Aerospace Engineering: A Contemporary Approach

Aerospace Engineering: A Contemporary Approach

Editor: Natalie Spagner

MURPHY & MOORE
www.murphy-moorepublishing.com

Published by Murphy & Moore Publishing
1 Rockefeller Plaza,
New York City, NY 10020, USA
www.murphy-moorepublishing.com

Aerospace Engineering: A Contemporary Approach
Edited by Natalie Spagner

International Standard Book Number: 978-1-63987-026-4 (Hardback)

Cataloging-in-Publication Data

Aerospace engineering : a contemporary approach / edited by Natalie Spagner.
p. cm.
Includes bibliographical references and index.
ISBN 978-1-63987-026-4
1. Aerospace engineering. 2. Aeronautics. 3. Astronautics. 4. Engineering. I. Spagner, Natalie.
TL545 .A37 2022
629.1--dc23

Contents

Preface

Aerospace engineering is a discipline of engineering that is related to the design, construction and development of aircraft and spacecrafts. It is classified into two main branches termed as aeronautical engineering and astronautical engineering. It uses tools from several other disciplines including aerodynamics, avionics, propulsion, materials science and structural analysis. Astronautical engineering deals with designing, developing and manufacturing of spacecrafts. Aeronautical engineering is concerned with the study, design, testing and manufacturing of flight capable machines and other aircraft. Few of the major elements of aerospace engineering are radar cross-section, fluid mechanics, electro-technology and solid mechanics. This book unfolds the innovative aspects of aerospace engineering which will be crucial for the progress of this field in the future. The topics included herein are significant and bound to provide incredible insights to readers. This book will serve as a resource guide to the readers in keeping pace with the rapid changes in this field.

Various studies have approached the subject by analyzing it with a single perspective, but the present book provides diverse methodologies and techniques to address this field. This book contains theories and applications needed for understanding the subject from different perspectives. The aim is to keep the readers informed about the progresses in the field; therefore, the contributions were carefully examined to compile novel researches by specialists from across the globe.

Indeed, the job of the editor is the most crucial and challenging in compiling all chapters into a single book. In the end, I would extend my sincere thanks to the chapter authors for their profound work. I am also thankful for the support provided by my family and colleagues during the compilation of this book.

Editor

1

Concise Neural Nonaffine Control of Air-Breathing Hypersonic Vehicles Subject to Parametric Uncertainties

Xiangwei Bu,[1] Qiong Wang,[2] Yan Zhao,[1] and Guangjun He[1]

[1]*Air and Missile Defense College, Air Force Engineering University, Xi'an 710051, China*
[2]*Science College, Air Force Engineering University, Xi'an 710051, China*

Correspondence should be addressed to Xiangwei Bu; buxiangwei1987@126.com

Academic Editor: Paul Williams

In this paper, a novel simplified neural control strategy is proposed for the longitudinal dynamics of an air-breathing hypersonic vehicle (AHV) directly using nonaffine models instead of affine ones. For the velocity dynamics, an adaptive neural controller is devised based on a minimal-learning parameter (MLP) technique for the sake of decreasing computational loads. The altitude dynamics is rewritten as a pure feedback nonaffine formulation, for which a novel concise neural control approach is achieved without backstepping. The special contributions are that the control architecture is concise and the computational cost is low. Moreover, the exploited controller possesses good practicability since there is no need for affine models. The semiglobally uniformly ultimate boundedness of all the closed-loop system signals is guaranteed via Lyapunov stability theory. Finally, simulation results are presented to validate the effectiveness of the investigated control methodology in the presence of parametric uncertainties.

1. Introduction

Air-breathing hypersonic vehicles (AHVs) are regarded as a promising technology for cost-efficient and reliable access to space. One of the major issues with respect to the developments of AHVs is to maintain controllability of the vehicle during various phases of flight [1–3]. However, flight control design for such vehicles is very challenging due to significant uncertainties, strong couplings, nonminimum phase characteristics, and significant flexible effects [4–7].

Recently, many classic control methodologies have been proposed for the longitudinal dynamics of AHVs. To guarantee the robustness with respect to parametric uncertainties, an adaptive sliding mode controller is addressed for an AHV [8, 9]. The starting point of that strategy [8, 9] is to linearize the vehicle model as an affine one, which is required for the control design. In [10], a continuous high-order sliding mode control scheme is investigated using an affine AHV model. In that study, the undesired chattering phenomenon is avoided owing to the continuous control inputs. For an AHV with actuator faults, an adaptive fault-tolerant controller is devised on the basis of a Takagi-Sugeno (T-S) fuzzy system [11]. In [12, 13], a linear parameter-varying (LPV) affine model is first constructed to describe the complex longitudinal dynamics of an AHV. Then, a nonfragile switching tracking control methodology is proposed [12, 13]. In [14], a compound control approach-combined trajectory linearization control (TLC) with active disturbance rejection control (ADRC) is presented for an AHV. Moreover, the robustness of that strategy is enhanced via linear-extended state observers [14]. For the longitudinal dynamics of an AHV, a new tracking control scheme is exploited by solving a system of linear algebraic equations [15]. In [16], a robust controller is explored for an AHV, and moreover, a new nonlinear disturbance observer (NDO) is developed for robustness enhancement.

It is noted that the altitude dynamics of AHVs can be rewritten as a strict feedback affine model based on strict assumptions, which makes backstepping control achievable [17–20]. In [1], a robust adaptive control method is studied via backstepping. However, the smooth parameter projection utilized in that paper may result in high computational loads because of too many adaptive parameters. Meanwhile, the problem of "explosion of terms" is ignored [1]. In [19], a new tracking differentiator is exploited to provide estimations of the time derivatives of virtual controllers while the problem of "explosion of terms" is eliminated. Based on the results of [21], a new NDO is constructed, which guarantees that the proposed backstepping controller can provide robust tracking of velocity and altitude reference trajectories in the presence of parametric uncertainties and external disturbances [19]. It has been demonstrated that neural networks (NNs) have the ability of global approximation [22–25]. Thus, improved backstepping control methodologies with robust performance can be achieved if the unknown nonlinearities are approximated by NNs [26–29]. To reduce the computational cost of neural approximation, the NNs are employed to approach the devised backstepping controllers rather than unknown functions while a nonsingular direct neural control strategy is addressed for an AHV [30]. Furthermore, a novel control scheme with low computational loads is exploited for an AHV based on a pure feedback affine model [31]. Meanwhile, the learning parameters are decreased via a minimal-learning parameter (MLP) method [31].

It is worth pointing out that all the above results are obtained based on affine models. That is, it requires that the AHV model has to be linear in the control inputs. However, in fact, the AHV model has no linear appearance of the control inputs [32, 33]. On the basis of the previous approaches, we have to first transform the AHV model into an affine formulation and then exploit controllers utilizing linearized or affine AHV models. In this paper, we propose a novel adaptive neural controller for the longitudinal dynamics of an AHV without using affine models. The vehicle dynamics is decomposed as the velocity subsystem and the altitude subsystem to be controlled separately. For both subsystems, simplified neural controllers with low computational loads are addressed via an MLP technique. By using a novel model transformation method, the altitude dynamics is rewritten as a pure feedback formulation. Hence, a simplified design is achieved. Finally, the effectiveness of the studied methodology is verified by simulation results. The special contributions of this paper are summarized as follows:

(1) Different from the existing studies, the exploited controller is directly derived from nonaffine models. Moreover, the strict requirements [30] that the unknown functions have to be strictly positive and bounded are released in this paper. Thereby, the proposed control scheme possesses excellent practicability and reliability.

(2) Compared with backstepping control strategies [18–20, 30], the explored controller exhibits a more concise structure and a lower computational load. In this study, there is no need of a complicated recursive design process of backstepping. In comparison with the neural backstepping control approach [30], much less controllers, NNs, and learning parameters are needed in this study.

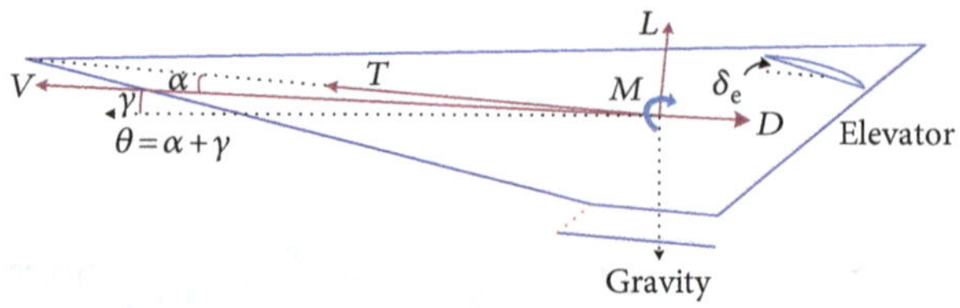

FIGURE 1: The geometry and force map of an AHV.

The remainder of this paper is structured as follows. The vehicle model and preliminaries are shown in Section 2. Section 3 presents the control design process. The simulation studies are made in Section 4, and the conclusions are proposed in Section 5.

2. AHV Model and Preliminaries

2.1. The AHV Model and Control Goal. In this paper, we consider a longitudinal dynamic model of an AHV developed by Bolender and Doman [32]. The AHV's longitudinal sketch and force map are shown in Figure 1. The equations of motion are constructed by applying Lagrange's equations. Moreover, the vehicle is viewed as a single flexible structure with mass-normalized mode shapes. Thus, the vehicle model contains flexible effects. The equations of motion are expressed as follows [33]:

$$\dot{V} = \frac{T\cos(\theta-\gamma) - D}{m} - g\sin\gamma, \tag{1}$$

$$\dot{h} = V\sin\gamma, \tag{2}$$

$$\dot{\gamma} = \frac{L + T\sin(\theta-\gamma)}{mV} - \frac{g\cos\gamma}{V}, \tag{3}$$

$$\dot{\theta} = Q, \tag{4}$$

$$\dot{Q} = \frac{M + \widetilde{\psi}_1\ddot{\eta}_1 + \widetilde{\psi}_2\ddot{\eta}_2}{I_{yy}}, \tag{5}$$

$$k_1\ddot{\eta}_1 = -2\zeta_1\omega_1\dot{\eta}_1 - \omega_1^2\eta_1 + N_1 - \widetilde{\psi}_1\frac{M}{I_{yy}} - \frac{\widetilde{\psi}_1\widetilde{\psi}_2\ddot{\eta}_2}{I_{yy}}, \tag{6}$$

$$k_2\ddot{\eta}_2 = -2\zeta_2\omega_2\dot{\eta}_2 - \omega_2^2\eta_2 + N_2 - \widetilde{\psi}_2\frac{M}{I_{yy}} - \frac{\widetilde{\psi}_2\widetilde{\psi}_1\ddot{\eta}_1}{I_{yy}}, \tag{7}$$

where the approximations of T, D, L, M, N_1, and N_2 are designed as follows [33]:

$$\begin{aligned}
T &\approx C_T^{\alpha^3}\alpha^3 + C_T^{\alpha^2}\alpha^2 + C_T^{\alpha}\alpha + C_T^0,\\
D &\approx \bar{q}S\left(C_D^{\alpha^2}\alpha^2 + C_D^{\alpha}\alpha + C_D^{\delta_e^2}\delta_e^2 + C_D^{\delta_e}\delta_e + C_D^0\right),\\
L &\approx \bar{q}S\left(C_L^{\alpha}\alpha + C_L^{\delta_e}\delta_e + C_L^0\right),\\
M &\approx z_T T + \bar{q}S\bar{c}\left[C_{M,\alpha}^{\alpha^2}\alpha^2 + C_{M,\alpha}^{\alpha}\alpha + C_{M,\alpha}^0 + c_e\delta_e\right],\\
N_1 &\approx N_1^{\alpha^2}\alpha^2 + N_1^{\alpha}\alpha + N_1^0,\\
N_2 &\approx N_2^{\alpha^2}\alpha^2 + N_2^{\alpha}\alpha + N_2^{\delta_e}\delta_e + N_2^0,\\
\bar{q} &= \frac{1}{2}\rho V^2, \quad \rho = \rho_0 \exp\left(-\frac{h-h_0}{h_s}\right),
\end{aligned} \tag{8}$$

with

$$\begin{aligned}
C_T^{\alpha^3} &= \beta_1(h,\bar{q})\Phi + \beta_2(h,\bar{q}),\\
C_T^{\alpha^2} &= \beta_3(h,\bar{q})\Phi + \beta_4(h,\bar{q}),\\
C_T^{\alpha} &= \beta_5(h,\bar{q})\Phi + \beta_6(h,\bar{q}),\\
C_T^{0} &= \beta_7(h,\bar{q})\Phi + \beta_8(h,\bar{q}).
\end{aligned} \tag{9}$$

The above model is composed of five rigid body states (i.e., V, h, γ, θ, and Q), four flexible states (i.e., η_1, $\dot{\eta}_1$, η_2, and $\dot{\eta}_2$), and two control inputs (i.e., Φ and δ_e). It is noticed that the control inputs do not occur explicitly in (1), (2), (3), (4), (5), (6), and (7). However, they appear through T, D, L, M, N_1, and N_2. For more detailed definitions of the model parameters and coefficients, the reader could refer to [33] or the Nomenclature.

The control goal pursued in this study is to devise adaptive neural controllers for an AHV without using affine modes such that velocity V and altitude h track their reference trajectories V_{ref} and h_{ref}, respectively. Meanwhile, all the closed-loop system signals are bounded and the proposed controller is robust to parametric uncertainties.

2.2. NN Description. To guarantee the controller's robustness, we employ the radial basis function NN (RBFNN) to approximate the lumped uncertainties of both subsystems of AHVs. The RBFNN is defined as the mapping relationship between the input vector $\mathbf{X} = [\ell_1, \ell_2, \dots, \ell_n]^T \in \mathbf{R}^n$ and the output $y \in \mathbf{R}$ [34].

$$y = \mathbf{W}^T\mathbf{h}(\mathbf{X}), \tag{10}$$

where $\mathbf{W} = [w_1, w_2, \dots, w_p]^T \in \mathbf{R}^p$ denotes a weight vector; p and n represent the node number and input number, respectively; $\mathbf{h}(\mathbf{X}) = [h_1(\mathbf{X}), h_2(\mathbf{X}), \dots, h_p(\mathbf{X})]^T \in \mathbf{R}^p$ with $h_j(\mathbf{X})$ is defined as follows:

$$h_j(\mathbf{X}) = \exp\left(-\frac{\|\mathbf{X}-\mathbf{c}\|^2}{2b_j^2}\right), \quad j = 1, 2, \dots, p, \tag{11}$$

where $\mathbf{c} = [c_1, c_2, \dots, c_n]^T \in \mathbf{R}^n$ and $\mathbf{b} = [b_{j1}, b_{j2}, \dots, b_{jn}]^T \in \mathbf{R}^n$ mean a center vector and a width vector of $h_j(\mathbf{X})$, respectively.

For an arbitrary continuous unknown function $F(\mathbf{X})$, it has be proved that there exists an ideal weight vector $\mathbf{W}^* = [w_1^*, w_2^*, \dots, w_p^*]^T \in \mathbf{R}^p$ such that [34]

$$F(\mathbf{X}) = \mathbf{W}^{*T}\mathbf{h}(\mathbf{X}) + \varepsilon, \quad |\varepsilon| \le \varepsilon_M, \tag{12}$$

where ε and ε_M are the approximation error and its upper bound, respectively. Noting that $\mathbf{W}^*$ is completely unknown, its elements $w_1^*, w_2^*, \dots, w_p^*$ are required to be adjusted adaptively. In what follows, we define $\varphi = \|\mathbf{W}^*\|^2$ and regulate φ rather than $w_1^*, w_2^*, \dots, w_p^*$ to achieve satisfactory neural approximation. In this way, the learning parameters are decreased greatly such that the computational burden is quite low. This is the so-called MLP scheme.

2.3. High-Order Tracking Differentiator. In the subsequent developments, we will employ a high-order tracking differentiator (HTD) [35] to estimate the newly defined states. This HTD is formulated as follows:

$$\begin{aligned}
\dot{\chi}_1 &= \chi_2,\\
\dot{\chi}_2 &= \chi_3,\\
\dot{\chi}_3 &= \chi_4,\\
\dot{\chi}_4 &= -R^4\left[a_1(\chi_1 - v) + a_2\frac{\chi_2}{R} + a_3\frac{\chi_3}{R^2} + a_4\frac{\chi_4}{R^3}\right],
\end{aligned} \tag{13}$$

where R, a_1, a_2, a_3, and a_4 are positive constants to be chosen; v is the input signal; χ_1, χ_2, χ_3, and χ_4 denote the states of the HTD and they stand for the estimations of v, $\dot{v}$, $\ddot{v}$, and $\dddot{v}$, respectively. The relative estimation errors are defined as follows: $e_1 = \chi_1 - v$, $e_2 = \chi_2 - \dot{v}$, $e_3 = \chi_3 - \ddot{v}$, and $e_4 = \chi_4 - \dddot{v}$. By choosing an infinitely large R for the HTD, we have

$$\begin{aligned}
&\lim_{R\to+\infty}(\chi_1 - v) = \lim_{R\to+\infty} e_1 = 0,\\
&\frac{d\lim_{R\to+\infty}(\chi_1 - v)}{dt} = \lim_{R\to+\infty}(\dot{\chi}_1 - \dot{v}) = \lim_{R\to+\infty}(\chi_2 - \dot{v}) = \lim_{R\to+\infty} e_2 = 0,\\
&\frac{d\lim_{R\to+\infty}(\chi_2 - \dot{v})}{dt} = \lim_{R\to+\infty}(\dot{\chi}_2 - \ddot{v}) = \lim_{R\to+\infty}(\chi_3 - \ddot{v}) = \lim_{R\to+\infty} e_3 = 0,\\
&\frac{d\lim_{R\to+\infty}(\chi_3 - \ddot{v})}{dt} = \lim_{R\to+\infty}(\dot{\chi}_3 - \dddot{v}) = \lim_{R\to+\infty}(\chi_4 - \dddot{v}) = \lim_{R\to+\infty} e_4 = 0.
\end{aligned} \tag{14}$$

Remark 1. It is obvious that the estimation errors e_i $(i = 1, 2, 3, 4)$ can converge to zero by designing an adequately large R. In choosing an infinitely large but bounded R for the THD in simulation study, we know that there exist positive constants $\bar{e}_i$ $(i = 1, 2, 3, 4)$ such that $\bar{e}_i \ge |e_i|$ $(i = 1, 2, 3, 4)$ [23].

2.4. Preliminaries. To start the control design, the following preliminaries are required.

Definition 1 (see [36]). For any compact set Ω_0, if there exist a positive parameter Σ and $I = I(\Sigma, Y_0(t_0))$ such that $\|Y_0(t)\| \le \Sigma$ for all $Y(t_0) \in \Omega_0$ and $t \ge t_0 + I$, then $Y_0(t)$ is semiglobally uniformly ultimately bounded.

Lemma 1 (implicit function theorem, see [36, 37]). *The implicit function $H : \mathbf{R}^l \times \mathbf{R}^r \to \mathbf{R}^l$ is continuously differentiable at each $(\boldsymbol{\varpi}, \boldsymbol{\sigma})$ of an open set $Y \subset \mathbf{R}^l \times \mathbf{R}^r$. Define $(\boldsymbol{\varpi}_0, \boldsymbol{\sigma}_0)$ as a point in Y for which $H(\boldsymbol{\varpi}_0, \boldsymbol{\sigma}_0) = 0$ and the Jacobian matrix $(\partial H/\partial \boldsymbol{\varpi})(\boldsymbol{\varpi}_0, \boldsymbol{\sigma}_0)$ is nonsingular. Then, there are neighborhoods $U \subset \mathbf{R}^l$ of $\boldsymbol{\varpi}_0$ and $G \subset \mathbf{R}^r$ of $\boldsymbol{\sigma}_0$ such that for each $\boldsymbol{\sigma} \in G$, the equation $H(\boldsymbol{\varpi}, \boldsymbol{\sigma}) = 0$ has a unique solution $\boldsymbol{\varpi} \in U$. Furthermore, the solution can be expressed as $\boldsymbol{\varpi} = g_0(\boldsymbol{\sigma})$, where $g_0(\cdot)$ is a continuously differentiable function at $\boldsymbol{\sigma} = \boldsymbol{\sigma}_0$.*

Remark 2. For the detailed proof of Lemma 1, the reader could refer to [36]. It is noted that if the implicit function $H(\boldsymbol{\varpi}, \boldsymbol{\sigma})$ satisfies Lemma 1, $H(\boldsymbol{\varpi}, \boldsymbol{\sigma})$ can be viewed as a function of $\boldsymbol{\sigma}$ since there exists a function $g_0(\cdot)$ such that $\boldsymbol{\varpi} = g_0(\boldsymbol{\sigma})$.

3. Controller Design

According to the discussions presented in [1, 7, 19, 20], we know that the vehicle dynamics can be reasonably decomposed as the velocity subsystem (i.e., (1)) and the altitude subsystem (i.e., (2), (3), (4), and (5)).

3.1. Velocity Controller Design. In this subsection, the objective is to devise a neural controller Φ for the velocity subsystem by applying a nonaffine model such that velocity V follows its reference command V_{ref}.

The velocity subsystem can be expressed as the following nonaffine model:

$$\dot{V} = f_V(V, \Phi), \tag{15}$$

where $f_V(V, \Phi)$ is a completely unknown and continuously differentiable function.

Assumption 1. The following inequality

$$\frac{\partial f_V(V, \Phi)}{\partial \Phi} > 0 \tag{16}$$

holds for all $(V, \Phi) \in \Omega_V \times \mathbf{R}$ with a controllability region Ω_V.

Remark 3. According to [33] and the response bounds of the rigid-body states, we conclude that Assumption 1 is satisfied.

Define velocity tracking error as follows:

$$\tilde{V} = V - V_{\text{ref}}. \tag{17}$$

Taking time derivative along (17) and substituting (15) lead to

$$\dot{\tilde{V}} = \dot{V} - \dot{V}_{\text{ref}} = k_V \Phi + F_V(V, \Phi) - \dot{V}_{\text{ref}}, \tag{18}$$

where $k_V > 0$ is a design parameter; $F_V(V, \Phi) = f_V(V, \Phi) - k_V \Phi$ denotes the lumped uncertainty of the velocity subsystem.

The control effort Φ is chosen as follows:

$$\Phi = k_V^{-1}(\Phi_0 - \Phi_1), \tag{19}$$

where $\Phi_0 = -k_{V1}\tilde{V} - k_{V2}\int_0^t \tilde{V} d\tau + \dot{V}_{\text{ref}}$ with $k_{V1} > 0$ and $k_{V2} > 0$; Φ_1 is a neural control signal designed to cancel $F_V(V, \Phi)$ by employing an RBFNN.

Invoking Lemma 1, we have that there exists Φ_1^* such that

$$H_1(V, \Phi_0, \Phi_1^*) \triangleq F_V\left(V, k_V^{-1}(\Phi_0 - \Phi_1^*)\right) - \Phi_1^* = 0. \tag{20}$$

Then, we obtain the following theorem.

Theorem 1. *Define*

$$k_V > \frac{1}{2}\frac{\partial f_V(V, \Phi)}{\partial \Phi}. \tag{21}$$

Then there exist a set $\Omega_V \subset \mathbf{R}$ and a unique Φ_1^* that is a function of V and Φ_0 such that $\Phi_1^*(V, \Phi_0)$ satisfies (20) for all $(V, \Phi_0) \in \Omega_V \times \mathbf{R}$.

Proof. The sufficient condition for the existence of Φ_1^* is that the following inequality holds [37, 38]:

$$\left|\frac{\partial F_V(V, \Phi)}{\partial \Phi_1^*}\right| < 1. \tag{22}$$

Considering (16), (19), and (21) and the fact that $F_V(V, \Phi) = f_V(V, \Phi) - k_V \Phi$, we obtain

$$\begin{aligned}\left|\frac{\partial F_V(V, \Phi)}{\partial \Phi_1^*}\right| &= \left|\frac{\partial}{\partial \Phi_1^*}[f_V(V, \Phi) - k_V \Phi]\right| \\ &= \left|\frac{\partial [f_V(V, \Phi) - k_V \Phi]}{\partial \Phi}\frac{\partial \Phi}{\partial \Phi_1^*}\right| \\ &= \left|\left[\frac{\partial f_V(V, \Phi)}{\partial \Phi} - k_V\right]\frac{1}{k_V}\right| \\ &= \left|\frac{1}{k_V}\frac{\partial f_V(V, \Phi)}{\partial \Phi} - 1\right| < 1.\end{aligned} \tag{23}$$

It is clear that Φ_1^* exists.

Furthermore, notice that

$$\begin{aligned}\frac{\partial}{\partial \Phi_1^*}H_1(V, \Phi_0, \Phi_1^*) &= \frac{\partial}{\partial \Phi_1^*}[F_V(V, \Phi^*) - \Phi_1^*] \\ &= \frac{\partial}{\partial \Phi^*}[f_V(V, \Phi^*) - k_V \Phi^*]\frac{\partial \Phi^*}{\partial \Phi_1^*} - 1 \\ &= \left[\frac{\partial}{\partial \Phi^*}f_V(V, \Phi^*) - k_V\right]\left(-\frac{1}{k_V}\right) - 1 \\ &= -\frac{1}{k_V}\frac{\partial f_V(V, \Phi^*)}{\partial \Phi^*},\end{aligned} \tag{24}$$

with $\Phi^* = k_V^{-1}(\Phi_0 - \Phi_1^*)$.

By (16) and (21), we know that $\partial H_1(V, \Phi_0, \Phi_1^*)/(\partial \Phi_1^*)$ is nonsingular. Thus, Theorem 1 holds. This is the end of proof.

Theorem 1 reveals that Φ is a function of V and Φ_0 for all $(V, \Phi_0) \in \Omega_V \times \mathbf{R}$ due to $\Phi = k_V^{-1}(\Phi_0 - \Phi_1)$. Thereby, $F_V(V, \Phi) = f_V(V, \Phi) - k_V \Phi$ also is a function of V and Φ_0. Define $\mathbf{X}_1 = [V, \Phi_0]^T \in \mathbf{R}^2$ as an input vector. Then we introduce an RBFNN to approximate $F_V(V, \Phi)$.

$$F_V(V, \Phi) = \mathbf{W}_1^{*T}\mathbf{h}_1(\mathbf{X}_1) + \iota_1, |\iota_1| \le \iota_{1M}, \tag{25}$$

where $\mathbf{W}_1^* = [w_{11}^*, w_{12}^*, \ldots, w_{1p_1}^*]^T \in \mathbf{R}^{p_1}$ is an ideal weight vector; ι_1 and ι_{1M} are the approximation error and its upper bound, respectively; $\mathbf{h}_1(\mathbf{X}_1) = [h_{11}(\mathbf{X}_1), h_{12}(\mathbf{X}_1), \ldots, h_{1p_1}(\mathbf{X}_1)]^T \in \mathbf{R}^{p_1}$, $h_{1j}(\mathbf{X}_1)$ $(j = 1, 2, \ldots, p_1)$ have similar formulations to (11).

Define $\varphi_1 = \|\mathbf{W}_1^*\|^2$ and choose

$$\Phi_1 = \frac{1}{2}\tilde{V}\hat{\varphi}_1 \mathbf{h}_1^T(\mathbf{X}_1)\mathbf{h}_1(\mathbf{X}_1), \tag{26}$$

where $\hat{\varphi}_1$ is the estimation of φ_1 with the following adaptive law:

$$\dot{\hat{\varphi}}_1 = \frac{\lambda_1}{2}\tilde{V}^2 \mathbf{h}_1^T(\mathbf{X}_1)\mathbf{h}_1(\mathbf{X}_1) - 2\hat{\varphi}_1, \tag{27}$$

where $\lambda_1 > 0$ is a design parameter.

Theorem 2. *Consider the closed-loop system consisting of plant (15) under Assumption 1 with controller (19) and adaptive law (27). Then all the signals involved are semiglobally uniformly ultimately bounded.*

Proof. Define

$$\tilde{\varphi}_1 = \hat{\varphi}_1 - \varphi_1. \tag{28}$$

Substituting (19), (25), and (26) into (18) yields

$$\begin{aligned}\dot{\tilde{V}} &= \Phi_0 - \Phi_1 + \mathbf{W}_1^{*T}\mathbf{h}_1(\mathbf{X}_1) + \iota_1 - \dot{V}_{\text{ref}} \\ &= -k_{V1}\tilde{V} - k_{V2}\int_0^t \tilde{V} d\tau - \frac{1}{2}\tilde{V}\hat{\varphi}_1 \mathbf{h}_1^T(\mathbf{X}_1)\mathbf{h}_1(\mathbf{X}_1) \\ &\quad + \mathbf{W}_1^{*T}\mathbf{h}_1(\mathbf{X}_1) + \iota_1.\end{aligned} \tag{29}$$

The Lyapunov function candidate is chosen as follows:

$$W_V = \frac{\tilde{V}^2}{2} + \frac{k_{V2}}{2}\left(\int_0^t \tilde{V} d\tau\right)^2 + \frac{\tilde{\varphi}_1^2}{2\lambda_1}. \tag{30}$$

Invoking (27), (28), and (29), the time derivative of W_V is derived as follows:

$$\begin{aligned}\dot{\mathbf{W}}_V &= \tilde{V}\dot{\tilde{V}} + k_{V2}\tilde{V}\int_0^t \tilde{V} d\tau + \frac{\tilde{\varphi}_1 \dot{\hat{\varphi}}_1}{\lambda_1} \\ &= -k_{V1}\tilde{V}^2 - \frac{1}{2}\tilde{V}^2 \varphi_1 \mathbf{h}_1^T(\mathbf{X}_1)\mathbf{h}_1(\mathbf{X}_1) + \tilde{V}\mathbf{W}_1^{*T}\mathbf{h}_1(\mathbf{X}_1) \\ &\quad + \tilde{V}\iota_1 - \frac{2\tilde{\varphi}_1\hat{\varphi}_1}{\lambda_1}.\end{aligned} \tag{31}$$

Due to $\tilde{\varphi}_1^2 + 2\tilde{\varphi}_1(\hat{\varphi}_1 - \tilde{\varphi}_1) + \varphi_1^2 = \tilde{\varphi}_1^2 + 2\tilde{\varphi}_1\varphi_1 + \varphi_1^2 = (\tilde{\varphi}_1 + \varphi_1)^2 \ge 0$, it follows that $2\tilde{\varphi}_1\hat{\varphi}_1 \ge \tilde{\varphi}_1^2 - \varphi_1^2$. Thus, the following inequality holds:

$$\begin{aligned}\dot{W}_V \le &- k_{V1}\tilde{V}^2 - \frac{\tilde{\varphi}_1^2}{\lambda_1} - \frac{1}{2}\tilde{V}^2 \varphi_1 \mathbf{h}_1^T(\mathbf{X}_1)\mathbf{h}_1(\mathbf{X}_1) \\ &+ \tilde{V}\mathbf{W}_1^{*T}\mathbf{h}_1(\mathbf{X}_1) + \tilde{V}\iota_1 + \frac{\varphi_1^2}{\lambda_1}.\end{aligned} \tag{32}$$

Notice that

$$\begin{aligned}\tilde{V}\mathbf{W}_1^{*T}\mathbf{h}_1(\mathbf{X}_1) &\le \frac{\tilde{V}^2}{2}\|\mathbf{W}_1^{*T}\mathbf{h}_1(\mathbf{X}_1)\|^2 + \frac{1}{2} \\ &= \frac{\tilde{V}^2}{2}\|\mathbf{W}_1^*\|^2\|\mathbf{h}_1(\mathbf{X}_1)\|^2 + \frac{1}{2} \\ &= \frac{\tilde{V}^2}{2}\varphi_1 \mathbf{h}_1^T(\mathbf{X}_1)\mathbf{h}_1(\mathbf{X}_1) + \frac{1}{2}, \\ \tilde{V}\iota_1 &\le \frac{1}{2}\tilde{V}^2\iota_{1M}^2 + \frac{1}{2}.\end{aligned} \tag{33}$$

Inequality (32) becomes

$$\dot{W}_V \le -\left(k_{V1} - \frac{1}{2}\iota_{1M}^2\right)\tilde{V}^2 - \frac{\tilde{\varphi}_1^2}{\lambda_1} + 1 + \frac{\varphi_1^2}{\lambda_1}. \tag{34}$$

Let $k_{V1} > \iota_{1M}^2/2$ and define the following compact sets:

$$\begin{aligned}\Omega_{\tilde{V}} &= \left\{\tilde{V}\,\middle|\,|\tilde{V}| \le \sqrt{\frac{1 + (\varphi_1^2/\lambda_1)}{k_{V1} - (1/2)\iota_{1M}^2}}\right\}, \\ \Omega_{\tilde{\varphi}_1} &= \left\{\tilde{\varphi}_1\,\middle|\,|\tilde{\varphi}_1| \le \sqrt{\frac{1 + (\varphi_1^2/\lambda_1)}{(1/\lambda_1)}}\right\}.\end{aligned} \tag{35}$$

It can be seen that $\dot{W}_V$ will be negative if $\tilde{V} \notin \Omega_{\tilde{V}}$ or $\tilde{\varphi}_1 \notin \Omega_{\tilde{\varphi}_1}$. Hence, $\tilde{V}$ and $\tilde{\varphi}_1$ are semiglobally uniformly ultimately bounded. This completes the proof.

3.2. Altitude Controller Design. In this subsection, we will explore a simplified neural controller δ_e for the altitude subsystem without using backstepping such that altitude h tracks its command h_{ref}.

Define altitude tracking error as $\tilde{h} = h - h_{ref}$. The command of γ is chosen as follows:

$$\gamma_d = \arcsin\left(\frac{-k_\gamma \tilde{h} + \dot{h}_{ref}}{V}\right), \quad (36)$$

where $k_\gamma > 0$ is a design parameter. If γ converges to γ_d, then we have that h converges to h_{ref} [26]. Hence, the subsequent control goal is to let γ converge to γ_d.

Define $x_1 = \gamma$, $x_2 = \theta$, $x_3 = Q$. Then, a nonaffine model can be derived from the rest of the altitude subsystem (i.e., (3), (4), and (5)).

$$\begin{aligned} \dot{x}_1 &= f_1(x_1, x_2), \\ \dot{x}_2 &= x_3, \\ \dot{x}_3 &= f_3(\mathbf{x}, \delta_e), \end{aligned} \quad (37)$$

with $\mathbf{x} = [x_1, x_2, x_3]^T$. In (37), $f_1(x_1, x_2)$ and $f_3(\mathbf{x}, \delta_e)$ are completely unknown and continuously differentiable functions.

Assumption 2. The following are inequalities:

$$\begin{aligned} \frac{\partial f_1(x_1, x_2)}{\partial x_2} &> 0, \\ \frac{\partial f_3(\mathbf{x}, \delta_e)}{\partial \delta_e} &> 0. \end{aligned} \quad (38)$$

The inequalities above are satisfied for all $(\mathbf{x}, \delta_e) \in \Omega_\mathbf{x} \times \mathbf{R}$ with a controllability region $\Omega_\mathbf{x}$.

Remark 4. Taking into account [33] and the response bounds of the rigid-body states, we know that Assumption 2 holds.

To avoid the complex design process of backstepping, the following model transformation is made.

Step 1. Define $z_1 = x_1 = \gamma$ and $z_2 = \dot{z}_1 = f_1(x_1, x_2)$. Considering (37), the time derivative of z_2 is given by the following:

$$\begin{aligned} \dot{z}_2 &= \frac{\partial f_1(x_1, x_2)}{\partial x_1}\dot{x}_1 + \frac{\partial f_1(x_1, x_2)}{\partial x_2}\dot{x}_2 \\ &= \frac{\partial f_1(x_1, x_2)}{\partial x_1} f_1(x_1, x_2) + \frac{\partial f_1(x_1, x_2)}{\partial x_2} x_3 \\ &\triangleq f_{h1}(\mathbf{x}). \end{aligned} \quad (39)$$

Step 2. Define $z_3 = \dot{z}_2 = f_{h1}(\mathbf{x})$. Invoking (37), the time derivative of z_3 is derived as follows:

$$\begin{aligned} \dot{z}_3 &= \frac{\partial f_{h1}(\mathbf{x})}{\partial x_1}\dot{x}_1 + \frac{\partial f_{h1}(\mathbf{x})}{\partial x_2}\dot{x}_2 + \frac{\partial f_{h1}(\mathbf{x})}{\partial x_3}\dot{x}_3 \\ &= \frac{\partial f_{h1}(\mathbf{x})}{\partial x_1} f_1(x_1, x_2) + \frac{\partial f_{h1}(\mathbf{x})}{\partial x_2} x_3 + \frac{\partial f_{h1}(\mathbf{x})}{\partial x_3} f_3(\mathbf{x}, \delta_e) \\ &\triangleq f_{h2}(\mathbf{x}, \delta_e). \end{aligned} \quad (40)$$

As a result, a pure feedback model with a nonaffine formulation is obtained:

$$\begin{aligned} \dot{z}_1 &= z_2, \\ \dot{z}_2 &= z_3, \\ \dot{z}_3 &= f_{h2}(\mathbf{x}, \delta_e), \end{aligned} \quad (41)$$

where $f_{h2}(\mathbf{x}, \delta_e)$ is an unknown continuous function.

Remark 5. From (37), (38), (39), and (40), we obtain that

$$\begin{aligned} \frac{\partial f_{h2}(\mathbf{x}, \delta_e)}{\partial \delta_e} &= \frac{\partial f_{h1}(\mathbf{x})}{\partial x_3}\frac{\partial f_3(\mathbf{x}, \delta_e)}{\partial \delta_e} \\ &= \frac{\partial f_1(x_1, x_2)}{\partial x_2}\frac{\partial f_3(\mathbf{x}, \delta_e)}{\partial \delta_e} > 0. \end{aligned} \quad (42)$$

Remark 6. Compared with (37), the expression of (41) is simpler. On the basis of (41), the complicated design process of backstepping can be eliminated.

Define flight-path angle tracking error e_0 and error function E as follows:

$$\begin{aligned} e_0 &= \gamma - \gamma_d = z_1 - \gamma_d, \\ E &= \left(\frac{d}{dt} + \mu\right)^3 \int_0^t e_0 d\tau = \ddot{e}_0 + 3\mu\dot{e}_0 + 3\mu^2 e_0 + \mu^3 \int_0^t e_0 d\tau, \end{aligned} \quad (43)$$

where μ is a positive constant to be chosen. Moreover, because the polynomial $(s + \mu)^3$ is Hurwitz, the boundedness of e_0 can be guaranteed if E is bounded.

Considering (41), the first three order time derivatives of e_0 are derived as follows:

$$\begin{aligned} \dot{e}_0 &= \dot{z}_1 - \dot{\gamma}_d = z_2 - \dot{\gamma}_d, \\ \ddot{e}_0 &= \dot{z}_2 - \ddot{\gamma}_d = z_3 - \ddot{\gamma}_d, \\ \dddot{e}_0 &= \dot{z}_3 - \dddot{\gamma}_d = k_h \delta_e + F_h(\mathbf{x}, \delta_e) - \dddot{\gamma}_d, \end{aligned} \quad (44)$$

where $k_h > 0$ is a design parameter; $F_h(\mathbf{x}, \delta_e) = f_{h2}(\mathbf{x}, \delta_e) - k_h \delta_e$ is a completely unknown function.

Thus, the time derivative of E is given as follows:

$$\begin{aligned}\dot{E} &= \dddot{e}_0 + 3\mu\ddot{e}_0 + 3\mu^2\dot{e}_0 + \mu^3 e_0 \\ &= k_h\delta_e + F_h(\mathbf{x}, \delta_e) - \dddot{\gamma}_d + 3\mu\ddot{e}_0 + 3\mu^2\dot{e}_0 + \mu^3 e_0.\end{aligned} \tag{45}$$

Note that the newly defined states z_2 and z_3 are unknown. From the above model transformation process, we have that $z_2 = \dot{\gamma}$ and $z_3 = \ddot{\gamma}$. Thus, let γ pass through the HTD presented in Section 2.3. We can get the estimations of z_2 and z_3, denoted by $\hat{z}_2$ and $\hat{z}_3$, respectively. Meanwhile, let γ_d be the input signal of the HTD. We can obtain the estimations of $\dot{\gamma}_d$, $\ddot{\gamma}_d$, and $\dddot{\gamma}_d$, represented by $\hat{\dot{\gamma}}_d$, $\hat{\ddot{\gamma}}_d$, and $\hat{\dddot{\gamma}}_d$, respectively. The relative estimation errors are defined as follows:

$$\begin{aligned} s_1 &= \hat{z}_2 - z_2, \\ s_2 &= \hat{z}_3 - z_3, \\ s_3 &= \hat{\dot{\gamma}}_d - \dot{\gamma}_d, \\ s_4 &= \hat{\ddot{\gamma}}_d - \ddot{\gamma}_d, \\ s_5 &= \hat{\dddot{\gamma}}_d - \dddot{\gamma}_d. \end{aligned} \tag{46}$$

By Remark 1, it is concluded that there exist positive constants $\bar{s}_i$ $(i = 1-5)$ such that $\bar{s}_i \geq |s_i|$ $(i = 1-5)$. Thus, (44) and (45) can be modified as follows:

$$\begin{aligned} \hat{\dot{e}}_0 &= \hat{z}_2 - \hat{\dot{\gamma}}_d, \\ \hat{\ddot{e}}_0 &= \hat{z}_3 - \hat{\ddot{\gamma}}_d, \\ \hat{\dddot{e}}_0 &= k_h\delta_e + F_h(\mathbf{x}, \delta_e) - \hat{\dddot{\gamma}}_d, \end{aligned} \tag{47}$$

$$\hat{\dot{E}} = k_h\delta_e + F_h(\mathbf{x}, \delta_e) - \hat{\dddot{\gamma}}_d + 3\mu\hat{\ddot{e}}_0 + 3\mu^2\hat{\dot{e}}_0 + \mu^3 e_0. \tag{48}$$

The control law is selected as follows:

$$\delta_e = k_h^{-1}(\delta_{e0} - \delta_{e1}), \tag{49}$$

where $\delta_{e0} = -k_{h1}\hat{E} + \hat{\dddot{\gamma}}_d - 3\mu\hat{\ddot{e}}_0 - 3\mu^2\hat{\dot{e}}_0 - \mu^3 e_0$; $k_h > 0$, $k_{h1} > 0$ are design parameters; δ_{e1} is a neural control signal devised to cancel $F_h(\mathbf{x}, \delta_e)$ by applying an RBFNN.

Considering Lemma 1, we know that there exists a δ_{e1}^* such that

$$H_2(\mathbf{x}, \delta_{e0}, \delta_{e1}^*) \triangleq F_h\left(\mathbf{x}, k_h^{-1}(\delta_{e0} - \delta_{e1}^*)\right) - \delta_{e1}^* = 0. \tag{50}$$

Theorem 3. *Define*

$$k_h > \frac{1}{2}\frac{\partial f_{h2}(\mathbf{x}, \delta_e)}{\partial \delta_e}. \tag{51}$$

Then, there exist a set $\Omega_{\mathbf{x}} \subset \mathbf{R}^3$ and a unique δ_{e1}^* that is a function of x and δ_e such that $\delta_{e1}^*(\mathbf{x}, \delta_{e0})$ satisfies (50) for all $(\mathbf{x}, \delta_{e0}) \in \Omega_{\mathbf{x}} \times \mathbf{R}$.

Proof. The sufficient condition for the existence of δ_{e1}^* is that the following inequality holds [37, 38]:

$$\left|\frac{\partial F_h(\mathbf{x}, \delta_e)}{\partial \delta_{e1}^*}\right| < 1. \tag{52}$$

Considering (42), (49), and (51), it follows that

$$\begin{aligned}\left|\frac{\partial F_h(\mathbf{x}, \delta_e)}{\partial \delta_{e1}^*}\right| &= \left|\frac{\partial[f_{h2}(\mathbf{x}, \delta_e) - k_h\delta_e]}{\partial \delta_{e1}^*}\right| \\ &= \left|\frac{\partial[f_{h2}(\mathbf{x}, \delta_e) - k_h\delta_e]}{\partial \delta_e}\frac{\partial \delta_e}{\partial \delta_{e1}^*}\right| \\ &= \left|\left[\frac{\partial f_{h2}(\mathbf{x}, \delta_e)}{\partial \delta_e} - k_h\right]\frac{1}{k_h}\right| \\ &= \left|\frac{1}{k_h}\frac{\partial f_{h2}(\mathbf{x}, \delta_e)}{\partial \delta_e} - 1\right| < 1.\end{aligned} \tag{53}$$

Thereby, δ_{e1}^* exists.

Furthermore,

$$\begin{aligned}\frac{\partial}{\partial \delta_{e1}^*}H_2(\mathbf{x}, \delta_{e0}, \delta_{e1}^*) &= \frac{\partial}{\partial \delta_{e1}^*}[F_h(\mathbf{x}, \delta_e^*) - \delta_{e1}^*] \\ &= \frac{\partial}{\partial \delta_{e1}^*}[f_{h2}(\mathbf{x}, \delta_e^*) - k_h\delta_e^*] - 1 \\ &= \frac{\partial}{\partial \delta_e^*}[f_{h2}(\mathbf{x}, \delta_e^*) - k_h\delta_e^*]\frac{\partial \delta_e^*}{\partial \delta_{e1}^*} - 1 \\ &= \left[\frac{\partial f_{h2}(\mathbf{x}, \delta_e^*)}{\partial \delta_e^*} - k_h\right]\left(-\frac{1}{k_h}\right) - 1 \\ &= -\frac{1}{k_h}\frac{\partial f_{h2}(\mathbf{x}, \delta_e^*)}{\partial \delta_e^*},\end{aligned} \tag{54}$$

with $\delta_e^* = k_h^{-1}(\delta_{e0} - \delta_{e1}^*)$. Obviously, $\partial H_2(\mathbf{x}, \delta_{e0}, \delta_{e1}^*)/(\partial \delta_{e1}^*)$ is nonsingular. Hence, Theorem 3 is satisfied. The proof is completed.

It is observed from Theorem 3 and Lemma 1 that δ_e can be treated as a function of $\mathbf{x}$ and δ_{e0} because of $\delta_e = k_h^{-1}(\delta_{e0} - \delta_{e1})$. Hence, $F_h(\mathbf{x}, \delta_e) = f_{h2}(\mathbf{x}, \delta_e) - k_h\delta_e$ can also be viewed as a function of $\mathbf{x}$ and δ_{e0}.

Define $\mathbf{X}_2 = [\mathbf{x}^T, \delta_{e0}]^T \in \mathbf{R}^4$ as an input vector. Then, we employ an RBFNN to approximate $F_h(\mathbf{x}, \delta_e)$.

$$F_h(\mathbf{x}, \delta_e) = \mathbf{W}_2^{*T}\mathbf{h}_2(\mathbf{X}_2) + \iota_2, \quad |\iota_2| \leq \iota_{2M}, \tag{55}$$

where $\mathbf{W}_2^* = [w_{21}^*, w_{22}^*, \ldots, w_{2p_2}^*]^T \in \mathbf{R}^{p_2}$ means an ideal weight vector; ι_2 and ι_{2M} represent the approximation error and its upper bound, respectively; $\mathbf{h}_2(\mathbf{X}_2) = [h_{21}(\mathbf{X}_2), h_{22}(\mathbf{X}_2), \ldots, h_{2p_2}(\mathbf{X}_2)]^T \in \mathbf{R}^{p_2}$, $h_{2j}(\mathbf{X}_2)$ $(j = 1, 2, \ldots, p_2)$ have similar formulations to (11).

δ_{e1} is chosen as follows:

$$\delta_{e1} = \frac{1}{2}\hat{E}\hat{\varphi}_2\mathbf{h}_2^T(\mathbf{X}_2)\mathbf{h}_2(\mathbf{X}_2), \tag{56}$$

where $\varphi_2 = \|\mathbf{W}_2^*\|^2$, $\hat{\varphi}_2$ is the estimation of φ_2 with the following adaptive law:

$$\dot{\hat{\varphi}}_2 = \frac{\lambda_2}{2}\hat{E}^2\mathbf{h}_2^T(\mathbf{X}_2)\mathbf{h}_2(\mathbf{X}_2) - 2\hat{\varphi}_2, \tag{57}$$

with $\lambda_2 > 0$.

Theorem 4. *Consider the closed-loop system consisting of plant (41) under Assumption 2 with controller (50) and adaptive law (58). Then all the signals involved are semi-globally uniformly ultimately bounded.*

Proof. Define

$$\tilde{\varphi}_2 = \hat{\varphi}_2 - \varphi_2. \tag{58}$$

Substituting (49), (55), and (56) into (48), we have

$$\dot{\hat{E}} = -k_{h1}\hat{E} - \frac{1}{2}\hat{E}\hat{\varphi}_2\mathbf{h}_2^T(\mathbf{X}_2)\mathbf{h}_2(\mathbf{X}_2) + \mathbf{W}_2^{*T}\mathbf{h}_2(\mathbf{X}_2) + \iota_2. \tag{59}$$

Design the following Lyapunov function candidate:

$$W_h = \frac{1}{2}\hat{E}^2 + \frac{\tilde{\varphi}_2^2}{2\lambda_2}. \tag{60}$$

Differentiating along (60) with respect to time and substituting (57), (58), and (59), we obtain

$$\begin{aligned}\dot{W}_h &= \hat{E}\dot{\hat{E}} + \frac{\tilde{\varphi}_2\dot{\hat{\varphi}}_2}{\lambda_2}\\ &= -k_{h1}\hat{E}^2 - \frac{1}{2}\hat{E}^2\varphi_2\mathbf{h}_2^T(\mathbf{X}_2)\mathbf{h}_2(\mathbf{X}_2) + \hat{E}\mathbf{W}_2^{*T}\mathbf{h}_2(\mathbf{X}_2)\\ &\quad + \iota_2\hat{E} - \frac{2\tilde{\varphi}_2\hat{\varphi}_2}{\lambda_2}.\end{aligned} \tag{61}$$

Notice that

$$\begin{aligned}\hat{E}\mathbf{W}_2^{*T}\mathbf{h}_2(\mathbf{X}_2) &\le \frac{\hat{E}^2}{2}\|\mathbf{W}_2^{*T}\mathbf{h}_2(\mathbf{X}_2)\|^2 + \frac{1}{2}\\ &= \frac{\hat{E}^2}{2}\|\mathbf{W}_2^*\|^2\|\mathbf{h}_2(\mathbf{X}_2)\|^2 + \frac{1}{2}\\ &= \frac{\hat{E}^2}{2}\varphi_2\mathbf{h}_2^T(\mathbf{X}_2)\mathbf{h}_2(\mathbf{X}_2) + \frac{1}{2}\,\frac{2\tilde{\varphi}_2\hat{\varphi}_2}{\lambda_2}\\ &\ge \frac{\tilde{\varphi}_2^2}{\lambda_2} - \frac{\varphi_2^2}{\lambda_2},\\ \hat{E}\iota_2 &\le \frac{1}{2}\hat{E}^2\iota_{2M}^2 + \frac{1}{2}.\end{aligned} \tag{62}$$

Thus, the following inequality is satisfied:

$$\dot{W}_h = -\left(k_{h1} - \frac{1}{2}\iota_{2M}^2\right)\hat{E}^2 - \frac{1}{\lambda_2}\tilde{\varphi}_2^2 + 1 + \frac{\varphi_2^2}{\lambda_2}. \tag{63}$$

Table 1: Initial trim conditions.

Item	Value	Units
V	7700	ft/s
h	85,000	ft
γ	0	deg
θ	1.6325	deg
Q	0	deg/s
η_1	0.9700	—
$\dot{\eta}_1$	0	—
η_2	0.7967	—
$\dot{\eta}_2$	0	—

Let $k_{h1} > \iota_{2M}^2/2$ and define the following compact sets:

$$\begin{aligned}\Omega_{\hat{E}} &= \left\{\hat{E}\,\middle|\,|\hat{E}| \le \sqrt{\frac{1 + (\varphi_2^2/\lambda_2)}{k_{h1} - (1/2)\iota_{2M}^2}}\right\},\\ \Omega_{\tilde{\varphi}_2} &= \left\{\tilde{\varphi}_2\,\middle|\,|\tilde{\varphi}_2| \le \sqrt{\frac{1 + (\varphi_2^2/\lambda_2)}{(1/\lambda_2)}}\right\}.\end{aligned} \tag{64}$$

It is clear that $\dot{W}_h$ will be negative if $\hat{E} \notin \Omega_{\hat{E}}$ or $\tilde{\varphi}_2 \notin \Omega_{\tilde{\varphi}_2}$. Therefore, $\hat{E}$ and $\tilde{\varphi}_2$ are semiglobally uniformly ultimately bounded. The proof is completed.

Remark 7. From (43), (44), (45), (46), (47), and (48), we get that

$$\begin{aligned}\hat{E} &= \ddot{e}_0 + 3\mu\dot{\hat{e}}_0 + 3\mu^2 e_0 + \mu^3\int_0^t e_0 d\tau\\ &= \ddot{e}_0 + s_2 - s_4 + 3\mu(\dot{e}_0 + s_1 - s_3) + 3\mu^2 e_0 + \mu^3\int_0^t e_0 d\tau\\ &= E + s_2 - s_4 + 3\mu(s_1 - s_3).\end{aligned} \tag{65}$$

It further follows that

$$|E| = |\hat{E} - s_2 + s_4 - 3\mu(s_1 - s_3)| \le |\hat{E}| + \bar{s}_2 + \bar{s}_4 + 3\mu(\bar{s}_1 + \bar{s}_3). \tag{66}$$

Thus, E and e_0 are also bounded. That is, γ can converge to γ_d.

Remark 8. Different from the affine control strategies [26–30], the proposed controller is derived from a nonaffine model such that it exhibits better practicability. Moreover, the prior information [30] that the unknown control gains approximated by NNs have to be strictly positive is not needed in this paper.

Remark 9. It is obvious that there is no need of the complicated design process of backstepping. Compared with [26, 30], the architecture of the developed controller is more concise. Meanwhile, the problem of "explosion of terms" is successfully eliminated.

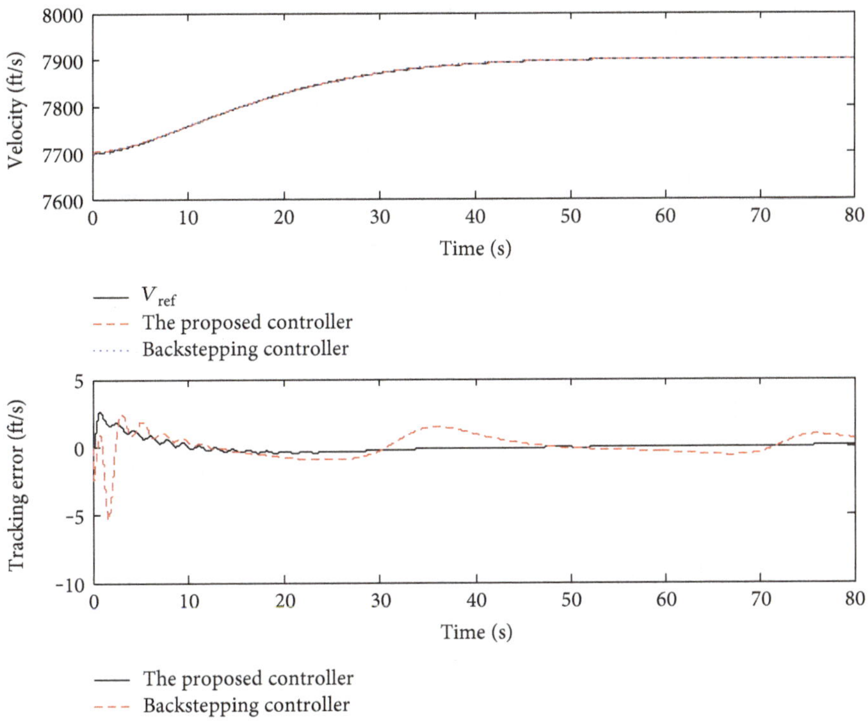

FIGURE 2: Velocity tracking performance.

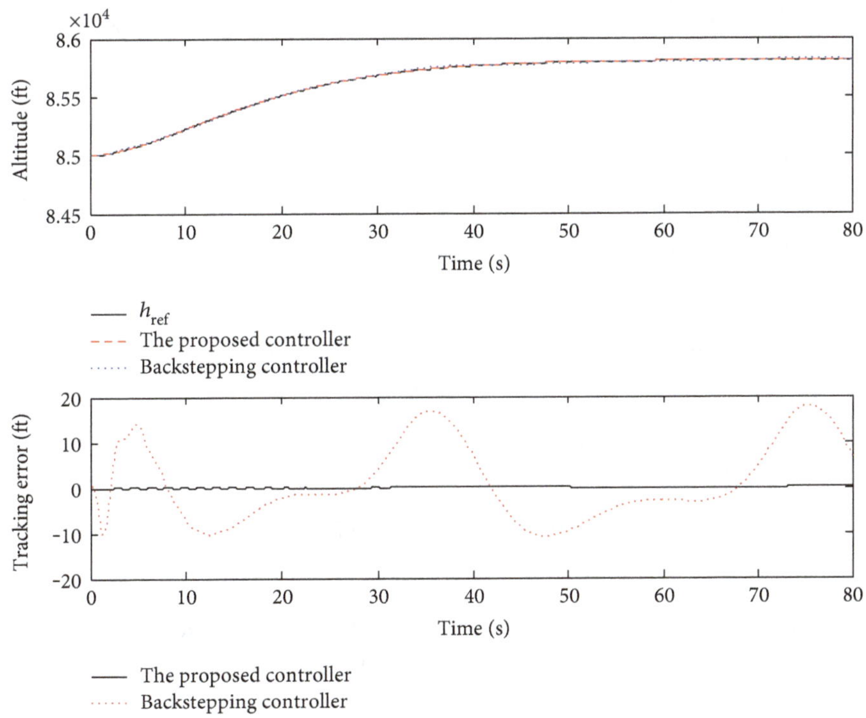

FIGURE 3: Altitude tracking performance.

Remark 10. Throughout this paper, only two NNs are applied to approach the lumped uncertainties of both subsystems. Furthermore, by the merit of the MLP scheme, there are totally only two learning parameters (i.e., $\widehat{\varphi}_1$ and $\widehat{\varphi}_2$) required for neural approximation. In comparison with [30], a lower computational burden design is achieved in this study.

Remark 11. If the input vectors of RBFNNs are chosen as $\mathbf{X}_1 = [V, \Phi]^T$ and $\mathbf{X}_2 = [\mathbf{x}^T, \delta_e]^T$, it will lead to a fixed-point

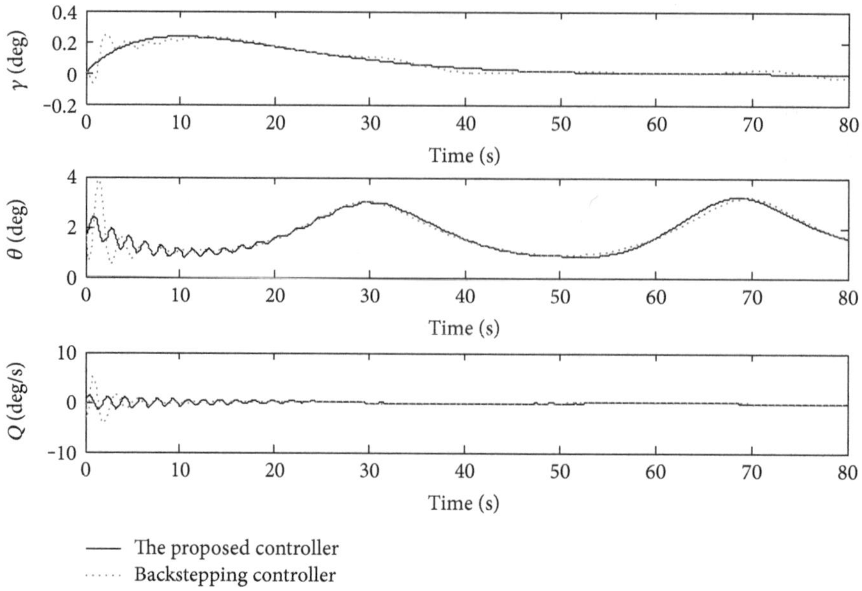

FIGURE 4: The attitude angles.

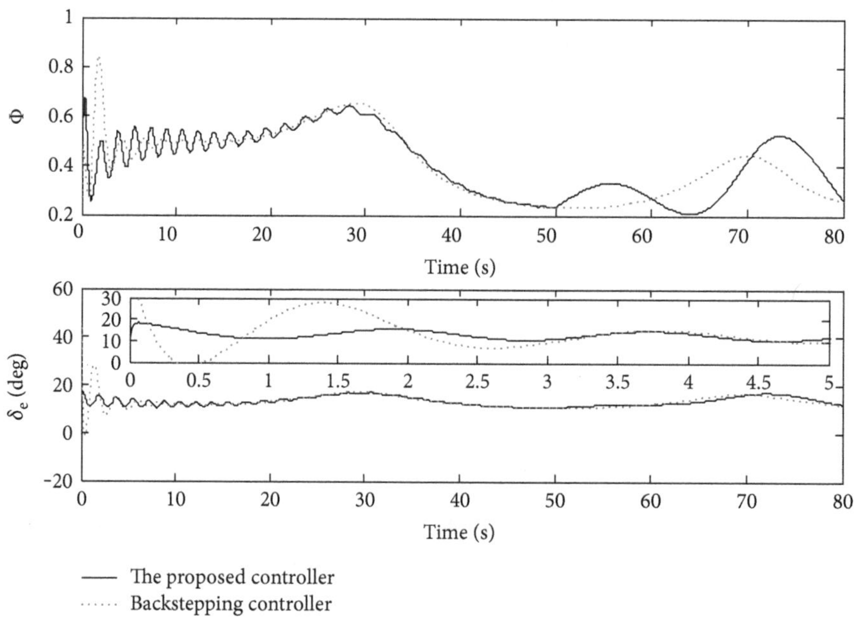

FIGURE 5: The control inputs.

problem, which will result in a computational burden. In this paper, such problem is avoided.

4. Simulation Results

In this section, we will make simulation studies to validate the tracking performance of the proposed control methodology. All the aerodynamic coefficients and parameters in (1), (2), (3), (4), (5), (6), and (7) are the same as [33]. The initial trim conditions of the vehicle are shown in Table 1. The reference trajectories of velocity and altitude are smoothed via the following filter:

$$\frac{0.1^2}{s^2 + 2 \times 0.9 \times 0.1 \times s + 0.1^2}. \quad (67)$$

The centers of RBFNNs are evenly spaced in $\mathbf{c}_1 \in [7700$ ft/s, 8700 ft/s$] \times [-0.1, 1]$, $\mathbf{c}_2 \in [-1$ deg, 1 deg$] \times [0$ deg, 5 deg$] \times [-5$ deg/s, 5 deg/s$] \times [0$ rad, 0.35 rad$]$. The weight vectors are chosen as $\mathbf{b}_1 = [33.33$ ft/s, $0.0367]^T$, $\mathbf{b}_2 = [0.0012$ rad, 0.0029 rad, 0.0058 rad/s, 0.0117 rad$]^T$. The node number is

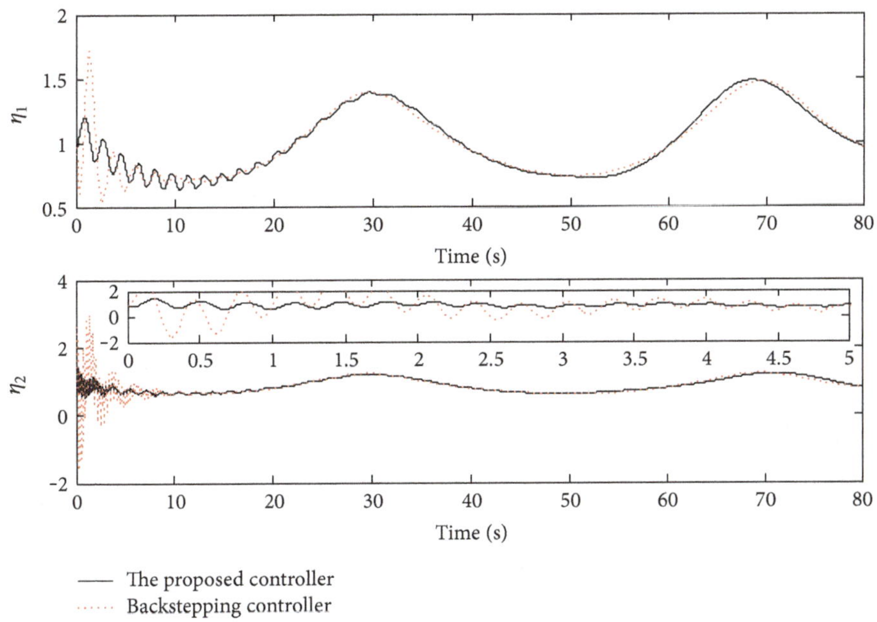

FIGURE 6: The flexible states.

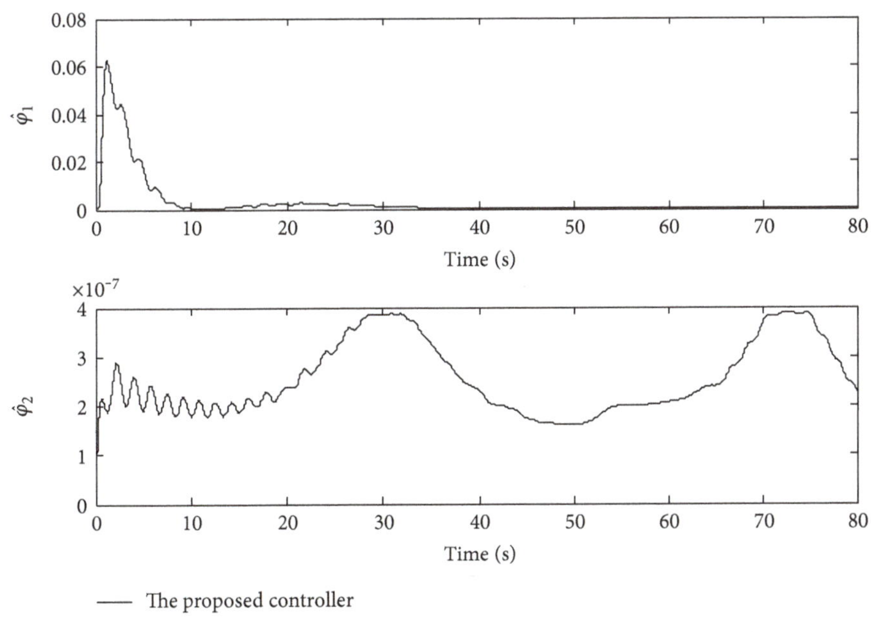

FIGURE 7: The estimations of φ_1 and φ_2.

set as $p_1 = p_2 = 20$. The design parameters are chosen as follows: $k_V = 0.9, k_{V1} = 0.2, k_{V2} = 0.8, k_\gamma = 2, k_h = 0.9, k_{h1} = 50$, $\mu = 7$, $\lambda_1 = \lambda_2 = 0.05$, $R = 0.05$, $a_1 = a_3 = 0.5, a_2 = a_4 = 0.1$. To show the superiority, the proposed controller is compared with a traditional backstepping control strategy presented in [19]. The following two cases are considered.

Case 1. In this case, the AHV is assumed to climb a maneuver from the initial trim conditions, listed in Table 1, to the final values $h = 85800$ ft and $V = 7900$ ft/s. To test the robustness of the designed controller, we define

$$C = C_0[1 + 0.4 \sin(0.05\pi t)], \tag{68}$$

where C means the value of uncertain coefficient and C_0 stands for the nominal value of C. With this definition, a parameter uncertainty up to 40% of the nominal value is considered.

The simulation results are shown in Figures 2, 3, 4, 5, 6, and 7. From Figures 2 and 3, it can be seen that the obtained velocity and altitude tracking performance is

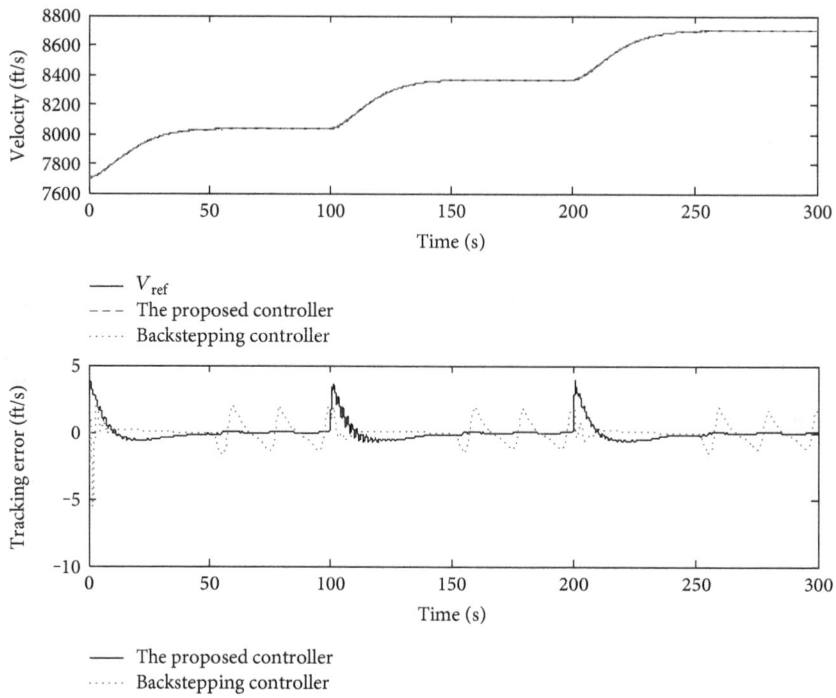

FIGURE 8: Velocity tracking and tracking error.

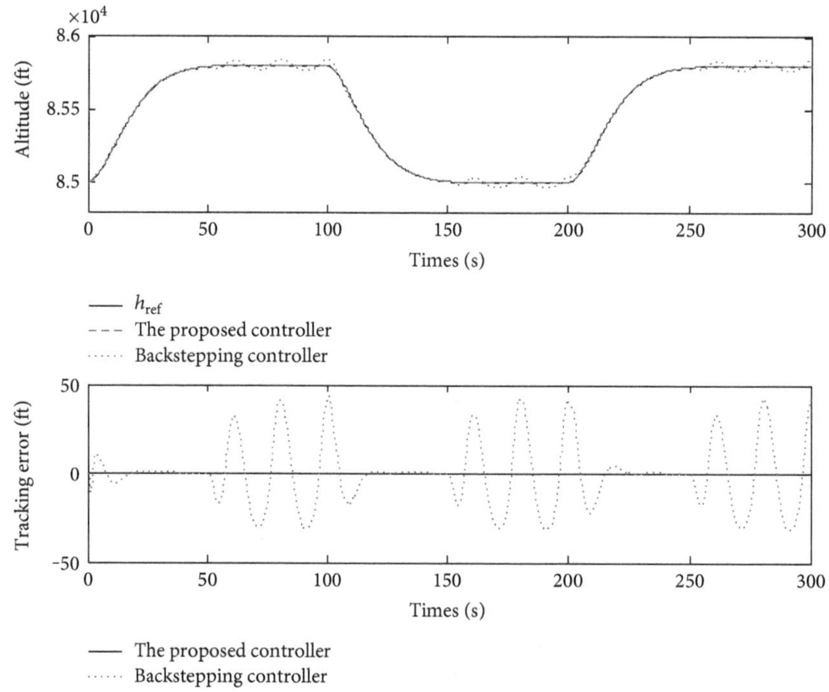

FIGURE 9: Altitude tracking and tracking error.

better than the one provided by the backstepping controller. Meanwhile, the attitude angles, the control inputs, and the flexible states, achieved by both control schemes, are smooth and without high-frequency chattering, as depicted in Figures 4, 5, and 6. Figure 7 reveals that the estimations of φ_1 and φ_2 are bounded.

Case 2. In this case, we take into consideration a more practical scenario. The altitude follows a square command with period 200 s and magnitude 800 ft. The velocity tracks the step command with 333.33 ft/s for every 100 seconds. The following condition of parametric uncertainty is considered.

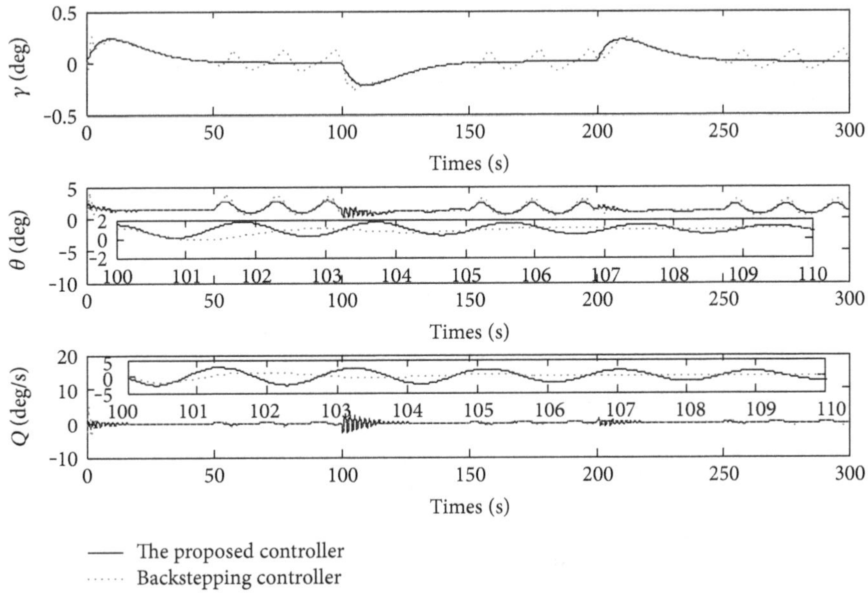

FIGURE 10: The flight-path angle, pitch angle, and pitch rate.

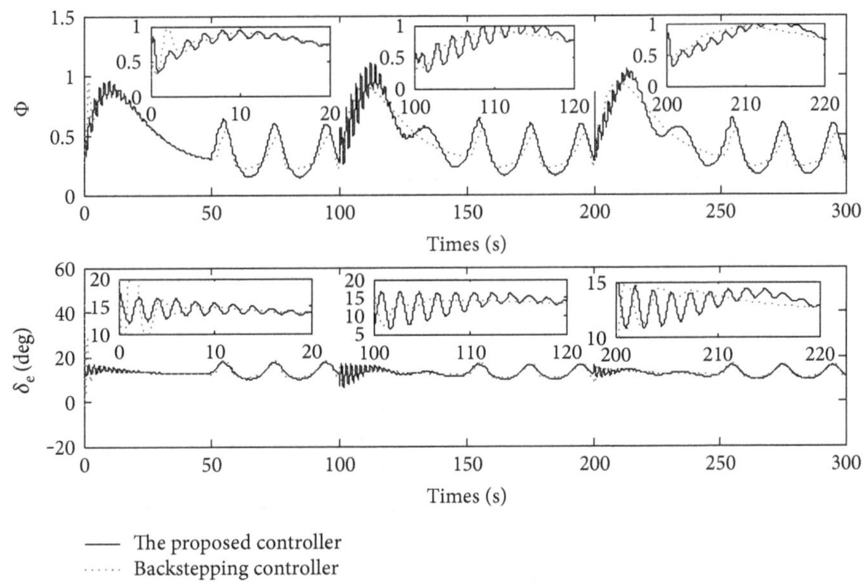

FIGURE 11: The control inputs.

$$C = \begin{cases} C_0, & 0\,\mathrm{s} \le t < 50\,\mathrm{s}, \\ C_0[1 + 0.4\sin(0.1\pi t)], & 50\,\mathrm{s} \le t < 100\,\mathrm{s}, \\ C_0, & 100\,\mathrm{s} \le t < 150\,\mathrm{s}, \\ C_0[1 + 0.4\sin(0.1\pi t)], & 150\,\mathrm{s} \le t < 200\,\mathrm{s}, \\ C_0, & 200\,\mathrm{s} \le t < 250\,\mathrm{s}, \\ C_0[1 + 0.4\sin(0.1\pi t)], & 250\,\mathrm{s} \le t \le 300\,\mathrm{s}. \end{cases} \quad (69)$$

The obtained simulation results, presented in Figures 8, 9, 10, 11, 12, and 13, indicate that the exploited control methodology can provide stable tracking of velocity and altitude commands in the presence of severe uncertainties. Moreover, it is observed from Figures 8 and 9 that compared with backstepping control, much smaller tracking errors are achieved by the explored control algorithm when the model parameters are uncertain. The attitude angles, the control inputs, and the flexible states, shown in Figures 10, 11, and 12, are even and can converge to stable values. Finally, the boundedness of $\hat{\varphi}_1$ and $\hat{\varphi}_2$ is presented in Figure 13.

5. Conclusions

This paper investigates the design of a concise neural controller for an AHV with parametric uncertainties. The vehicle

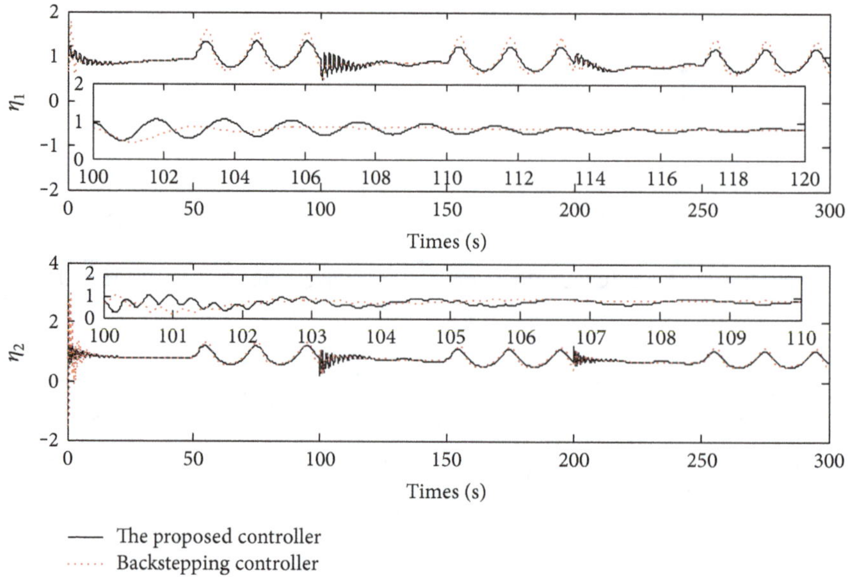

FIGURE 12: The flexible states.

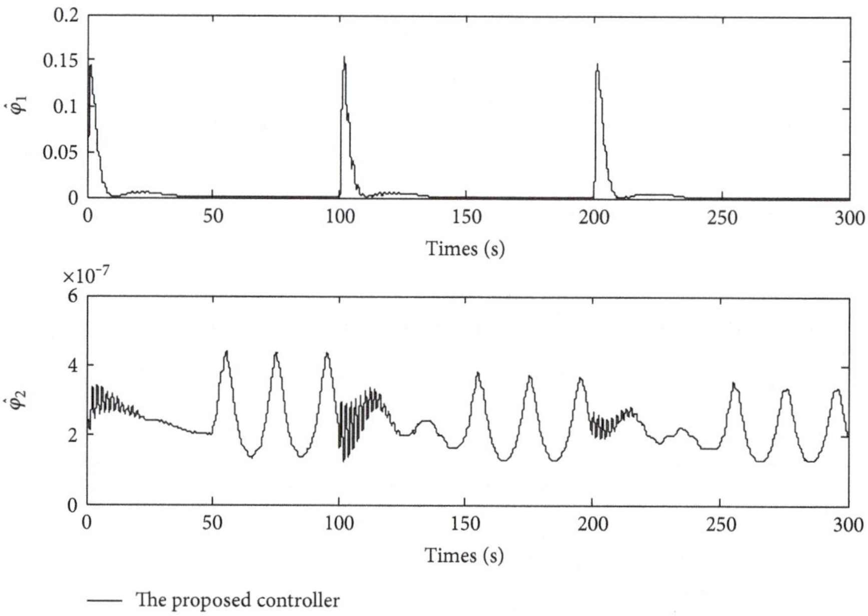

FIGURE 13: The estimations of φ_1 and φ_2.

dynamics is decomposed into the respective velocity and altitude subsystems that are controlled separately. For both subsystems, simplified neural controllers are devised directly employing nonaffine models. Specially, a novel control scheme with a concise structure is proposed for the altitude dynamics without utilizing backstepping. Throughout this paper, only two NNs are applied and only two learning parameters are required for neural approximation. It is proved that all the closed-loop system signals are semiglobally uniformly ultimately bounded. Finally, the superiority of the design in command tracking over the traditional backstepping control is well validated by simulation results in the presence of model uncertainties.

Nomenclature

m: Vehicle mass
ρ: Density of air
$\bar{q}$: Dynamic pressure
S: Reference area
h: Altitude
V: Velocity
γ: Flight-path angle
θ: Pitch angle
α: Angle of attack ($\alpha = \theta - \gamma$)
Q: Pitch rate
T: Thrust

D: Drag
L: Lift
M: Pitching moment
I_{yy}: Moment of inertia
$\bar{c}$: Aerodynamic chord
z_T: Thrust moment arm
Φ: Fuel equivalence ratio
δ_e: Elevator angular deflection
N_i: ith generalized force
$N_i^{\alpha_j}$: jth order contribution of α to N_i
N_i^0: Constant term in N_i
$N_2^{\delta_e}$: Contribution of δ_e to N_2
$\beta_i(h, \bar{q})$: ith trust fit parameter
η_i: ith generalized elastic coordinate
ζ_i: Damping ratio for elastic mode η_i
ω_i: Natural frequency for elastic mode η_i
$C_D^{\alpha^i}$: ith order coefficient of α in D
$C_D^{\delta_e^i}$: ith order coefficient of δ_e in D
C_D^0: Constant coefficient in D
$C_L^{\alpha^i}$: ith order coefficient of α in L
$C_L^{\delta_e}$: Coefficient of δ_e contribution in L
C_L^0: Constant coefficient in L
$C_{M,\alpha}^{\alpha^i}$: ith order coefficient of α in M
$C_{M,\alpha}^0$: Constant coefficient in M
$C_T^{\alpha^i}$: ith order coefficient of α in T
C_T^0: Constant coefficient in T
h_0: Nominal altitude for air density approximation
ρ_0: Air density at the altitude h_0
$\tilde{\psi}_i$: Constrained beam coupling constant for η_i
c_e: Coefficient of δ_e in M
$1/h_s$: Air density decay rate
$\mathbf{R}^n$: n-dimensional Euclidean space
$\mathbf{R}$: The set of all real numbers
$||\bullet||$: The 2-norm of a vector
$|\bullet|$: The absolute value of a scalar.

Conflicts of Interest

The authors declare that they have no conflicts of interest.

Acknowledgments

This work was supported by the National Natural Science Foundation of China (Grant nos. 61603410 and 61573374) and Young Talent Fund of University Association for Science and Technology in Shaanxi, China (Grant no. 20170107).

References

[1] L. Fiorentini, A. Serrani, M. A. Bolender, and D. B. Doman, "Nonlinear robust adaptive control of flexible air-breathing hypersonic vehicles," *Journal of Guidance, Control, and Dynamics*, vol. 32, no. 2, pp. 401–416, 2009.

[2] C. Mu, Z. Ni, C. Sun, and H. He, "Air-breathing hypersonic vehicle tracking control based on adaptive dynamic programming," *IEEE Transactions on Neural Networks and Learning Systems*, vol. 28, no. 3, pp. 584–598, 2017.

[3] X. W. Bu, X. Y. Wu, J. Q. Huang, and D. Z. Wei, "A guaranteed transient performance-based adaptive neural control scheme with low-complexity computation for flexible air-breathing hypersonic vehicles," *Nonlinear Dynamics*, vol. 84, no. 4, pp. 2175–2194, 2016.

[4] C. Mu, C. Sun, and W. Xu, "Fast sliding mode control on air-breathing hypersonic vehicles with transient response analysis," *Proceedings of the Institution of Mechanical Engineers, Part I: Journal of Systems and Control Engineering*, vol. 230, no. 1, pp. 23–34, 2015.

[5] X. W. Bu, X. Y. Wu, J. Q. Huang, and D. Z. Wei, "Robust estimation-free prescribed performance back-stepping control of air-breathing hypersonic vehicles without affine models," *International Journal of Control*, vol. 89, no. 11, pp. 2185–2200, 2016.

[6] H. B. Sun, S. H. Li, and C. Y. Sun, "Finite time integral sliding mode control of hypersonic vehicles," *Nonlinear Dynamics*, vol. 73, no. 1-2, pp. 229–244, 2013.

[7] X. W. Bu, X. Y. Wu, J. Q. Huang, Z. Ma, and R. Zhang, "Minimal-learning-parameter based simplified adaptive neural back-stepping control of flexible air-breathing hypersonic vehicles without virtual controllers," *Neurocomputing*, vol. 175, pp. 816–825, 2016.

[8] J. M. Wang, Y. J. Wu, and X. M. Dong, "Recursive terminal sliding mode control for hypersonic flight vehicle with sliding mode disturbance observer," *Nonlinear Dynamics*, vol. 81, no. 3, pp. 1489–1510, 2015.

[9] X. X. Hu, L. G. Wu, C. H. Hu, and H. J. Gao, "Adaptive sliding mode tracking control for a flexible air-breathing hypersonic vehicle," *Journal of the Franklin Institute*, vol. 349, no. 2, pp. 559–577, 2012.

[10] J. Wang, Q. Zong, R. Su, and B. L. Tian, "Continuous high order sliding mode controller design for a flexible air-breathing hypersonic vehicle," *ISA Transactions*, vol. 53, no. 3, pp. 690–698, 2014.

[11] Q. K. Shen, B. Jiang, and V. Cocquempot, "Fuzzy logic system-based adaptive fault-tolerant control for near-space vehicle attitude dynamics with actuator faults," *IEEE Transactions on Fuzzy Systems*, vol. 21, no. 2, pp. 289–300, 2013.

[12] Y. Q. Huang, C. Y. Sun, C. S. Qian, J. M. Zhang, and L. Wang, "Polytopic LPV modeling and gain-scheduled switching control for a flexible air-breathing hypersonic vehicle," *Journal of Systems Engineering and Electronics*, vol. 24, no. 1, pp. 118–127, 2003.

[13] Y. Q. Huang, C. Y. Sun, C. S. Qian, and L. Wang, "Non-fragile switching tracking control for a flexible air-breathing hypersonic vehicle based on polytopic LPV model," *Chinese Journal of Aeronautics*, vol. 26, no. 4, pp. 948–959, 2013.

[14] X. L. Shao and H. L. Wang, "Active disturbance rejection based trajectory linearization control for hypersonic reentry vehicle with bounded uncertainties," *ISA Transactions*, vol. 54, pp. 27–38, 2015.

[15] H. F. Sun, Z. L. Yang, and B. Meng, "Tracking control of a class of non-linear systems with applications to cruise control of air-breathing hypersonic vehicles," *International Journal of Control*, vol. 88, no. 5, pp. 885–896, 2015.

[16] J. Yang, S. H. Li, C. Y. Sun, and L. Guo, "Nonlinear-disturbance-observer-based robust flight control for airbreathing hypersonic vehicles," *IEEE Transactions on Aerospace and Electronic Systems*, vol. 49, no. 2, pp. 1263–1275, 2013.

[17] X. W. Bu, X. Y. Wu, Z. Ma, R. Zhang, and J. Q. Zhang, "Novel auxiliary error compensation design for the adaptive neural control of a constrained flexible air-breathing hypersonic vehicle," *Neurocomputing*, vol. 171, pp. 313–324, 2016.

[18] D. X. Gao, S. X. Wang, and H. J. Zhang, "A singularly perturbed system approach to adaptive neural back-stepping control design of hypersonic vehicles," *Journal of Intelligent & Robotic Systems*, vol. 73, no. 1–4, pp. 249–259, 2014.

[19] X. W. Bu, X. Y. Wu, R. Zhang, Z. Ma, and J. Q. Huang, "Tracking differentiator design for the robust backstepping control of a flexible air-breathing hypersonic vehicle," *Journal of the Franklin Institute*, vol. 352, no. 4, pp. 1739–1765, 2015.

[20] Q. Zong, F. Wang, R. Su, and S. K. Shao, "Robust adaptive backstepping tracking control for a flexible air-breathing hypersonic vehicle subject to input constraint," *Proceedings of the Institution of Mechanical Engineers Part G: Journal of Aerospace Engineering*, vol. 229, no. 1, pp. 10–25, 2014.

[21] X. W. Bu, X. Y. Wu, Y. X. Chen, and R. Y. Bai, "Design of a class of new nonlinear disturbance observers based on tracking differentiators for uncertain dynamic systems," *International Journal of Control, Automation, and Systems*, vol. 13, no. 3, pp. 595–602, 2015.

[22] D. Wang and C. Mu, "A novel neural optimal control framework with nonlinear dynamics: closed-loop stability and simulation verification," *Neurocomputing*, vol. 266, pp. 353–360, 2017.

[23] X. W. Bu, X. Y. Wu, M. Y. Tian, J. Q. Huang, R. Zhang, and Z. Ma, "High-order tracking differentiator based adaptive neural control of a flexible air-breathing hypersonic vehicle subject to actuators constraints," *ISA Transactions*, vol. 58, pp. 237–247, 2015.

[24] D. Wang, C. Mu, and D. Liu, "Data-driven nonlinear near-optimal regulation based on iterative neural dynamic programming," *Acta Automatica Sinica*, vol. 43, no. 3, pp. 366–375, 2017.

[25] X. W. Bu, X. Y. Wu, D. Z. Wei, and J. Q. Huang, "Neural-approximation-based robust adaptive control of flexible air-breathing hypersonic vehicles with parametric uncertainties and control input constraints," *Information Sciences*, vol. 346-347, pp. 29–43, 2016.

[26] Q. Zong, F. Wang, B. L. Tian, and R. Su, "Robust adaptive dynamic surface control design for a flexible air-breathing hypersonic vehicle with input constraints and uncertainty," *Nonlinear Dynamics*, vol. 78, no. 1, pp. 289–315, 2014.

[27] W. A. Butt, L. Yan, and A. S. Kendrick, "Adaptive integral dynamic surface control of a hypersonic flight vehicle," *International Journal of Systems Science*, vol. 46, no. 10, pp. 1717–1728, 2015.

[28] B. Xu, "Robust adaptive neural control of flexible hypersonic flight vehicle with dead-zone input nonlinearity," *Nonlinear Dynamics*, vol. 80, no. 3, pp. 1509–1520, 2015.

[29] B. Xu and Y. Zhang, "Neural discrete back-stepping control of hypersonic flight vehicle with equivalent prediction model," *Neurocomputing*, vol. 154, pp. 337–346, 2015.

[30] X. W. Bu, X. Y. Wu, Z. Ma, and R. Zhang, "Nonsingular direct neural control of air-breathing hypersonic vehicle via back-stepping," *Neurocomputing*, vol. 153, pp. 164–173, 2015.

[31] B. Xu, Y. H. Fan, and S. M. Zhang, "Minimal-learning-parameter technique based adaptive neural control of hypersonic flight dynamics without back-stepping," *Neurocomputing*, vol. 164, pp. 201–209, 2015.

[32] M. A. Bolender and D. B. Doman, "Nonlinear longitudinal dynamical model of an air-breathing hypersonic vehicle," *Journal of Spacecraft and Rockets*, vol. 44, no. 2, pp. 374–387, 2007.

[33] J. T. Parker, A. Serrani, S. Yurkovich, M. A. Bolender, and D. B. Doman, "Control-oriented modeling of an air-breathing hypersonic vehicle," *Journal of Guidance, Control, and Dynamics*, vol. 30, no. 3, pp. 856–869, 2007.

[34] R. M. Sanner and J. E. Slotine, "Gaussian networks for direct adaptive control," *IEEE Transactions on Neural Networks*, vol. 3, no. 6, pp. 837–863, 1992.

[35] X. Wang, Z. Chen, and Z. Yuan, "Design and analysis for new discrete tracking-differentiators," *Applied Mathematics-A Journal of Chinese Universities Series B*, vol. 18, no. 2, pp. 214–222, 2003.

[36] S. S. Ge, C. C. Hang, T. Lee, and T. Zhang, *Stable Adaptive Neural Network Control*, Kluwer Academic, USA, 2002.

[37] J. H. Park, S. H. Huh, S. H. Kim, S. J. Seo, and G. T. Park, "Direct adaptive controller for nonaffine nonlinear systems using self-structuring neural networks," *IEEE Transactions on Neural Networks*, vol. 16, no. 2, pp. 414–422, 2005.

[38] A. J. Calise, N. Hovakimyan, and M. Idan, "Adaptive output feedback control of nonlinear systems using neural networks," *Automatica*, vol. 37, no. 1, pp. 1201–1211, 2001.

2

Nonlinear Dynamics of the High-Speed Rotating Plate

Minghui Yao, Li Ma, and Wei Zhang

Beijing Key Laboratory of Nonlinear Vibrations and Strength of Mechanical Structures, College of Mechanical Engineering, Beijing University of Technology, Beijing 100124, China

Correspondence should be addressed to Minghui Yao; merry_mingming@163.com

Academic Editor: Christopher J. Damaren

High speed rotating blades are crucial components of modern large aircraft engines. The rotating blades under working condition frequently suffer from the aerodynamic, elastic and inertia loads, which may lead to large amplitude nonlinear oscillations. This paper investigates nonlinear dynamic responses of the blade with varying rotating speed in supersonic airflow. The blade is simplified as a pre-twist and presetting cantilever composite plate. Warping effect of the rectangular cross-section of the plate is considered. Based on the first-order shear deformation theory and von-Karman nonlinear geometric relationship, nonlinear partial differential dynamic equations of motion for the plate are derived by using Hamilton's principle. Galerkin approach is applied to discretize the partial differential governing equations of motion to ordinary differential equations. Asymptotic perturbation method is exploited to derive four-degree-of-freedom averaged equation for the case of 1 : 3 internal resonance-1/2 sub-harmonic resonance. Based on the averaged equation, numerical simulation is used to analyze the influence of the perturbation rotating speed on nonlinear dynamic responses of the blade. Bifurcation diagram, phase portraits, waveforms and power spectrum prove that periodic motion and chaotic motion exist in nonlinear vibration of the rotating cantilever composite plate.

1. Introduction

High-speed rotating blades are essential components of modern large aircraft engines. Blades are designed with large aspect ratios and thin-wall structure in order to raise operational efficiency. In the actual working condition, the rotating blades are frequently subjected to the aerodynamic, elastic, and inertia loads. Various types of excitation lead to large amplitude nonlinear parametric vibrations of the blades, which can result in the resonance phenomena and undesirable disasters, especially when the rotating blades operate with high speed and huge centrifugal force. According to the investigation, the vibration failure of the aircraft engine is more than 60% of the total failure, while the vibration failure of the blade accounts for more than 70% of the total vibration failure. Resonance and flutter produced by the forced vibration and self-excited vibration are the main reason leading to the blade failure. So, it is very important that a reasonable model is established to accurately predict nonlinear vibration characteristics and other complex dynamics of the blade.

In recent years, considerable attention has been given to the studies on the vibration characteristics of rotating blades. Lin et al. [1] deduced governing differential equations and general elastic boundary conditions of a nonuniform pretwist Timoshenko beam by using Hamilton's principle. Yao et al. [2] investigated the nonlinear dynamic responses of the thin-walled rotating cantilever beam. Sarkar and Ganguli [3] assumed the modal function as polynomials satisfying all four boundary conditions and discussed free vibration of the nonhomogeneous Timoshenko beam. Georgiades et al. [4] derived equations of motion of a rotating composite Timoshenko beam by utilizing Hamilton's principle. Xie et al. [5, 6] numerically investigated the effect of symmetric and asymmetric shroud gaps, rotational speeds, and the aerodynamic force amplitude on dynamic characteristics of the rotating Euler-Bernoulli beam with a mass point at the free end. Huang and Kuang [7] investigated the effect of a near root local blade crack on the stability of a bladed disk, in which an individual blade is modeled simply as a cantilever Euler-Bernoulli beam.

Currently, plenty of theoretical analyses are focused on the beam model of the blade, such as the simple Euler-Bernoulli beam and the pretwist Timoshenko beam. But, theoretical studies on the plate model of the blade are still few. Young and Liou [8] established equations of a rotating cantilever plate with a time-varying speed and numerically investigated the effect of the Coriolis force on boundaries of the unstable regions. Yoo and Kim [9] derived linear equations of motion and analyzed free vibration of rotating cantilever plates. Sinha and Turner [10] derived governing partial differential equations of motion for a rotating pretwist plate to investigate the static and dynamic frequencies of the blade. Li and Zhang [11] presented a dynamic model of a functionally graded rectangular plate undergoing large overall motions. Kouchakzadeh et al. [12] analyzed the nonlinear aeroelasticity of a laminated composite plate in supersonic airflow by using the classical plate theory. Malgaca et al. [13] tried to control the vibration of the rotating blade at different speeds by utilizing root-embedded piezoelectric materials. Kam et al. [14] proposed a procedure to analyze the structural failure of a composite wind blade subjected to quasi-static loads. Tang and Chen [15] presented the solvability conditions of nonlinear partial differential equations for in-plane moving plates of two cases, in which the one case is without internal resonance and another case is under internal resonances, respectively. Sun et al. [16] used a quadratic layerwise theory and a new dynamic model to study dynamic behaviors for a multilayer pretwist rotating blade. Wang et al. [17] established a time-dependent nonlinear model of a flexible blade-rotor-bearing system by using the Lagrange method. Shariyat et al. [18] performed the nonlinear dynamic analysis of rectangular composite plates. Zhang and Li [19] adopted the Lagrangian method to acquire dynamic equations of motion for the pretwist and predeformed rotating cantilever plate subjected to the harmonic aerodynamic force. Mendonça et al. [20] considered internal damping in the shaft to study dynamic behaviors of the rotors mounted on composite shafts. Banichuk et al. [21, 22] investigated the stability and bifurcation of the rotating blade under different conditions.

Based on the shallow shell theory and the Ritz method, Leissa et al. [23] explored frequencies and mode shapes of turbomachinery blades with the coupling of bending and twist. Kee and Kim [24] assumed blades as the moderately thick open cylindrical shell models. Sun et al. [25] applied the general shell theory to investigate the influence of parameters on natural frequency and damping characteristics of the shell model blade. Sinha and Zylka [26] simplified the rotating pretwisted turbomachinery cantilevered airfoil as an anisotropic shell and derived the free vibration equations of motion for the transverse deflection of the shell including the warping effect. Volker and Joachim [27] analyzed the aeroelastic phenomena of twenty compressor blades simplified as the spring-damper models mounted on the hub. Ekici et al. [28] proposed a nonlinear harmonic balance method to compute the unsteady self-excited aerodynamic of asymmetry turbomachinery blades. Farhadi and Hosseini-Hashemi [29] studied aeroelastic behaviors of a rotating thick plate in the supersonic airflow. Lachenal et al. [30] presented the design, analysis, and realization of a zero stiffness twist morphing wind turbine blade subjected to the gust loads.

Dynamic modeling of the rotating blade requires an accurate expression of the aerodynamic force. But, the physical mechanism of aerodynamic interaction has been mainly investigated by experimental methods. Models have been proposed to deal with aerodynamic interaction problems, which is done with by the piston theory widely. Ashley and Zartarian [31] firstly proposed the quasi-steady piston theory to deal with aerodynamic interaction problems. Navazi and Haddadpour [32] studied the thermal stability of the functionally graded plate subjected to the aerodynamic load obtained from the first-order piston theory. According to large deformation geometric relationship, the piston theory and the quasi-static thermal stress theory, Yuan and Qiu [33] established the aerodynamic model of a composite stiffened panel and used Hamilton's principle to derive the equations of motion for the system. Yang et al. [34] applied a modified local piston theory to analyze aeroelastic behaviors of curved panels.

Although extensive studies have been carried out on rotating cantilever beams, studies on nonlinear dynamic responses of the pretwist, presetting rotating cantilever plates are still few. In this paper, nonlinear dynamic behaviors of the blade with varying rotating speeds under the supersonic airflow are investigated. Considering the shear deformation and the warping effect, equations of motion for the cantilever plate are derived by using Hamilton's principle. The Galerkin approach is applied to discretize the partial differential governing equations of motion to ordinary differential equations. The asymptotic perturbation method is exploited to obtain averaged equations of the system in the case of 1 : 3 internal resonance-1/2 subharmonic resonance. Based on the averaged equations, numerical simulation is applied to investigate the bifurcation and chaotic dynamics of the rotating cantilever composite plate. In order to analyze the internal resonance, we choose the perturbation rotating speed as the controlling parameter to investigate nonlinear behaviors of the pretwist and presetting rotating cantilever plate. From the results of the numerical simulation, it is found that the system performs periodic and chaotic motions under specific conditions. It is observed that the perturbation rotating speed has a significant influence on the nonlinear dynamic behaviors of the rotating plate. Since we can control the responses of the system from the chaotic motions to the periodic motions by changing the perturbation rotating speed, we can control the large amplitude nonlinear vibrations of the blade.

2. Equations of Motion for the Rotating Cantilevered Blade

The schematic diagram of the rotating cantilever blade is shown in Figure 1. The blade is simplified as a pretwist and

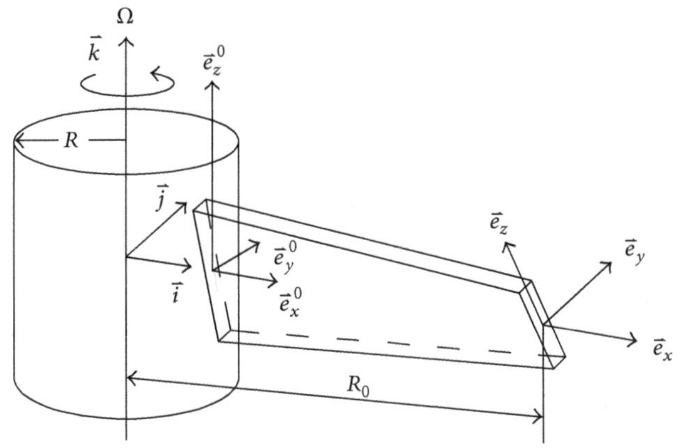

FIGURE 1: The model of the pretwist rotating cantilever rectangular plate.

presetting rotating cantilever plate in the following dynamical analysis. Rotating speed of the blade is $\Omega(t)=\Omega_0+\Omega_1\cos(\tilde{\omega}t)$, where Ω_0 is the steady-state rotating speed and $\Omega_1\cos(\tilde{\omega}t)$ is the periodic perturbation. The shape of the blade is a rectangular plate, which is characterized by the span length L, the chord length C, and the thickness h. The cross-section of the cantilever plate is shown in Figure 2. The plate is clamped to a rigid hub of radius r. There are a presetting angle θ_r in the fixed end and a pretwist angle θ_R at the free end. $(\vec{\mathbf{i}},\vec{\mathbf{j}},\vec{\mathbf{k}})$ is the inertial frame and the origin of it is in the center of the hub. At the edge of the hub, there is the rotating coordinate system $(\vec{\mathbf{e}}_x^0,\vec{\mathbf{e}}_y^0,\vec{\mathbf{e}}_z^0)$. We define that the spanwise direction is $\vec{\mathbf{e}}_x^0$, the chordwise direction is $\vec{\mathbf{e}}_y^0$, and the thickness direction is $\vec{\mathbf{e}}_z^0$. Another local triad system $(\hat{e}_x,\hat{e}_y,\hat{e}_z)$ attached to the free end of the plate is introduced on the plate, which is called the sectional coordinate system.

The twist rate θ' of the blade is expressed as

$$\theta'=\frac{(\theta_R-\theta_r)}{(R-r)}, \tag{1a}$$

and the twist angle $\theta(x)$ of the blade is written as

$$\theta(x)=\theta_r+\theta' x. \tag{1b}$$

Based on the Kirchhoff hypothesis, the first-order shear deformation theory including the warping effect is considered to establish the displacement field. Displacements of any point along x, y, and z directions can be expressed by the displacement of the neutral plane of the plate as follows [35]:

$$u(x,y,z,t)=u_0(x,y,t)+z\varphi_x(x,y,t)-z\left(\theta'\right)^2 x\omega_0, \tag{2a}$$

$$v(x,y,z,t)=v_0(x,y,t)+z\varphi_y(x,y,t), \tag{2b}$$

$$w(x,y,z,t)=w_0(x,y,t), \tag{2c}$$

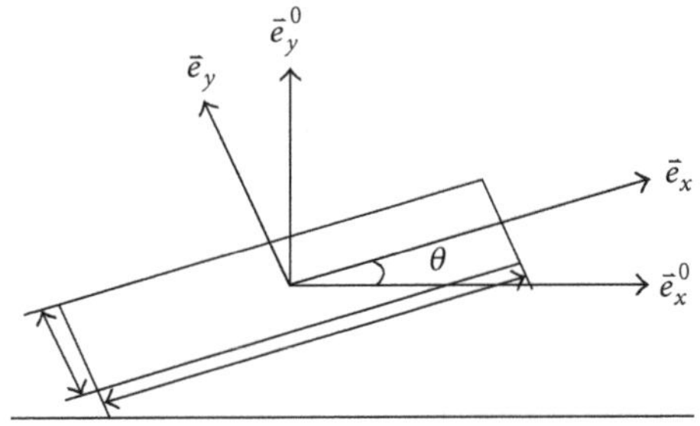

FIGURE 2: The section of the rotating cantilever plate.

where

$$\phi_x=\gamma_{xz}-\frac{\partial w_0}{\partial x},\qquad \phi_y=\gamma_{yz}-\frac{\partial w_0}{\partial y}. \tag{2d}$$

Nonlinear strains of the von Karman plate theory are given as

$$\begin{Bmatrix}\varepsilon_{xx}\\ \varepsilon_{yy}\\ \gamma_{yz}\\ \gamma_{xz}\\ \gamma_{xy}\end{Bmatrix}=\{\varepsilon^0\}+z\{\varepsilon^1\}=\begin{Bmatrix}\dfrac{\partial u_0}{\partial x}+\dfrac{1}{2}\left(\dfrac{\partial w_0}{\partial x}\right)^2\\ \dfrac{\partial v_0}{\partial y}+\dfrac{1}{2}\left(\dfrac{\partial w_0}{\partial y}\right)^2\\ \dfrac{\partial w_0}{\partial y}+\varphi_y\\ \dfrac{\partial w_0}{\partial x}+\varphi_x\\ \dfrac{\partial u_0}{\partial y}+\dfrac{\partial v_0}{\partial x}+\dfrac{\partial w_0}{\partial x}\dfrac{\partial w_0}{\partial y}\end{Bmatrix}+z\begin{Bmatrix}\dfrac{\partial\varphi_x}{\partial x}-\left(\theta'\right)^2 w_0-\left(\theta'\right)^2 xw_{0,x}\\ \dfrac{\partial\varphi_y}{\partial y}\\ 0\\ 0\\ \dfrac{\partial\varphi_x}{\partial y}+\dfrac{\partial\varphi_y}{\partial x}\end{Bmatrix}, \tag{3}$$

where $\{\varepsilon^0\}$ is the membrane strain and $\{\varepsilon^1\}$ is the flexural strain.

For the advanced fiber-reinforced composite material blade, the constitutive relation of the composite plate is expressed as follows:

$$\begin{Bmatrix}\sigma_{xx}\\ \sigma_{yy}\\ \sigma_{xy}\end{Bmatrix}=\begin{bmatrix}\bar{Q}_{11}&\bar{Q}_{12}&\bar{Q}_{16}\\ \bar{Q}_{12}&\bar{Q}_{22}&\bar{Q}_{26}\\ \bar{Q}_{16}&\bar{Q}_{26}&\bar{Q}_{66}\end{bmatrix}\begin{Bmatrix}\varepsilon_{xx}\\ \varepsilon_{yy}\\ \gamma_{xy}\end{Bmatrix}, \tag{4a}$$

$$\begin{Bmatrix}\sigma_{yz}\\ \sigma_{xz}\end{Bmatrix}=k\begin{bmatrix}\bar{Q}_{44}&\bar{Q}_{45}\\ \bar{Q}_{45}&\bar{Q}_{55}\end{bmatrix}\begin{Bmatrix}\gamma_{yz}\\ \gamma_{xz}\end{Bmatrix}, \tag{4b}$$

where

$$\begin{aligned} Q_{11} &= \frac{E_1}{1-\gamma_{12}\gamma_{21}}, \\ Q_{12} &= \frac{\gamma_{12}E_2}{1-\gamma_{12}\gamma_{21}}, \\ Q_{22} &= \frac{E_2}{1-\gamma_{12}\gamma_{21}}, \\ Q_{66} &= G_{12}, \\ Q_{44} &= G_{23}, \\ Q_{55} &= G_{13}, \end{aligned} \tag{4c}$$

where E_1 is the longitudinal modulus of the fiber, E_2 is the transverse modulus, γ_{12} and γ_{21} are the major Poisson's ratio, respectively, and G_{12}, G_{23}, and G_{13} are the shear modulus, respectively.

In-plane force resultants $\mathbf{N}_{xx}$, $\mathbf{N}_{yy}$, and $\mathbf{N}_{xy}$; the moment resultants $\mathbf{M}_{xx}$, $\mathbf{M}_{yy}$, and $\mathbf{M}_{xy}$; and the transverse forces $\mathbf{Q}_x$ and $\mathbf{Q}_y$ are written in matrices as follows:

$$\begin{Bmatrix} \{\mathbf{N}\} \\ \{\mathbf{M}\} \end{Bmatrix} = \begin{bmatrix} [A] & [B] \\ [B] & [D] \end{bmatrix} \begin{Bmatrix} \{\varepsilon^0\} \\ \{\varepsilon^1\} \end{Bmatrix}, \tag{5a}$$

$$\begin{Bmatrix} \mathbf{Q}_y \\ \mathbf{Q}_x \end{Bmatrix} = k \begin{bmatrix} A_{44} & A_{45} \\ A_{45} & A_{55} \end{bmatrix} \begin{Bmatrix} \frac{\partial w_0}{\partial y} + \phi_y \\ \frac{\partial w_0}{\partial x} + \phi_x \end{Bmatrix}, \tag{5b}$$

where k represents the shear coefficient, A_{ij} is called the extensional stiffness, D_{ij} is the bending stiffness, and B_{ij} is the bending-extensional coupling stiffness, which are defined as

$$[A_{ij},\ B_{ij},\ D_{ij}] = \int_{-h/2}^{h/2} Q_{ij}[1,\ z,\ z^2]d_z, \tag{6a}$$

I_0, I_1, and I_2 are mass moments of inertia, which are defined as

$$[I_0,\ I_1,\ I_2] = \int_{-h/2}^{h/2} \rho[1,\ z,\ z^2]d_z. \tag{6b}$$

The rotating plate mounted in the hub is rotating with the variable speed. The displacement of a random point P without deformation on the plate is expressed as

$$\vec{r}(x,y,z,t) = (R_0 + x)\vec{\mathbf{e}}_x + y\vec{\mathbf{e}}_y + z\vec{\mathbf{e}}_z, \tag{7a}$$

and the deformation displacement of a random point P on the rotating plate is written as

$$\vec{R}(x,y,z,t) = (x + R_0 + u)\vec{\mathbf{e}}_x + (y+v)\vec{\mathbf{e}}_y + (z+w)\vec{\mathbf{e}}_z. \tag{7b}$$

where u, v, and w denote displacement components along x, y, and z directions, respectively.

When the blade rotates with a constant angular velocity ω, the transient angle Θ can be written as

$$\Theta = \int_0^t [\Omega_0 + \Omega_1 \cos \tilde{\omega} t] d_t. \tag{8}$$

The instantaneous direction of the local unit vector $(\vec{\mathbf{e}}_x, \vec{\mathbf{e}}_y, \vec{\mathbf{e}}_z)$ for any typical point on the blade with respect to the stationary global Cartesian unit vector $(\vec{\mathbf{i}}, \vec{\mathbf{j}}, \vec{\mathbf{k}})$ is given by

$$\begin{Bmatrix} \vec{\mathbf{e}}_x^{\,0} \\ \vec{\mathbf{e}}_y^{\,0} \\ \vec{\mathbf{e}}_z^{\,0} \end{Bmatrix} = \begin{bmatrix} \cos\Theta & \sin\Theta & 0 \\ -\sin\Theta & \cos\Theta & 0 \\ 0 & 0 & 0 \end{bmatrix} \begin{Bmatrix} \vec{\mathbf{i}} \\ \vec{\mathbf{j}} \\ \vec{\mathbf{k}} \end{Bmatrix}, \tag{9a}$$

$$\begin{Bmatrix} \vec{\mathbf{e}}_x \\ \vec{\mathbf{e}}_y \\ \vec{\mathbf{e}}_z \end{Bmatrix} = \begin{bmatrix} 1 & 0 & 0 \\ 0 & \cos(\beta x + \theta_r) & \sin(\beta x + \theta_r) \\ 0 & -\sin(\beta x + \theta_r) & \cos(\beta x + \theta_r) \end{bmatrix} \begin{Bmatrix} \vec{\mathbf{e}}_x^{\,0} \\ \vec{\mathbf{e}}_y^{\,0} \\ \vec{\mathbf{e}}_z^{\,0} \end{Bmatrix}. \tag{9b}$$

The speed and the acceleration of any random point P on the plate are described as follows:

$$\begin{aligned} \dot{R} = {} & [\dot{u}\cos\Theta - (x + R_0 + u)\sin\Theta\cdot\Omega - \dot{v}\cos\theta\sin\Theta \\ & - (y+v)\cos\theta\cos\Theta\cdot\Omega + \dot{w}\sin\theta\sin\Theta \\ & + (z+w)\sin\theta\cos\Theta\cdot\Omega]\vec{\mathbf{i}} + [\dot{u}\sin\Theta \\ & + (x+R_0+u)\cos\Theta\cdot\Omega + \dot{v}\cos\theta\cos\Theta \\ & - (y+v)\cos\theta\sin\Theta\cdot\Omega - \dot{w}\sin\theta\cos\Theta \\ & + (z+w)\sin\theta\sin\Theta\cdot\Omega]\vec{\mathbf{j}} + (\dot{v}\sin\theta + \dot{w}\cos\theta)\vec{\mathbf{k}}, \end{aligned} \tag{10a}$$

$$\begin{aligned} \ddot{R} = {} & [\ddot{u}\cos\Theta - 2\dot{u}\sin\Theta\cdot\Omega - (x+R_0+u)\cos\Theta\cdot\Omega^2 \\ & - (x+R_0+u)\sin\Theta\cdot\dot{\Omega} - \ddot{v}\cos\theta\sin\Theta - 2\dot{v}\cos\theta\cos\Theta \\ & \cdot\Omega + (y+v)\cos\theta\sin\Theta\cdot\Omega^2 - (y+v)\cos\theta\cos\Theta\cdot\dot{\Omega} \\ & + \ddot{w}\sin\theta\sin\Theta + 2\dot{w}\sin\theta\cos\Theta\cdot\Omega - (z+w)\sin\theta\sin\Theta \\ & \cdot\Omega^2 + (z+w)\sin\theta\cos\Theta\cdot\dot{\Omega}]\vec{\mathbf{i}} + [\ddot{u}\sin\Theta + 2\dot{u}\cos\Theta \\ & \cdot\Omega - (x+R_0+u)\sin\Theta\cdot\Omega^2 + (x+R_0+u)\cos\Theta\cdot\dot{\Omega} \\ & + \ddot{v}\cos\theta\cos\Theta - 2\dot{v}\cos\theta\sin\Theta\cdot\Omega - (y+v)\cos\theta\cos\Theta \\ & \cdot\Omega^2 - (y+v)\cos\theta\sin\Theta\cdot\dot{\Omega} - \ddot{w}\sin\theta\cos\Theta \\ & + 2\dot{w}\sin\theta\sin\Theta\cdot\Omega + (z+w)\sin\theta\cos\Theta\cdot\Omega^2 \\ & + (z+w)\sin\theta\sin\Theta\cdot\dot{\Omega}]\vec{\mathbf{j}} + (\ddot{v}\sin\theta + \ddot{w}\cos\theta)\vec{\mathbf{k}}, \end{aligned} \tag{10b}$$

where

$$\theta = \theta(x) = \theta_r + \theta' x. \tag{10c}$$

The variation of the kinetic energy is given by

$$
\begin{aligned}
\delta K &= -\int_V \rho \ddot{R} \delta R d_V \\
&= \int_s \Big\{ I_0 \{ \ddot{u}_0 \cos\Theta - \ddot{v}_0 \cos\theta \sin\Theta + \ddot{w}_0 \sin\theta \sin\Theta - 2(\dot{u}_0 \sin\Theta + \dot{v}_0 \cos\theta \cos\Theta \\
&\quad - \dot{w}_0 \sin\theta \cos\Theta) \cdot \Omega - [(x + R_0 + u_0)\cos\Theta - (y + v_0)\cos\theta \sin\Theta + w_0 \sin\theta \sin\Theta] \cdot \Omega^2 \\
&\quad - [(x + R_0 + u_0)\sin\Theta - (y + v_0)\cos\theta \cos\Theta - w_0 \sin\theta \cos\Theta] \cdot \dot{\Omega}\} \cdot \delta u_0 \\
&\quad + I_0 \{ \ddot{u}_0 \sin\Theta - \ddot{v}_0 \cos\theta \cos\Theta - \ddot{w}_0 \sin\theta \cos\Theta + 2(\dot{u}_0 \cos\Theta + \dot{v}_0 \cos\theta \sin\Theta \\
&\quad + \dot{w}_0 \sin\theta \sin\Theta) \cdot \Omega - [(x + R_0 + u_0)\sin\Theta + (y + v_0)\cos\theta \cos\Theta - w_0 \sin\theta \cos\Theta] \cdot \Omega^2 \\
&\quad + [(x + R_0 + u_0)\cos\Theta - (y + v_0)\cos\theta \sin\Theta - w_0 \sin\theta \sin\Theta] \cdot \dot{\Omega}\} \cdot \delta v_0 \\
&\quad + \Big\{ I_0\left(\ddot{v}_0 \sin\theta + \ddot{w}_0 \cos\theta\right) - I_2 \theta'^2 x \Big\{ \left(\ddot{\phi}_x - \theta'^2 x \ddot{w}_0\right) \cos\Theta - \ddot{\phi}_y \cos\theta \sin\Theta \\
&\quad - 2\left[\dot{\phi}_x - \theta'^2 x \dot{w}_0 + \dot{\phi}_y \cos\theta \cos\Theta\right] \cdot \Omega - \left[\left(\phi_x - \theta'^2 x w_0\right) \cos\Theta - \phi_y \cos\theta \sin\Theta + \sin\theta \sin\Theta\right] \\
&\quad \cdot \Omega^2 - \left[\left(\phi_x - \theta'^2 x w_0\right) \sin\Theta + \phi_y \cos\theta \cos\Theta - \sin\theta \cos\Theta\right] \cdot \dot{\Omega} \Big\}\Big\} \delta w_0 \\
&\quad + I_2 \Big\{ \left(\ddot{\phi}_x - \theta'^2 x \ddot{w}_0\right) \cos\Theta - \ddot{\phi}_y \cos\theta \sin\Theta - 2\left[\dot{\phi}_x - \theta'^2 x \dot{w}_0 + \dot{\phi}_y \cos\theta \cos\Theta\right] \cdot \Omega - \Big[\left(\phi_x - \theta'^2 x w_0\right) \cos\Theta \\
&\quad - \phi_y \cos\theta \sin\Theta + \sin\theta \sin\Theta\Big] \cdot \Omega^2 - \left[\left(\phi_x - \theta'^2 x w_0\right) \sin\Theta + \phi_y \cos\theta \cos\Theta - \sin\theta \cos\Theta\right] \cdot \dot{\Omega} \Big\} \delta\phi_x \\
&\quad + I_2 \Big\{ \left(\ddot{\phi}_x - \theta'^2 x \ddot{w}_0\right) \sin\Theta + \ddot{\phi}_y \cos\theta \cos\Theta + 2\left[\left(\dot{\phi}_x - \theta'^2 x \dot{w}_0\right) \cos\Theta - \dot{\phi}_y \cos\theta \sin\Theta\right] \cdot \Omega \\
&\quad - \left[\left(\phi_x - \theta'^2 x w_0\right) \sin\Theta + \phi_y \cos\theta \cos\Theta - \sin\theta \cos\Theta\right] \cdot \Omega^2 + \Big[\left(\phi_x - \theta'^2 x w_0\right) \cos\Theta \\
&\quad - \phi_y \cos\theta \sin\Theta + \sin\theta \sin\Theta\Big] \cdot \dot{\Omega} \Big\} \delta\phi_y \Big\} d_x d_y.
\end{aligned} \tag{11}
$$

A large centrifugal force will be generated when the compressor blade of the aircraft engine operates with the high rotating speed. Components of centrifugal forces along the spanwise and chordwise directions are described by F_s and F_c, which can be expressed as

$$F_s = \int_x^L \rho \Omega^2 (r + x) d_x, \tag{12a}$$

$$F_c = \int_y^{\pm C/2} \rho \Omega^2 y \sin^2\theta d_y. \tag{12b}$$

The variation of the potential energy δU consists the variation of the strain energy δU_1 and the variation of the centrifugal potential energy δU_2, namely, $\delta U = \delta U_1 + \delta U_2$, where

$$
\begin{aligned}
\delta U_1 = \int_{\Omega_0} \Big\{ \int_{-h/2}^{h/2} &\Big[\sigma_{xx}\left(\delta\varepsilon_{xx}^{(0)} + z\delta\varepsilon_{xx}^{(1)}\right) \\
&+ \sigma_{yy}\left(\delta\varepsilon_{yy}^{(0)} + z\delta\varepsilon_{yy}^{(1)}\right) + \sigma_{xy}\left(\delta\gamma_{xy}^{(0)} + z\delta\gamma_{xy}^{(1)}\right) \\
&+ \sigma_{xz}\delta\gamma_{xz}^{(0)} + \sigma_{yz}\delta\gamma_{yz}^{(0)}\Big] d_z \Big\} d_x d_y,
\end{aligned} \tag{13a}
$$

$$
\begin{aligned}
\delta U_2 = \Big\{ \int_{-h/2}^{h/2} &F_s \frac{1}{2} \delta \left[\left(\left(\frac{\partial w}{\partial x}\right)^2 + \left(\frac{\partial v}{\partial x}\right)^2\right)\right. \\
&\left. + F_c \frac{1}{2} \delta \left(\left(\frac{\partial w}{\partial y}\right)^2 + \left(\frac{\partial u}{\partial y}\right)^2\right)\right] d_z \Big\} d_x d_y.
\end{aligned} \tag{13b}
$$

Aerodynamic forces of the blade derived by the first-order piston theory are written as

$$
\begin{aligned}
\Delta p_y &= C_\infty \rho_\infty \left(\frac{\partial v}{\partial t} + U_y^t \frac{\partial v}{\partial x}\right) \\
&= C_\infty \rho_\infty \left(\dot{v} + P \sin\theta \frac{\partial v}{\partial x}\right),
\end{aligned} \tag{14a}
$$

$$
\begin{aligned}
\Delta p_z &= C_\infty \rho_\infty \left(\frac{\partial w}{\partial t} + U_z^t \frac{\partial w}{\partial x}\right) \\
&= C_\infty \rho_\infty \left(\dot{w} + P \cos\theta \frac{\partial w}{\partial x}\right),
\end{aligned} \tag{14b}
$$

where ρ_∞ is the air density, C_∞ is the speed of sound, P is the speed of the airflow, and $U_y^t = P \sin\theta$, $U_z^t = P \cos\theta$.

The variation of the virtual work of aerodynamic forces is expressed as follows:

$$\begin{aligned}\delta W &= \int_S \Delta p_z \delta w d_x d_y + \int_S - \Delta p_y \delta v d_x d_z \\ &= \int_S \left(\Delta p_z \delta w + \frac{\partial z}{\partial y} \Delta p_y \delta v \right) d_x d_y.\end{aligned} \tag{15}$$

Nonlinear dynamic equations of motion for the rotating blade are established by using Hamilton's principle.

$$\int_0^t (\delta K - \delta U + \delta W) d_t = 0. \tag{16}$$

Substituting (11), (13a), (13b), and (15) into (16), the nonlinear partial governing equations for nonlinear vibration of the rotating blade expressed in terms of displacement variables can be derived as follows:

$$\begin{aligned}\delta u_0 :\ & A_{11}\frac{\partial^2 u_0}{\partial x^2} + 2A_{16}\frac{\partial^2 u_0}{\partial x \partial y} + A_{66}\frac{\partial^2 u_0}{\partial y^2} + A_{16}\frac{\partial^2 v_0}{\partial x^2} + A_{26}\frac{\partial^2 v_0}{\partial y^2} + (A_{12}+A_{66})\frac{\partial^2 v_0}{\partial x \partial y} - 2B_{11}\theta'^2\frac{\partial w_0}{\partial x} - B_{11}\theta'^2 x\frac{\partial^2 w_0}{\partial x^2} \\ & + A_{11}\frac{\partial w_0}{\partial x}\frac{\partial^2 w_0}{\partial x^2} + A_{66}\frac{\partial w_0}{\partial x}\frac{\partial^2 w_0}{\partial y^2} + 2A_{16}\frac{\partial w_0}{\partial x}\frac{\partial^2 w_0}{\partial x \partial y} + A_{16}\frac{\partial w_0}{\partial y}\frac{\partial^2 w_0}{\partial x^2} + A_{26}\frac{\partial w_0}{\partial y}\frac{\partial^2 w_0}{\partial y^2} + (A_{12}+A_{66})\frac{\partial w_0}{\partial y}\frac{\partial^2 w_0}{\partial x \partial y} \\ & - B_{16}\theta'^2 x\frac{\partial^2 w_0}{\partial x \partial y} + B_{66}\frac{\partial^2 \phi_x}{\partial y^2} + 2B_{16}\frac{\partial^2 \phi_x}{\partial x \partial y} + B_{11}\frac{\partial^2 \phi_x}{\partial x^2} + B_{16}\frac{\partial^2 \phi_y}{\partial x^2} + B_{26}\frac{\partial^2 \phi_y}{\partial y^2} + (B_{12}+B_{66})\frac{\partial^2 \phi_y}{\partial x \partial y} - B_{16}\theta'^2\frac{\partial^2 w_0}{\partial y} \\ = & I_o\{(\ddot{u}_0 \cos\Theta - \ddot{v}_0 \cos\theta \sin\Theta + \ddot{w}_0 \sin\theta \sin\Theta) - 2(\dot{u}_0 \sin\Theta + \dot{v}_0 \cos\theta \cos\Theta - w_0 \sin\theta \cos\Theta)\cdot\Omega \\ & - [(x+R_0+u_0)\cos\Theta - (y+v_0)\cos\theta \sin\Theta + w_0 \sin\theta \sin\Theta]\cdot\Omega^2 - [(x+R+u_0)\sin\Theta + (y+v_0)\cos\theta \sin\Theta + w_0 \sin\theta \cos\Theta]\cdot\dot{\Omega}\} \\ & + 2I_0 y \sin^2\theta\frac{\partial w_0}{\partial y}\cdot\Omega^2 + 2I_0 y \sin^2\theta\frac{\partial u_0}{\partial y}\cdot\Omega^2 - I_0 \sin^2\theta\left(\frac{C^2}{4} - y^2\right)\frac{\partial^2 w_0}{\partial y^2}\cdot\Omega^2 - I_0 \sin^2\theta\left(\frac{C^2}{4} - y^2\right)\frac{\partial^2 u_0}{\partial y^2}\cdot\Omega^2,\end{aligned} \tag{17}$$

$$\begin{aligned}\delta v_0 :\ & A_{16}\frac{\partial^2 u_0}{\partial x^2} + A_{26}\frac{\partial^2 u_0}{\partial y^2} + (A_{12}+A_{66})\frac{\partial^2 u_0}{\partial x \partial y} + A_{66}\frac{\partial^2 v_0}{\partial x^2} + A_{22}\frac{\partial^2 v_0}{\partial y^2} + 2A_{26}\frac{\partial^2 v_0}{\partial x \partial y} - 2B_{16}\theta'^2\frac{\partial w_0}{\partial x} \\ & - B_{16}\theta'^2 x\frac{\partial^2 w_0}{\partial x^2} + A_{16}\frac{\partial w_0}{\partial x}\frac{\partial^2 w_0}{\partial x^2} + A_{26}\frac{\partial w_0}{\partial x}\frac{\partial^2 w_0}{\partial y^2} + (A_{12}+A_{66})\frac{\partial w_0}{\partial x}\frac{\partial^2 w_0}{\partial x \partial y} + A_{22}\frac{\partial w_0}{\partial y}\frac{\partial^2 w_0}{\partial y^2} \\ & + A_{66}\frac{\partial w_0}{\partial y}\frac{\partial^2 w_0}{\partial x^2} + 2A_{26}\frac{\partial w_0}{\partial y}\frac{\partial^2 w_0}{\partial x \partial y} + B_{16}\frac{\partial^2 \phi_x}{\partial x^2} + B_{26}\frac{\partial^2 \phi_x}{\partial y^2} v + (B_{12}+B_{66})\frac{\partial^2 \phi_x}{\partial x \partial y} + B_{66}\frac{\partial^2 \phi_y}{\partial x^2} \\ & + B_{22}\frac{\partial^2 \phi_y}{\partial y^2} + 2B_{26}\frac{\partial^2 \phi_y}{\partial x \partial y} - B_{12}\theta'^2 x\frac{\partial^2 w_0}{\partial x \partial y} - B_{12}\theta'^2\frac{\partial w_0}{\partial y} + C_\infty \rho_\infty\left[\left(\dot{v}_0 + z\dot{\phi} y\right) + p \sin\theta\left(\frac{\partial v_0}{\partial x} + z\frac{\partial \phi_y}{\partial x}\right)\right]\frac{\partial z}{\partial y} \\ = & I_0\{(\ddot{u}_0 \sin\Theta + \ddot{v}_0 \cos\theta \cos\Theta - \ddot{w}_0 \sin\theta \cos\Theta) + 2(\dot{u}_0 \cos\Theta - \dot{v}_0 \cos\theta \sin\Theta + \dot{w}_0 \sin\theta \sin\Theta)\cdot\Omega - [(x+R_0+u_0)\sin\Theta \\ & + (y+v_0)\cos\theta \cos\Theta - w_0 \sin\theta \cos\Theta]\cdot\Omega^2 + [(x+R_0+u_0)\cos\Theta - (y+v_0)\cos\theta \sin\Theta + w_0 \sin\theta \sin\Theta]\cdot\dot{\Omega}\} \\ & - I_0(r+x)\frac{\partial w_0}{\partial x}\cdot\Omega^2 - I_0(r+x)\frac{\partial v_0}{\partial x}\cdot\Omega^2 + I_0\left[r(L-x) + \frac{1}{2}(L^2 - x^2)\right]\frac{\partial^2 w_0}{\partial x^2}\cdot\Omega^2 - I_0\left[r(L-x) + \frac{1}{2}(L^2 - x^2)\right]\frac{\partial^2 v_0}{\partial x^2}\cdot\Omega^2,\end{aligned} \tag{18}$$

$$\begin{aligned}\delta w_0 :\ & \frac{3}{2}A_{11}\left(\frac{\partial w_0}{\partial x}\right)^2\frac{\partial^2 w_0}{\partial x^2} + \left(A_{66} + \frac{1}{2}A_{12}\right)\left(\frac{\partial w_0}{\partial x}\right)^2\frac{\partial^2 w_0}{\partial y^2} + 3A_{16}\left(\frac{\partial w_0}{\partial x}\right)^2\frac{\partial^2 w_0}{\partial x \partial y} + 3A_{16}\frac{\partial w_0}{\partial x}\frac{\partial w_0}{\partial y}\frac{\partial^2 w_0}{\partial x^2} + 3A_{26}\frac{\partial w_0}{\partial x}\frac{\partial w_0}{\partial y}\frac{\partial^2 w_0}{\partial y^2} \\ & + (2A_{12}+3A_{66})\frac{\partial w_0}{\partial x}\frac{\partial w_0}{\partial y}\frac{\partial^2 w_0}{\partial x \partial y} - 3B_{11}\theta'^2 x\frac{\partial w_0}{\partial x}\frac{\partial^2 w_0}{\partial x^2} - 4B_{16}\theta'^2 x\frac{\partial w_0}{\partial x}\frac{\partial^2 w_0}{\partial x \partial y} - 3B_{16}\theta'^2 x\frac{\partial w_0}{\partial x}\frac{\partial w_0}{\partial y} + \frac{3}{2}A_{22}\left(\frac{\partial w_0}{\partial y}\right)^2\frac{\partial^2 w_0}{\partial y^2} \\ & + \left(A_{66} + \frac{1}{2}A_{12}\right)\left(\frac{\partial w_0}{\partial y}\right)^2\frac{\partial^2 w_0}{\partial x^2} + 3A_{26}\left(\frac{\partial w_0}{\partial y}\right)^2\frac{\partial^2 w_0}{\partial x \partial y} - 2B_{12}\theta' x\frac{\partial^2 w_0}{\partial y \partial x}\frac{\partial w_0}{\partial y} - 2B_{16}\theta'^2 x\frac{\partial^2 w_0}{\partial x^2}\frac{\partial w_0}{\partial y} + A_{11}\frac{\partial^2 u_0}{\partial x^2}\frac{\partial w_0}{\partial x} + 2A_{16}\frac{\partial^2 u_0}{\partial x \partial y}\frac{\partial w_0}{\partial x} \\ & + A_{66}\frac{\partial^2 u_0}{\partial y^2}\frac{\partial w_0}{\partial x} + A_{16}\frac{\partial^2 v_0}{\partial x^2}\frac{\partial w_0}{\partial x} + A_{26}\frac{\partial^2 v_0}{\partial y^2}\frac{\partial w_0}{\partial x} + (A_{12}+A_{66})\frac{\partial^2 v_0}{\partial x \partial y}\frac{\partial w_0}{\partial x} - 2B_{11}\left(\frac{\partial w_0}{\partial x}\right)^2 + B_{66}\frac{\partial^2 \phi_x}{\partial y^2}\frac{\partial w_0}{\partial x} + 2B_{16}\frac{\partial^2 \phi_x}{\partial x \partial y}\frac{\partial w_0}{\partial x} \\ & + B_{11}\frac{\partial^2 \phi_x}{\partial x^2}\frac{\partial w_0}{\partial x} + B_{16}\frac{\partial^2 \phi_y}{\partial x^2}\frac{\partial w_0}{\partial x} + B_{26}\frac{\partial^2 \phi_y}{\partial y^2}\frac{\partial w_0}{\partial x} + (B_{12}+B_{66})\frac{\partial^2 \phi_y}{\partial x \partial y}\frac{\partial w_0}{\partial x} + A_{16}\frac{\partial^2 u_0}{\partial x^2}\frac{\partial w_0}{\partial y} + A_{26}\frac{\partial^2 u_0}{\partial y^2}\frac{\partial w_0}{\partial y} + (A_{12}+A_{66})\frac{\partial^2 u_0}{\partial x \partial y}\frac{\partial w_0}{\partial y} \\ & + A_{66}\frac{\partial^2 v_0}{\partial x^2}\frac{\partial w_0}{\partial y} + A_{22}\frac{\partial^2 v_0}{\partial y^2}\frac{\partial w_0}{\partial y} + 2A_{26}\frac{\partial^2 v_0}{\partial x \partial y}\frac{\partial w_0}{\partial y} + B_{16}\frac{\partial^2 \phi_x}{\partial x^2}\frac{\partial w_0}{\partial y} + B_{26}\frac{\partial^2 \phi_x}{\partial y^2}\frac{\partial w_0}{\partial y} + (B_{12}+B_{16})\frac{\partial^2 \phi_x}{\partial x \partial y}\frac{\partial w_0}{\partial y} + B_{66}\frac{\partial^2 \phi_y}{\partial x^2}\frac{\partial w_0}{\partial y}\end{aligned}$$

$$
\begin{aligned}
&+ B_{22}\frac{\partial^2\phi_y}{\partial y^2}\frac{\partial w_0}{\partial y} + 2B_{26}\frac{\partial^2\phi_y}{\partial x\partial y}\frac{\partial w_0}{\partial y} + B_{12}\theta'^2\left(\frac{\partial w_0}{\partial y}\right)^2 + 2A_{16}\frac{\partial u_0}{\partial x}\frac{\partial^2 w_0}{\partial x\partial y} + 2A_{26}\frac{\partial v_0}{\partial y}\frac{\partial^2 w_0}{\partial x\partial y} + 2A_{66}\frac{\partial u_0}{\partial y}\frac{\partial^2 w_0}{\partial x\partial y} + 2A_{66}\frac{\partial^2 v_0}{\partial x}\frac{\partial^2 w_0}{\partial x\partial y}\\
&+ 2B_{16}\frac{\partial\phi_x}{\partial x}\frac{\partial^2 w_0}{\partial x\partial y} - 2B_{16}\theta'^2 w_0\frac{\partial^2 w_0}{\partial x\partial y} + 2B_{26}\frac{\partial\phi_y}{\partial y}\frac{\partial^2 w_0}{\partial x\partial y} + 2B_{66}\frac{\partial\phi_x}{\partial y}\frac{\partial^2 w_0}{\partial x\partial y} + 2B_{66}\frac{\partial\phi_y}{\partial x}\frac{\partial^2 w_0}{\partial x\partial y} + A_{11}\frac{\partial u_0}{\partial x}\frac{\partial^2 w_0}{\partial x^2} + A_{12}\frac{\partial v_0}{\partial y}\frac{\partial^2 w_0}{\partial x^2}\\
&+ A_{16}\frac{\partial u_0}{\partial y}\frac{\partial^2 w_0}{\partial x^2} + A_{16}\frac{\partial^2 v_0}{\partial x}\frac{\partial^2 w_0}{\partial x^2} + B_{11}\frac{\partial\phi_x}{\partial x}\frac{\partial^2 w_0}{\partial x^2} - B_{11}\theta'^2 w_0\frac{\partial^2 w_0}{\partial x^2} + B_{12}\frac{\partial^2\phi_y}{\partial y}\frac{\partial^2 w_0}{\partial x^2} + B_{16}\frac{\partial^2\phi_x}{\partial y}\frac{\partial^2 w_0}{\partial x^2} + B_{16}\frac{\partial\phi_y}{\partial x}\frac{\partial^2 w_0}{\partial x^2}\\
&+ A_{12}\frac{\partial u_0}{\partial x}\frac{\partial^2 w_0}{\partial y^2} + A_{22}\frac{\partial v_0}{\partial y}\frac{\partial^2 w_0}{\partial y^2} + A_{26}\frac{\partial u_0}{\partial y}\frac{\partial^2 w_0}{\partial y^2} + A_{26}\frac{\partial v_0}{\partial x}\frac{\partial^2 w_0}{\partial y^2} + B_{12}\frac{\partial\phi_x}{\partial x}\frac{\partial^2 w_0}{\partial y^2} - B_{12}\theta'^2 w_0\frac{\partial^2 w_0}{\partial y^2} - B_{12}\theta'^2 x\frac{\partial w_0}{\partial x}\frac{\partial^2 w_0}{\partial y^2}\\
&+ B_{22}\frac{\partial\phi_y}{\partial y}\frac{\partial^2 w_0}{\partial y^2} + B_{26}\frac{\partial\phi_x}{\partial y}\frac{\partial^2 w_0}{\partial y^2} + B_{26}\frac{\partial\phi_y}{\partial x}\frac{\partial^2 w_0}{\partial y^2} - B_{11}\theta'^2 x\frac{\partial^2 u_0}{\partial x^2} - B_{12}\theta'^2 x\frac{\partial^2 v_0}{\partial x\partial y} - B_{16}\theta'^2 x\frac{\partial^2 u_0}{\partial x\partial y} - B_{16}\theta'^2 x\frac{\partial^2 v_0}{\partial x^2}\\
&- D_{11}\theta'^2 x\frac{\partial^2\phi_x}{\partial x^2} - 2D_{11}\theta'^4 x\frac{\partial^2 w_0}{\partial x} - D_{11}\theta'^4 x\frac{\partial^2 u_0}{\partial x^2} - D_{12}\theta'^2 x\frac{\partial^2\phi_y}{\partial x\partial y} - D_{16}\theta'^2 x\frac{\partial^2\phi_x}{\partial x\partial y} - D_{16}\theta'^2 x\frac{\partial^2\phi_y}{\partial x^2} + kA_{45}\frac{\partial^2 w_0}{\partial x\partial y} + kA_{45}\frac{\partial\phi_y}{\partial x}\\
&+ kA_{55}\frac{\partial^2 w_0}{\partial x^2} + kA_{55}\frac{\partial\phi_x}{\partial x} + kA_{44}\frac{\partial^2 w_0}{\partial y^2} + kA_{44}\frac{\partial\phi_y}{\partial y} + kA_{45}\frac{\partial^2 w_0}{\partial x\partial y} + kA_{45}\frac{\partial^2\phi_x}{\partial y} + C_\infty\rho_\infty\dot{w}_0 + C_\infty\rho_\infty P\cos\theta\frac{\partial w_0}{\partial x}\\
&= I_0(\ddot{v}_0\sin\theta + \ddot{w}_0\cos\theta) - I_2\theta'^2 x\Big\{\left(\ddot{\varphi}_x - \theta'^2 x\ddot{w}_0\right)\cos\Theta - \ddot{\varphi}_y\cos\theta\sin\Theta - 2\left(\dot{\varphi}_x - \theta'^2 x\dot{w}_0 + \dot{\varphi}_y\cos\theta\cos\Theta\right)\cdot\Omega\\
&- \left[\left(\varphi_x - \theta'^2 x w_0\right)\cos\Theta - \varphi_y\cos\theta\sin\Theta + \sin\theta\sin\Theta\right]\cdot\Omega^2 - \left[\left(\varphi_x - \theta'^2 x W_0\right)\sin\Theta + \varphi_y\cos\theta\cos\Theta - \sin\theta\cos\Theta\right]\cdot\dot{\Omega}\Big\}\\
&- 2I_0 y\sin^2\theta\frac{\partial w_0}{\partial y}\cdot\Omega^2 - 2I_0 y\sin^2\theta\frac{\partial u_0}{\partial y}\cdot\Omega^2 + I_0\sin^2\theta\left(\frac{C^2}{4} - y^2\right)\frac{\partial^2 w_0}{\partial y^2}\cdot\Omega^2 + I_0\sin^2\theta\left(\frac{C^2}{4} - y^2\right)\frac{\partial^2 u_0}{\partial y^2}\cdot\Omega^2\\
&- I_0(r + x)\frac{\partial w_0}{\partial x}\cdot\Omega^2 - I_0(r + x)\frac{\partial v_0}{\partial x}\cdot\Omega^2 + I_0\left[r(L - x) + \frac{1}{2}(L^2 - x^2)\right]\frac{\partial^2 w_0}{\partial x^2}\cdot\Omega^2 - I_0\left[r(L - x) + \frac{1}{2}(L^2 - x^2)\right]\frac{\partial^2 v_0}{\partial x^2}\cdot\Omega^2,
\end{aligned}
\tag{19}
$$

$$
\begin{aligned}
\delta\phi_x:\ & B_{11}\frac{\partial^2 u_0}{\partial x^2} + 2B_{16}\frac{\partial^2 u_0}{\partial x\partial y} + B_{66}\frac{\partial^2 u_0}{\partial y^2} + B_{16}\frac{\partial^2 v_0}{\partial x^2} + (B_{12} + B_{66})\frac{\partial^2 v_0}{\partial x\partial y} + B_{26}\frac{\partial^2 v_0}{\partial y^2} + B_{11}\frac{\partial w_0}{\partial x}\frac{\partial^2 w_0}{\partial x^2}\\
&+ B_{11}\frac{\partial w_0}{\partial x}\frac{\partial^2 w_0}{\partial x^2} + B_{16}\frac{\partial w_0}{\partial y}\frac{\partial^2 w_0}{\partial x^2} + B_{12}\frac{\partial w_0}{\partial y}\frac{\partial^2 w_0}{\partial x\partial y} + 2B_{16}\frac{\partial w_0}{\partial x}\frac{\partial^2 w_0}{\partial x\partial y} + B_{66}\frac{\partial w_0}{\partial y}\frac{\partial^2 w_0}{\partial x\partial y} + B_{26}\frac{\partial w_0}{\partial y}\frac{\partial^2 w_0}{\partial y^2}\\
&+ B_{66}\frac{\partial w_0}{\partial x}\frac{\partial^2 w_0}{\partial y^2} - 2D_{11}\theta'^2\frac{\partial w_0}{\partial x} - D_{11}\theta'^2 x\frac{\partial^2 w_0}{\partial x^2} - D_{16}\theta'^2 x\frac{\partial^2 w_0}{\partial x\partial y} + D_{11}\frac{\partial^2\phi_x}{\partial x^2} + 2D_{16}\frac{\partial^2\phi_x}{\partial x\partial y} + D_{66}\frac{\partial^2\phi_x}{\partial y^2}\\
&+ D_{16}\frac{\partial^2\phi_y}{\partial x^2} + (D_{12} + D_{66})\frac{\partial^2\phi_y}{\partial x\partial y} + D_{26}\frac{\partial^2\phi_y}{\partial y^2} - kA_{45}\phi_y - kA_{55}\frac{\partial w_0}{\partial x} - kA_{45}\frac{\partial w_0}{\partial y} - kA_{55}\phi_x - kA_{45}\phi_y\\
&= I_2\Big\{\left(\ddot{\phi}_x - \theta'^2\ddot{w}_0 x\right)\cos\Theta - \ddot{\phi}_y\cos\theta\sin\Theta - 2\left(\dot{\phi}_x - \theta'^2 x\dot{w}_0 + \dot{\phi}_y\cos\theta\cos\Theta\right)\cdot\Omega\\
&- \left[\left(\phi_x - \theta'^2 x w_0\right)\cos\Theta - \phi_y\cos\theta\sin\Theta + \sin\theta\sin\Theta\right]\cdot\Omega^2 - \left[\left(\phi_x - \theta'^2 x w_0\right)\sin\Theta + \phi_y\cos\theta\cos\Theta - \sin\theta\sin\Theta\right]\cdot\dot{\Omega}\Big\},
\end{aligned}
\tag{20}
$$

$$
\begin{aligned}
\delta\phi_y:\ & B_{16}\frac{\partial^2 u_0}{\partial x^2} + (B_{12} + B_{66})\frac{\partial^2 u_0}{\partial x\partial y} + B_{26}\frac{\partial^2 u_0}{\partial y^2} + B_{66}\frac{\partial^2 v_0}{\partial x^2} + B_{22}\frac{\partial^2 v_0}{\partial y^2} + 2B_{26}\frac{\partial^2 v_0}{\partial x\partial y} + B_{16}\frac{\partial w_0}{\partial x}\frac{\partial^2 w_0}{\partial x^2} + 2B_{26}\frac{\partial w_0}{\partial y}\frac{\partial^2 w_0}{\partial x\partial y}\\
&+ B_{66}\frac{\partial w_0}{\partial y}\frac{\partial^2 w_0}{\partial x^2} + (B_{66} + B_{12})\frac{\partial w_0}{\partial x}\frac{\partial^2 w_0}{\partial x\partial y} + B_{22}\frac{\partial w_0}{\partial y}\frac{\partial^2 w_0}{\partial y^2} + B_{26}\frac{\partial w_0}{\partial x}\frac{\partial^2 w_0}{\partial y^2} - 2D_{16}\theta'^2\frac{\partial w_0}{\partial x} - D_{16}\theta'^2 x\frac{\partial^2 w_0}{\partial x^2}\\
&- D_{12}\theta'^2 x\frac{\partial^2 w_0}{\partial x\partial y} + D_{16}\frac{\partial^2\phi_x}{\partial x^2} + (D_{12} + D_{66})\frac{\partial^2\phi_x}{\partial x\partial y} + D_{26}\frac{\partial^2\phi_x}{\partial y^2} + D_{66}\frac{\partial^2\phi_y}{\partial x^2} + D_{22}\frac{\partial^2\phi_y}{\partial y^2} + 2D_{26}\frac{\partial^2\phi_y}{\partial x\partial y} - kA_{44}\frac{\partial w_0}{\partial y}\\
&- kA_{45}\frac{\partial w_0}{\partial x} - kA_{45}\phi_x - kA_{44}\phi_y - D_{12}\theta'^2 x\frac{\partial^2 w_0}{\partial x\partial y} + C_\infty\rho_\infty\left(\dot{w}_0 + P\cos\theta\cdot\frac{\partial w_0}{\partial x}\right)\cdot z\frac{\partial z}{\partial y}\\
&= I_2\Big\{\left[\left(\ddot{\phi}_x - \theta'^2 x\ddot{w}_0\right)\sin\Theta + \ddot{\phi}_y\cos\theta\cos\Theta + 2\left(\dot{\phi}_x - \theta'^2 x\dot{w}_0\right)\cos\Theta - \dot{\phi}_y\cos\theta\sin\Theta\right]\cdot\Omega\\
&- \left[\left(\phi_y - \theta'^2 x\ddot{w}_0\right)\sin\Theta + \phi_y\cos\theta\cos\Theta - \sin\theta\cos\Theta\right]\cdot\Omega^2 + \left[\left(\phi_x - \theta'^2 x w_0\right)\cos\Theta - \phi_y\cos\theta\sin\Theta + \sin\theta\sin\Theta\right]\cdot\dot{\Omega}\Big\}.
\end{aligned}
\tag{21}
$$

The boundary conditions are given as follows:

$$\begin{aligned}
&x=0: u=v=w=\phi_x=\phi_y=0,\\
&x=L: \mathbf{N}_{xy}=\mathbf{N}_{xx}=\mathbf{M}_{xx}=\mathbf{M}_{xy}=\mathbf{Q}_x=0,\\
&y=0: \mathbf{N}_{xy}=\mathbf{N}_{yy}=\mathbf{M}_{yy}=\mathbf{M}_{xy}=\mathbf{Q}_y=0,\\
&y=C: \mathbf{N}_{xy}=\mathbf{N}_{yy}=\mathbf{M}_{yy}=\mathbf{M}_{xy}=\mathbf{Q}_y=0.
\end{aligned} \tag{22}$$

The dimensionless variables and parameters are introduced as follows:

$$\begin{aligned}
\bar{u}_0&=\frac{u_0}{L},\\
\bar{v}_0&=\frac{v_0}{C},\\
\bar{w}_0&=\frac{w_0}{h},\\
\bar{x}&=\frac{x}{L},\\
\bar{y}&=\frac{y}{C},\\
\bar{\phi}_x&=\phi_x,\\
\bar{\phi}_y&=\phi_y,\\
\bar{\theta}&=\theta,\\
\bar{\theta}'&=L\theta',\\
\bar{r}&=\frac{r}{L},\\
\overline{A_{ij}}&=\frac{(LC)^{1/2}}{Eh^2},\\
\overline{B_{ij}}&=\frac{(LC)^{1/2}}{Eh^3}B_{ij},\\
\overline{D_{ij}}&=\frac{(LC)^{1/2}}{Eh^4}D_{ij},\\
\overline{I_i}&=\frac{1}{\rho(LC)^{(1+i)/2}},\\
\bar{\Omega}&=\Omega\sqrt{\frac{\rho CL}{E}},\\
\bar{t}&=t\sqrt{\frac{E}{\rho CL}},\\
\bar{C}_\infty&=C_\infty\sqrt{\frac{\rho}{E}},\\
\bar{p}&=p\sqrt{\frac{\rho}{E}},\\
\bar{\rho}_\infty&=\frac{\rho_\infty L^2}{\rho A}.
\end{aligned} \tag{23}$$

Substituting (23) into (17), (18), (19), (20), and (21), we obtain the dimensionless governing equations of motion for the blade.

Based on the practical working condition of the blade, and theoretical and numerical studies given by Dowell et al. [36, 37], it is known that vibrations of the first two order modes for the blade play an important role during vibration. The Galerkin approach is applied to obtain ordinary differential equations of motion in the dimensionless form. The Galerkin approach is derived by the Taylor expansion method, which is a mathematically convergent method. The first two order mode functions can produce convergent results. Mode functions of the plate $u(x,y,t)$, $v(x,y,t)$, $w(x,y,t)$, $\phi_x(x,y,t)$, and $\phi_y(x,y,t)$ are given as follows based on a combination method of beam functions

$$\begin{aligned}
u(x,y,t)&=u_1(t)\sin\frac{\pi x}{2L}\cos\frac{\pi y}{C}+u_2(t)\sin\frac{3\pi x}{2L}\cos\frac{2\pi y}{C},\\
v(x,y,t)&=v_1(t)\sin\frac{\pi x}{2L}\sin\frac{\pi y}{C}+v_2(t)\sin\frac{3\pi x}{2L}\sin\frac{2\pi y}{C},\\
w(x,y,t)&=w_1(t)X_1(x)Y_1(y)+w_2(t)X_2(x)Y_2(y),\\
\phi_x(x,y,t)&=\phi_{x1}(t)\sin\frac{\pi x}{2L}\cos\frac{\pi y}{C}+\phi_{x2}(t)\sin\frac{\pi x}{L}\cos\frac{2\pi y}{C},\\
\phi_y(x,y,t)&=\phi_{y1}(t)\sin\frac{\pi y}{C}\left(1-\cos\frac{\pi x}{2L}\right)\\
&\quad+\phi_{y2}(t)\sin\frac{2\pi y}{C}\left(1-\cos\frac{\pi x}{L}\right),
\end{aligned} \tag{24a}$$

where

$$X_j=\cosh\frac{k_jx}{a}-\cos\frac{k_jx}{a}-\beta_j\left(\sinh\frac{k_jx}{a}-\sin\frac{k_jx}{a}\right), \tag{24b}$$

$$Y_i=\cosh\frac{k_iy}{b}+\cos\frac{k_iy}{b}-\alpha_i\left(\sinh\frac{k_iy}{b}+\sin\frac{k_iy}{b}\right), \tag{24c}$$

where k_i are coefficients related to frequencies, $k_i^4=k_j^4=\omega^2\rho A/EJ$, κ_i and λ_j are coefficients of beam functions as follows:

$$\kappa_i=\frac{\cosh k_ib-\cos k_ib}{\sinh k_ib-\sin k_ib}\quad(i=3,4,5,\dots), \tag{24d}$$

$$\lambda_j=\frac{\cosh k_ja-\cos k_ja}{\sinh k_ja-\sin k_ja}\quad(j=1,2,3,\dots). \tag{24e}$$

Substitute (24a), (24b), (24c), (24d), and (24e) into the dimensionless governing equations of motion and express u, v, φ_x, and φ_y with w. Ordinary differential equations of transverse vibration for the first mode and the second mode of the rotating blade in the dimensionless form are derived through the Galerkin method as follows:

$$\begin{aligned}&\ddot{w}_1 + \left[\alpha_1 + f_{11}\Omega_1\sin(\tilde{\omega}t) + f_{12}\Omega_1\cos(\tilde{\omega}t) + f_{13}\Omega_1^2\cos^2(\tilde{\omega}t)\right]\\&\cdot w_1 + [\alpha_2 + f_{14}\Omega_1\cos(\tilde{\omega}t)]\dot{w}_1 + \alpha_3 {w_1}^2 + \alpha_4 w_1 w_2\\&+ \alpha_5 w_1 {w_2}^2 + \alpha_6 {w_1}^2 w_2 + \alpha_7 {w_1}^3 + \alpha_8 {w_2}^2 + \alpha_9 {w_2}^3\\&+ \left[\alpha_{10} + f_{15}\Omega_1\cos(\tilde{\omega}t) + f_{16}\Omega_1\sin(\tilde{\omega}t) + f_{17}\Omega_1^2\cos^2(\tilde{\omega}t)\right]\\&\cdot w_2 + [\alpha_{11} + f_{18}\Omega_1\cos(\tilde{\omega}t)]\dot{w}_2 + \alpha_{12}\ddot{w}_2\\&= \alpha_{13} + f_{19}\Omega_1\cos(\tilde{\omega}t) + f_{20}\Omega_1\sin(\tilde{\omega}t) + f_{21}\Omega_1^2\cos^2(\tilde{\omega}t),\end{aligned} \tag{25a}$$

$$\begin{aligned}&\ddot{w}_2 + \left[\beta_1 + f_{31}\sin(\tilde{\omega}t) + f_{32}\Omega_1\cos(\tilde{\omega}t) + f_{33}\Omega_1^2\cos^2(\tilde{\omega}t)\right]\\&\cdot w_2 + [\beta_{14} + f_{35}\Omega_1\cos(\tilde{\omega}t)]\dot{w}_2 + \beta_3 w_2^2 + \beta_4 w_1 w_2\\&+ \beta_5 w_1^2 w_2 + \beta_6 w_2^2 w_1 + \beta_7 w_2^3 + \beta_8 w_1^2 + \beta_9 w_1^3\\&+ \left[\beta_{10} + f_{35}\Omega_1\cos(\tilde{\omega}t) + f_{36}\Omega_1\sin(\tilde{\omega}t)\right.\\&\left.+ f_{37}\Omega_1^2\cos^2(\tilde{\omega}t)\right] w_1 + [\beta_{11} + f_{38}\Omega_1\cos(\tilde{\omega}t)]\dot{w}_1 + \beta_{12}\ddot{w}_1\\&= \beta_{13} + f_{39}\Omega_1\cos(\tilde{\omega}t) + f_{40}\Omega_1\sin(\tilde{\omega}t) + f_{41}\Omega_1^2\cos^2(\tilde{\omega}t),\end{aligned} \tag{25b}$$

where expressions of coefficients α_i, β_i, and f_{ij} are described in the appendix.

3. Perturbation Analysis

There are both square nonlinear terms and cubic terms in (25a) and (25b), so the asymptotic perturbation method is exploited to derive average equations. The small parameter ε is introduced. The scale transformation is performed as follows:

$$\begin{aligned}\alpha_1 &\longrightarrow \varepsilon^2\alpha_1,\\ f_{11} &\longrightarrow \varepsilon^2 f_{11},\\ f_{12} &\longrightarrow \varepsilon^2 f_{12},\\ f_{13} &\longrightarrow \varepsilon^2 f_{13},\\ \alpha_2 &\longrightarrow \varepsilon^2\alpha_2,\\ f_{14} &\longrightarrow \varepsilon^2 f_{14},\\ \alpha_{10} &\longrightarrow \varepsilon^2\alpha_{10},\\ f_{15} &\longrightarrow \varepsilon^2 f_{15},\\ f_{16} &\longrightarrow \varepsilon^2 f_{16},\\ f_{17} &\longrightarrow \varepsilon^2 f_{17},\\ \alpha_{11} &\longrightarrow \varepsilon^2\alpha_{11},\\ f_{18} &\longrightarrow \varepsilon^2 f_{18},\\ \alpha_{12} &\longrightarrow \varepsilon^2\alpha_{12},\\ \alpha_{13} &\longrightarrow \varepsilon^2\alpha_{13},\\ f_{19} &\longrightarrow \varepsilon^2 f_{19},\\ f_{20} &\longrightarrow \varepsilon^2 f_{20},\\ f_{21} &\longrightarrow \varepsilon^2 f_{21},\\ \beta_1 &\longrightarrow \varepsilon^2\beta_1,\\ f_{31} &\longrightarrow \varepsilon^2 f_{31},\\ f_{32} &\longrightarrow \varepsilon^2 f_{32},\\ f_{33} &\longrightarrow \varepsilon^2 f_{33},\\ \beta_2 &\longrightarrow \varepsilon^2\beta_2,\\ f_{34} &\longrightarrow \varepsilon^2 f_{34},\\ \beta_{10} &\longrightarrow \varepsilon^2\beta_{10},\\ f_{37} &\longrightarrow \varepsilon^2 f_{37},\\ \beta_{11} &\longrightarrow \varepsilon^2\beta_{11},\\ f_{38} &\longrightarrow \varepsilon^2 f_{38},\\ \beta_{12} &\longrightarrow \varepsilon^2\beta_{12},\\ \beta_{13} &\longrightarrow \varepsilon^2\beta_{13},\\ f_{39} &\longrightarrow \varepsilon^2 f_{39},\\ f_{40} &\longrightarrow \varepsilon^2 f_{40},\\ f_{41} &\longrightarrow \varepsilon^2 f_{41}.\end{aligned} \tag{26}$$

The case of 1:3 internal resonance-1/2 sub-harmonic resonance for the rotating plate is considered. Relationship between ω_1 and ω_2 is expressed as

$$\begin{aligned}3\omega_1 &= \omega_2,\\ \omega_1 &= \frac{1}{2}\tilde{\omega} + \varepsilon^2\sigma_1,\\ \omega_2 &= \frac{3}{2}\tilde{\omega} + \varepsilon^2\sigma_2,\\ \tilde{\omega} &= \Omega,\end{aligned} \tag{27}$$

where σ_1 and σ_2 are two detuning parameters.

Substituting (26) and (27) into (25a) and (25b), we obtain the following equation

$$\begin{aligned}&\ddot{w}_1 + \left[\frac{\Omega^2}{4} + \Omega\sigma_1\varepsilon^2 + \varepsilon^4\sigma_1^2\right] w_1\\&+ \left[\varepsilon^2\alpha_2 + \varepsilon^2 f_{14}\Omega_1\cos(\tilde{\omega}t)\right]\dot{w}_1 + \alpha_3 {w_1}^2\\&+ \alpha_4 w_1 w_2 + \alpha_5 w_1 w_2^2 + \alpha_6 {w_1}^2 w_2 + \alpha_7 {w_1}^3\\&+ \alpha_8 {w_2}^2 + \alpha_9 {w_2}^3 + \left[\varepsilon^2\alpha_{10} + \varepsilon^2 f_{15}\Omega_1\cos(\tilde{\omega}t)\right.\\&\left.+ \varepsilon^2 f_{16}\Omega_1\sin(\tilde{\omega}t) + \varepsilon^2 f_{17}\Omega_1^2\cos^2(\tilde{\omega}t)\right] w_2\\&+ \left[\varepsilon^2\alpha_{11} + \varepsilon^2 f_{18}\Omega_1\cos(\tilde{\omega}t)\right]\dot{w}_2 + \varepsilon^2\alpha_{12}\ddot{w}_2\\&= \varepsilon^2\alpha_{13} + \varepsilon^2 f_{19}\Omega_1\cos(\tilde{\omega}t) + \varepsilon^2 f_{20}\Omega_1\sin(\tilde{\omega}t)\\&+ \varepsilon^2 f_{21}\Omega_1^2\cos^2(\tilde{\omega}t),\end{aligned} \tag{28a}$$

$$\begin{aligned}&\ddot{w}_2 + \left[\frac{9}{4}\Omega^2 + 3\Omega\sigma_2\varepsilon^2 + \varepsilon^4\sigma_2^2\right] w_2\\&+ \left[\varepsilon^2\beta_{14} + \varepsilon^2 f_{35}\Omega_1\cos(\tilde{\omega}t)\right]\dot{w}_2 + \beta_3 w_2^2\\&+ \beta_4 w_1 w_2 + \beta_5 {w_1}^2 w_2 + \beta_6 w_2^2 w_1 + \beta_7 w_2^3\\&+ \beta_8 w_1^2 + \beta_9 w_1^3 + \left[\varepsilon^2\beta_{10} + \varepsilon^2 f_{35}\Omega_1\cos(\tilde{\omega}t)\right.\\&\left.+ \varepsilon^2 f_{36}\Omega_1\sin(\tilde{\omega}t) + \varepsilon^2 f_{37}\Omega_1^2\cos^2(\tilde{\omega}t)\right] w_1\\&+ \left[\varepsilon^2\beta_{11} + \varepsilon^2 f_{38}\Omega_1\cos(\tilde{\omega}t)\right]\dot{w}_1 + \varepsilon^2\beta_{12}\ddot{w}_1\\&= \varepsilon^2\beta_{13} + \varepsilon^2 f_{39}\Omega_1\cos(\tilde{\omega}t) + \varepsilon^2 f_{40}\Omega_1\sin(\tilde{\omega}t)\\&+ \varepsilon^2 f_{41}\Omega_1^2\cos^2(\tilde{\omega}t).\end{aligned} \tag{28b}$$

The time scale transformation is introduced as

$$\tau = \varepsilon^q t, \tag{29}$$

where q is the positive rational number, which will be determined in the following deduction.

Functions $x(t)$ and $y(t)$ are the solutions of (28a) and (28b), which can be expressed as a power series of small parameter ε.

$$x(t) = \sum_{n=-\infty}^{\infty} \varepsilon^{r_n}\psi_n(\tau,\varepsilon)e^{-\text{in}(\Omega/2)t}, \tag{30a}$$

$$y(t) = \sum_{n=-\infty}^{\infty} \varepsilon^{r_n}\phi_n(\tau,\varepsilon)e^{-\text{in}(3\Omega/2)t}. \tag{30b}$$

In (30a) and (30b), if $n \neq 0$, then $r_n = |n|$, otherwise $r_0 = r$. In the following derivation, we will fix r_n.

The real part of $x(t)$ and $y(t)$ are written as follows:

$$\psi_n(\tau,\varepsilon) = \psi_{-n}{}^*(\tau,\varepsilon), \tag{31a}$$

$$\phi_n(\tau,\varepsilon) = \phi_{-n}{}^*(\tau,\varepsilon), \tag{31b}$$

where the item with the asterisk on the right-hand side is the complex conjugate of the item on the left-hand side.

Solutions of (28a) and (28b) can be rewritten more explicitly as follows:

$$\begin{aligned} x(t) &= \varepsilon^r\psi_0(\tau,\varepsilon) + \varepsilon\psi_1(\tau,\varepsilon)e^{-i(\Omega/2)t} \\ &+ \varepsilon^2\psi_2(\tau,\varepsilon)e^{-i\Omega t} + \varepsilon^3\psi_3(\tau,\varepsilon)e^{-i(3\Omega/2)t} \\ &+ \varepsilon^4\psi_4(\tau,\varepsilon)e^{-i(4\Omega/2)t} + \text{cc} + o(\varepsilon^5), \end{aligned} \tag{32a}$$

$$\begin{aligned} y(t) &= \varepsilon^r\phi_0(\tau,\varepsilon) + \varepsilon\phi_1(\tau,\varepsilon)e^{-i(3\Omega/2)t} \\ &+ \varepsilon^2\phi_2(\tau,\varepsilon)e^{-3i\Omega t} + \varepsilon^3\phi_3(\tau,\varepsilon)e^{-i(9\Omega/2)t} \\ &+ \varepsilon^4\phi_4(\tau,\varepsilon)e^{-i6\Omega t} + \text{cc} + o(\varepsilon^5), \end{aligned} \tag{32b}$$

where cc denotes terms of the complex conjugate of functions on the right-hand side.

Solutions of (28a) and (28b) can be expressed as harmonic functions, whose coefficients are related to small parameter ε, as shown in (32a) and (32b). Functions $\psi_n(\tau,\varepsilon)$ and $\varphi_n(\tau,\varepsilon)$ are given by

$$\psi_n(\tau,\varepsilon) = \sum_{i=0}^{+\infty} \varepsilon^i\psi_n{}^{(i)}(\tau), \tag{33a}$$

$$\phi_n(\tau,\varepsilon) = \sum_{i=0}^{+\infty} \varepsilon^i\phi_n{}^{(i)}(\tau). \tag{33b}$$

We suppose when $\varepsilon \to 0$, the limits of $\psi_n(\tau,\varepsilon)$ and $\phi_n(\tau,\varepsilon)$ exist. Meanwhile, we define $\psi_n^{(0)} = \psi_n$ and $\phi_n^{(0)} = \phi_n$ for $n \neq 1$ and $\psi_1^{(0)} = \psi$ and $\phi_1^{(0)} = \phi$ for $n = 1$. When $n = 2$, we obtain the derivative results as follows:

$$\frac{d}{d_t}\left(\psi_n e^{-\text{in}(\Omega/2)t}\right) = \left(-\text{in}\frac{\Omega}{2}\psi_n + \varepsilon^q\frac{d\psi_n}{d\tau}\right)e^{-\text{in}(\Omega/2)t}, \tag{34a}$$

$$\frac{d}{d_t}\left(\phi_n e^{-\text{in}(3\Omega/2)t}\right) = \left(-\text{in}\frac{3}{2}\Omega\phi_n + \varepsilon^q\frac{d\phi_n}{d\tau}\right)e^{-\text{in}(3\Omega/2)t}. \tag{34b}$$

In order to solve the coefficients of $\psi_n(\tau,\varepsilon)$ and $\phi_n(\tau,\varepsilon)$, substituting (30a) and (30b) into (28a) and (28b), we obtain equations for each harmonic with order n and a fixed order of approximation on the perturbation parameter ε.

When $n = 0$, we get the following expressions:

$$\begin{aligned} &\frac{1}{4}\Omega^2\varepsilon^r\psi_0 + 2\alpha_8\phi_1\phi_1^*\varepsilon^2 + 2\alpha_3\psi_1\psi_1^*\varepsilon^2 \\ &\quad - f_{20}\Omega_1\sin(\tilde{\omega}t)\varepsilon^2 - f_{19}\Omega_1\cos(\tilde{\omega}t)\varepsilon^2 \\ &\quad - f_{21}\Omega_1^2\cos^2(\tilde{\omega}t)\varepsilon^2 - \alpha_{13}\varepsilon^2 = 0, \end{aligned} \tag{35a}$$

$$\begin{aligned} &\frac{9}{4}\Omega^2\varepsilon^r\phi_0 + 2\beta_3\phi_1\phi_1^*\varepsilon^2 + 2\beta_8\psi_1\psi_1^*\varepsilon^2 \\ &\quad - \beta_{13}\varepsilon^2 - f_{40}\Omega_1\sin(\tilde{\omega}t)\varepsilon^2 - f_{39}\Omega_1\cos(\tilde{\omega}t)\varepsilon^2 \\ &\quad - f_{41}\Omega_1^2\cos^2(\tilde{\omega}t)\varepsilon^2 = 0. \end{aligned} \tag{35b}$$

The solvability of (35a) and (35b) requires $r = 2$. Thus, we obtain the following expressions:

$$\begin{aligned} \psi_0 = &-\frac{1}{\Omega^2}(8\alpha_3\psi_1\psi_1^* + 8\alpha_8\phi_1\phi_1^* - 4f_{20}\Omega_1\sin(\tilde{\omega}t) \\ &- 4f_{19}\Omega_1\cos(\tilde{\omega}t) - 4f_{21}\Omega_1^2\cos^2(\tilde{\omega}t) - 4\alpha_{13}), \end{aligned} \tag{36a}$$

$$\begin{aligned} \phi_0 = &-\frac{4}{9\Omega^2}(8\beta_3\phi_1\phi_1^* + 8\beta_8\psi_1\psi_1^* - 4\beta_{13} - 4f_{40}\Omega_1\sin(\tilde{\omega}t) \\ &- 4f_{39}\Omega_1\cos(\tilde{\omega}t) - 4f_{41}\Omega_1^2\cos^2(\tilde{\omega}t)). \end{aligned} \tag{36b}$$

When $n = 2$, considering derivative forms of (34a) and (34b) yields

$$\alpha_3\psi_1^2 + \alpha_4\psi_1^*\phi_1 - \frac{3}{4}\Omega^2\psi_2 = 0, \tag{37a}$$

$$\beta_3\phi_1^2 - \frac{27}{4}\Omega^2\phi_2 = 0. \tag{37b}$$

Then, the following equations are obtained

$$\psi_2 = \frac{4}{3}\frac{\alpha_3\psi_1^2 + \alpha_4\psi_1^*\phi_1}{\Omega^2}, \tag{38a}$$

$$\phi_2 = \frac{4}{27}\frac{\beta_3\phi_1^2}{\Omega^2}. \tag{38b}$$

When $n = 1$, for the balance of the nonlinear and linear terms, we must make $q = 2$. Equations can be derived as follows:

$$\begin{aligned} &-I\Omega D_1(\psi_1) + 2\alpha_3\psi_0\psi_1 + \alpha_4\phi_0\psi_1 + \alpha_4\phi_1\psi_2^* + 3\alpha_7\psi_1^2\psi_1^* \\ &\quad + 2\alpha_3\psi_2\psi_1^* + \alpha_6\phi_1\psi_1^* + \Omega\sigma_1\psi_1 - \frac{1}{2}I\Omega\cos(\tilde{\omega}t)f_{14}\Omega_1\psi_1 \\ &\quad + 2\alpha_5\psi_1\phi_1\phi_1^* - \frac{1}{2}I\Omega\alpha_2\psi_1 = 0, \end{aligned} \tag{39a}$$

$$\begin{aligned} &-3I\Omega D_1(\psi_1) + 2\beta_3\phi_0\phi_1 + \beta_4\phi_1\psi_0 + \beta_9\psi_1^3 + 3\Omega\sigma_2\phi_1 \\ &\quad + 2\beta_3\phi_2\phi_1^* + 3\beta_7\phi_1^2\phi_1^* + 2\beta_8\psi_1\psi_2 - \frac{3}{2}I\Omega\cos(\tilde{\omega}t)f_{34}\Omega_1\phi_1 \\ &\quad + 2\beta_5\psi_1\phi_1\psi_1^* - \frac{3}{2}I\Omega\beta_2\phi_1 = 0. \end{aligned} \tag{39b}$$

Substitute (36a) and (36b) and (38a) and (38b) into (39a) and (39b), differential equations about ψ_1 and ϕ_1 are derived as

$$D_1(\psi_1) = (g_1 + \Omega_1 g_2 - I\sigma_1 + Ih_3 + I\Omega_1 h_1 + I\Omega_1^2 h_2)\psi_1 + Ih_4\psi_1\phi_1\phi_1^* + Ih_5\psi_1^{*2}\phi_1 + Ih_6\psi_1^*\psi_1^2, \quad (40a)$$

$$D_1(\varphi_1) = (g_3 + \Omega_1 g_4 - I\sigma_2 + Ik_3 + I\Omega_1 k_1 + I\Omega_1^2 k_2)\varphi_1 + Ik_4\varphi_1\psi_1\psi_1^{*2} + Ik_5\varphi_1^*\varphi_1^2 + Ik_6\psi_1^3. \quad (40b)$$

Make ψ_1 and ϕ_1 as follows:

$$\psi_1 = x_1 + ix_2, \quad (41a)$$

$$\phi_1 = x_3 + ix_4. \quad (41b)$$

Substitute (41a) and (41b) into (40a) and (40b), averaged equation in the Cartesian form is obtained as

$$\dot{x}_1 = (g_1 + \Omega_1 g_2)x_1 + (\sigma_1 - h_3 - \Omega_1 h_1 - \Omega_1^2 h_2)x_2 - h_6 x_2(x_1^2 + x_2^2) - h_4 x_2(x_3^2 + x_4^2) - h_5 x_4(x_1^2 - x_2^2) + 2h_5 x_1 x_2 x_3, \quad (42a)$$

$$\dot{x}_2 = -(\sigma_1 - h_3 - \Omega_1 h_1 - \Omega_1^2 h_2)x_1 + (g_1 + \Omega_1 g_2)x_2 + h_6 x_1(x_1^2 + x_2^2) + h_4 x_1(x_3^2 + x_4^2) + h_5 x_3(x_1^2 - x_2^2) + 2h_5 x_1 x_2 x_4, \quad (42b)$$

$$\dot{x}_3 = (g_3 + \Omega_1 g_4)x_1 + (\sigma_2 - k_3 - \Omega_1 k_1 - \Omega_1^2 k_2)x_2 - k_4 x_4(x_1^2 + x_2^2) - k_5 x_4(x_3^2 + x_4^2) - k_6 x_2(3x_1^2 - x_2^2), \quad (42c)$$

$$\dot{x}_4 = -(\sigma_2 - k_3 - \Omega_1 k_1 - \Omega_1^2 k_2)x_1 + (g_3 + \Omega_1 g_4)x_2 + k_4 x_3(x_1^2 + x_2^2) + k_5 x_3(x_3^2 + x_4^2) + k_6 x_1(x_1^2 - 3x_2^2), \quad (42d)$$

where expressions of coefficients σ_i, g_i, h_i, and k_i are described in the appendix.

4. Numerical Simulation of Nonlinear Vibrations

4.1. Comparison about Frequencies. In this section, based on the kinetic energy and the potential energy of the blade deduced in the previous section, the Chebyshev polynomial and Ritz method are used to calculate the frequency of the model, which are compared with the frequency obtained by finite element method.

To validate the model proposed, we compare frequencies theoretically calculated with frequencies obtained by the finite element method. Geometrical parameters and physical properties of the compressor blade are selected as the spanwise length $L = 0.0534$ m, the chordwise length $C = 0.03773$ m, the mounting angle $\theta = 34.49°$, the hub radius $r = 0.0845$ m, the density $\rho = 7800\,\text{kg/m}^3$, the Poisson's ratio $\gamma = 0.3$, the elastic modulus $E = 1.96$ GPa, the inlet velocity $v_{in} = 39.0496$ m/s, the inlet mass flow $M_{in} = 2.557$ kg/s, the real flow velocity $V_{in} = 78.472$ m/s. Table 1 shows frequencies of different orders obtained by two methods above. When the rotating speed increases from 0 rpm to 15000 rpm, frequencies calculated by those two methods are in good agreement. So, the model is reasonable.

TABLE 1: Frequency comparison.

	1st		2nd		3rd	
	FEM	C-R	FEM	C-R	FEM	C-R
0 rpm	1130.4	1076.6	4038.4	3117.2	6288.6	6253.9
30000 rpm	1133.4	1080.6	4039.1	3118.7	6291.8	6257.7
6000 rpm	1142.2	1092.8	4041	3123.2	6301.2	6269.2
9000 rpm	1156.6	1112.7	4044.2	3130.6	6316.9	6288.3
12000 rpm	1176.6	1140	4048.7	3141	6338.8	6314.9
15000 rpm	1201.7	1156.2	4054.5	3169.9	6366.9	6341

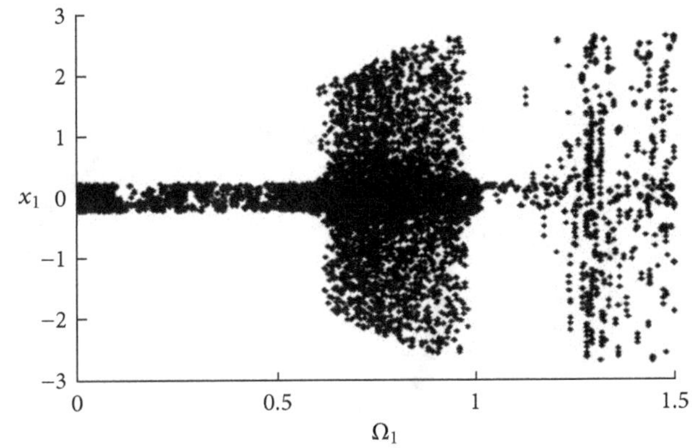

FIGURE 3: Bifurcation diagram on the plane (x_1, Ω_1) when the perturbation rotating speed Ω_1 varies in the interval $\Omega_1 = 0\sim1.5$. The initial condition is $x_{10} = 0.2$, $x_{20} = 0.101$, $x_{30} = -0.26$, and $x_{40} = 1.2$.

4.2. Numerical Simulation. Based on (42a), (42b), (42c), and (42d), we consider the influences of dimensionless parameters on nonlinear vibration behaviors of the rotating cantilever blade. Numerical simulation is utilized to investigate nonlinear dynamics of the rotating cantilever composite plate subjected to the aerodynamic force and the centrifugal force. Nonlinear oscillation of the system is investigated by choosing the perturbation rotating speed Ω_1 as the controlling parameter. When other parameters and the initial condition do not vary, we only change Ω_1 to detect the influence of the periodic perturbation rotating speed on vibration of the rotating cantilever composite blade. The Runge-Kutta algorithm and the Poincare map theory are used to construct numerical results of the bifurcation diagram, which describes the vibration law of the displacement x_1, when Ω_1 changes in a certain region.

When the parameter Ω_1 is chosen in the interval $\Omega_1 = 0 \sim 1.5$, and other parameters and the initial condition are, respectively, chosen as $\sigma_1 = 3.68$, $\sigma_2 = 1.88$, $g_1 = -0.0015$, $g_2 = 0.0012$, $g_3 = -0.0015$, $g_4 = 0.0012$, $h_1 = 5.95$, $h_2 = 0.29$,

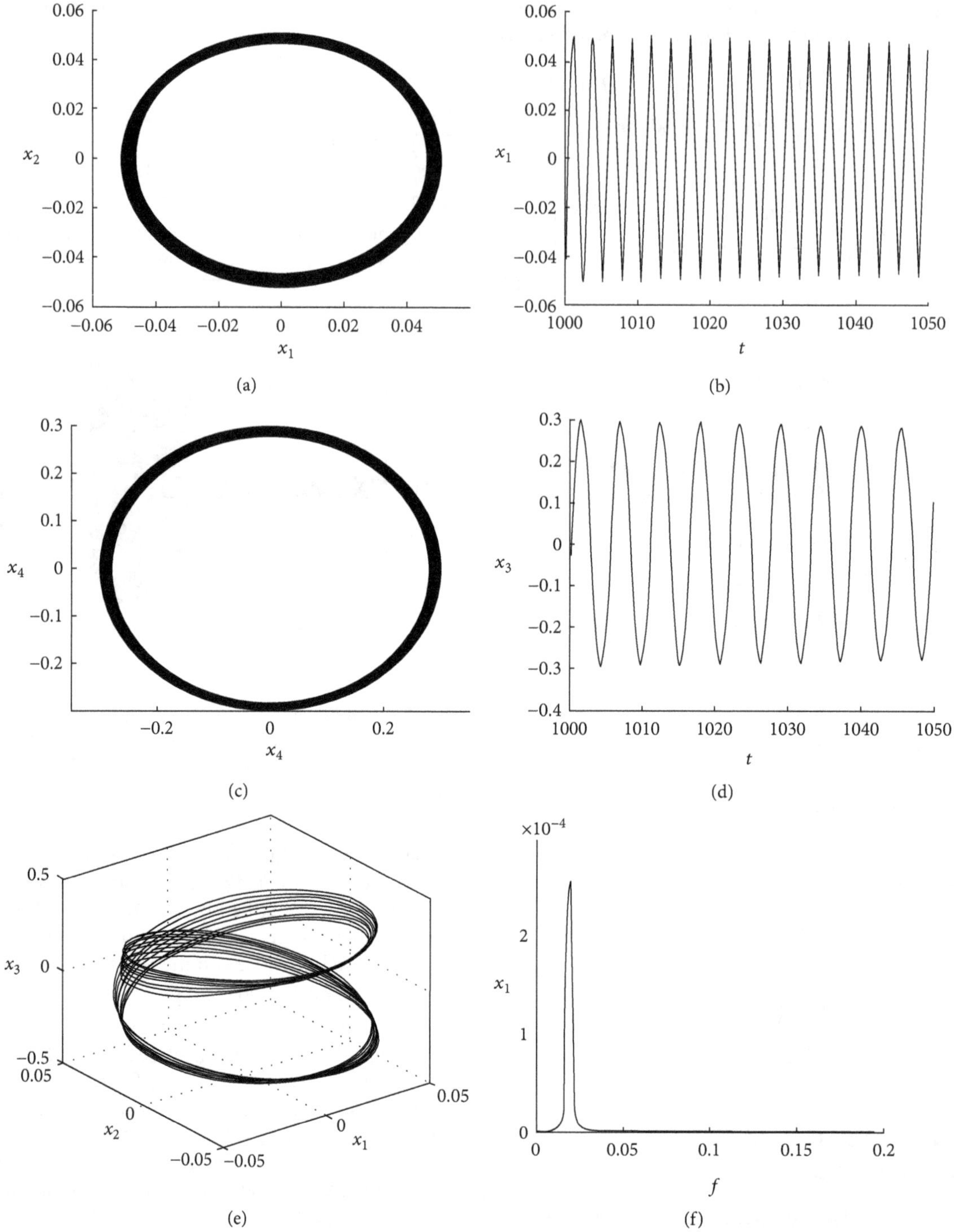

FIGURE 4: The multiperiodic motion of the blade with varying rotating speeds when $\Omega_1 = 0.076$. (a) The phase portrait on the plane (x_1, x_2); (b) the waveform on the plane (t, x_1); (c) the phase portrait on the plane (x_3, x_4); (d) the waveform on the plane (t, x_3); (e) the phase portrait in the three-dimensional space (x_1, x_2, x_3); and (f) the power spectrum on the plane (f, x_1).

$h_3 = 1.2$, $h_4 = -3.57$, $h_5 = 6.4$, $h_6 = 6.39$, $k_1 = 8.25$, $k_2 = 0.2925$, $k_3 = 0.18$, $k_4 = 19.9$, $k_5 = -1.4$, $k_6 = 1.25$, $x_{10} = 0.2$, $x_{20} = 0.101$, $x_{30} = -0.26$, and $x_{40} = 1.2$, we obtain Figure 3. Figure 3 describes the bifurcation diagram on the plane (x_1, Ω_1) when Ω_1 varies in the interval $(0, 1.5)$. It can be observed that periodic motion and chaotic motion appear in the vibration of the blade. With the increase of the perturbation rotating speed Ω_1, Figure 3 presents the following law: periodic motion → chaotic motion → periodic motion → chaotic motion → periodic motion.

The chaotic and periodic responses can be identified by several conventional criteria. Thus, based on the bifurcation diagram, the waveform, phase portraits, and the power spectrum are utilized to further verify the existence of the chaotic and periodic motion of the blade. Figures 4 and 5 illustrate that the system occurs multiperiodic motion when the

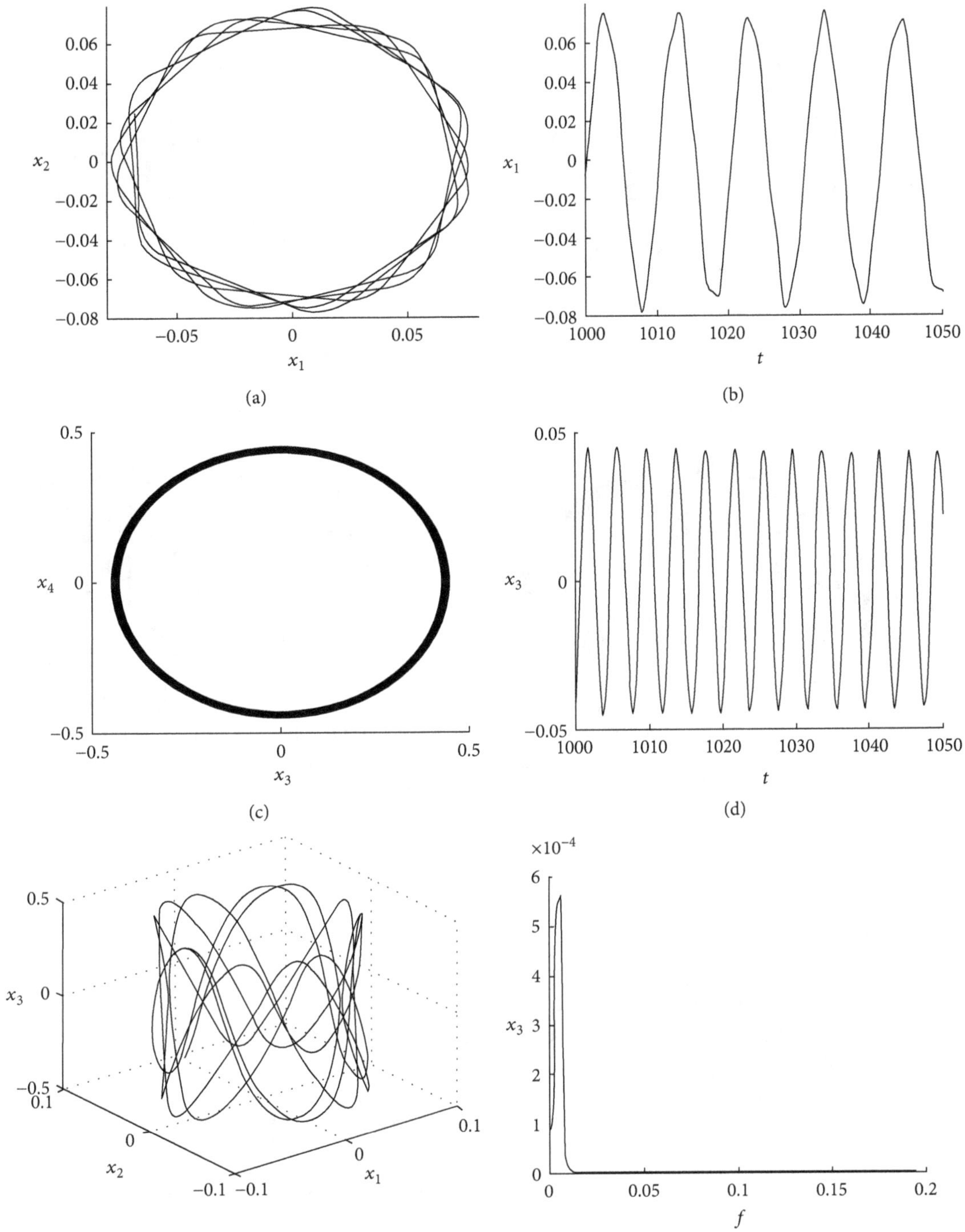

FIGURE 5: The multiperiodic motion of the blade with varying rotating speeds when $\Omega_1 = 0.412$. (a) The phase portrait on the plane (x_1, x_2); (b) the waveform on the plane (t, x_1); (c) the phase portrait on the plane (x_3, x_4); (d) the waveform on the plane (t, x_3); (e) the phase portrait in the three-dimensional space (x_1, x_2, x_3); and (f) the power spectrum on the plane (f, x_1).

perturbation rotating speed Ω_1 is chosen as $\Omega_1 = 0.076$ and $\Omega_1 = 0.412$, respectively. Figures 4(a) and 4(c) represent the phase portraits on the planes (x_1, x_2) and (x_3, x_4), respectively. Figures 4(b) and 4(d) give the waveform on the plane (t, x_1) and (t, x_3), respectively. Figure 4(e) represents the three-dimensional phase portrait in the space (x_1, x_2, x_3). Figure 4(f) describes the power spectrum on the plane (f, x_1). It can be shown from Figure 4 that the amplitude of the first-order mode is larger than that of the second-order mode. Figure 6 indicates that the chaotic motion of the system appears when $\Omega_1 = 0.8$. Figures 7 and 8 illustrate that multiperiodic motion of the rotating composite plate occur, when the perturbation rotating speed Ω_1 continues to increase and is selected as $\Omega_1 = 0.93$ and $\Omega_1 = 1.1$,

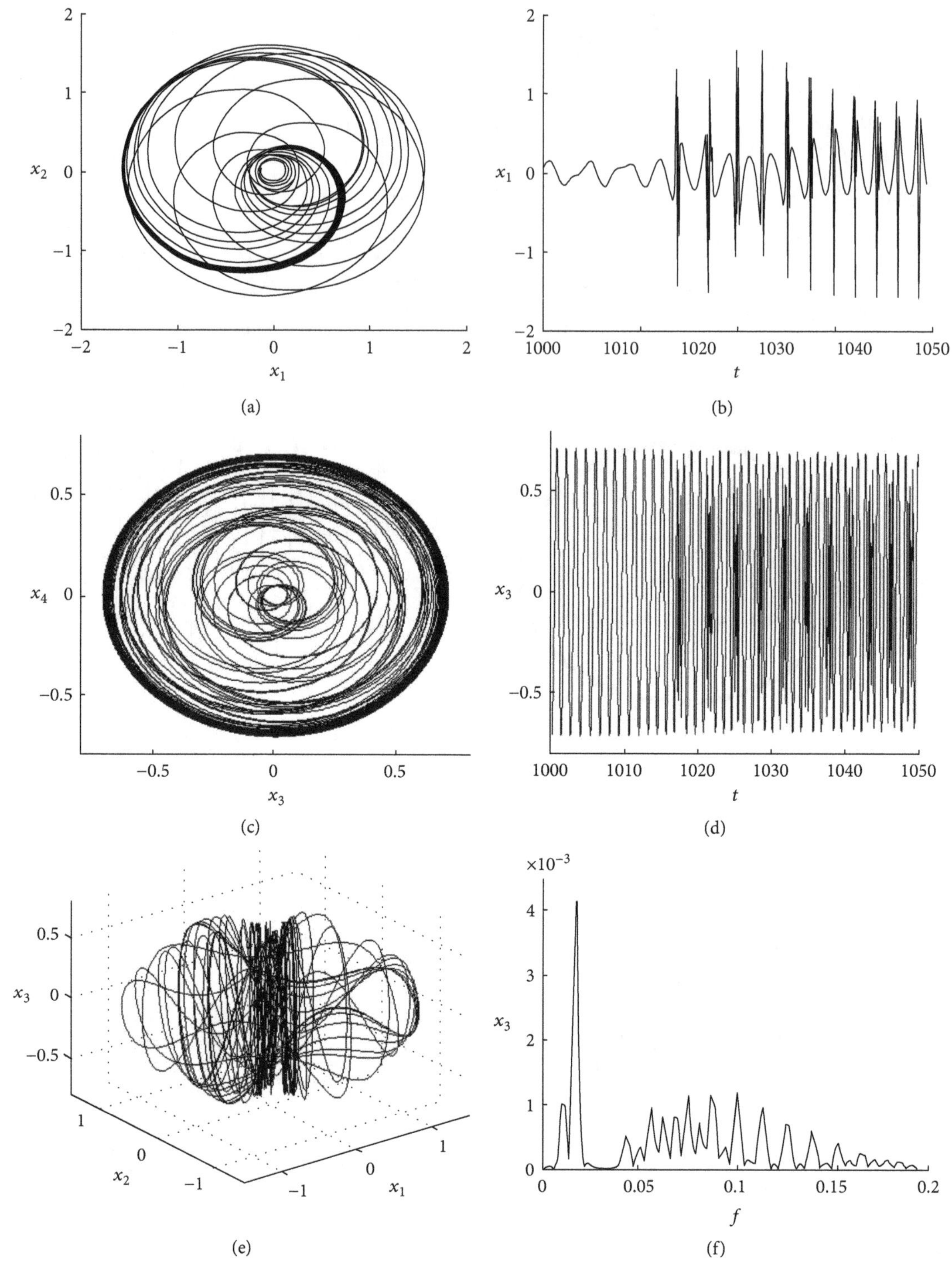

FIGURE 6: The chaotic motion of the blade with varying rotating speeds when $\Omega_1 = 0.8$. (a) The phase portrait on the plane (x_1, x_2); (b) the waveform on the plane (t, x_1); (c) the phase portrait on the plane (x_3, x_4); (d) the waveform on the plane (t, x_3); (e) the phase portrait in the three-dimensional space (x_1, x_2, x_3); and (f) the power spectrum on the plane (f, x_1).

respectively. Figure 9 shows that the chaotic motion of the rotating cantilever composite plate exists when $\Omega_1 = 1.201$. Figure 10 illustrates that the system exhibits multiperiodic motion when Ω_1 increases to $\Omega_1 = 1.33$.

Numerical simulation indicates that the system performs complex dynamic behaviors, such as multiperiodic motion and chaotic motion, when Ω_1 changes in a certain region. The perturbation rotating speed Ω_1 affects the dynamic behaviors of the system significantly.

5. Conclusions

In this paper, nonlinear vibration behaviors of the high rotating cantilever plate subjected to the centrifugal force

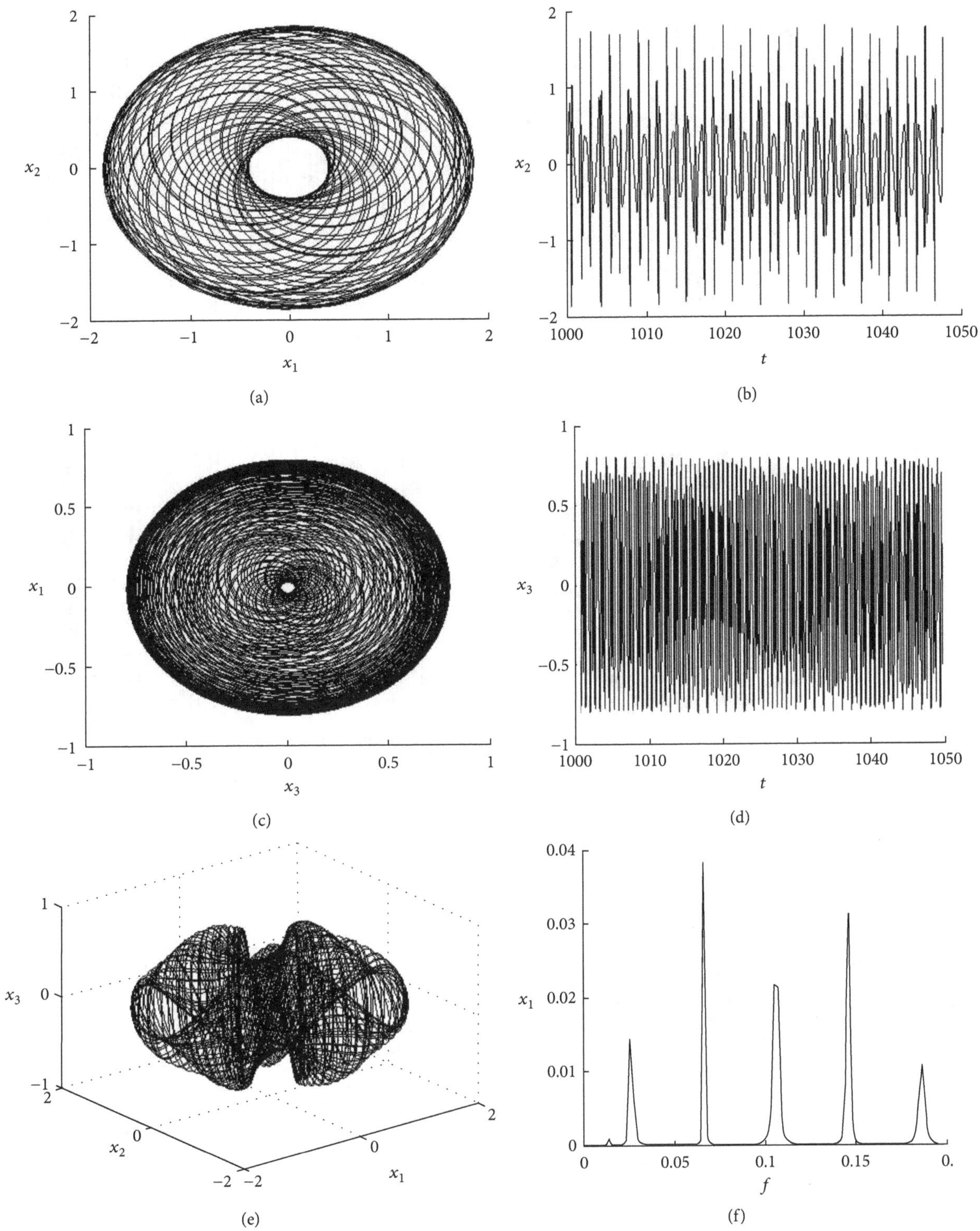

Figure 7: The multiperiodic motion of the blade with varying rotating speeds when $\Omega_1 = 0.93$. (a) The phase portrait on the plane (x_1, x_2); (b) the waveform on the plane (t, x_1); (c) the phase portrait on the plane (x_3, x_4); (d) the waveform on the plane (t, x_3); (e) the phase portrait in the three-dimensional space (x_1, x_2, x_3); (f) the power spectrum on the plane (f, x_1).

and the aerodynamic force are investigated. Nonlinear partial differential governing equations of the rotating blade are derived by utilizing Hamilton's principle. The Galerkin method is applied to discretize the nonlinear partial differential governing equations. The method of asymptotic perturbation is adopted to obtain the four-dimension nonlinear averaged equation of motion for the blade in the case of 1 : 3 internal resonance-1/2 subharmonic resonance. Numerical simulations are used to study nonlinear dynamic responses of the blade.

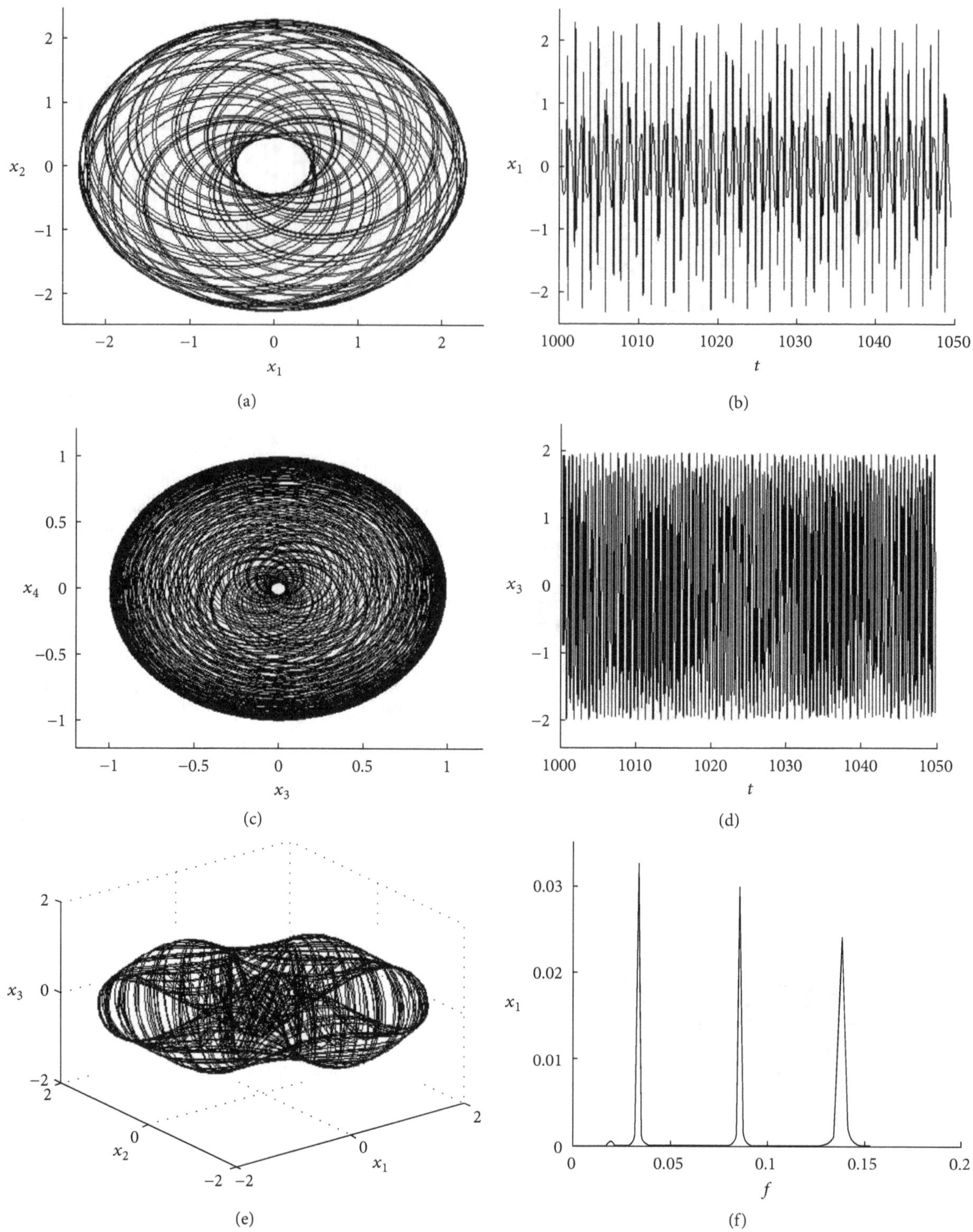

FIGURE 8: The multiperiodic motion of the blade with varying rotating speeds when $\Omega_1 = 1.1$. (a) The phase portrait on the plane (x_1, x_2); (b) the waveform on the plane (t, x_1); (c) the phase portrait on the plane (x_3, x_4); (d) the waveform on the plane (t, x_3); (e) the phase portrait in the three-dimensional space (x_1, x_2, x_3); and (f) the power spectrum on the plane (f, x_1).

The bifurcation diagram, phase portrait, and power spectrum are conducted to demonstrate that periodic motions and chaotic motions occur in nonlinear vibrations of the rotating blade under certain conditions. In this paper, the perturbation rotating speed is selected as the controlling parameter. Numerical simulation shows that nonlinear dynamic motions of the blade are sensitive to the perturbation rotating speed. The periodic motion and chaotic motion of the blade appear alternately. Periodic motions and chaotic motions exist in the averaged equations. It is well known that

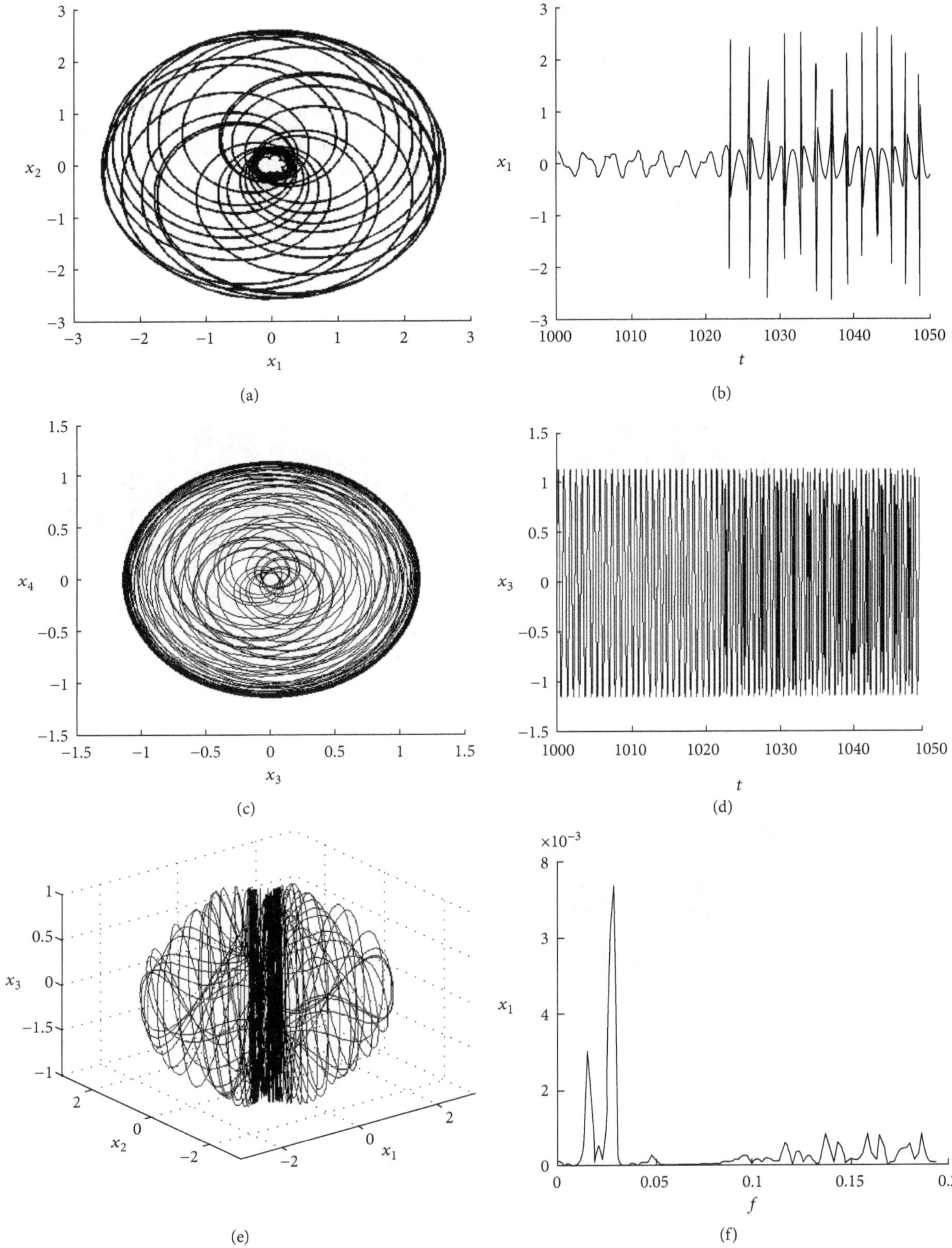

FIGURE 9: The chaotic motion of the blade with varying rotating speeds when $\Omega_1 = 1.201$. (a) the phase portrait on the plane (x_1, x_2); (b) the waveform on the plane (t, x_1); (c) the phase portrait on the plane (x_3, x_4); (d) the waveform on the plane (t, x_3); (e) the phase portrait in the three-dimensional space (x_1, x_2, x_3); and (f) the power spectrum on the plane (f, x_1).

periodic motions and chaotic motions in the averaged equations can lead to amplitude-modulated periodic and chaotic vibrations in the original system under certain conditions. Therefore, amplitude-modulated periodic and chaotic motions occur in the rotating blade. Occurrence of the chaotic motion means that the rotating cantilever blade

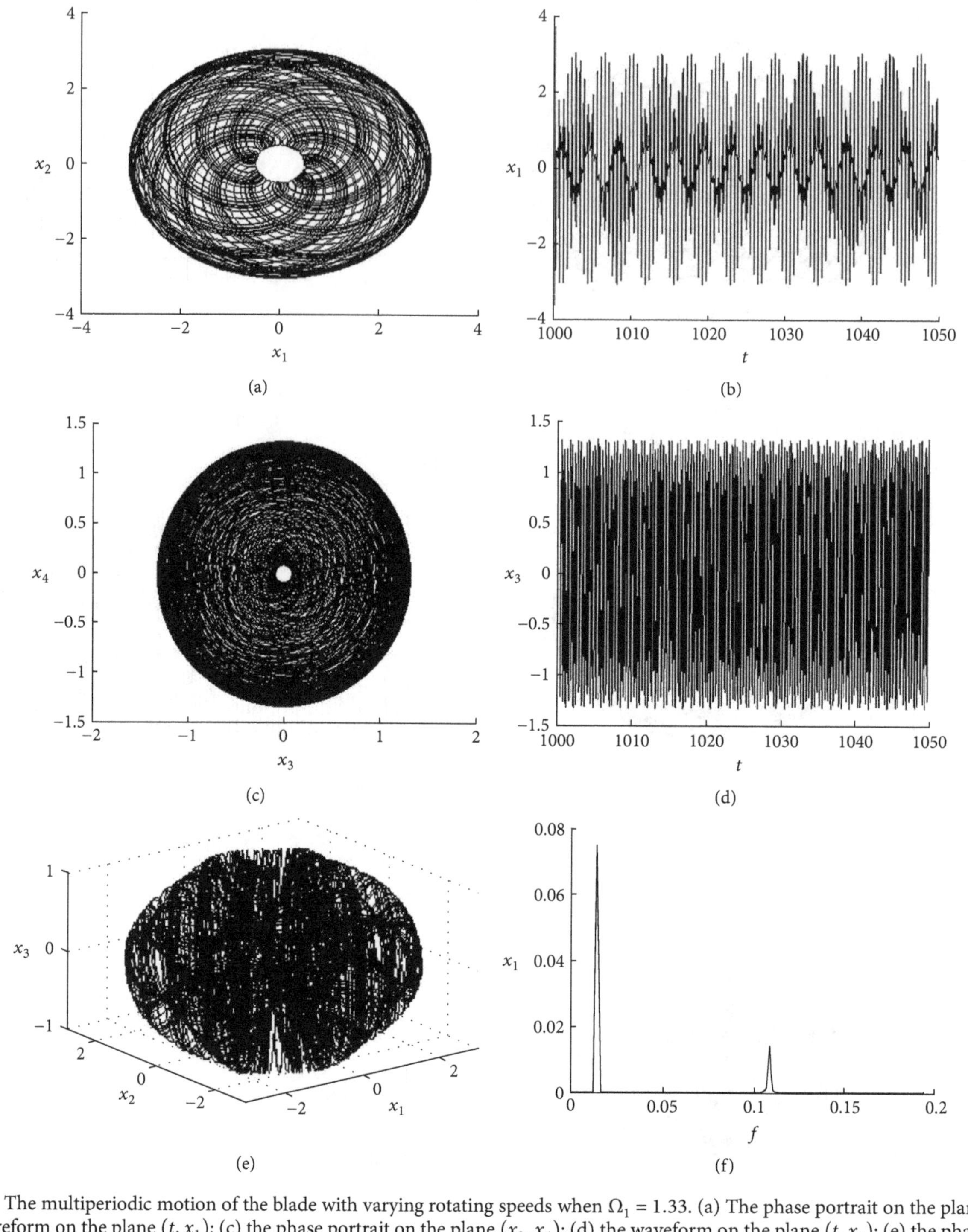

Figure 10: The multiperiodic motion of the blade with varying rotating speeds when $\Omega_1 = 1.33$. (a) The phase portrait on the plane (x_1, x_2); (b) the waveform on the plane (t, x_1); (c) the phase portrait on the plane (x_3, x_4); (d) the waveform on the plane (t, x_3); (e) the phase portrait in the three-dimensional space (x_1, x_2, x_3); and (f) the power spectrum on the plane (f, x_1).

may perform large amplitude nonlinear vibration, which leads to the damage of the system. We can control the responses of the system from the chaotic motions to the periodic motions by changing the perturbation rotating speed. Finally, the large-amplitude vibrations of the blade can be controlled. Thus, study on nonlinear behavior of the blade is of great significance.

Analytical results of this paper can be widely used in the practical aircraft engines. Large amplitude nonlinear vibration produced by the forced vibration and self-excited vibration occasionally leads to the undesirable disaster of the blade. Since chaotic motion is the large amplitude nonlinear vibration, we can adjust the perturbation rotating speed to control the responses of the system in order to avoid the appearance of the chaotic motion. Moreover, the possibility of aircraft engine failure caused by nonlinear vibration is reduced to ensure the safety of the aeroplane.

Appendix

Coefficients in Equations

Coefficients represented in (25a) and (25b) are given as follows:

$$\alpha_1 = \frac{1}{\alpha_0}\left[-1.963c_{97} + \left(0.653\Omega_0^2 + 0.33c_{106}p_{10}\right)\cos\Theta\right] + \frac{h_8}{\alpha_0},$$

$$\alpha_2 = \frac{1}{\alpha_0}(0.653c_{104} - 0.949c_{96} + 0.33\Omega_0 c_{103}p_{10}),$$

$$\alpha_3 = \frac{h_1}{\alpha_0},$$

$$\alpha_4 = \frac{h_2}{\alpha_0},$$

$$\alpha_5 = \frac{h_3}{\alpha_0},$$

$$\alpha_6 = \frac{h_4}{\alpha_0},$$

$$\alpha_7 = \frac{h_5}{\alpha_0},$$

$$\alpha_8 = \frac{h_6}{\alpha_0},$$

$$\alpha_9 = \frac{h_7}{\alpha_0},$$

$$\alpha_{10} = \frac{1}{\alpha_0}\left[-3.401c_{97} + \left(1.131\Omega_0^2 + 0.33c_{106}p_{11}\right)\cos\Theta\right] + \frac{h_9}{\alpha_0},$$

$$\alpha_{11} = \frac{1}{\alpha_0}(1.131c_{104} - 1.644c_{96} + 0.33\Omega_0 c_{103}p_{11}),$$

$$\alpha_{12} = \frac{1}{\alpha_0}(0.329c_{100}p_{11} + 1.131c_{101})\cos\Theta + \frac{1}{\alpha_0}1.71c_{99},$$

$$\alpha_{13} = \frac{1}{\alpha_0}\left[(0.563c_{109}\sin\Theta + 0.3297c_{109}p_{12}\cos\Theta)\Omega_0^2\right] + \frac{h_{10}}{\alpha_0},$$

$$f_{11} = \frac{1}{\alpha_0}[(-0.626c_{111} - 0.33c_{110}p_{10})\tilde{\omega}\sin\Theta],$$

$$f_{12} = \frac{1}{\alpha_0}[(1.305c_{107} - 0.659c_{106}p_{10})\Omega_0\cos\Theta],$$

$$f_{13} = \frac{1}{\alpha_0}[(0.33c_{106}p_{10} + 0.653c_{107})\cos\Theta],$$

$$f_{14} = \frac{1}{\alpha_0}(0.653c_{104} + 0.33c_{103}p_{10}),$$

$$f_{15} = \frac{1}{\alpha_0}[(-1.131c_{111}\tilde{\omega} - 0.33c_{110}p_{11})\sin\Theta],$$

$$f_{16} = \frac{1}{\alpha_0}[(2.261c_{107} - 0.659c_{106}p_{11})\Omega_0\cos\Theta],$$

$$f_{17} = \frac{1}{\alpha_0}[(0.33c_{106}p_{11} + 1.131c_{107})\cos\Theta],$$

$$f_{18} = \frac{1}{\alpha_0}(1.131c_{104} + 0.33c_{103}p_{11}),$$

$$f_{19} = \frac{1}{\alpha_0}[(1.127c_{109}\sin\Theta + 0.6595c_{106}p_{12}\cos\Theta)\Omega_0],$$

$$f_{20} = \frac{1}{\alpha_0}[(-0.563c_{113}\cos\Theta - 0.3297c_{110}p_{12}\sin\Theta)\tilde{\omega}],$$

$$f_{21} = \frac{1}{\alpha_0}(-0.563c_{109}\sin\Theta - 0.3297c_{106}p_{12}\cos\Theta),$$

$$\beta_1 = \frac{1}{\beta_0}\left[-1.963c_{97} + \left(0.653\Omega_0^2 + 0.33c_{106}p_{10}\right)\cos\Theta\right] + \frac{h_{18}}{\beta_0},$$

$$\beta_2 = \frac{1}{\beta_0}[(1.131c_{104} + 0.571c_{103}p_{10} - 0.262c_{105}p_4)\Omega_0 - 0.547c_{105}p_1 - 1.644c_{96}],$$

$$\beta_3 = \frac{h_{11}}{\beta_0},$$

$$\beta_4 = \frac{h_{12}}{\beta_0},$$

$$\beta_5 = \frac{h_{13}}{\beta_0},$$

$$\beta_6 = \frac{h_{14}}{\beta_0},$$

$$\beta_7 = \frac{h_{15}}{\beta_0},$$

$$\beta_8 = \frac{h_{16}}{\beta_0},$$

$$\beta_9 = \frac{h_{17}}{\beta_0},$$

$$\beta_{10} = \frac{1}{\beta_0}\left[-3.401c_{97} + \left(1.131\Omega_0^2 + 0.33c_{106}p_{11}\right)\cos\Theta\right] + \frac{h_{19}}{\beta_0},$$

$$\beta_{11} = \frac{1}{\beta_0}(2.593c_{104} + 0.571c_{103}p_{11})\Omega_0,$$

$$\beta_{12} = \frac{1}{\beta_0}[(-0.468p_{17} + 0.111p_{14})c_{98} + 3.921c_{99} + (0.571c_{100}p_{11} + 2.593c_{101})\cos\Theta, + (-0.262c_{102}p_5 - 0.547c_{102}p_2)\sin\Theta],$$

$$\beta_{13} = \frac{1}{\beta_0}\left[(0.9756c_{109} + 0.262c_{108}p_{16} - 0.547c_{112}p_3)\Omega_0^2\sin\Theta + 0.571c_{106}p_{12}\Omega_0^2\cos\Theta\right] + \frac{h_{20}}{\beta_0},$$

$$f_{31} = \frac{1}{\beta_0}[(-0.626c_{111} - 0.33c_{110}p_{10})\tilde{\omega}\sin\Theta],$$

$$f_{32} = \frac{1}{\beta_0}[(1.305c_{107} - 0.659c_{106}p_{10})\Omega_0\cos\Theta],$$

$$f_{33} = \frac{1}{\beta_0}[(0.33c_{106}p_{10} + 0.653c_{107})\cos\Theta],$$

$$f_{34} = \frac{1}{\beta_0}(1.131c_{104} + 0.571c_{103}p_{10} - 0.262c_{105}p_4 - 0.547c_{105}p_1),$$

$$f_{35} = \frac{1}{\beta_0}[(-1.131c_{111}\tilde{\omega} - 0.33c_{110}p_{11})\sin\Theta],$$

$$f_{36} = \frac{1}{\beta_0}[(2.261c_{107} - 0.659c_{106}p_{11})\Omega_0\cos\Theta],$$

$$f_{37} = \frac{1}{\beta_0}[(0.33c_{106}p_{11} + 1.131c_{107})\cos\Theta],$$

$$f_{38} = \frac{1}{\beta_0}(2.593c_{104} + 0.571c_{103}p_{11} - 0.262c_{105}p_5 - 0.547c_{105}p_2),$$

$$f_{39} = \frac{1}{\beta_0}[(-1.0947c_{108} + 0.524c_{108}p_6 + 1.951c_{109})\Omega_0 \sin\Theta + 1.142c_{106}p_{12}\Omega_0\cos\Theta],$$

$$f_{40} = \frac{1}{\beta_0}[(0.547c_{112}p_3 + 0.262c_{112}p_6 - 0.976c_{113})\tilde{\omega}\cos\Theta - 0.571c_{110}p_{12}\tilde{\omega}\sin\Theta],$$

$$f_{41} = \frac{1}{\beta_0}[(0.9756c_{109} - 0.262c_{108}p_6 - 0.547c_{108}p_3)\sin\Theta - 0.571c_{106}p_{12}\cos\Theta], \tag{A.1}$$

where

$$\alpha_0 = (0.33c_{100}p_{10} + 0.653c_{101})\cos\Theta + 0.987c_{99},$$

$$\beta_0 = (0.111p_{13} + 0.468)c_{98} + 1.7099c_{99} + (0.571c_{100}p_{10} + 1.131c_{101})\cos\Theta + (-0.262c_{102}p_4 - 0.547c_{102}p_1)\sin\Theta,$$

$$c_{96} = \frac{\bar{C}_\infty\bar{\rho}_\infty}{\bar{A}_{11}}\frac{2L^2}{3h^4}A,$$

$$c_{97} = \frac{\bar{C}_\infty\bar{\rho}_\infty}{\bar{A}_{11}}\frac{2L\sqrt{LC}}{3h^4}A\cos\bar{\theta},$$

$$c_{98} = \frac{2\bar{I}_0}{3\bar{A}_{11}}\frac{CL^4}{h^5}\sin\bar{\theta},$$

$$c_{99} = \frac{\bar{I}_0}{\bar{A}_{11}}\frac{2L^4}{3h^4}\cos\bar{\theta},$$

$$c_{100} = -\frac{\bar{I}_2}{\bar{A}_{11}}\frac{2CL^4}{3h^5}\bar{\theta}'^2,$$

$$c_{101} = \frac{\bar{I}_2}{\bar{A}_{11}}\frac{2CL^3}{3h^4}\bar{\theta}'^4,$$

$$c_{102} = \frac{\bar{I}_2}{\bar{A}_{11}}\frac{2CL^4}{3h^5}\bar{\theta}'^2\cos\bar{\theta},$$

$$c_{103} = -\frac{4\bar{I}_2}{3\bar{A}_{11}}\frac{CL^4}{h^5}\bar{\theta}'^2$$

$$c_{104} = \frac{4\bar{I}_2}{3\bar{A}_{11}}\frac{CL^3}{h^4}\bar{\theta}'^4,$$

$$c_{105} = -\frac{4\bar{I}_2}{3\bar{A}_{11}}\frac{CL^4}{h^5}\bar{\theta}'^2\cos\bar{\theta},$$

$$c_{106} = -\frac{\bar{I}_2}{\bar{A}_{11}}\frac{2CL^4}{3h^5}\bar{\theta}'^2,$$

$$c_{107} = \frac{\bar{I}_2}{\bar{A}_{11}}\frac{2CL^3}{3h^4}\bar{\theta}'^4,$$

$$c_{108} = \frac{\bar{I}_2}{\bar{A}_{11}}\frac{2CL^4}{3h^5}\bar{\theta}'^2\cos\bar{\theta},$$

$$c_{109} = -\frac{\bar{I}_2}{\bar{A}_{11}}\frac{2CL^4}{3h^5}\bar{\theta}'^2\sin\bar{\theta},$$

$$c_{110} = -\frac{\bar{I}_2}{\bar{A}_{11}}\frac{2CL^4}{3h^5}\bar{\theta}'^2,$$

$$c_{111} = \frac{\bar{I}_2}{\bar{A}_{11}}\frac{2CL^3}{3h^4}\bar{\theta}'^2,$$

$$c_{112} = -\frac{\bar{I}_2}{\bar{A}_{11}}\frac{2CL^4}{3h^5}\bar{\theta}'^2\cos\bar{\theta},$$

$$c_{113} = \frac{\bar{I}_2}{\bar{A}_{11}}\frac{2CL^4}{3h^5}\bar{\theta}'^2\sin\bar{\theta},$$

$$p_1 = \frac{s_{71}}{n_5},$$

$$s_{71} = (h_{71}t_3 + h_{72}t_9 + h_{73}t_{15} + h_{74}t_{21} + h_{76}m_{43}) \cdot m_{53}m_{61} + h_{71}t_3m_{51} + h_{71}t_4m_{61} + h_{72}t_9m_{51} + h_{72}t_{10}m_{61} + h_{73}t_{15}m_{51} + h_{73}t_{16}m_{61} + h_{74}t_{21}m_{51} + h_{74}t_{22}m_{61} + h_{76}m_{43}m_{51} + h_{76}m_{44}m_{61} + h_{77}m_{53}m_{61} + h_{71}t_1 + h_{72}t_7 + h_{73}t_{13} + h_{74}t_{19} + h_{76}m_{41} + h_{77}m_{51},$$

$$p_2 = \frac{s_{72}}{n_5},$$

$$s_{72} = (h_{71}t_3 + h_{72}t_9 + h_{73}t_{15} + h_{74}t_{21} + h_{76}m_{43}) \cdot m_{53}m_{62} + h_{71}t_3m_{52} + h_{71}t_4m_{62} + h_{72}t_9m_{52} + h_{72}t_{10}m_{62} + h_{73}t_{15}m_{52} + h_{73}t_{16}m_{62} + h_{74}t_{21}m_{52} + h_{74}t_{22}m_{62} + h_{76}m_{43}m_{52} + h_{76}m_{44}m_{62} + h_{77}m_{53}m_{62} + h_{71}t_2 + h_{72}t_8 + h_{73}t_{14} + h_{74}t_{20} + h_{76}m_{42} + h_{77}m_{52} + h_{75},$$

$$p_3 = \frac{s_{73}}{n_5},$$

$$s_{73} = (h_{71}t_3 + h_{72}t_9 + h_{73}t_{15} + h_{74}t_{21} + h_{76}m_{43}) \cdot m_{53}m_{64} + h_{71}t_3m_{55} + h_{71}t_4m_{64} + h_{72}t_9m_{55} + h_{72}t_{10}m_{64} + h_{73}t_{15}m_{55} + h_{73}t_{16}m_{64} + h_{74}t_{21}m_{55} + h_{74}t_{22}m_{64} + h_{76}m_{43}m_{55} + h_{76}m_{44}m_{64} + h_{77}m_{53}m_{64} + h_{71}t_6 + h_{72}t_{12} + h_{73}t_{14} + h_{73}t_{18} + h_{74}m_{24} + h_{76}m_{46} + h_{77}m_{55},$$

$$n_5 = (h_{71}t_3 + h_{72}t_9 + h_{73}t_{15} + h_{74}t_{21})m_{53}m_{63} + h_{76}m_{43}m_{53}m_{63} + h_{71}m_{54}t_3 + h_{71}m_{63}t_4 + h_{72}m_{54}t_9 + h_{72}m_{63}t_{10} + h_{73}m_{54}t_{15} + h_{73}m_{63}t_{16} + h_{74}m_{54}t_{21} + h_{74}m_{63}t_{22} + h_{76}m_{43}m_{54} + h_{76}m_{44}m_{63} + h_{77}m_{53}m_{63} + h_{71}t_5 + h_{72}t_{11} + h_{73}t_{17} + h_{74}t_{23} + h_{76}m_{45} + h_{77}m_{54} + h_{78},$$

$$p_4 = m_{63}m_{71} + m_{61},$$
$$p_5 = m_{63}m_{72} + m_{62},$$
$$p_6 = m_{63}m_{73} + m_{64}$$
$$p_{10} = m_{43}m_{53}m_{63}m_{71} + m_{43}m_{53}m_{61} + m_{43}m_{54}m_{71} + m_{44}m_{63}m_{71} + m_{43}m_{51} + m_{44}m_{61} + m_{45}m_{71} + m_{41},$$
$$p_{11} = m_{43}m_{53}m_{63}m_{72} + m_{43}m_{53}m_{62} + m_{43}m_{54}m_{72} + m_{44}m_{63}m_{72} + m_{43}m_{52} + m_{44}m_{62} + m_{45}m_{72} + m_{42},$$
$$p_{12} = m_{43}m_{53}m_{63}m_{73} + m_{43}m_{53}m_{64} + m_{43}m_{54}m_{73} + m_{44}m_{63}m_{73} + m_{43}m_{55} + m_{44}m_{64} + m_{45}m_{73} + m_{46},$$
$$\begin{aligned}p_{16} = {} & m_{21}m_{33}m_{43}(m_{53}p_4 + m_{54}p_1 + m_{51}) + m_{21}m_{33}m_{45}p_1 \\ & + m_{21}m_{34}(m_{53}p_4 + m_{54}p_1) + m_{24}m_{43}(m_{54}p_1 + m_{53}p_4) \\ & + m_{21}m_{33}m_{41} + m_{21}m_{34}m_{51} + m_{21}m_{35}p_4 + m_{21}m_{36}p_1 \\ & + m_{24}m_{43}m_{51} + m_{24}m_{44}p_4 + m_{24}m_{45}p_1 + m_{25}m_{53}p_4 \\ & + m_{25}m_{54}p_1 + m_{21}m_{31} + m_{24}m_{41} + m_{25}m_{51} + m_{26}p_4 \\ & + m_{27}p_1 + m_{22},\end{aligned}$$
$$\begin{aligned}p_{17} = {} & m_{21}m_{33}m_{43}(m_{53}p_5 + m_{54}p_2 + m_{52}) + m_{21}m_{33} \\ & \cdot (m_{44}p_5 + m_{45}p_2) + m_{21}m_{34}(m_{53}p_5 + m_{54}p_2) \\ & + m_{24}m_{43}(m_{53}p_5 + m_{54}p_2) + m_{21}m_{33}m_{42} + m_{21}m_{34}m_{52} \\ & + m_{21}m_{35}p_5 + m_{21}m_{36}p_2 + m_{24}m_{43}m_{52} + m_{24}m_{44}p_5 \\ & + m_{24}m_{45}p_2 + m_{25}m_{53}p_5 + m_{25}m_{54}p_2 + m_{21}m_{32} \\ & + m_{24}m_{42} + m_{25}m_{52} + m_{26}p_5 + m_{27}p_2 + m_{23}.\end{aligned} \quad (A.2)$$

Coefficients represented in (42a), (42b), (42c), and (42d) are as follows:

$$g_1 = -\frac{1}{2}\alpha_2,$$
$$g_2 = -\frac{1}{2}\cos(\tilde{\omega}t)f_{14},$$
$$g_3 = -\frac{1}{2}\beta_2,$$
$$g_4 = -\frac{1}{2}\cos(\tilde{\omega}t)f_{34},$$
$$h_1 = -\frac{I}{\Omega^3}\left[8\alpha_3 f_{19}\cos(\tilde{\omega}t) + \frac{4}{9}\alpha_4 f_{40}\sin(\tilde{\omega}t) + 8\alpha_3 f_{20}\sin(\tilde{\omega}t) + \frac{4}{9}\alpha_4 f_{39}\cos(\tilde{\omega}t)\right],$$
$$h_2 = -\frac{I}{\Omega^3}\left[8\alpha_3 f_{21}\cos^2(\tilde{\omega}t) + \frac{4}{9}\alpha_4 f_{41}\cos^2(\tilde{\omega}t)\right],$$
$$h_3 = -\frac{I}{\Omega^3}\left[8\alpha_3\alpha_{13} + \frac{4}{9}\alpha_4\beta_{13}\right],$$
$$h_4 = -\frac{2I\alpha_5}{\Omega} + \frac{I}{\Omega^3}\left[16\alpha_3\alpha_8 - \frac{4}{3}\alpha_4^2 + \frac{8}{9}\alpha_4\beta_3\right],$$
$$h_5 = -\frac{I\alpha_6}{\Omega} - \frac{4I\alpha_3\alpha_4}{\Omega^3},$$
$$h_6 = -\frac{3I\alpha_7}{\Omega} + \frac{I}{\Omega^3}\left[\frac{40}{3}\alpha_3^2 + \frac{8}{9}\alpha_4\beta_8\right],$$
$$k_1 = -\frac{I}{3\Omega^3}\left[4\beta_4 f_{19}\cos(\tilde{\omega}t) + \frac{8}{9}\beta_3 f_{40}\sin(\tilde{\omega}t) + 4\beta_4 f_{20} \cdot \sin(\tilde{\omega}t) + \frac{8}{9}\beta_3 f_{39}\cos(\tilde{\omega}t)\right],$$
$$k_2 = -\frac{I}{3\Omega^3}\left[4\beta_4 f_{21}\cos^2(\tilde{\omega}t) + \frac{8}{9}\beta_3 f_{41}\cos^2(\tilde{\omega}t)\right],$$
$$k_3 = -\frac{I}{3\Omega^3}\left[4\beta_4\alpha_{13} + \frac{8}{9}\beta_3\beta_{13}\right],$$
$$k_4 = -\frac{2I\beta_5}{3\Omega} + \frac{I}{3\Omega^3}\left[8\alpha_3\beta_4 - \frac{8}{3}\alpha_4\beta_8 + \frac{16}{9}\beta_8\beta_3\right],$$
$$k_5 = -\frac{I\beta_7}{\Omega} + \frac{8I\beta_4\alpha_8}{3\Omega^3} + \frac{40I\beta_3^2}{81\Omega^3},$$
$$k_6 = -\frac{I\beta_9}{\Omega} - \frac{8\beta_8\alpha_3}{9\Omega^3}. \quad (A.3)$$

Nomenclature

Ω:	Rotating speed
Ω_0:	Steady-state rotating speed
Ω_1:	Periodic perturbation amplitude of the rotating speed
$\tilde{\omega}$:	Angular velocity of the blade
L:	Plate spanwise length
C:	Plate chordwise length
h:	Plate thickness
r:	Radius of the rigid hub
θ_r:	Presetting angle
θ_R:	Pre-twist angle
$(\vec{\mathbf{i}}, \vec{\mathbf{j}}, \vec{\mathbf{k}})$:	Inertial frame
$(\vec{\mathbf{e}}_x^0, \vec{\mathbf{e}}_y^0, \vec{\mathbf{e}}_z^0)$:	Rotating coordinate system
$(\hat{e}_x, \hat{e}_y, \hat{e}_z)$:	Local triad system
θ':	Twist rate
θ:	Twist angle
u_0, v_0, w_0:	Components of displacement of a point at the neutral plane
u, v, w:	Components of displacement of a random point
ϕ_x, ϕ_y:	Angular displacement
$\varepsilon_{xx}, \varepsilon_{yy}, \gamma_{yz}, \gamma_{xz}, \gamma_{xy}$:	Components of strains of a point
ε^0:	Membrane strain
ε^1:	Flexural strain
κ_i, λ_j:	Coefficients of beam functions
$\sigma_{xx}, \sigma_{yy}, \sigma_{yz}, \sigma_{xz}, \sigma_{xy}$:	Components of stresses of a point
Q_{ij}:	Plane stress-reduced stiffnesses
σ_1, σ_2:	Detuning parameters
E_1, E_2:	Longitudinal modulus and transverse modulus of the fiber
γ_{12}, γ_{21}:	Poisson's ratio
G_{12}, G_{23}, G_{13}:	Shear modulus
$\mathbf{N}_{xx}, \mathbf{N}_{yy}, \mathbf{N}_{xy}$:	In-plane force resultants
$\mathbf{M}_{xx}, \mathbf{M}_{yy}, \mathbf{M}_{xy}$:	Moment resultants

$\mathbf{Q}_x$, $\mathbf{Q}_y$: Transverse forces
k: Shear coefficient
A_{ij}: Extensional stiffness
D_{ij}: Bending stiffness
B_{ij}: Bending-extensional coupling stiffness
I_0, I_1, I_2: Mass moments of inertia
$\vec{r}$: Displacement of a point without deformation
Θ: Rotating angle of the blade
$\dot{R}$, $\ddot{R}$: Speed and acceleration
K: Kinetic energy
F_s, F_c: Components of the centrifugal force
ρ: Density of the blade
U: Potential energy
U_1: Strain energy
U_2: Centrifugal potential energy
ρ_∞: Air density
C_∞: Speed of sound
P: Speed of the airflow
W: Virtual work of aerodynamic forces
δ: The variation
ε: Small parameter
ω_1, ω_2: The first two mode frequencies
τ: Time scale transformation
q: Positive rational number
v_{in}: Inlet velocity
M_{in}: Inlet mass flow
V_{in}: Real flow velocity.

Conflicts of Interest

The authors declare that there are no conflicts of interest regarding the publication of this paper.

Acknowledgments

The authors gratefully acknowledge the support of the National Natural Science Foundation of China (NNSFC) through Grant no. 11372015.

References

[1] S. M. Lin, W. R. Wang, and S. Y. Lee, "The dynamic analysis of nonuniformly pretwisted Timoshenko beams with elastic boundary conditions," *International Journal of Mechanical Sciences*, vol. 43, no. 10, pp. 2385–2405, 2001.

[2] M. H. Yao, Y. P. Chen, and W. Zhang, "Nonlinear vibrations of blade with varying rotating speed," *Nonlinear Dynamics*, vol. 68, no. 4, pp. 487–504, 2012.

[3] K. Sarkar and R. Ganguli, "Analytical test functions for free vibration analysis of rotating non-homogeneous Timoshenko beams," *Meccanica*, vol. 49, no. 6, pp. 1469–1477, 2014.

[4] F. Georgiades, J. Latalski, and J. Warminski, "Equations of motion of rotating composite beam with a nonconstant rotation speed and an arbitrary preset angle," *Meccanica*, vol. 49, no. 8, pp. 1833–1858, 2014.

[5] H. Ma, F. Xie, H. Nai, and B. Wen, "Vibration characteristics analysis of rotating shrouded blades with impacts," *Journal of Sound and Vibration*, vol. 378, pp. 92–108, 2016.

[6] F. Xie, H. Ma, C. Cui, and B. Wen, "Vibration response comparison of twisted shrouded blades using different impact models," *Journal of Sound and Vibration*, vol. 397, pp. 171–191, 2017.

[7] B. W. Huang and J. H. Kuang, "Variation in the stability of a rotating blade disk with a local crack defect," *Journal of Sound and Vibration*, vol. 294, no. 3, pp. 486–502, 2006.

[8] T. H. Young and G. T. Liou, "Coriolis effect on the vibration of a cantilever plate with time-varying rotating speed," *Journal of Vibration and Acoustics*, vol. 114, no. 2, pp. 232–241, 1992.

[9] H. Yoo and S. K. Kim, "Free vibration analysis of rotating cantilever plates," *AIAA Journal*, vol. 40, no. 11, pp. 2188–2196, 2002.

[10] S. K. Sinha and K. E. Turner, "Natural frequencies of a pretwisted blade in a centrifugal force field," *Journal of Sound and Vibration*, vol. 330, no. 11, pp. 2655–2681, 2011.

[11] L. Li and D. G. Zhang, "Free vibration analysis of rotating functionally graded rectangular plates," *Composite Structures*, vol. 136, pp. 493–504, 2016.

[12] M. A. Kouchakzadeh, M. Rasekh, and H. Haddadpour, "Panel flutter analysis of general laminated composite plates," *Composite Structures*, vol. 92, no. 12, pp. 2906–2915, 2010.

[13] L. Malgaca, H. Al-Qahtani, and M. Sunar, "Vibration control of rotating blades using root-embedded piezoelectric materials," *Arabian Journal for Science and Engineering*, vol. 40, no. 5, pp. 1487–1495, 2015.

[14] T. Y. Kam, H. M. Su, and C. Y. Huang, "Quasi-static buckling and first-ply failure loads of shear web reinforced glass-fabric composite wind blades," *Composite Structures*, vol. 160, pp. 1225–1235, 2017.

[15] Y. Q. Tang and L. Q. Chen, "Nonlinear free transverse vibrations of in-plane moving plates: without and with internal resonances," *Journal of Sound and Vibration*, vol. 330, no. 1, pp. 110–126, 2011.

[16] J. Sun, I. L. Arteaga, and L. Kari, "Dynamic modeling of a multilayer rotating blade via quadratic layerwise theory," *Composite Structures*, vol. 99, no. 5, pp. 276–287, 2013.

[17] L. Wang, D. Q. Cao, and W. Huang, "Nonlinear coupled dynamics of flexible blade–rotor–bearing systems," *Tribology International*, vol. 43, no. 4, pp. 759–778, 2010.

[18] M. Shariyat, M. Moradi, and S. Samaee, "Enhanced model for nonlinear dynamic analysis of rectangular composite plates with embedded SMA wires, considering the instantaneous local phase changes," *Composite Structures*, vol. 109, pp. 106–118, 2014.

[19] B. Zhang and Y. Li, "Nonlinear vibration of rotating pre-deformed blade with thermal gradient," *Nonlinear Dynamics*, vol. 86, no. 1, pp. 459–478, 2016.

[20] W. R. D. P. Mendonça, E. C. D. Medeiros, and A. L. R. Pereira, "The dynamic analysis of rotors mounted on composite shafts with internal damping," *Composite Structures*, vol. 167, pp. 50–62, 2017.

[21] X. Zhang, F. Chen, B. Zhang, and T. Jing, "Local bifurcation analysis of a rotating blade," *Applied Mathematical Modelling*, vol. 40, no. 5-6, pp. 4023–4031, 2016.

[22] N. Banichuk, A. Barsuk, J. Jeronen, P. Neittaanmaki, and T. Tuovinen, "On bifurcation analysis of implicitly given functionals in the theory of elastic stability," *Mathematical Modeling and Optimization of Complex Structures*, vol. 40, pp. 175–188, 2015.

[23] A. W. Leissa, J. K. Lee, and A. J. Wang, "Rotating blade vibration analysis using shells," *Journal of Engineering for Power*, vol. 104, no. 2, pp. 296–302, 1982.

[24] Y. J. Kee and J. H. Kim, "Vibration characteristics of initially twisted rotating shell type composite blades," *Composite Structures*, vol. 64, no. 2, pp. 151–159, 2004.

[25] J. Sun, I. L. Arteaga, and L. Kari, "General shell model for a rotating pretwisted blade," *Journal of Sound and Vibration*, vol. 332, no. 22, pp. 5804–5820, 2013.

[26] S. K. Sinha and R. P. Zylka, "Vibration analysis of composite airfoil blade using orthotropic thin shell bending theory," *International Journal of Mechanical Sciences*, vol. 121, pp. 90–105, 2017.

[27] C. Volker and B. Joachim, "Numerical investigation of non-linear fluid-structure interaction in vibrating compressor blades," *Journal of Turbomachinery*, vol. 123, no. 2, pp. 402–408, 2000.

[28] K. Ekici, R. E. Kielb, and K. C. Hall, "Aerodynamic asymmetry analysis of unsteady flows in turbomachinery," *Journal of Turbomachinery*, vol. 132, no. 1, article 011006, 2009.

[29] S. Farhadi and S. H. Hosseini-Hashemi, "Aeroelastic behavior of cantilevered rotating rectangular plates," *International Journal of Mechanical Sciences*, vol. 53, no. 4, pp. 316–328, 2011.

[30] X. Lachenal, S. Daynes, and P. M. Weaver, "A zero torsional stiffness twist morphing blade as a wind turbine load alleviation device," *Smart Materials & Structures*, vol. 22, no. 6, article 065016, 2013.

[31] A. Holt and Z. Garabed, "Piston theory-a new aerodynamic tool for the aeroelastician," *Journal of the Aeronautical Sciences*, vol. 23, no. 12, pp. 1109–1118, 1956.

[32] H. M. Navazi and H. Haddadpour, "Aero-thermoelastic stability of functionally graded plates," *Composite Structures*, vol. 80, no. 4, pp. 580–587, 2007.

[33] K. H. Yuan and Z. P. Qiu, "Nonlinear flutter analysis of stiffened composite panels in supersonic flow," *Science China Physics, Mechanics and Astronomy*, vol. 53, no. 2, pp. 336–344, 2010.

[34] Z. Yang, J. Zhou, and Y. Gu, "Integrated analysis on static/dynamic aeroelasticity of curved panels based on a modified local piston theory," *Journal of Sound and Vibration*, vol. 333, no. 22, pp. 5885–5897, 2014.

[35] J. N. Reddy, *Mechanics of Laminated Composite Plates and Shells Theory and Analysis*, CRC Press, Boca Raton, FL, USA, 2000.

[36] E. Dowell and W. Ye, "Limit cycle oscillation of a fluttering cantilever plate," *AIAA Journal*, vol. 29, no. 11, pp. 1929–1936, 1991.

[37] D. Tang, E. H. Dowell, and K. C. Hall, "Limit cycle oscillations of a cantilevered wing in low subsonic flow," *AIAA Journal*, vol. 37, no. 3, pp. 364–371, 1999.

3

Quantification of Limit Cycle Oscillations in Nonlinear Aeroelastic Systems with Stochastic Parameters

C. C. Cui, J. K. Liu, and Y. M. Chen

Department of Mechanics, Sun Yat-sen University, No. 135 Xingang Road West, Guangzhou 510275, China

Correspondence should be addressed to Y. M. Chen; chenymao@mail.sysu.edu.cn

Academic Editor: Shaoping Wang

This paper presents an analytical feature for limit cycle oscillation (LCO) in the nonlinear aeroelastic system of an airfoil, with major emphasis on its applications in LCO quantification. The nonlinear stiffness is modeled as the product of the pth power of vibration displacement and the qth power of velocity, with its coefficient as a stochastic parameter. One interesting finding is that the LCO amplitude is directly proportional to the $1/(1-p-q)$th power of the coefficient, whereas the frequency is independent of the coefficient. Based on this feature, the statistics and distribution functions of the LCO amplitude are obtained semianalytically, which are validated by Monte Carlo simulations. In addition, we discuss the possible influences of the nonlinear stiffness on flutter suppression of the airfoil subjected to Gaussian white noises. Surprisingly, increasing the nonlinear stiffness alone does not necessarily reduce the vibration amplitude as expected. Instead, it may sometimes induce disastrous subcritical LCOs with much higher vibration amplitudes.

1. Introduction

The nonlinear aeroelastic system of an airfoil is a typical self-excited system, which can exhibit lots of nonlinear dynamical behaviors such as bifurcations, limit cycle oscillations (LCO), and chaotic responses [1, 2]. Quantification of airfoil LCOs via analytical and/or semianalytical techniques has been an active area of research for many years. It has stimulated the interests and curiosities of many researchers [3–9]. Due to the design and manufacturing errors, uncertainties are usually inevitable in airfoil aeroelastic systems [10]. They usually happen to system parameters as stochastic [11] or uncertain-but-bounded variables [12]. As uncertainties are included, it will become much more cumbersome to investigate nonlinear aeroelastic responses such as LCOs, bifurcations, and chaos.

Recent years have witnessed an increasing amount of research works focusing on the stochastic analysis of airfoil LCO. Expansion technique has been widely used to propose solution approaches on this topic. Mathematically, these methods aim at transforming a stochastic problem into a deterministic one, according to appropriate representations of the output variables or stochastic processes. Attar and Dowell [13] employed Monte Carlo simulation and a response surface method, respectively, to map the output characteristics of a delta wing into two input parameters. They considered two variables, that is, the thickness and modulus of elasticity. Millman et al. [14] also applied the response surface mapping to simulate the bifurcations of LCOs. Another technique, called stochastic Galerkin method, was proposed by Xiu et al. [15] to study LCOs arising in nonlinear systems with uncertainties. Pettit and Beran proposed methods for quantification of LCOs by using polynomial chaos [16] and Wiener-Haar expansion [17], respectively. Additionally, Millman et al. [18] applied Fourier chaos expansion to investigate the bifurcation behaviors of an airfoil. The Gegenbauer polynomial can also be applied in the orthogonal decomposition of stochastic nonlinear flutter system, as suggested by Wu et al. [19].

Aside from the above-mentioned approaches based on expansion technique, there are some other approaches developed to solve nonlinear stochastic aeroelastic systems. The stochastic collocation method was applied by Deng et al. [20] to solve the nonlinear aeroelastic system of an airfoil with a control surface. In this method, deterministic solutions are obtained at a series of discrete values for stochastic parameters. Based on the attained solutions, the stochastic analysis can be carried out using an interpellation method. Chen et al. [21] applied the incremental harmonic method to solve a nonlinear aeroelastic system with uncertain-but-bounded parameters. In addition, the B-spline stochastic projection technique was utilized by Millman et al. to quantify uncertainties [22]. It was also extended, by Beran et al. [23], to the treatment of aerodynamic nonlinearities governed by discrete Euler equations. More recently, the normal form theory was adopted to investigate the flutter boundary of an airfoil [24].

As mentioned above, a lot of numerical analysis has been made to better understand the effect of uncertain parameters on LCOs as well as other responses. Analytical or semianalytical prediction on aeroelastic responses is of fundamental significance to aeroelastics research; for example, Balakrishnan and Tuffaha established an explicit formula to determine aeroelastic divergence speed [25]. Nevertheless, it is rare to find more analytical and even semianalytical results. This study is motivated by an analytical dependence of the airfoil LCO on the nonlinear stiffness coefficient. More specifically, this parametric dependence regards the analytical relationship between the amplitude (frequency) and the coefficient of the nonlinear coefficient. It will be proved and extended, in Section 2, to general self-excited systems with multiple nonlinearities. Based on these features, we will present an algorithm to quantify the main statics of LCOs arising in the aeroelastic system with the nonlinear coefficient as a stochastic parameter. Furthermore, we will discuss the effects of the nonlinear coefficients on LCO suppression when the airfoil is subjected to Gaussian white noises.

2. Parametric Dependence of LCO

The nonlinear aeroelastic system [1, 2] of an airfoil is usually modeled as a set of ordinary differential equations. To elucidate the parametric dependence of the LCO, the system is rewritten as the following self-excited oscillator:

$$\mathbf{M}\ddot{\mathbf{x}} + \mathbf{C}\dot{\mathbf{x}} + \mathbf{K}\mathbf{x} + c\mathbf{F}(\mathbf{x}, \dot{\mathbf{x}}) = \mathbf{0} \tag{1}$$

in which the superscript denotes the differentiation with respect to time t, $\mathbf{x}$ is the unknown vector of dimension n, and $\mathbf{M}$, $\mathbf{C}$, and $\mathbf{K}$ are the mass, damping, and stiffness matrixes, respectively. The coefficient, c, is a nonzero constant. The nonlinear term, $\mathbf{F}(\mathbf{x}, \dot{\mathbf{x}})$, is given as a continuous function in one component of $\mathbf{x}$:

$$\mathbf{F}(\mathbf{x}, \dot{\mathbf{x}}) = \left[0, \ldots, 0, (\mathbf{x}_i)^p (\dot{\mathbf{x}}_i)^q, 0, \ldots, 0\right]^{\mathrm{T}} \tag{2}$$

with p and q as given constants. The superscript "T" denotes the transpose and $\mathbf{x}_i$ the ith component of vector $\mathbf{x}$. The major result regarding the relationship between LCO and the nonlinear coefficient is given as follows:

(I) The LCO amplitude is directly proportional to $c^{1/(1-p-q)}$.

(II) The LCO frequency is independent of c.

Table 1: The frequency and amplitude of the van der Pol equation with $\varepsilon = 1$, by Delamotte's and 100th-order HAM solution, respectively.

Solution	a	ω
	$\delta = 1$	
Delamotte	2.0086198609	0.94295584744161
HAM	2.00861986087484	0.94295584744161
	$\delta = 0.25$	
HAM	4.01723972174969	0.94295584744161
	$\delta = 0.01$	
HAM	20.08619860874844	0.94295584744161

Denote the LCO solution as $\mathbf{x} = \mathbf{y}$ when $c = 1$, with the angular frequency as ω. Under the transformation $\tau = \omega t$, (1) becomes

$$\omega^2 \mathbf{M}\mathbf{x}'' + \omega \mathbf{C}\mathbf{x}' + \mathbf{K}\mathbf{x} + c\mathbf{F}\left(\mathbf{x}, \omega\mathbf{x}'\right) = \mathbf{0}, \tag{3}$$

where the superscript denotes the differentiation with respect to τ. Considering that $\mathbf{y}$ is a solution of (3) as $c = 1$, we have

$$\omega^2 \mathbf{M}\mathbf{y}'' + \omega \mathbf{C}\mathbf{y}' + \mathbf{K}\mathbf{y} + \mathbf{F}\left(\mathbf{y}, \omega\mathbf{y}'\right) = \mathbf{0}. \tag{4}$$

For any nonzero constant λ, substituting $\mathbf{x} = \lambda\mathbf{y}$ into the left side of (3), we obtain

$$\omega^2 \mathbf{M}\mathbf{y}'' + \omega \mathbf{C}\mathbf{y}' + \mathbf{K}\mathbf{y} + \lambda^{p+q-1} c\mathbf{F}\left(\mathbf{y}, \omega\mathbf{y}'\right). \tag{5}$$

Comparing (4) and (5), we know that as long as $\lambda^{p+q-1}c = 1$, $\mathbf{x} = \lambda\mathbf{y}$ is also a solution to (3); that is,

$$\lambda = c^{1/(1-p-q)}. \tag{6}$$

As the coefficient c varies, one of the LCOs can be expressed as $c^{1/(1-p-q)}\mathbf{y}$, which implies that the amplitude is in direct proportion to $c^{1/(1-p-q)}$ and that the frequency is independent of c.

We take the famous van der Pol equation as an illustration

$$\ddot{x} + x - \varepsilon\dot{x} + \delta x^2 \dot{x} = 0. \tag{7}$$

As is well known, as both ε and δ are positive constants, this system has a stable LCO. In this example, we have $p = 2$ and $q = 1$. According to the above properties, the amplitude of the LCO is in inverse proportion to $\sqrt{\delta}$.

We examine the above-mentioned properties with the help of some highly accurate solutions. Tables 1 and 2 present the results obtained by Delamotte [26], Buonomo [27], and the homotopy analysis method (HAM) [28, 29], respectively. Here, Delamotte's and Buonomo's solutions are provided, respectively, to validate the HAM ones. With such excellent agreement between these solutions, it is reasonable to consider the HAM results to be precise and use it as a benchmark for comparison in examining the above-mentioned features. One can observe that the frequency remains the same as nonlinear coefficient δ varies. On the other hand, as δ increases (or decreases) by 100 times, the amplitudes decreases (or increases) 10 times. That means the amplitude is inversely proportional to $\sqrt{100} = 10$.

Table 2: The frequency and amplitude of the van der Pol equation with $\varepsilon = 0.3$, by Buonomo's and 100th-order HAM solution, respectively.

Solution	a	ω
	$\delta = 0.3$	
Buonomo	Not given	0.994419844392168
HAM	2.000922385554212	0.994419844392168
	$\delta = 1.2$	
HAM	1.000461192777106	0.994419844392168
	$\delta = 30$	
HAM	0.2000922385554212	0.994419844392168

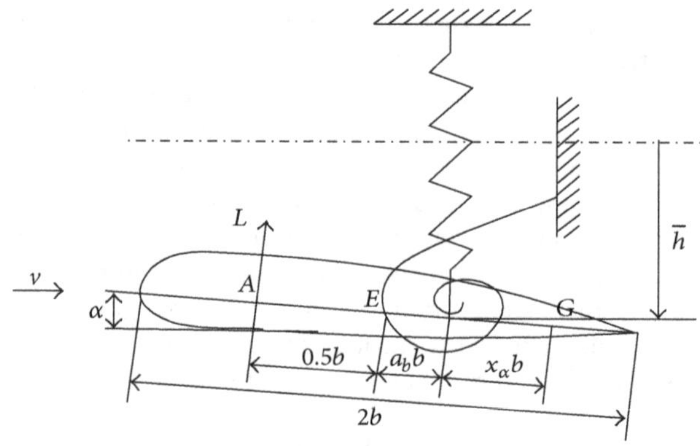

Figure 1: Sketch of a two-dimensional airfoil. b is the length of mid-chord, ab is the distance between the elastic axis and the mid-chord, and xb is the distance between the mass center and the elastic axis.

3. Quantification of LCOs in Nonlinear Aeroelastic Systems

The physical model presented in Figure 1 is a two-dimensional airfoil, oscillating in pitch and plunge, which has been employed by many authors [5–9].

The pitch angle about the elastic axis is denoted by α, positive with the nose up; the plunge deflection is denoted by $\bar{h}$, positive in the downward direction. In terms of nondimensional time $t = \omega_\alpha t_1$ (t_1 is the real time) and nondimensional plunge displacement $\xi = \bar{h}/b$, the coupled motions of the airfoil in incompressible unsteady flow can be described as [30, 31]

$$\begin{aligned} &C_0\ddot{\xi} + C_1\ddot{\alpha} + C_2\dot{\xi} + C_3\dot{\alpha} + C_4\xi + C_5\alpha + C_6 w_1 + C_7 w_2 \\ &\quad + C_8 w_3 + C_9 w_4 + C_{10}\xi + \eta\xi^3 = 0, \\ &d_0\ddot{\xi} + d_1\ddot{\alpha} + d_2\dot{\xi} + d_3\dot{\alpha} + d_4\xi + d_5\alpha + d_6 w_1 + d_7 w_2 \\ &\quad + d_8 w_3 + d_9 w_4 + d_{10}\xi + \delta\alpha^3 = 0, \end{aligned} \tag{8}$$

where η and δ are the coefficients of the cubic stiffness; and w_i's are given as

$$\begin{aligned} w_1 &= \int_0^t e^{-\varepsilon_1(t-\sigma)}\alpha(\sigma)\,d\sigma, \\ w_2 &= \int_0^t e^{-\varepsilon_2(t-\sigma)}\alpha(\sigma)\,d\sigma, \\ w_3 &= \int_0^t e^{-\varepsilon_1(t-\sigma)}\xi(\sigma)\,d\sigma, \\ w_4 &= \int_0^t e^{-\varepsilon_2(t-\sigma)}\xi(\sigma)\,d\sigma, \end{aligned} \tag{9}$$

in which $\varepsilon_1 = 0.335$ and $\varepsilon_2 = 0.0455$. The coefficients and more details about the model are stated in Li et al. [30] or Liu and Dowell [31]. As one of the nonlinear stiffness coefficients (η and δ) is given as 0, the system contains a single nonlinearity. The nonlinear coefficients are given as bounded random parameters

$$\begin{aligned} \eta &= \eta_0 + \eta_1 v, \\ \delta &= \delta_0 + \delta_1 v, \end{aligned} \tag{10}$$

where η_0, δ_0 and η_1, δ_1 are positive constants and $\eta_0 > \eta_1$, $\delta_0 > \delta_1$; $v \in [-1, 1]$ is given as a stochastic parameter with the probability function as $f(v)$.

Case 1 (pitch nonlinearity $\eta = 0$, $\delta \neq 0$). The system parameters in (8) are given in Table 3, and the value of nondimensional wind speed U varies. Numerical solutions are obtained by the Runge-Kutta method. Using analytical techniques developed for linear flutter analysis, the critical flutter speed is found to be $U = U_f = 6.0385$. As U increases beyond U_f, a stable LCO arises. As U/U_f increases further, a secondary Hopf bifurcation arises following a jump for LCO amplitude [31].

We now turn to calculate the statistics of the LCO amplitudes. Since we consider a cubic nonlinearity such that $p = 3$ and $q = 0$, the extremes of the LCOs, denoted as H for ξ and A for α, respectively, can be expressed as

$$\begin{aligned} H &= \frac{H(1)}{\sqrt{\delta}}, \\ A &= \frac{A(1)}{\sqrt{\delta}}, \end{aligned} \tag{11}$$

where $H(1)$ and $A(1)$ are the extremes as δ (or η) equals 1. According to (11), the statistics and distribution functions of A and H can be determined by the following procedures. Note that this special case was revealed by Chen and Liu using the homotopy analysis method [32].

Mathematically, the mean and standard deviation of A can be, respectively, given as

$$EA = \int_{-1}^{1} A(\delta) f(v)\,dv = A(1)\int_{-1}^{1} \frac{f(v)}{\sqrt{\delta_0 + \delta_1 v}}dv, \tag{12}$$

$$\begin{aligned} DA &= \sqrt{\int_{-1}^{1} A^2(\delta) f(v)\,dv - (EA)^2} \\ &= \sqrt{A^2(1)\int_{-1}^{1} \frac{f(v)}{(\delta_0 + \delta_1 v)}dv - (EA)^2}. \end{aligned} \tag{13}$$

TABLE 3: Parameter values in system (8).

Parameter	c_0	c_1	c_2	c_3	c_4	c_5	c_6	
Value	1.01	0.255	0.01	0.02	0.0022	0.0122	$1.4E-4$	
Parameter	d_0	d_1	d_2	d_3	d_4	d_5	d_6	
Value	1.02	1.015	0	0.04	0	0	0	
Parameter	c_7	c_8	c_9	c_{10}	d_7	d_8	d_9	d_{10}
Value	0.0014	$-6.8E-6$	$-6E-8$	$0.0625U^{-2}$	0	0	0	U^{-2}

TABLE 4: Comparisons between the mean and standard derivation by the presented method and by MCS.

	EA	DA	EH	DH
		$U = 1.5U_f$, $\delta_0 = 100$, $\delta_1 = 10$		
MCS, 5000 samples	0.1230	$3.0929E-3$	0.3179	$8.0184E-3$
Presented method	0.1233	$3.0927E-3$	0.3195	$8.0945E-3$
Relative difference	0.2%	0.007%	0.5%	0.09%
		$U = 2U_f$, $\delta_0 = 100$, $\delta_1 = 50$		
MCS, 5000 samples	0.2003	0.0271	0.6342	0.0858
Presented method	0.2009	0.0272	0.6315	0.0856
Relative difference	0.3%	0.4%	0.4%	0.2%

A bounded probability density function with monopeak and symmetrically distributed within $[-1, 1]$, named as arc-like probability density function, is defined as follows [10]:

$$f(v) = \begin{cases} \dfrac{2\sqrt{1-v^2}}{\pi}, & |v| \le 1 \\ 0, & |v| > 1. \end{cases} \tag{14}$$

Since the amplitude of the LCO can be semianalytically expressed as monotonous function of δ, the minimum ($A_{\min}$) and maximum ($A_{\max}$) of A can be easily determined as δ varies in a bounded region. Moreover, the probable density function of $A(\delta)$ can be obtained according to the probability theory as

$$g(a) = f[\delta(a)]\left|\delta'(a)\right|, \quad A_{\min} \le a \le A_{\max}. \tag{15}$$

And, the distribution function can then be integrated as

$$G(a) = \int_{A_{\min}}^{a} g(a)\,da, \quad A_{\min} \le a \le A_{\max}. \tag{16}$$

Table 4 shows the comparisons between the solutions provided by the presented method (i.e., (12) and (13)) and Monte Carlo simulation (MCS) results. The relative discrepancies are less than 0.5% when 5000 samples are taken into account. The distribution functions of A and H are plotted in Figures 2 and 3, respectively. Excellent agreement can also be observed. Note that the distribution curves for A and H are similar to each other. That is because the functions of A and H can be expressed by the same functions in δ.

As U/U_f increases beyond 2.1, there are three extremes for pitch displacement, as revealed by Liu and Dowell [31].

TABLE 5: The mean and standard derivation of A of the three extremes obtained by the presented method and by MCS with 5000 samples, respectively. Parameters are given as $U/U_f = 2.5$, $\delta_0 = 100$, and $\delta_1 = 50$.

	EA	DA
MCS	0.2881	0.0390
Presented method	0.2880	0.0390
MCS	0.1839	0.0249
Presented method	0.1839	0.0249
MCS	0.0759	0.0103
Presented method	0.0760	0.0103

Table 5 shows the results obtained by the presented method for the means and derivations. They are both in excellent agreement with the MCS results. The distribution functions for the extremes are plotted in Figure 4. It also shows that the results of the presented method agree well with the MCS ones.

Case 2 (plunge nonlinearity $\eta \ne 0$, $\delta = 0$). As a cubic plunge stiffness is considered, the bifurcation chart obtained by the incremental harmonic balance method is plotted in Figure 5. Note that the critical flutter speed depends on the linear coefficients rather than nonlinear coefficients. Different from the first case, a subcritical Hopf bifurcation is detected at U_f. As U/U_f decreases from 1 to 0.68 or so, there exists an unstable LCO. The LCO gains its stability at $U/U_f = 0.68$, and a stable LCO always arises again as U/U_f increases from 0.68.

Table 6 presents the comparisons of the results by the presented method and MCS ones. The distribution functions

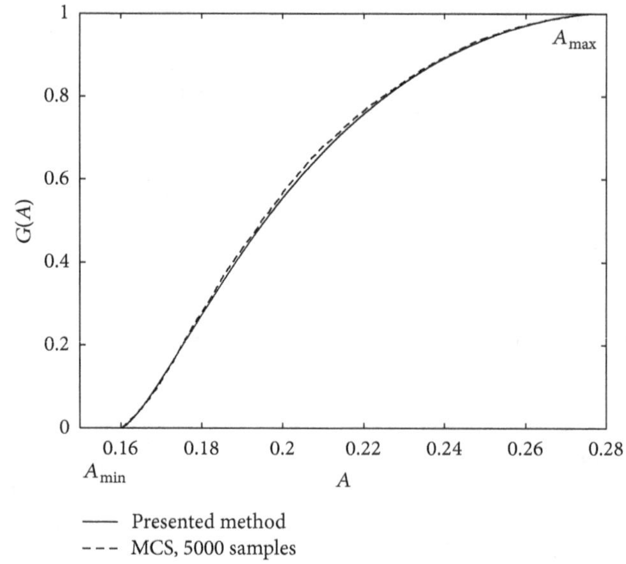

FIGURE 2: Distribution functions of A obtained by the presented method and by MCS, respectively. Parameters are given as $U = 2U_f$, $\delta_0 = 100$, and $\delta_1 = 50$.

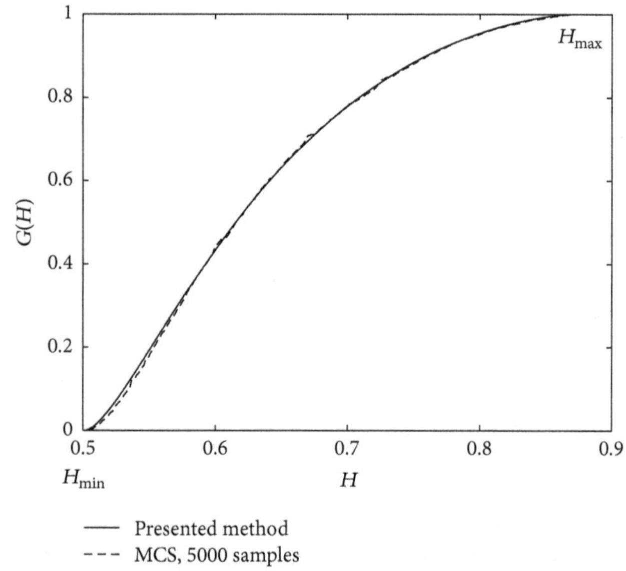

FIGURE 3: Distribution functions of H obtained by the presented method and by MCS, respectively. Parameters are given as $U/U_f = 2$, $\delta_0 = 100$, and $\delta_1 = 50$.

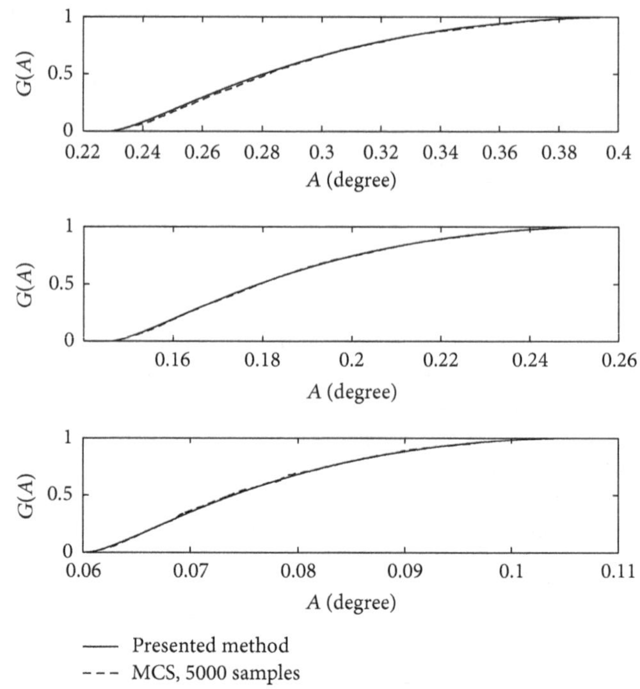

FIGURE 4: The distribution functions for the three extremes of A with $U/U_f = 2.5$, $\delta_0 = 100$, and $\delta_1 = 50$.

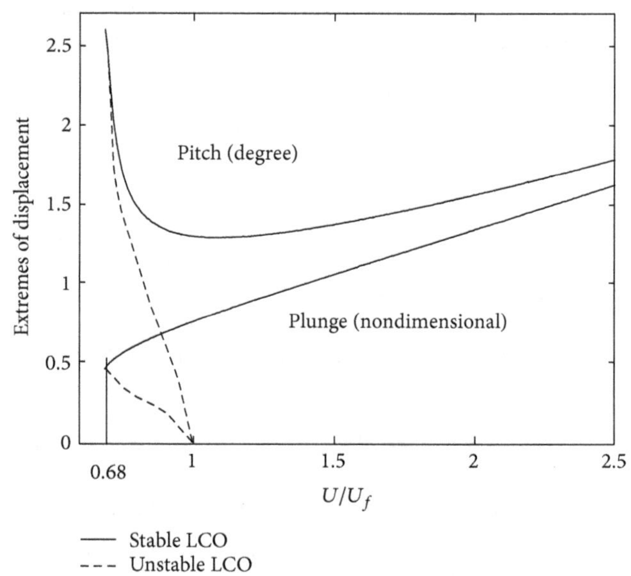

FIGURE 5: Bifurcation chart of the aeroelastic system with $\delta = 0$ and $\eta = 80$.

are also obtained for both the pitch and plunge amplitudes, respectively. Excellent agreement can be also observed in Figure 6.

4. Influence of Nonlinear Stiffness to LCO Suppression

For the aeroelastic system with a cubic stiffness, the LCO amplitude is inversely proportional to the square root of the nonlinear coefficient. Accordingly, one would like to reduce the LCO amplitude by raising the nonlinear stiffness. As an illustrative example, Figure 7 shows that the amplitude reduces to a large extent as the nonlinear coefficient increases from 8 to 100. The initial condition (IC) is defined as $[\alpha(0), \dot{\alpha}(0), \xi(0), \dot{\xi}(0)]$. The single fixed point of system (8) is the origin, which is unstable as U is larger than U_f. In such a case, the motions converge to the stable LCO regardless of ICs. As expected, increasing the cubic stiffness can suppress the LCO for the case of $U/U_f > 1$.

TABLE 6: Comparisons of the mean and standard derivation by the presented method with those by MCS.

	EA	*DA*	*EH*	*DH*
		$U = 1U_f, \delta_0 = 100, \delta_1 = 80$		
MCS, 5000 samples	1.2605	0.3273	0.7330	0.1902
Presented method	1.2646	0.3295	0.7339	0.1912
Relative difference	0.3%	0.7%	0.1%	0.5%
		$U = 2U_f, \delta_0 = 100, \delta_1 = 80$		
MCS, 4000 samples	1.3349	0.3494	1.0236	0.2640
Presented method	1.3364	0.3482	1.0269	0.2676
Relative difference	0.1%	0.3%	0.3%	1.3%

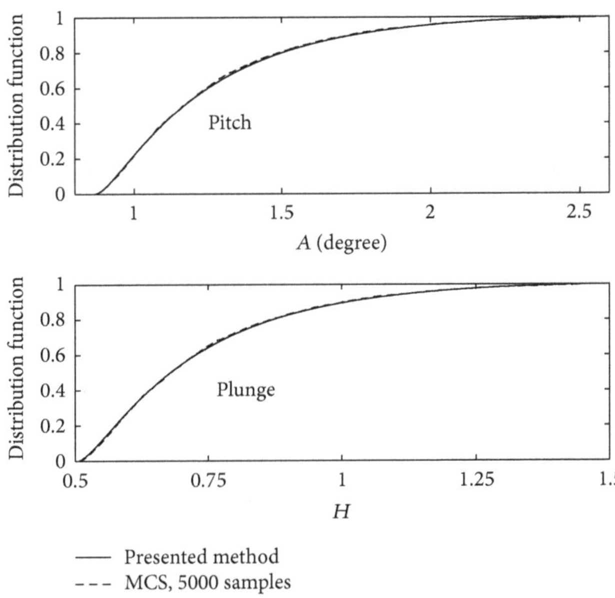

FIGURE 6: The distribution functions for the three extremes of A and H with $U/U_f = 1$, $\delta_0 = 100$, and $\delta_1 = 80$.

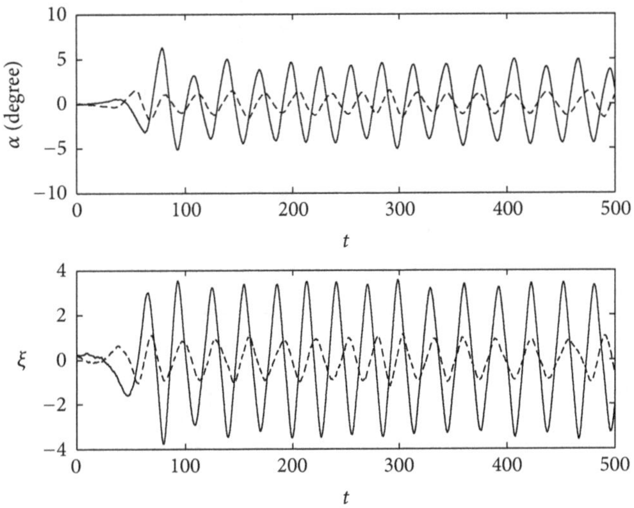

FIGURE 7: Time histories for system (8) with IC $= [1^{\circ}, 0, 0, 0]$. Parameters are $U/U_f = 1.5$ and $\eta = 8$ (solid lines) or $\eta = 100$ (dashed lines).

When $U/U_f < 1$, however, increasing the cubic stiffness does not necessarily lead to decreasing of the LCO amplitude as expected. According to Figure 5, two LCOs coexist when U/U_f is located between 0.68 and 1, with the lower one being unstable and the other stable. Note that the single fixed point is also stable. That implies the motions may converge to either the fixed point or the stable LCO. By choosing different ICs, the convergent region of the LCO can be determined according to time histories obtained by time marching integration. As Figure 8 shows, there is critical boundary in the $\alpha(0) - h(0)$ plane (when $\dot{\alpha}(0) = \dot{\xi}(0) = 0$). The vibration approaches the stable LCO when the IC is given within the shadow region; otherwise it diminishes to the fixed point.

Generally, it is useful for vibration suppression to enlarge the convergent region of the fixed point because it implies that the motions being convergent as noises happen is more probable. As mentioned above, the smaller the nonlinear coefficient is, the larger the convergent region of the fixed point becomes. Figure 9 shows the time histories of system (8) with the same value for U but with different values for the nonlinear coefficient. As $\eta = 8$, the motion converges to the fixed point. If $\eta = 100$ is chosen, the motion approaches a LCO finally. In this case, increasing the cubic stiffness results in larger amplitude vibration rather than vibration suppression.

Consider Gaussian white noises in both the pitching and plunging DOFs, such as

$$\begin{aligned} &C_0\ddot{\xi} + C_1\ddot{\alpha} + C_2\dot{\xi} + C_3\dot{\alpha} + C_4\xi + C_5\alpha + C_6w_1 + C_7w_2 \\ &\quad + C_8w_3 + C_9w_4 + C_{10}\xi + \eta\xi^3 = 0.1\varepsilon_1(t), \\ &d_0\ddot{\xi} + d_1\ddot{\alpha} + d_2\dot{\xi} + d_3\dot{\alpha} + d_4\xi + d_5\alpha + d_6w_1 + d_7w_2 \\ &\quad + d_8w_3 + d_9w_4 + d_{10}\xi + \delta\alpha^3 = 0.1\varepsilon_2(t) \end{aligned} \tag{17}$$

in which the perturbations $\varepsilon_i(t)$'s are given as white Gaussian noises with mean values as 1. The time histories, as plotted in Figure 10, show that the vibration amplitude for $\eta = 8$ is smaller than that for $\eta = 100$. Actually, the vibrations resulted

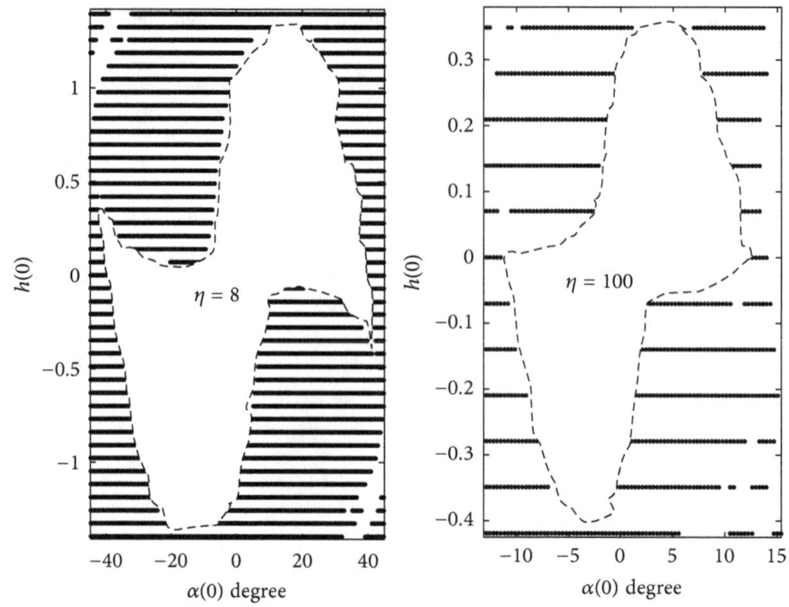

FIGURE 8: The convergent region for the stable LCO and fixed point of system (8) with $U/U_f = 0.7$; shadow region denotes convergent points for the LCO while the blank for the fixed point.

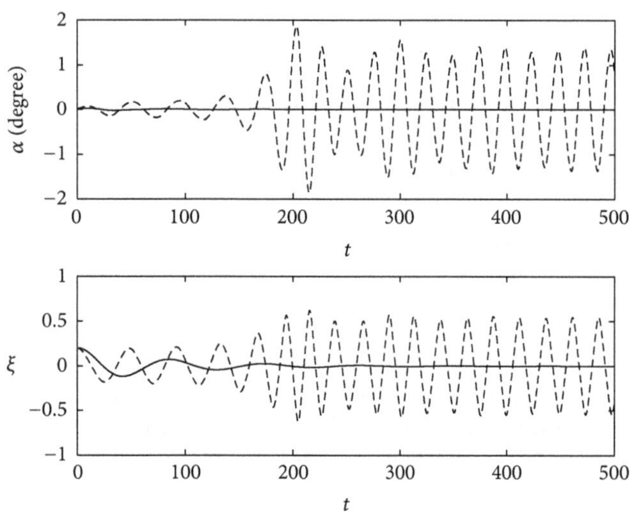

FIGURE 9: Time histories for system (8) with IC $= [1^{\circ}, 0, 0, 0]$. Parameters are $U/U_f = 0.8$ and $\eta = 8$ (solid lines) or $\eta = 100$ (dashed lines).

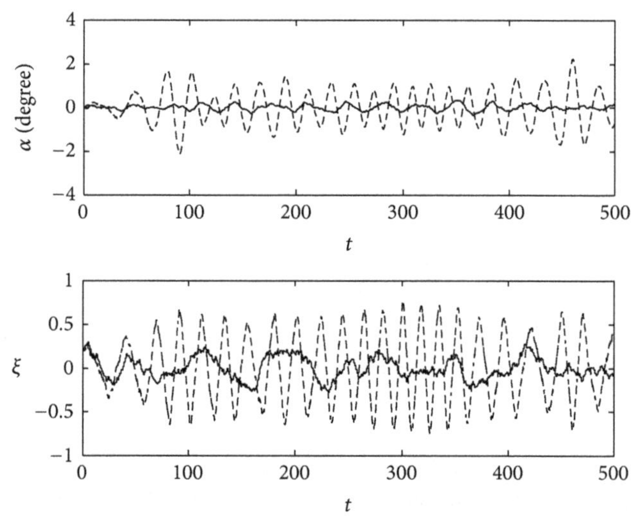

FIGURE 10: Time histories for system (17) with IC $= [1^{\circ}, 0, 0, 0]$. Parameters are $U/U_f = 0.8$ and $\eta = 8$ (solid lines) or $\eta = 100$ (dashed lines).

from the white noises as $\eta = 8$. The vibration is too small to cause the vibration to reach the convergent region of the stable LCO. For the case of $\eta = 100$, on the other hand, the motions are in essence LCOs subjected to noises.

5. Conclusions

We have investigated the LCO of an airfoil aeroelastic system with a nonlinear stiffness. The nonlinearity is expressed as a product of pth power of vibration displacement and qth power of the velocity. Interestingly, the LCO amplitude is directly proportional to $1/(1 - p - q)$th power of the nonlinear coefficient, while the frequency is independent of this coefficient.

Special emphasis has been placed on the applications of the above-mentioned feature. An illustrative application is in LCO quantification for the nonlinear aeroelastic system with the nonlinear coefficient as a stochastic parameter. The statistics and distribution functions of the LCO amplitudes can be semianalytically determined. Excellent agreement between the obtained results and the MCS ones validates the feasibility and efficiency of the applications in LCO quantification.

In addition, the influence of the nonlinear stiffness on LCO suppression has been discussed in detail. When a stable LCO exists and the fixed point is unstable, that is, for the case when the wind speed increases beyond the critical flutter speed, the LCO can be suppressed by means of increasing the cubic stiffness. Below the flutter speed, however, increasing the cubic stiffness cannot reduce (as expected) but possibly result in larger magnitude of vibration.

As is well known, parameters in nonlinear dynamical systems are closely related to system stability, bifurcations, chaos, and so on. Therefore, parametric study is one of major tasks in nonlinear dynamics analysis. This study could also be considered to be a preliminary and useful attempt at the dependence of vibration on system parameters (especially nonlinear coefficients) of self-excited systems. The simple yet efficient applications imply that the feature revealed in this paper could be more applicable in more nonlinear systems.

Competing Interests

The authors declare that they have no competing interests.

Acknowledgments

This work is supported by the National Natural Science Foundation of China (11272361 and 11572356) and Guangdong Province Natural Science Foundation (1414050000412).

References

[1] E. H. Dowell, R. Clark, D. Cox et al., *A modern course in aeroelasticity*, vol. 116 of *Solid Mechanics and Its Applications*, Kluwer Academic Publishers, Dordrecht, Netherlands, 4th edition, 2004.

[2] B. H. K. Lee, S. J. Price, and Y. S. Wong, "Nonlinear aeroelastic analysis of airfoils: bifurcation and chaos," *Progress in Aerospace Sciences*, vol. 35, no. 3, pp. 205–334, 1999.

[3] S. A. Eftekhari, F. Bakhtiari-Nejad, and E. H. Dowell, "Damage detection of an aeroelastic panel using limit cycle oscillation analysis," *International Journal of Non-Linear Mechanics*, vol. 58, pp. 99–110, 2014.

[4] D. C. Li, S. J. Guo, and J. W. Xiang, "Aeroelastic dynamic response and control of an airfoil section with control surface nonlinearities," *Journal of Sound and Vibration*, vol. 329, no. 22, pp. 4756–4771, 2010.

[5] X.-Z. Xu, Y.-K. Gao, and W.-G. Zhang, "Iterative learning control of a nonlinear aeroelastic system despite gust load," *International Journal of Aerospace Engineering*, vol. 2015, Article ID 237804, 13 pages, 2015.

[6] H. Dai, X. Yue, J. Yuan, and D. Xie, "A fast harmonic balance technique for periodic oscillations of an aeroelastic airfoil," *Journal of Fluids and Structures*, vol. 50, pp. 231–252, 2014.

[7] M. R. Chiarelli and S. Bonomo, "Aeroelastic analysis of wings in the transonic regime: planform's influence on the dynamic instability," *International Journal of Aerospace Engineering*, vol. 2016, 13 pages, 2016.

[8] F. X. Chen, J. K. Liu, and Y. M. Chen, "Flutter analysis of an airfoil with nonlinear damping using equivalent linearization," *Chinese Journal of Aeronautics*, vol. 27, no. 1, pp. 59–64, 2014.

[9] R. Vasconcellos and A. Abdelkefi, "Nonlinear dynamical analysis of an aeroelastic system with multi-segmented moment in the pitch degree-of-freedom," *Communications in Nonlinear Science and Numerical Simulation*, vol. 20, no. 1, pp. 324–334, 2015.

[10] C. L. Pettit, M. R. Hajj, and P. S. Beran, "A stochastic approach for modeling incident gust effects on flow quantities," *Probabilistic Engineering Mechanics*, vol. 25, no. 1, pp. 153–162, 2010.

[11] T. Fang, X. L. Leng, and C. Q. Song, "Chebyshev polynomial approximation for dynamical response problem of random system," *Journal of Sound and Vibration*, vol. 266, no. 1, pp. 198–206, 2003.

[12] X. Wang and Z. Qiu, "Interval finite element analysis of wing flutter," *Chinese Journal of Aeronautics*, vol. 21, no. 2, pp. 134–140, 2008.

[13] P. J. Attar and E. H. Dowell, "A stochastic analysis of the limit cycle behavior of a nonlinear aeroelastic model using the response surface method," in *Proceedings of the 46th AIAA/ASME/ASCE/AHS/ASC Structures, Structural Dynamics and Materials Conference, Structures, Structural Dynamics, and Materials and Co-Located Conferences*, AIAA 2005-1986, Austin, Tex, USA, June 2005.

[14] D. R. Millman, P. I. King, R. C. Maple et al., "Predicting uncertainty propagation in a highly nonlinear system with a stochastic projection method," in *Proceedings of the 45th AIAA/ASCE/AHS/ASC Structures, Structural Dynamics, and Materials Conference*, AIAA 2004-1613, Palm Springs, Calif, USA, April 2004.

[15] D. Xiu, I. G. Kevrekidis, and R. Ghanem, "An equation-free, multiscale approach to uncertainty quantification," *Computing in Science and Engineering*, vol. 7, no. 3, pp. 16–23, 2005.

[16] C. L. Pettit and P. S. Beran, "Polynomial chaos expansion applied to airfoil limit cycle oscillations," AIAA 2004-1691, 2004.

[17] C. L. Pettit and P. S. Beran, "Wiener-Haar expansion of airfoil limit cycle oscillations," AIAA 2005-1985, 2005.

[18] D. R. Millman, P. I. King, and P. S. Beran, "Airfoil pitch-and-plunge bifurcation behavior with fourier chaos expansions," *Journal of Aircraft*, vol. 42, no. 2, pp. 376–384, 2005.

[19] C. Wu, H. Zhang, and T. Fang, "Flutter analysis of an airfoil with bounded random parameters in incompressible flow via Gegenbauer polynomial approximation," *Aerospace Science and Technology*, vol. 11, no. 7-8, pp. 518–526, 2007.

[20] J. Deng, C. Anton, and Y. S. Wong, "Stochastic collocation method for secondary bifurcation of a nonlinear aeroelastic system," *Journal of Sound and Vibration*, vol. 330, no. 13, pp. 3006–3023, 2011.

[21] Y. M. Chen, J. K. Liu, and G. Meng, "Incremental harmonic balance method for nonlinear flutter of an airfoil with uncertain-but-bounded parameters," *Applied Mathematical Modelling*, vol. 36, no. 2, pp. 657–667, 2012.

[22] D. R. Millman, P. I. King, R. C. Maple, P. S. Beran, and L. K. Chilton, "Uncertainty quantification with a B-spline stochastic projection," *AIAA Journal*, vol. 44, no. 8, pp. 1845–1853, 2006.

[23] P. S. Beran, C. L. Pettit, and D. R. Millman, "Uncertainty quantification of limit-cycle oscillations," *Journal of Computational Physics*, vol. 217, no. 1, pp. 217–247, 2006.

[24] M. Ghommem, M. R. Hajj, and A. H. Nayfeh, "Uncertainty analysis near bifurcation of an aeroelastic system," *Journal of Sound and Vibration*, vol. 329, no. 16, pp. 3335–3347, 2010.

[25] A. V. Balakrishnan and A. Tuffaha, "The transonic dip in aeroelastic divergence speed—an explicit formula," *Journal of the Franklin Institute*, vol. 349, no. 1, pp. 59–73, 2012.

[26] B. Delamotte, "Nonperturbative (but approximate) method for solving differential equations and finding limit cycles," *Physical Review Letters*, vol. 70, no. 22, pp. 3361–3364, 1993.

[27] A. Buonomo, "On the periodic solution of the van der Pol equation for small values of the damping parameter," *International Journal of Circuit Theory and Applications*, vol. 26, no. 1, pp. 39–52, 1998.

[28] S.-J. Liao, "An analytic approximate approach for free oscillations of self-excited systems," *International Journal of Non-Linear Mechanics*, vol. 39, no. 2, pp. 271–280, 2004.

[29] Y. M. Chen and J. K. Liu, "A study of homotopy analysis method for limit cycle of van der Pol equation," *Communications in Nonlinear Science and Numerical Simulation*, vol. 14, no. 5, pp. 1816–1821, 2009.

[30] D. Li, S. Guo, and J. Xiang, "Study of the conditions that cause chaotic motion in a two-dimensional airfoil with structural nonlinearities in subsonic flow," *Journal of Fluids and Structures*, vol. 33, pp. 109–126, 2012.

[31] L. P. Liu and E. H. Dowell, "The secondary bifurcation of an aeroelastic airfoil motion: effect of high harmonics," *Nonlinear Dynamics*, vol. 37, no. 1, pp. 31–49, 2004.

[32] Y. M. Chen and J. K. Liu, "Homotopy analysis method for limit cycle oscillations of an airfoil with cubic nonlinearities," *Journal of Vibration and Control*, vol. 16, no. 2, pp. 163–179, 2010.

4

Evaluation of Assembly Gap from 3D Laser Measurements via FEA Simulation

Xueshu Liu,[1] Junhao Chang,[2] Yuxing Yang,[2] Yiqi Wang,[2] Yongjie Bao,[2] and Hang Gao[2]

[1]*School of Automotive Engineering, Dalian University of Technology, Dalian 116024, China*
[2]*School of Mechanical Engineering, Dalian University of Technology, Dalian 116024, China*

Correspondence should be addressed to Xueshu Liu; liuxs@dlut.edu.cn

Academic Editor: Santiago Hernández

Due to the compliance and geometrical defects of composite parts, gaps exist between assembly components after the preassembly. Assembly requirements impose to fill these gaps to eliminate any unexpected internal stresses. Although it is identified as a problematic and expensive nonadded value stage, a gap measurement is still needed. This paper develops a numerical process for gap prediction before the assembly step. After the assembly components are scanned in a specified configuration, finite element meshes are created using the scanned data and the shape variations of each component caused by constraints and forces in different configurations are evaluated by finite element analysis. Assembly gaps are finally assessed by assembling the simulated preassembly shapes of all components. The feasibility of the proposed method is proved by an experiment.

1. Introduction

Due to the excellent properties, carbon fiber reinforced plastic (CFRP) has been widely used in aircraft structures in recent years. However, because of the complicated manufacturing processes and anisotropic property of CFRP as well, it is almost impossible to manufacture a composite product without geometrical deviation including process-induced deformation and thickness variations at present, especially for large ones. As a result, gaps are formed at the faying surfaces during the assembly process and prestress would be consequently introduced by the assembly forces, which would result in the development of cracks of delamination. To avoid such problems, assembly gaps should be filled to eliminate the undesirable prestress. Unfortunately, in the aviation industry, the usage of shim is carefully restricted to avoid increasing the weight of structure and lowering the load capacity of aircraft as a result. On the other hand, a deficiency of shim will affect the mechanical performance of the structure as reported in [1–6]. Therefore, the accurate measurement of assembly gap plays an important role in the shimming process.

Nevertheless, the assessment of the assembly gap of CFRP structures is not trivial. First, the aircraft structures are commonly designed as a closed or semiclosed region such as the wing box. So not all the gaps can be reached during assembly when a direct measurement method is adopted. Moreover, for accurate shimming, the gaps should be represented in 3D, which is a mission impossible for direction measurement. Second, even assembly gaps can be evaluated by numerical methods like virtual assembly after scanning all the components and rebuilding their actual geometries; high-accuracy assessment of assembly gap is still a challenge for aircraft structures. Because some components are flexible since their thicknesses are usually much less than their length and width like the aircraft plate, they tend to deform under their own weight, which results in the shape variations when their space orientation change. In addition, the measurement shape is normally different from that in assembly configuration because the assembly process generates deformations due to the application of forces and restrictions at fixation points and the requirement of prestress assembly as well. Therefore, the difference between the measurement shape and assembly shape has to be taken into consideration so as to accurately evaluate the assembly gaps and remove any possible prestress generated during the assembly process.

So far, many efforts on the shape prediction of flexible part in different configurations have been done. For example, Ascione and Polini gave a solution to the problem of nonrigid freeform inspection by a coordinate measuring machine [7]. But it is expensive and time-consuming to manufacture such equipment. So the recent trends in inspection methods for compliant components are referred to as the simulated displacement method [8]. For instance, Abenhaim et al. proposed the iterative displacement inspection algorithm for the inspection of flexible parts [9]. Gentilini and Shimada reported a method for the shape inspection of a flexible assembly part by virtually mounting it into the assembly [10]. First, the dense measured mesh is smoothed and reduced to become suitable for finite element analysis (FEA). Then, the specific displacement boundary conditions are defined and applied for the FE simulation of the assembly process. After that, quality inspection of the simulated post-assembly shape can be done using visualization tools. Radvar-Esfahlan and Tahan presented a method for nonrigid part inspection which was based on simulation using FEA [11]. An approach based on only partial views of a part was reported in [12]. The authors focused on the regions where inspection is required and used the concept of characteristic points which implies more simplicity for calculations and for the surface acquisition process. Instead of FEA, a spring-mass model was defined and used to simulate deformations in [13]. To improve the reliability of the FE analysis results, Abenhaim et al. proposed a boundary displacement constrained optimization using FEA [14–16], and Karganroudi et al. introduced an algorithm to automatically filter out sample points that are close to defects [17]. Thiebaut et al. presented a method to obtain the form deviation of flexible components from part measurements independently of the assembly configuration of use [18].

The abovementioned contributions focus on the differences of a flexible part while in different configurations. One more step goes ahead, and the assembly gaps can be evaluated by a virtual method. Lacroix et al. represented a numerical process to predict an assembly gap before an assembly step from component measurements [19]. The main issue relates to the integration of measuring data into finite element meshes created by using a nominal model. Different from them, in this study, a new procedure for assembly gap prediction is proposed. Starting from the measurement data (usually point cloud), finite element meshes of the flexible part are created and the deformations caused by forces and restrictions in different configurations are evaluated by using FEA. After the assembly shapes of all components are obtained, virtual assembly technology is employed to assess the assembly gaps. Finally, the accuracy of this method is validated by an experiment.

2. Principle of the Method

Assembly component is always measured in a given configuration, using an over-constrained measuring set-up. So the measured shape S^{m} corresponds to the theoretical shape S^{d}, added to the form deviation E^{f} and the deformation under the combined effect of the measuring set-up and the component's own weight E^{mc}, which can be expressed by

$$S^{\mathrm{m}} = S^{\mathrm{d}} + E^{\mathrm{f}} + E^{\mathrm{mc}}. \quad (1)$$

Note that the term $(S^{\mathrm{d}} + E^{\mathrm{f}})$ represents the free state of a component, which is the shape it should have in the state of absence of gravity and any constraints.

Similarly, the shape of a component in assembly configuration S^{a} can be given by

$$S^{\mathrm{a}} = S^{\mathrm{d}} + E^{\mathrm{f}} + E^{\mathrm{ac}}, \quad (2)$$

where E^{ac} indicates the deformation under the combined effect of the assembly set-up and the component's own weight.

Note that S^{d} and E^{f} was kept constant no matter in what kind of configuration the component involves. So the relationship between S^{m} and S^{a} is given by

$$S^{\mathrm{a}} = S^{\mathrm{m}} - E^{\mathrm{mc}} + E^{\mathrm{ac}}. \quad (3)$$

The deformations caused by forces and restrictions in different configurations, namely, E^{mc} and E^{ac}, can be assessed by FEA. After all the shapes of components in assembly configuration are obtained, assemble them virtually and gaps can be clearly defined.

3. Mesh Generation from Point Cloud

After scanning a flexible part in a specified configuration with a laser scanner, a point cloud $\tilde{P}$ representing surfaces of the part is obtained. Postprocessing steps are usually needed because $\tilde{P}$ suffers from noise, which includes physical phenomena and hardware-related issues [10]. Such noise can be removed by using mean curvature flow, normal field diffusion algorithms, and so on. After noise removal, a clean point cloud $P = \{p_i \in R^3 \mid_{i=1,2,\ldots}\}$ is got and used for assembly shape prediction by FEA.

To create finite element meshes from P for simulation, a procedure was proposed and shown in Figure 1. First, principal component analysis is performed on P and the normal direction $\mathbf{n}$ of P is computed. A plane $\Phi(\mathbf{n}, \widehat{p})$ is then defined by $\mathbf{n}$ and $\widehat{p} \in R^3$ (Figure 1(b)). A map $f : P \rightarrow \bar{P}$, which is given by (4), is used to project P onto Φ and $\bar{P} = \{f(p_i) \mid_{\forall p_i \in P}\}$ is yielded.

$$f : \mathbf{p}_i + \frac{\mathbf{n} \cdot \overrightarrow{p_i \widehat{p}} \cdot \mathbf{n}}{\|\mathbf{n}\|^2}, \quad (4)$$

where $\mathbf{p}_i$ indicates the Euclidean vector $\overrightarrow{op_i}$ and o represents the point (0,0,0) in R^3.

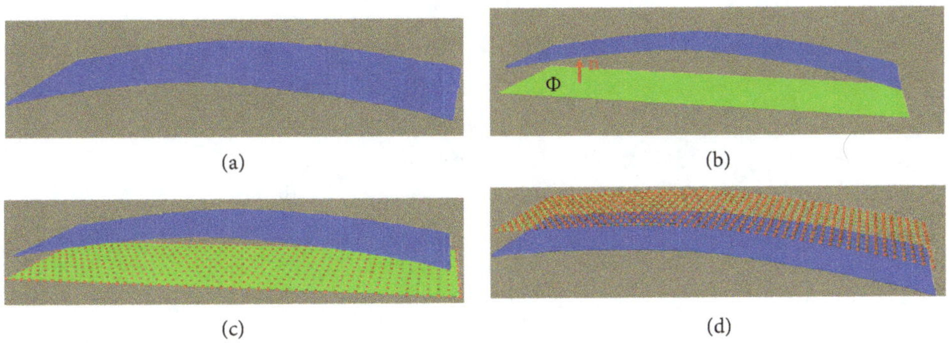

FIGURE 1: Procedure of mesh generation from point cloud. (a) Point cloud P. (b) Projection plane Φ. (c) Sampling points Q (red). (d) Mesh model.

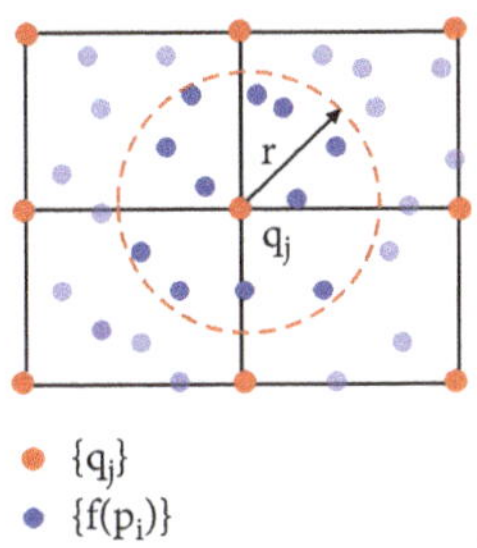

FIGURE 2: Illustration of average distance computation.

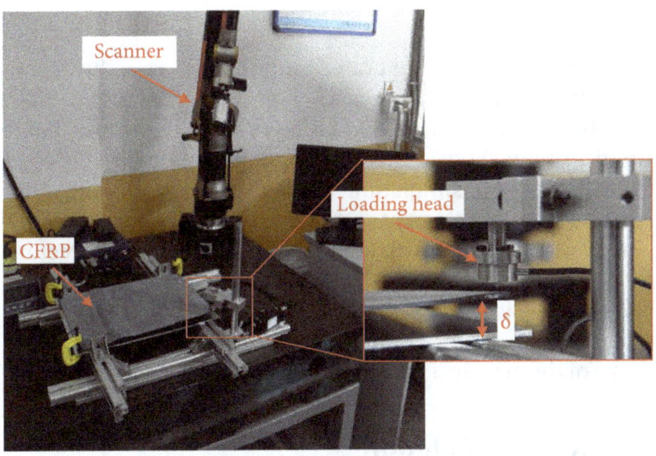

FIGURE 3: Experiment set-up.

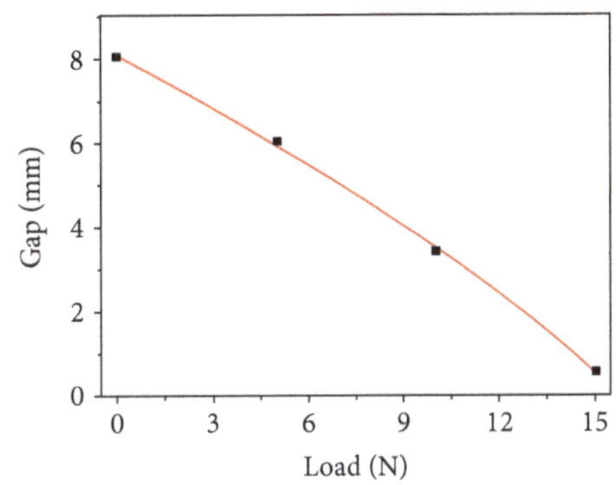

FIGURE 4: Relationship between load and gap size δ.

TABLE 1: Mechanical properties of T700/YPH-25.

Property	Value	Unit
E_1	143	GPa
E_2	8.0	GPa
G12	4.4	GPa
G23	2.8	GPa
μ12	0.34	
μ23	0.50	

The distance $D(p_i)$ from $p_i \in P$ to Φ is computed by

$$D(p_i) = \frac{\mathbf{n} \cdot \overrightarrow{p_i \hat{p}}}{\|\mathbf{n}\|}. \tag{5}$$

Next, sampling points $Q = \{q_j \in \Phi \mid_{j=1,2,\ldots}\}$ are generated on Φ with respect to user-defined parameters (Figure 1(c), red points). An average distance $H(q_j)$ of $q_j \in Q$ is computed by

$$H(q_j) = \frac{\sum \lambda_i D(\forall p_i \in U)}{\sum \lambda_i}, \tag{6}$$

where $U = \{p_m \in P \mid d(f(p_m), q_j) < r\}$, $d(,)$ represents the Euclidean distance between two points, r is a user-defined parameter, and λ_i indicates the coefficient of weight given by

$$\lambda_i = 1 - \frac{d\left(f(p_i), q_j\right)}{2r}. \tag{7}$$

The thinking behind (7) is that the less the distance from $f(p_i)$ to q_j is, the larger the weight coefficient is. Considering the definition of U, it is clear that $\lambda_i \in [0.5, 1]$.

An example for $H(q_j)$ computation is given in Figure 2 where the red dots represent $\{q_j\}$ and the blue ones indicate $\{f(p_i)\}$. After a q_j is specified (the center), candidates used for computation are determined with respect to the definition of U (dark blue dots inside the red dashed circle). To ensure the accuracy of computation, the number of

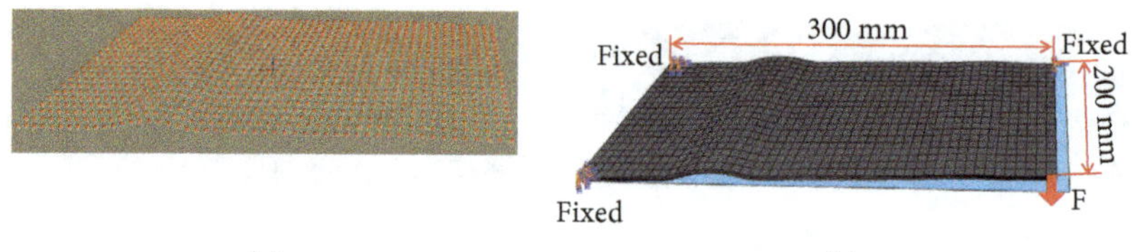

FIGURE 5: Finite element model creation. (a) 2D mesh model. (b) FE model.

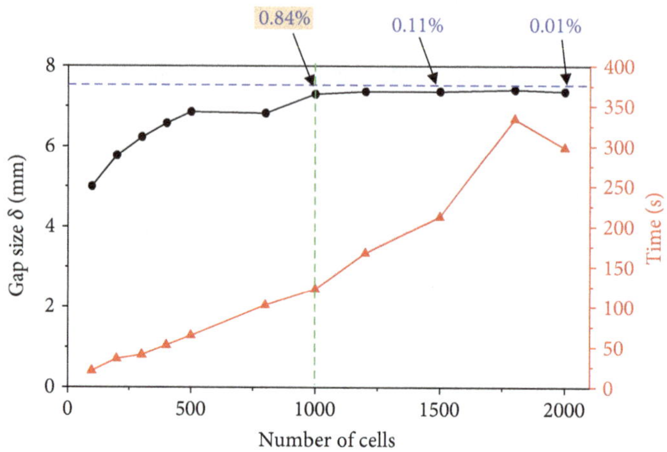

FIGURE 6: Influence of mesh size on computational accuracy and time.

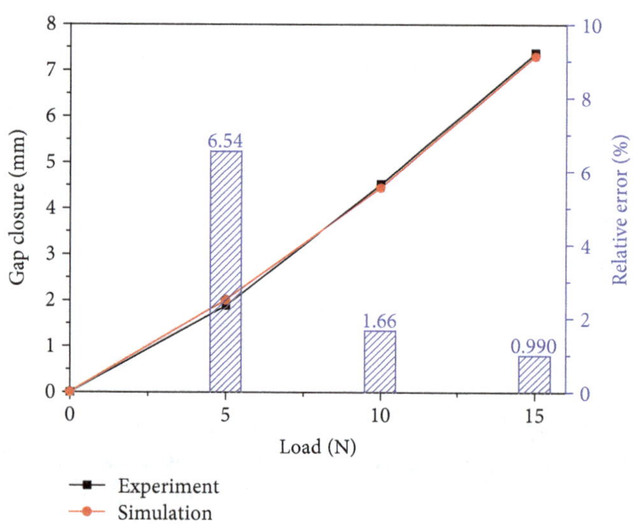

FIGURE 7: Comparison of simulation and experiment.

candidates $\#U$ is compared to a user-defined variable δ. In case $\#U < \delta$, the parameter r is enlarged by 1.5 times ($r = 1.5r$) until $\#U \geq \delta$ is met. In this study, $\delta = 20$.

All the sampling points Q are then mapped by $g : Q \rightarrow V$, which is given by (8), to generate vertices $V = \{g(q_j) \mid_{\forall q_j \in Q}\}$ used as nodes of finite element meshes.

$$g : \mathbf{q_j} + \frac{\mathbf{n} \cdot H\left(q_j\right)}{\|\mathbf{n}\|}. \tag{8}$$

V meshes can be generated and used for assembly shape prediction by FEA. Figure 1(d) gives such an example. Note that the meshes were moved far away from the source point cloud so as to show clearly. The meshes can be directly used as an input of a simulation when the plate is modeled by two-dimensional elements in a finite element method. Alternatively, they can be extruded along their normal direction with a specified thickness so as solid elements are created and the plate is then modeled by three-dimensional elements. After the boundary conditions and loads are applied, the shape of the plate in a specified configuration can be predicted.

4. Method Validation

For all numerical approaches to assembly gap prediction, the key point is if the compliance and contacts in assemblies are considered. So to validate the proposed method, an experiment is conducted as shown in Figure 3. A rectangular CFRP specimen (300 mm × 200 mm) with form deviation is fixed on a metallic plate by constraining its three corners with clamps. A loading equipment is used to deform the CFRP specimen, and the load head is set right above the free corner. The distance δ from the free corner to the metallic plate is measured 3 times for each configuration, and the averaged value is used for simulation result validation. A laser scanner is used to scan the CFRP specimen after the load is applied. Note that considering the size of the specimen deformation caused by gravity is ignored in this study.

Loads of 0 N, 5 N, 10 N, and 15 N were applied, and the relationship between the load and δ is shown in Figure 4. It can be seen that with the increasing of load, the distance δ

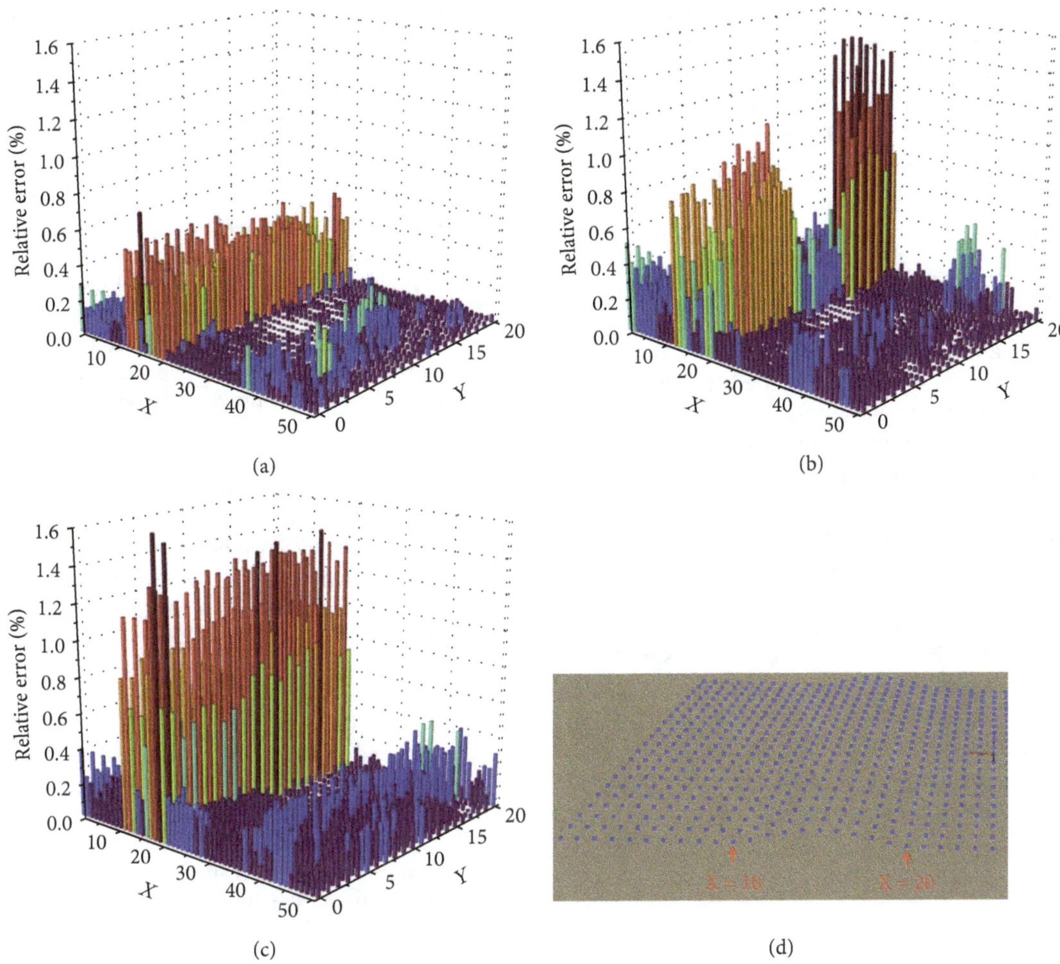

Figure 8: Comparison of simulation and experiment. (a) 5 N. (b) 10 N. (c) 15 N. (d) Arc area.

decreases and the relationship between them is approximately linear. The specimen was scanned 4 times and used for the proposed method validation.

The CFRP specimen was made by T700/YPH-25 carbon fiber/epoxy prepregs, which has nominal ply thickness of 0.15 mm. Mechanical properties of the material are given in Table 1. The lay-up sequence of the specimen is [45/90/0/90 /0/90/0/90/0/90/−45/0/90/0/90/−45/90/90/0], and the nominal thickness is 2.85 mm.

Starting from the point cloud taken under 0 N load and following the procedure given in the previous section, two-dimensional meshes are obtained as shown in Figure 5(a). After extruding the mesh 2.85 mm thickness, solid elements are got and used for assembly shape prediction (Figure 5(b)).

It is well known that the solution precision of FEA is sensitive to mesh size. Generally speaking, the less the mesh size is, the higher the accuracy of the solution will be and the more time the computation will take. In addition, a mesh aspect ratio, i.e., the ratio of the longest to the shortest side in a mesh, should be as possible as small to ensure the best results. So to get mesh-independent simulation result and acceptable solution precision as well, an experiment is done and the influences of mesh size (number of meshes) on computational accuracy and time are shown in Figure 6. Note that the mesh aspect ratio is kept no more than 3 for linear computation. Considering computational accuracy and time, the CFRP specimen is meshed into 1000 meshes.

After applying the boundary conditions and loads, simulations were carried out. The simulation results are compared to experimental ones, and the comparison result is given in Figure 7. It can be seen that when the applied load is 5 N, the relative error between the simulation result and the experimental one is 6.54%. When the load increases up to 15 N, the relative error decreases to 0.99%, which proves the feasibility of the proposed method.

In addition, after the simulations were carried out, the results were output and compared to the experimental ones as shown in Figure 8, where the coordinates X and Y are used to locate the nodes used to create the meshes (red points in Figure 5(a)) and distance (Z-axis) indicates the deviation from the experimental results to the simulation ones at each node.

It can be seen that with the increasing of load, the deviations increase and the largest deviation appeared in the arc area ($10 < X < 20$, Figure 8(d)). This may be mainly due to the meshing rules used in this study. Generally speaking, intensive meshes should be used for the arc area so as to capture the rapid changing of stress. So adaptive meshes will bring us a more accurate result and reduce the deviation as a result. In addition, the reconstruction of experimental result inevitably introduces errors and may consequently enlarge the deviation. Nevertheless, compared to the initial gap size, the deviations are relatively low and the computation results are acceptable for assembly gap assessment.

5. Conclusion

A numerical approach to assembly gap prediction of composite structures is presented in this paper. First, assembly components are scanned in a specified configuration and finite element meshes are then generated by using the scanned data. Next, the shape variations of each component are evaluated by finite element analysis after considering the differences between constraints and forces presented in scanning and assembly configurations. Finally, the assembly gaps are assessed by assembling the simulated preassembly shapes of all components. The feasibility of the proposed method is proved by an experiment.

Being a new method, there are still rooms for modification and improvement. First, the meshing rules should be updated so as to minimize the deviation of simulation results for components with irregular shapes. In addition, the mesh generation procedure will be optimized in the future such that the proposed method can be used for large complex structure analysis without taking too much time.

Conflicts of Interest

The authors declare that there is no conflict of interest regarding the publication of this paper.

Acknowledgments

This work was supported by the Major State Basic Research Development Program (grant number 2014CB046504), the National Natural Science Foundation of China (grant number 51475073), and the Fundamental Research Funds for the Central Universities (DUT17JC19). The authors would like to acknowledge the above financial supports.

References

[1] C. Hühne, A.-K. Zerbst, G. Kuhlmann, C. Steenbock, and R. Rolfes, "Progressive damage analysis of composite bolted joints with liquid shim layers using constant and continuous degradation models," *Composite Structures*, vol. 92, no. 2, pp. 189–200, 2010.

[2] J. X. Dhôte, A. J. Comer, W. F. Stanley, and T. M. Young, "Study of the effect of liquid shim on single-lap joint using 3d digital image correlation," *Composite Structures*, vol. 96, pp. 216–225, 2013.

[3] J. X. Dhôte, A. J. Comer, W. F. Stanley, and T. M. Young, "Investigation into compressive properties of liquid shim for aerospace bolted joints," *Composite Structures*, vol. 109, pp. 224–230, 2014.

[4] Y. Zhai, D. Li, X. Li, and L. Wang, "An experimental study on the effect of joining interface condition on bearing response of single-lap, countersunk composite-aluminum bolted joints," *Composite Structures*, vol. 134, pp. 190–198, 2015.

[5] N. M. Chowdhury, J. Wang, W. K. Chiu, and P. Chang, "Experimental and finite element studies of thin bonded and hybrid carbon fibre double lap joints used in aircraft structures," *Composites Part B: Engineering*, vol. 85, pp. 233–242, 2016.

[6] L. Liu, J. Zhang, K. Chen, H. Wang, and M. Liu, "Experimental and numerical analysis of the mechanical behavior of composite-to-titanium bolted joints with liquid shim," *Aerospace Science and Technology*, vol. 49, pp. 167–172, 2016.

[7] R. Ascione and W. Polini, "Measurement of nonrigid freeform surfaces by coordinate measuring machine," *The International Journal of Advanced Manufacturing Technology*, vol. 51, no. 9-12, pp. 1055–1067, 2010.

[8] G. N. Abenhaim, A. Desrochers, and A. Tahan, "Nonrigid parts' specification and inspection methods: notions, challenges, and recent advancements," *The International Journal of Advanced Manufacturing Technology*, vol. 63, no. 5-8, pp. 741–752, 2012.

[9] G. N. Abenhaim, A. S. Tahan, A. Desrochers, and R. Maranzana, "A novel approach for the inspection of flexible parts without the use of special fixtures," *Journal of Manufacturing Science and Engineering*, vol. 133, no. 1, article 011009, 2011.

[10] I. Gentilini and K. Shimada, "Predicting and evaluating the post-assembly shape of thin-walled components via 3d laser digitization and fea simulation of the assembly process," *Computer-Aided Design*, vol. 43, no. 3, pp. 316–328, 2011.

[11] H. Radvar-Esfahlan and S.-A. Tahan, "Nonrigid geometric metrology using generalized numerical inspection fixtures," *Precision Engineering*, vol. 36, no. 1, pp. 1–9, 2012.

[12] A. Jaramillo, F. Prieto, and P. Boulanger, "Fast dimensional inspection of deformable parts from partial views," *Computers in Industry*, vol. 64, no. 9, pp. 1076–1081, 2013.

[13] A. Jaramillo, F. Prieto, and P. Boulanger, "Deformable part inspection using a spring-mass system," *Computer-Aided Design*, vol. 45, no. 8-9, pp. 1128–1137, 2013.

[14] G. N. Abenhaim, A. Desrochers, A. S. Tahan, and J. Bigeon, "A finite-element boundary condition setting method for the virtual mounting of compliant components," *Journal of Computing and Information Science in Engineering*, vol. 15, no. 4, article 041005, 2015.

[15] G. N. Abenhaim, A. Desrochers, A. S. Tahan, and J. Bigeon, "A virtual fixture using a fe-based transformation model embedded into a constrained optimization for the dimensional inspection of nonrigid parts," *Computer-Aided Design*, vol. 62, pp. 248–258, 2015.

[16] V. Sabri, S. A. Tahan, X. T. Pham, D. Moreau, and S. Galibois, "Fixtureless profile inspection of non-rigid parts using the numerical inspection fixture with improved definition of displacement boundary conditions," *The International Journal*

of Advanced Manufacturing Technology, vol. 82, no. 5-8, pp. 1343–1352, 2016.

[17] S. Sattarpanah Karganroudi, J.-C. Cuillière, V. Francois, and S.-A. Tahan, "Automatic fixtureless inspection of non-rigid parts based on filtering registration points," *The International Journal of Advanced Manufacturing Technology*, vol. 87, no. 1-4, pp. 687–712, 2016.

[18] F. Thiebaut, C. Lacroix, L. Andolfatto, and C. Lartigue, "Evaluation of the shape deviation of non rigid parts from optical measurements," *International Journal of Advanced Manufacturing Technology*, vol. 88, no. 5-8, pp. 1937–1944, 2017.

[19] C. Lacroix, L. Mathieu, F. Thiebaut, M. Douilly, and H. Falgarone, "Numerical process based on measuring data for gap prediction of an assembly," *Procedia CIRP*, vol. 27, pp. 97–102, 2015.

5

Reliability Analysis of the Chatter Stability during Milling using a Neural Network

Sen Hu, Xianzhen Huang, Yimin Zhang, and Chunmei Lv

School of Mechanical Engineering and Automation, Northeastern University, Shenyang 110819, China

Correspondence should be addressed to Xianzhen Huang; xzhhuang83@gmail.com

Academic Editor: Christopher J. Damaren

The parameters of a system have the randomness generally in the process of milling, which influences the stability of the milling. This paper uses the neural network to get a comprehensive analysis of the influences of random factors in milling and proposes a method for reliability analysis of the regenerative chatter stability in milling. Dynamic model of milling regenerative chatter is established, and stability lobe diagram is obtained by the full-discretization method (FDM). The neural network is applied to approximate the functional relationship of the limit axial cutting depth; then the reliability is computed with the Monte Carlo simulation method (MCSM) and the moment method (MM), respectively. Finally, the results of an example are used to demonstrate the efficiency and accuracy of the proposed method.

1. Introduction

High-speed milling processes are widely employed in the aerospace, automobile, ship building, and energy industry, among other fields, because they offer a high precision, high surface quality, and high material removal rate. However, in recent years, with the rapid development of modern industry and the ever increasing pressure of market competition, mechanical products have shown a tendency to become more and more special. Yet, complex shapes, thin-walled structures, the use of materials that are difficult to process, and high accuracy requirements all increase the probability of the occurrence of vibrations during product processing. This will ultimately affect the processing quality and reduce production efficiency. In more serious cases, production accidents might occur and threaten the safety of the personnel and thus the security of the enterprise. Accordingly, research on the origin of vibrations during milling processes has become extremely important.

Chatter is a kind of self-excited vibration, which is often caused by a regenerative feedback occurring in the system. Four different types of chatter vibrations can be distinguished according to the different feedback modes: regenerative chatter [1], mode-coupling chatter [2], frictional chatter [3], and thermal chatter [4]. In milling processes, regenerative chatter is the main cause of instabilities during the machining process, effectively reducing the quality of the produced parts [5].

Various methods have been reported in the literature to predict the chatter stability. In general, they can be classified into two categories: numerical methods [6–8] and analytical (or semianalytical) methods [9–22]. Sridhar et al. [6] proposed a numerical algorithm to analyze the milling stability. Smith and Tlusty [7] suggested using peak-to-peak (PTP) diagrams for analyzing the milling stability in the time-domain. Campomanes and Altintas [8] later improved the time-domain simulation model and proposed using the ratio of the dynamic uncut chip thickness to the static uncut chip thickness as an instability criterion. Based on the theory of periodic differential equations, Minis and Yanushevsky [9] employed the Nyquist criterion to calculate the stable boundary of the milling process. Altintaş and Budak [10] introduced the zeroth-order approximation (ZOA) method. This method has a high computational efficiency, but it cannot be used to predict the flip bifurcation in low radial immersion milling. Merdol and Altintas [11] then applied the

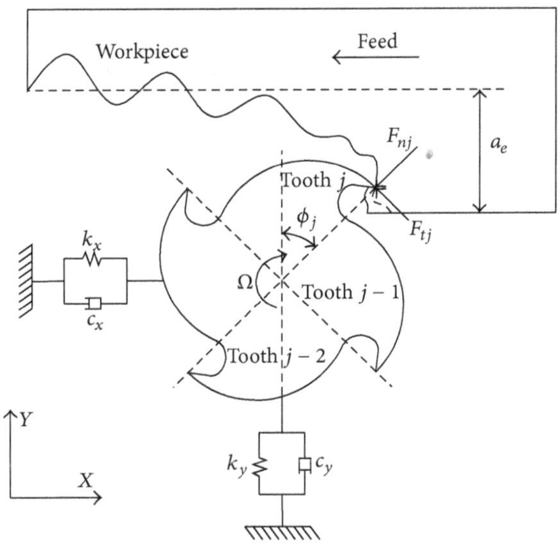

FIGURE 1: Dynamic model of the milling process.

multifrequency method to solve this problem. Bayly et al. [12] proposed the temporal finite element analysis (TFEA) method. The TFEA method can be applied to simultaneously predict the milling stability and the machining error. Insperger and Stépán [13] presented the semidiscretization method (SDM) for milling stability analysis. To improve the calculation accuracy, they then proposed the first-order SDM [14]. Ding et al. [15] established the full-discretization method (FDM) for milling stability analysis, which is based on the immediate integration method. The FDM can also simultaneously predict the milling stability and the machining error. Another kind of FDM is presented by Li et al. [16]. All the time-dependent terms are discretized in this method, and the transition matrix is obtained by numerical iteration method. Based on the TFEA method, Ding et al. [17] used integral equation technique to calculate the response of dynamic milling system and proposed the numerical integration method. Subsequently, they proposed a time-domain semianalytical method for stability analysis of milling in the framework of the differential quadrature method [18]. Besides, there are some other methods, such as [19–22]. Albertelli et al. [23] built a model for the stability analysis of a heavy milling machine and optimized both the planning process and the tool selection process.

The above-mentioned methods can be considered as a foundation for the analysis and the control of milling chatter. However, these methods are based on fixed cutting parameters. But in the actual machining process, the cutting parameters may vary and therefore must be treated as uncertain and random variables [24]. For example, variations in the working environment (such as voltage fluctuations) can make the feed of the workpiece random. Consequentially, the resulting measurement errors, machining errors, and other errors will lead to variations of the geometric dimensions of the tool. Kurdi et al. [25] studied the uncertainty in stability and surface location error in high-speed milling. Duncan et al. [26] described a procedure for adding uncertainty bounds to the analytical milling stability limits by the Monte Carlo simulation method (MCSM). Graham et al. [27] utilized the edge theorem and the zero exclusion condition to develop a robust chatter stability model based on the analytical chatter stability milling model. The reliability of milling chatter system was calculated using the first-order second-moment method (FOSMM) and compared to the Monte Carlo simulation method (MCSM) [28]. Lin [29] described the reliability of the cutting tools in the high-speed machining by normal distribution model.

In this study, we analyzed the milling stability using the FDM, and an artificial neural network was employed to analyze the effect of random factors on the milling stability. Furthermore, a method for calculating the reliability of the milling stability was developed and is presented in this paper.

2. Milling Chatter Stability Analysis

The dynamic model typically used for describing milling processes is shown in Figure 1. In this figure, ϕ_j is the angular position of jth tooth, Ω is the spindle speed (r/min), a_e is the radial cutting depth, and F_{tj} and F_{nj} are the tangential and normal components of the cutting force for jth tooth, respectively.

The governing equation of the dynamic milling system shown in Figure 1 can be expressed as [30]

$$\mathbf{M}\ddot{\mathbf{q}}(t) + \mathbf{C}\dot{\mathbf{q}}(t) + \mathbf{K}\mathbf{q}(t) = \mathbf{K}_c(t)\left[\mathbf{q}(t) - \mathbf{q}(t-T)\right], \quad (1)$$

where $\mathbf{M}$, $\mathbf{C}$, and $\mathbf{K}$ represent the modal mass matrix, damping matrix, and stiffness matrix, respectively. $\mathbf{q}(t)$ is the modal vector of the cutter. T is the time delay: that is, $T = 60(N\Omega)^{-1}$, where N is the number of cutting teeth. $\mathbf{K}_c(t)$ denotes the cutting coefficient matrix which varies periodically with time: that is, $\mathbf{K}_c(t) = \mathbf{K}_c(t+T)$, which can be written as

$$\mathbf{K}_c(t) = -a_p \begin{bmatrix} h_{xx}(t) & h_{xy}(t) \\ h_{yx}(t) & h_{yy}(t) \end{bmatrix}, \quad (2)$$

where a_p is the axial cutting depth and $h_{xx}(t)$, $h_{xy}(t)$, $h_{yx}(t)$, and $h_{yy}(t)$ are the cutting coefficients given by

$$\begin{aligned} h_{xx}(t) &= \sum_{j=1}^{N} w(\phi_j(t))\, s\phi_j \left[K_t c\phi_j + K_n s\phi_j\right], \\ h_{xy}(t) &= \sum_{j=1}^{N} w(\phi_j(t))\, c\phi_j \left[K_t c\phi_j + K_n s\phi_j\right], \\ h_{yx}(t) &= \sum_{j=1}^{N} w(\phi_j(t))\, s\phi_j \left[-K_t s\phi_j + K_n c\phi_j\right], \\ h_{yy}(t) &= \sum_{j=1}^{N} w(\phi_j(t))\, c\phi_j \left[-K_t s\phi_j + K_n c\phi_j\right], \end{aligned} \quad (3)$$

where $s\phi_j = \sin(\phi_j(t))$, $c\phi_j = \cos(\phi_j(t))$. K_n and K_t are the linearized normal and tangential cutting coefficients, respectively. $\phi_j(t)$ is defined by

$$\phi_j(t) = \left(\frac{2\pi\Omega}{60}\right)t + \frac{2\pi(j-1)}{N} \quad (4)$$

and $w(\phi_j)$ is the window function given by

$$w(\phi_j) = \begin{cases} 1, & \text{if } \phi_s \le \phi_j \le \phi_e \\ 0, & \text{otherwise,} \end{cases} \quad (5)$$

where ϕ_s and ϕ_e are the start and exit angles of jth tooth, respectively. For climb (down) milling, $\phi_s = \arccos(2r_c - 1)$ and $\phi_e = \pi$; for conventional (up) milling, $\phi_s = 0$ and $\phi_e = \arccos(1 - 2r_c)$, where r_c is the radial immersion ratio which is defined by $r_c = a_e/D$, where D is the cutter diameter.

Assuming that $\mathbf{p}(t) = \mathbf{M}\dot{\mathbf{q}}(t) + \mathbf{C}\mathbf{q}(t)/2$ and $\mathbf{x}(t) = [\mathbf{q}(t)\ \ \mathbf{p}(t)]^T$, (1) can be transformed into the state-space form

$$\dot{\mathbf{x}}(t) = \mathbf{A}\mathbf{x}(t) + \mathbf{B}(t)(\mathbf{x}(t) - \mathbf{x}(t-T)), \quad (6)$$

where

$$\mathbf{A} = \begin{bmatrix} -\dfrac{\mathbf{M}^{-1}\mathbf{C}}{2} & \mathbf{M}^{-1} \\ \dfrac{\mathbf{C}\mathbf{M}^{-1}\mathbf{C}}{4} - \mathbf{K} & -\dfrac{\mathbf{C}\mathbf{M}^{-1}}{2} \end{bmatrix}, \quad \mathbf{B}(t) = \begin{bmatrix} 0 & 0 \\ \mathbf{K}_c(t) & 0 \end{bmatrix}. \quad (7)$$

In the FDM [15], the first step is to discretize the time by dividing the period T into m equal parts; that is, $T = m\Delta t$ (with $m \in Z$). For a given time, (6) can be reformulated as

$$\dot{\mathbf{y}}(t) = \mathbf{A}\mathbf{y}(t) + \widetilde{\mathbf{B}}(t)(\widetilde{\mathbf{y}}(t) - \widetilde{\mathbf{y}}(t-T)), \quad t \in [t_j, t_{j+1}], \quad (8)$$

where $t_j = j\Delta t$ (with $j \in Z$). The periodic-coefficient item $\widetilde{\mathbf{B}}(t)$, the state item $\widetilde{\mathbf{y}}(t)$, and the time delay item $\widetilde{\mathbf{y}}(t-T)$ can be determined through linear approximation:

$$\begin{aligned} \widetilde{\mathbf{B}}(t) &= \mathbf{B}_{j+1} + \frac{\mathbf{B}_{j+1} - \mathbf{B}_j}{\Delta t}(t - t_{j+1}), \\ \widetilde{\mathbf{y}}(t) &= \mathbf{y}_{j+1} + \frac{\mathbf{y}_{j+1} - \mathbf{y}_j}{\Delta t}(t - t_{j+1}), \\ \widetilde{\mathbf{y}}(t-T) &= \mathbf{y}_{j-m+1} + \frac{\mathbf{y}_{j-m+1} - \mathbf{y}_{j-m}}{\Delta t}(t - t_{j+1}) \end{aligned} \quad (9)$$

with $\mathbf{B}_j = \mathbf{B}(t_j)$, $\mathbf{y}_j = \mathbf{y}(t_j)$. Treating (8) as an ordinary differential equation (ODE) and solving over the discretization period $[t_j, t_{j+1}]$ yield

$$\mathbf{y}_{j+1} = (\mathbf{\Phi}_0 + \mathbf{F}_j)\mathbf{y}_j + \mathbf{F}_{j+1}\mathbf{y}_{j+1} - \mathbf{F}_{j+1}\mathbf{y}_{j-m+1} - \mathbf{F}_j\mathbf{y}_{j-m}, \quad (10)$$

where

$$\mathbf{F}_j = \frac{\mathbf{\Phi}_3}{\Delta t^2}\mathbf{B}_j + \left(\frac{\mathbf{\Phi}_2}{\Delta t} - \frac{\mathbf{\Phi}_3}{\Delta t^2}\right)\mathbf{B}_{j+1}, \quad (11)$$

$$\mathbf{F}_{j+1} = \left(\frac{\mathbf{\Phi}_2}{\Delta t} - \frac{\mathbf{\Phi}_3}{\Delta t^2}\right)\mathbf{B}_j + \left(\mathbf{\Phi}_1 - 2\frac{\mathbf{\Phi}_2}{\Delta t} + \frac{\mathbf{\Phi}_3}{\Delta t^2}\right)\mathbf{B}_{j+1}, \quad (12)$$

$$\begin{aligned} \mathbf{\Phi}_0 &= e^{\mathbf{A}\Delta t}, \\ \mathbf{\Phi}_1 &= \mathbf{A}^{-1}(\mathbf{\Phi}_0 - \mathbf{I}), \\ \mathbf{\Phi}_2 &= \mathbf{A}^{-1}(\Delta t\mathbf{\Phi}_0 - \mathbf{\Phi}_1), \\ \mathbf{\Phi}_3 &= \mathbf{A}^{-1}(\Delta t^2\mathbf{\Phi}_0 - 2\mathbf{\Phi}_2). \end{aligned} \quad (13)$$

In (13), $\mathbf{I}$ is the identity matrix.

According to (10), a discrete map can be defined as

$$\mathbf{z}_{j+1} = \mathbf{D}_j\mathbf{z}_j, \quad (14)$$

where

$$\mathbf{z}_j = \operatorname{col}\left(\mathbf{y}_j, \mathbf{y}_{j-1}, \dots, \mathbf{y}_{j+1-m}, \mathbf{y}_{j-m}\right),$$

$$\mathbf{D}_j = \begin{bmatrix} \left[\mathbf{I} - \mathbf{F}_{j+1}\right]^{-1}\left(\mathbf{\Phi}_0 + \mathbf{F}_j\right) & \mathbf{0} & \cdots & \mathbf{0} & -\left[\mathbf{I} - \mathbf{F}_{j+1}\right]^{-1}\mathbf{F}_{j+1} & -\left[\mathbf{I} - \mathbf{F}_{j+1}\right]^{-1}\mathbf{F}_j \\ \mathbf{I} & \mathbf{0} & \cdots & \mathbf{0} & \mathbf{0} & \mathbf{0} \\ \mathbf{0} & \mathbf{I} & \cdots & \mathbf{0} & \mathbf{0} & \mathbf{0} \\ \vdots & \vdots & \ddots & \vdots & \vdots & \vdots \\ \mathbf{0} & \mathbf{0} & \cdots & \mathbf{I} & \mathbf{0} & \mathbf{0} \\ \mathbf{0} & \mathbf{0} & \cdots & \mathbf{0} & \mathbf{I} & \mathbf{0} \end{bmatrix}. \quad (15)$$

The transition matrix can be written as follows:

$$\mathbf{\Phi} = \mathbf{D}_{m-1}\mathbf{D}_{m-2}\cdots\mathbf{D}_1\mathbf{D}_0. \quad (16)$$

According to Floquet theory [31], the chatter stability depends on the eigenvalues of the transition matrix $\mathbf{\Phi}$: if the moduli of all the eigenvalues of the transition matrix $\mathbf{\Phi}$ are less than unity, the system is stable; otherwise, it is unstable; that is,

$$\max\left(\left|\Lambda\left(\mathbf{\Phi}\right)\right|\right) \begin{cases} < 1 & \text{stable} \\ = 1 & \text{stable boundary} \\ > 1 & \text{unstable.} \end{cases} \quad (17)$$

When the cutting system is in its critical stable state, its axial cutting depth is called the limiting axial cutting depth. The above formulas show that the limiting axial cutting depth is determined by the normal cutting force coefficient K_n, the tangential cutting force coefficient K_t, the modal mass m, the modal damping coefficient c, the modal stiffness coefficient k, the radial immersion ratio r_c, the spindle speed Ω, and the number of cutting teeth N; that is,

$$a_{\text{plim}} = \left\{a_p \mid \max\left(\left|\Lambda\left(\mathbf{\Phi}\left(K_n, K_t, m, c, k, r_c, N, \Omega\right)\right)\right|\right) = 1\right\}. \quad (18)$$

3. Reliability Analysis of the Chatter Stability

Mechanical reliability refers to the ability of mechanical products to complete the required functions under the specified conditions over a specified time period [32–34]. Measuring this ability with probability is reliability. The reliability of the chatter stability refers to the probability that the actual axial cutting depth is less than the limiting axial cutting depth. According to (18), there is no explicit expression for the limiting axial cutting depth. Therefore, the first step is to use a neural network to approximate the functional relationship of the limiting axial cutting depth.

3.1. Establishing a Back-Propagation Neural Network. An artificial neural network usually exhibits a good self-learning ability and a good fault tolerance and can be used for nonlinear mapping. Thus, it is well suited to deal with large amounts of data and to perform complex nonlinear mapping tasks. To date, back-propagation (BP) neural networks are among the most essential and most perfect artificial neural networks.

The design of a BP neural network mainly includes the design of the input layer, the output layer, the hidden layer, and the transformation function.

3.1.1. Design of the Input and the Output Layer. According to the analysis of the chatter stability based on the FDM, the input layer was selected to consist of 9 neurons, which represent the tangential cutting force coefficient K_t, the normal cutting force coefficient K_n, the modal damping coefficient c_x, the modal stiffness coefficient k_x, and the modal mass m_x in the x-direction, as well as the modal damping coefficient c_y, the modal stiffness coefficient k_y, and the modal mass m_y in the y-direction, and the radial immersion ratio r_c, respectively. For the output layer, a single-neuron structure was adopted, which represents the limiting axial cutting depth a_{plim}.

3.1.2. Design of the Hidden Layer. According to Hecht-Nielsen [35, 36], a BP neural network with three layers is enough to complete the general mapping. In a three-layer network, the approximate relationship between the number of neurons in the hidden layer n_2 and the number of neurons in the input layer n_1 is given by

$$n_2 = 2 \times n_1 + 1. \quad (19)$$

Therefore, the number of neurons in the hidden layer was selected to be 19.

3.1.3. Data Normalization. The conversion function of the network is a sigmoid function. Both ends of the output curve of this function are smooth, but in the middle part the curve varies drastically. Therefore, it is necessary to normalize the input and output variables by replacing the original data with corresponding values in the interval $(0, 1)$ to achieve a faster convergence rate.

The resulting topological structure of the BP neural network model used in this study is shown in Figure 2.

The basic algorithm for calculating the milling chatter stability using a BP neural network is as follows. Firstly, n

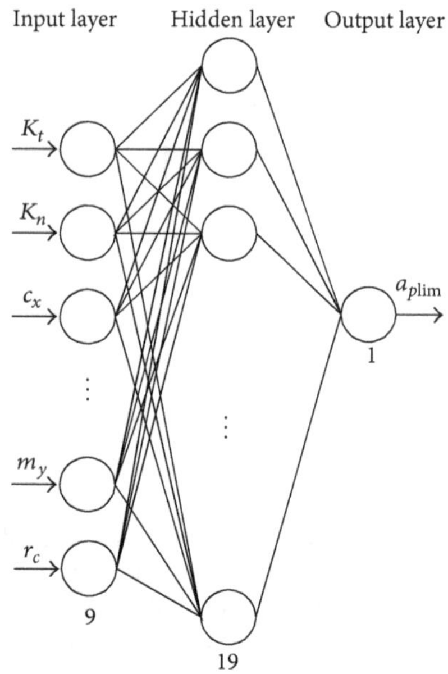

FIGURE 2: Topological structure of the BP neural network model used in this study.

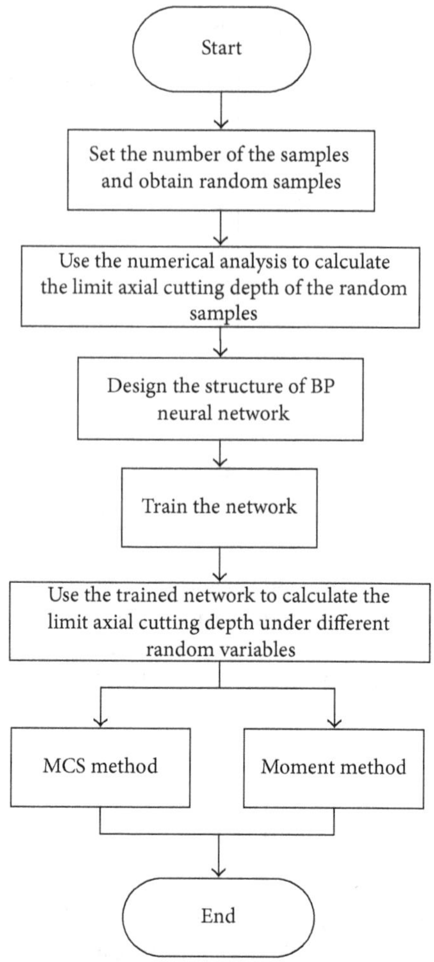

FIGURE 3: Flowchart for calculating the chatter stability reliability utilizing the BP neural network.

groups of random samples $\mathbf{x}_i$ $(i = 1, 2, \ldots, n)$ are generated from the basic variables (such as the kinetic parameters K_n, K_t, c_x, c_y, k_x, k_y, m_x, m_y). Then, the limiting axial cutting depth $a_{plim}(\mathbf{x}_i)$ of the random samples can be obtained through numerical analysis. The network grid will be trained using the data $\{\mathbf{x}_i, a_{plim}(\mathbf{x}_i)\}$ $(i = 1, 2, \ldots, n)$. The trained network can be used to calculate the limiting axial cutting depth for different values of the basic variables. On the basis of this algorithm, either the Monte Carlo simulation method (MCSM) or the moment method (MM) can be used to determine the reliability of the milling chatter stability. The described algorithm is illustrated in Figure 3.

3.2. The Monte Carlo Simulation Method. The MCSM offers a wide universality and is usually fairly easy to implement. Furthermore, as the number of simulations increases, its accuracy will improve.

The basic idea of the MCSM is as follows. Firstly, random samples in the space of the input variables are generated, and then the structural response value is calculated by plugging the random samples into the system model. The probabilistic nature and statistical regulation of the whole system are determined by the probabilistic nature and statistical regulation of the random samples and their responses. The basic steps of the MCSM are as follows:

(1) Based on the distribution characteristics of the basic variables, n groups of random samples $\mathbf{x}_i$ $(i = 1, 2, \ldots, n)$ are generated.

(2) The structural response value $g(\mathbf{x}_i)$ is then calculated by plugging n groups of random samples into the reliability analysis model.

(3) The number of samples in the failure domain is counted and denoted as n_f.

(4) The probability of failure is defined as the ratio of the number of samples in the failure region to the total number of samples: that is,

$$P_f \approx \frac{n_f}{n}. \tag{20}$$

3.3. Moment Method. Although the MCSM is straightforward to implement as it only requires repetitive executions of deterministic simulations, typically a large number of simulations are needed because the solution statistics converge relatively slowly. Therefore, the MCSM is often used as a standard to verify the accuracy of other methods.

In practical engineering, the first-order second-moment method (FOSMM) is the most simple and most commonly used method, but it will generate a large error for higher degree-nonlinear functions. In addition to the gradient of the nonlinear function, the second-order second-moment method (SOSMM) also considers the concavity and curvature of the limit state surface in the vicinity of the design point. Therefore, using the SOSMM can improve the accuracy of the reliability analysis.

Table 1: Dynamic parameters of the simulated milling system and mean values and standard deviations of the corresponding probability distributions.

Random variable	Mean value	Standard deviation	Probability distribution type
K_t	6×10^8 N/m^2	3×10^7 N/m^2	Normal
K_n	2×10^8 N/m^2	1×10^7 N/m^2	Normal
c_x, c_y	5.089 N·s/m	5.089×10^{-2} N·s/m	Normal
k_x, k_y	1.34×10^6 N/m	1.34×10^4 N/m	Normal
m_x, m_y	0.03993 kg	0.0003993 kg	Normal
r_c	0.6	0.006	Normal

The value of the reliability index β corresponds to the distance from the origin to the hyperplane in the standard normal space, which is the geometric meaning of the reliability index β. Therefore, it is first necessary to normalize the random variable $\mathbf{x}$; that is,

$$\mathbf{u} = \frac{\mathbf{x} - \mu_\mathbf{X}}{\sigma_\mathbf{X}}. \quad (21)$$

The reliability index β and the design point $\mathbf{u}^*$ can be obtained using the FOSMM. The second-order Taylor expansion of the function $g(\mathbf{u})$ at the design point $\mathbf{u}^*$ is

$$\begin{aligned} \tilde{g}(\mathbf{u}) &\approx g(\mathbf{u}^*) + \nabla g(\mathbf{u}^*)^T (\mathbf{u} - \mathbf{u}^*) \\ &\quad + \frac{1}{2} (\mathbf{u} - \mathbf{u}^*)^T \nabla^2 g(\mathbf{u}^*) (\mathbf{u} - \mathbf{u}^*), \end{aligned} \quad (22)$$

where $\nabla^2 g(\mathbf{u}^*)$ is the second-order derivative of the function.

The two ends of (22) were then divided by $|\nabla g(\mathbf{u}^*)|$:

$$\tilde{g}(\mathbf{u}) \approx \boldsymbol{\alpha}^T (\mathbf{u} - \mathbf{u}^*) + \frac{1}{2} (\mathbf{u} - \mathbf{u}^*)^T \mathbf{B} (\mathbf{u} - \mathbf{u}^*), \quad (23)$$

where

$$\begin{aligned} \boldsymbol{\alpha} &= \frac{\nabla g(\mathbf{u}^*)}{|\nabla g(\mathbf{u}^*)|}, \\ \mathbf{B} &= \frac{\nabla^2 g(\mathbf{u}^*)}{|\nabla g(\mathbf{u}^*)|}. \end{aligned} \quad (24)$$

In order to consider the curvature of the limit state surface, the variables must be converted from the standard normal space $\mathbf{u}$ to the space $\mathbf{y}$. In the space $\mathbf{y}$, the design point $\mathbf{y}^*$ falls on the coordinate axis ($\mathbf{y}^* = \{0, 0, \ldots, \beta\}^T$). An orthogonal matrix $\mathbf{H}$ can then be constructed so that $\mathbf{y} = \mathbf{H}\mathbf{u}$. Now, (23) can be rewritten as

$$\beta - y_n + \frac{1}{2} \begin{pmatrix} y_{n-1} \\ y_n - \beta \end{pmatrix} \mathbf{R} \begin{pmatrix} y_{n-1} \\ y_n - \beta \end{pmatrix} = 0, \quad (25)$$

where $\mathbf{y}_{n-1} = [y_1, y_2, \ldots, y_{n-1}]^T$ and $\mathbf{R} = \mathbf{H}\mathbf{B}\mathbf{H}^T$.

Solving the above equation and retaining the first two orders of the solution,

$$y_n = \beta + \frac{1}{2} y_n^T \mathbf{R}_{n-1} y_{n-1}, \quad (26)$$

where $\mathbf{R}_{n-1}$ is a new matrix consisting of $n - 1$ rows and columns of the matrix $\mathbf{R}$. The new approximation of (26) can be obtained by calculating the eigenvalue of the matrix $\mathbf{R}_{n-1}$, denoted as

$$y_n = \beta + \frac{1}{2} \sum_{i=1}^{n-1} w_i y_i^2, \quad (27)$$

where w_i is the eigenvalue of the matrix $\mathbf{R}_{n-1}$ and also the curvature of the limit state surface at the design point.

Thus, the failure domain F can be approximated by $\tilde{F} = \{\tilde{g}(y) < 0\}$. Finally, the probability of failure can be estimated using the method proposed by Breitung [37].

$$P_f = P(\tilde{F}) = \Phi(-\beta) \prod_{i=2}^{n-1} (1 + w_i \beta)^{-1/2}. \quad (28)$$

4. Numerical Example

For this example, we assumed that the dynamic characteristics in the x-direction and the y-direction of the milling machine system are balanced and symmetrical. The number of the cutter teeth is $N = 2$, and the dynamic parameters of the system and mean values and standard deviations of the corresponding probability distributions are listed in Table 1.

The value of the standard deviation of the random parameters should be obtained through the test or the statistical analysis of the experimental data. If the experimental data are not available, the standard deviations of mechanical property parameters could be determined by the variation coefficient [38]. If the random variables are affected by a large number of independent factors, they are generally subject to normal distribution [39].

Substitute mean values of the dynamic parameters K_t, K_n, c_x, c_y, k_x, k_y, m_x, m_y, and r_c into (17). Then, stability lobes of the milling system can be obtained. To study the effects of variations of variable X_i ($i = 1, 2, \ldots, 9$) on the stability boundaries, the stability lobes are obtained when $x_i = \mu_{X_i} - 3\sigma_{X_i}$ and $x_i = \mu_{X_i} + 3\sigma_{X_i}$ (other variables are equal to their mean values). All the results are depicted in Figure 4. From the results, the variances of K_t, k_x, k_y, m_x, and m_y have significant influences on the stability boundaries.

The critical axial cutting depth of the system is found to be lowest for a spindle speed Ω of 10740 r/min; that is, $\min(a_{\text{plim}}) = 9.33 \times 10^{-5}$ m. At this speed, the possibility of the occurrence of chatter is highest, so we selected this speed to

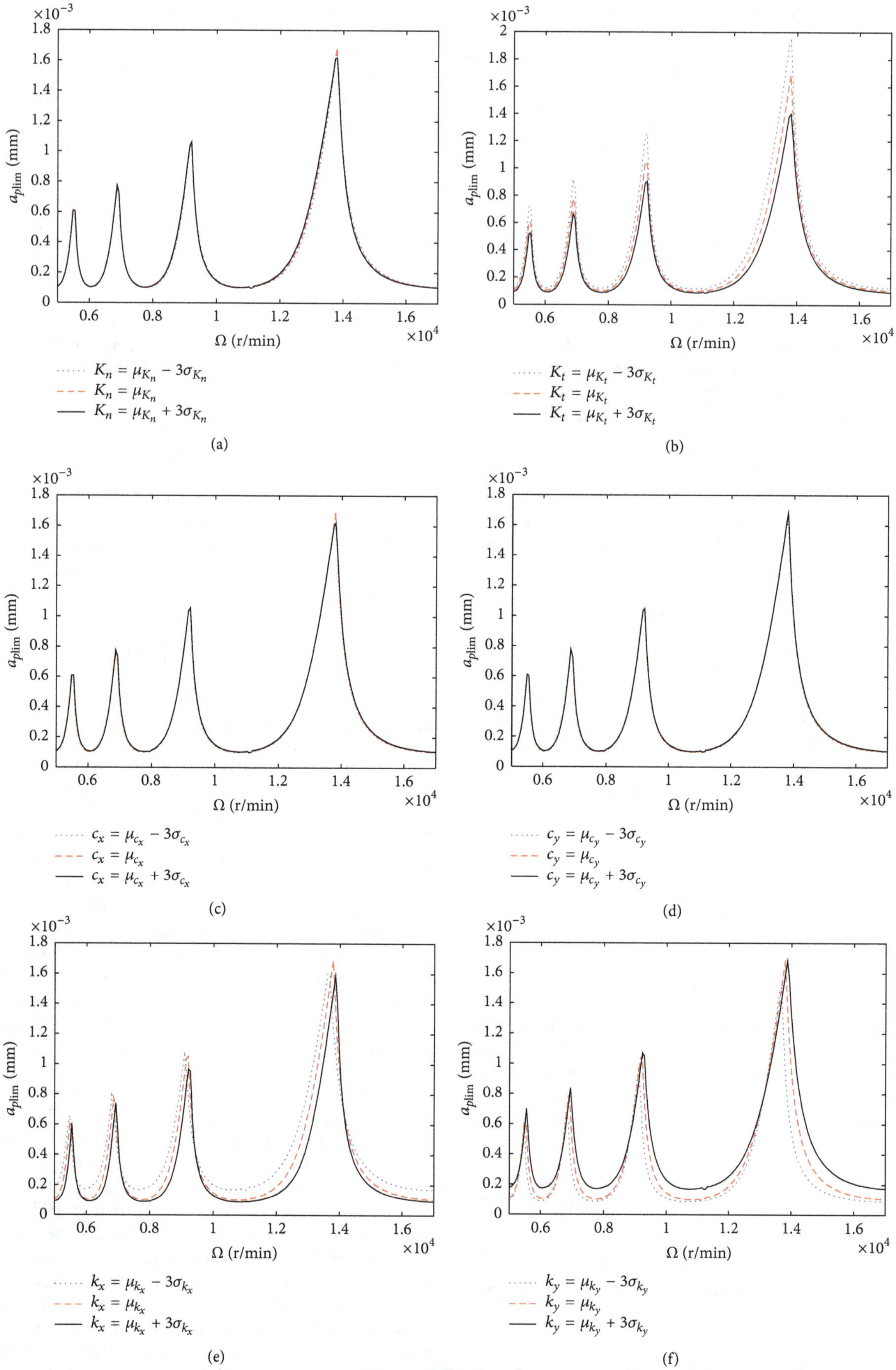

FIGURE 4: Continued.

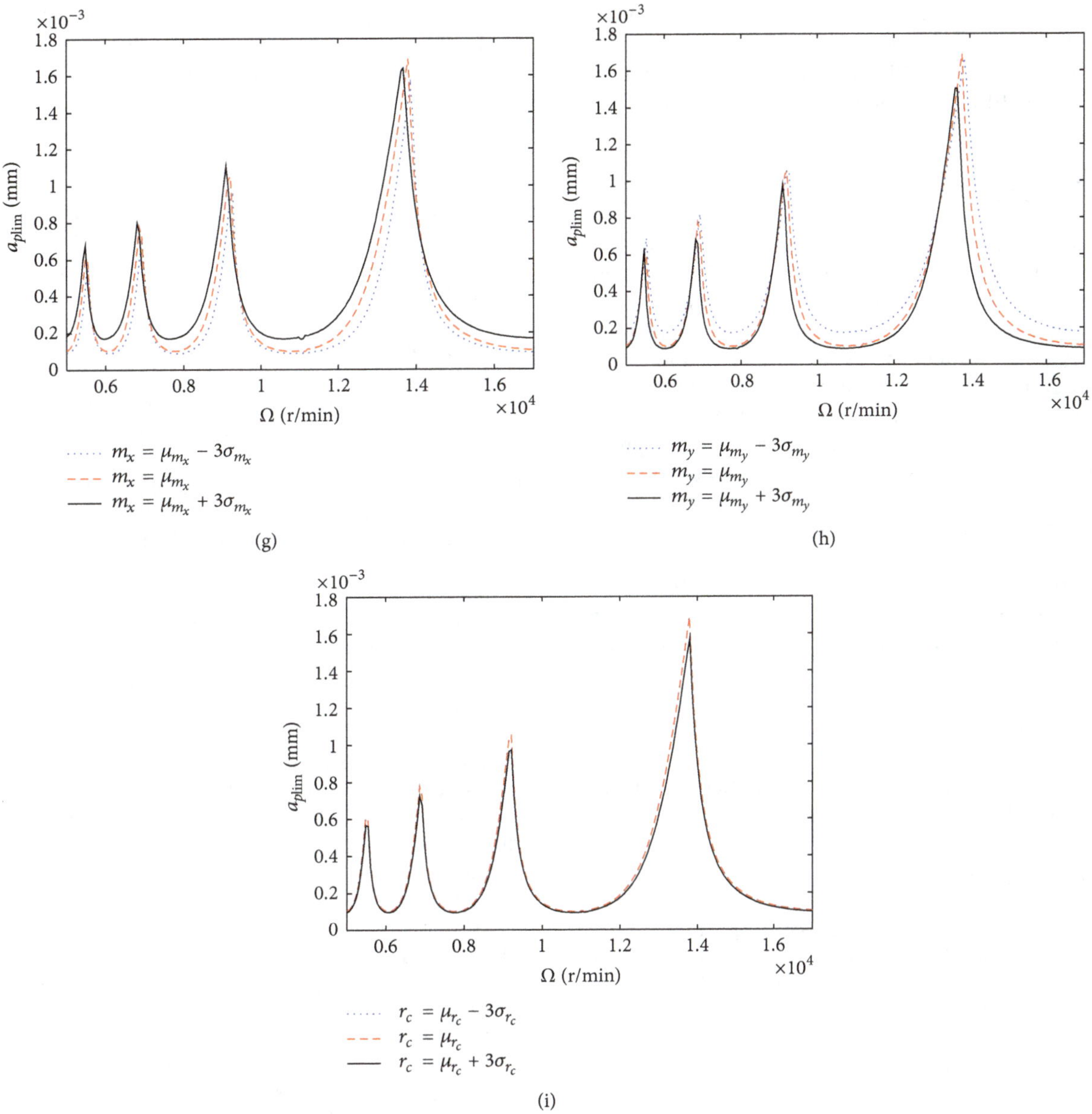

FIGURE 4: The effects of different variables on the stability boundaries.

TABLE 2: Comparison of the reliabilities obtained using the different methods.

	BP-AFOSM	BP-SOSMM	BP-MCSM	MCSM
Reliability	0.6349	0.7070	0.6879	0.6874

calculate the reliability of the chatter stability. The functional expression of the limiting axial cutting depth is obtained using the BP neural network. The actual axial cutting depth is considered to be equal to the minimum of the critical axial cutting depth of 9.33×10^{-5} m. Then, the MCSM and the MM are used to determine the reliability of the milling chatter stability. The direct MCSM is employed for comparison to demonstrate the validity of the methods based on the BP neural network. The number of direct MCSM samples is 1×10^5, and it costs about 22 hours (0.8 seconds are needed for each deterministic analysis). However, in the proposed method, the number of training samples of the BP neural network is 500, and about 400 seconds is needed for the method based on the BP neural network. After fitting the functional relationship of the limit axial cutting depth, the reliability is computed with the MCSM and MM, respectively. The MCSM needs about 153 seconds, and the AFOSMM needs about 6.4 seconds, and the SOSMM needs about 12 seconds. Therefore, the time consumed by the method based on the BP neural network is significantly less than that of the direct MCSM. The reliabilities obtained using the different methods are compared in Table 2. The accuracy of the methods based on the BP neural network is found to be acceptable. The accuracy of the SOSMM is much higher than that of the advanced first-order second-moment method

(AFOSMM), but it is still lower than MCSM. In addition, smaller values could be chosen for the actual axial cutting depth to enhance the reliability. For instance, when choosing an actual axial cutting depth of $a_p = 8 \times 10^{-5}$ m, the reliability will increase to $R = 0.9935$.

5. Conclusions

Vibrations are among the main obstacles limiting the accuracy and reliability of milling processes, and chatter is considered the most important vibration in a milling machine. In this paper, the stability of regenerative chatter that might occur during the milling process was analyzed, and a reliability analysis of the chatter stability was performed utilizing a BP neural network. The effect of random factors on the stability of the milling process was analyzed. This approach is more suitable for practical engineering applications than theoretical calculations based on fixed cutting parameters. The reliability of the stability of the milling process was calculated for different spindle speeds using both Monte Carlo simulations and the moment method. The results are considered to be helpful for improving the machining accuracy and machining efficiency of high-speed milling processes.

Competing Interests

The authors declare that there is no conflict of interests regarding the publication of this paper.

Acknowledgments

The authors gratefully acknowledge the support of National Natural Science Foundation of China (51575094, 51135003, and 51305071), Program for New Century Excellent Talents in University of Ministry of Education of China (NCET-13-0103), and Fundamental Research Funds for the Central Universities (N140304003).

References

[1] S. Smith and J. Tlusty, "Update on high-speed milling dynamics," *Journal of Engineering for Industry*, vol. 112, no. 2, pp. 142–149, 1990.

[2] J. Tlusty and F. Ismail, "Basic non-linearity in machining chatter," *CIRP Annals—Manufacturing Technology*, vol. 30, no. 1, pp. 299–304, 1981.

[3] M. Wiercigroch and A. M. Krivtsov, "Frictional chatter in orthogonal metal cutting," *Philosophical Transactions of the Royal Society A: Mathematical, Physical and Engineering Sciences*, vol. 359, no. 1781, pp. 713–738, 2001.

[4] M. A. Davies and T. J. Burns, "Thermomechanical oscillations in material flow during high-speed machining," *Philosophical Transactions of the Royal Society A: Mathematical, Physical and Engineering Sciences*, vol. 359, no. 1781, pp. 821–846, 2001.

[5] Y. Altintas, G. Stépán, D. Merdol, and Z. Dombovari, "Chatter stability of milling in frequency and discrete time domain," *CIRP Journal of Manufacturing Science & Technology*, vol. 1, no. 1, pp. 35–44, 2008.

[6] R. Sridhar, R. E. Hohn, and G. W. Long, "A stability algorithm for the general milling process: contribution to machine tool chatter research—7," *Journal of Engineering for Industry*, vol. 90, no. 2, pp. 330–334, 1968.

[7] S. Smith and J. Tlusty, "Efficient simulation programs for chatter in milling," *CIRP Annals—Manufacturing Technology*, vol. 42, no. 1, pp. 463–466, 1993.

[8] M. L. Campomanes and Y. Altintas, "An improved time domain simulation for dynamic milling at small radial immersions," *Journal of Manufacturing Science & Engineering*, vol. 125, no. 3, pp. 416–422, 2003.

[9] I. Minis and R. Yanushevsky, "A new theoretical approach for the prediction of machine tool chatter in milling," *Journal of Engineering for Industry*, vol. 115, no. 1, pp. 1–8, 1993.

[10] Y. Altintaş and E. Budak, "Analytical prediction of stability lobes in milling," *CIRP Annals—Manufacturing Technology*, vol. 44, no. 1, pp. 357–362, 1995.

[11] S. D. Merdol and Y. Altintas, "Multi frequency solution of chatter stability for low immersion milling," *Journal of Manufacturing Science & Engineering*, vol. 126, no. 3, pp. 459–466, 2004.

[12] P. V. Bayly, J. E. Halley, B. P. Mann, and M. A. Davies, "Stability of interrupted cutting by temporal finite element analysis," *Journal of Manufacturing Science and Engineering*, vol. 125, no. 2, pp. 220–225, 2003.

[13] T. Insperger and G. Stépán, "Semi-discretization method for delayed systems," *International Journal for Numerical Methods in Engineering*, vol. 55, no. 5, pp. 503–518, 2002.

[14] T. Insperger, G. Stépán, and J. Turi, "On the higher-order semi-discretizations for periodic delayed systems," *Journal of Sound & Vibration*, vol. 313, no. 1-2, pp. 334–341, 2008.

[15] Y. Ding, L. M. Zhu, X. J. Zhang, and H. Ding, "A full-discretization method for prediction of milling stability," *International Journal of Machine Tools & Manufacture*, vol. 50, no. 5, pp. 502–509, 2010.

[16] M. Li, G. Zhang, and Y. Huang, "Complete discretization scheme for milling stability prediction," *Nonlinear Dynamics*, vol. 71, no. 1-2, pp. 187–199, 2013.

[17] Y. Ding, L. M. Zhu, X. J. Zhang, and H. Ding, "Numerical integration method for prediction of milling stability," *Journal of Manufacturing Science & Engineering*, vol. 133, no. 3, Article ID 031005, pp. 391–407, 2011.

[18] Y. Ding, L. M. Zhu, X. J. Zhang, and H. Ding, "Stability analysis of milling via the differential quadrature method," *Journal of Manufacturing Science and Engineering*, vol. 135, no. 4, Article ID 044502, 7 pages, 2013.

[19] F. M. Asl and A. G. Ulsoy, "Analysis of a system of linear delay differential equations," *Journal of Dynamic Systems, Measurement, and Control*, vol. 125, no. 2, pp. 215–223, 2003.

[20] S. Yi, P. W. Nelson, and A. G. Ulsoy, "Delay differential equations via the matrix Lambert W function and bifurcation analysis: application to machine tool chatter," *Mathematical Biosciences and Engineering*, vol. 4, no. 2, pp. 355–368, 2007.

[21] E. A. Butcher, O. A. Bobrenkov, E. Bueler, and P. Nindujarla, "Analysis of milling stability by the chebyshev collocation method: algorithm and optimal stable immersion levels," *Journal of Computational and Nonlinear Dynamics*, vol. 4, no. 3, pp. 1–12, 2009.

[22] E. A. Butcher, H. Ma, E. Bueler, V. Averina, and Z. Szabo, "Stability of linear time-periodic delay-differential equations via Chebyshev polynomials," *International Journal for Numerical Methods in Engineering*, vol. 59, no. 7, pp. 895–922, 2004.

[23] P. Albertelli, N. Cau, G. Bianchi, and M. Monno, "The effects of dynamic interaction between machine tool subsystems on cutting process stability," *The International Journal of Advanced Manufacturing Technology*, vol. 58, no. 9–12, pp. 923–932, 2012.

[24] A. Iglesias, J. Munoa, and J. Ciurana, "Optimisation of face milling operations with structural chatter using a stability model based process planning methodology," *International Journal of Advanced Manufacturing Technology*, vol. 70, no. 1–4, pp. 559–571, 2014.

[25] M. H. Kurdi, T. L. Schmitz, R. T. Haftka, and B. P. Mann, "A numerical study of uncertainty in stability and surface location error in high-speed milling," *ASME International Mechanical Engineering Congress & Exposition*, vol. 16, no. 1, pp. 387–395, 2005.

[26] G. S. Duncan, M. H. Kurdi, T. L. Schmitz, and J. P. Snyder, "Uncertainty propagation for selected analytical milling stability limit analyses," *Transactions of the North American Manufacturing Research Institute of SME*, vol. 34, pp. 17–24, 2006.

[27] E. Graham, M. Mehrpouya, and S. S. Park, "Robust prediction of chatter stability in milling based on the analytical chatter stability," *Journal of Manufacturing Processes*, vol. 15, no. 4, pp. 508–517, 2013.

[28] Y. Liu, L. L. Meng, K. Liu, and Y. M. Zhang, "Chatter reliability of milling system based on first-order second-moment method," *International Journal of Advanced Manufacturing Technology*, 2016.

[29] W. S. Lin, "The reliability analysis of cutting tools in the HSM processes," *Archives of Materials Science & Engineering*, vol. 30, no. 2, pp. 97–100, 2008.

[30] J. Gradišek, M. Kalveram, T. Insperger et al., "On stability prediction for milling," *International Journal of Machine Tools and Manufacture*, vol. 45, no. 7-8, pp. 769–781, 2005.

[31] M. Farkas, *Periodic Motions*, vol. 104 of *Applied Mathematical Sciences*, Springer, New York, NY, USA, 1994.

[32] L. Changyou, W. Wei, Z. Yimin, G. Song, L. Zhenyuan, and Q. Changshuai, "Indexing accuracy reliability sensitivity analysis of power tool turret," *Maintenance and Reliability*, vol. 17, no. 1, pp. 27–34, 2014.

[33] Y. Chu, Z. Yuan, and J. Chen, "Research on dynamic reliability of a jet pipe servo valve based on generalized stochastic Petri nets," *International Journal of Aerospace Engineering*, vol. 2015, Article ID 171642, 8 pages, 2015.

[34] Y. Pang, H.-Z. Huang, N.-C. Xiao, Y. Liu, and Y.-F. Li, "Possibilistic reliability analysis of repairable system with omitted or delayed failure effects," *Maintenance and Reliability*, vol. 14, no. 3, pp. 195–202, 2012.

[35] R. Hecht-Nielsen, "Theory of the backpropagation neural network," *Neural Networks*, vol. 1, no. 1, pp. 65–93, 1988.

[36] R. Hecht-Nielsen, "Kolmogorov's mapping neural network existence theorem," in *Proceedings of the 1st IEEE International Conference on Neural Network*, pp. 11–14, San Diego, Calif, USA, 1987.

[37] K. Breitung, "Asymptotic approximations for multinormal integrals," *Journal of Engineering Mechanics*, vol. 110, no. 3, pp. 357–366, 1984.

[38] Y. Zhang, X. He, Q. Liu, and B. Wen, "Robust reliability design of Banjo flange with arbitrary distribution parameters," *Journal of Pressure Vessel Technology*, vol. 127, no. 4, pp. 408–413, 2005.

[39] P. D. T. O'Connor and A. Kleyner, *Practical Reliability Engineering*, John Wiley & Sons, Chichester, UK, 5th edition, 2012.

6

A Highly Accurate Approach for Aeroelastic System with Hysteresis Nonlinearity

C. C. Cui,[1,2] S. X. Xie,[3] X. C. Huang,[4] J. K. Liu,[2] and Y. M. Chen[2]

[1]*State Key Laboratory of Complex Electromagnetic Environment Effects on Electronics and Information System, P.O. Box 085 Luoyang, China*
[2]*Department of Mechanics, Sun Yat-sen University, No. 135 Xingang Road West, Guangzhou 510275, China*
[3]*Basic Education Department, Jiangxi Industry Polytechnic College, Qinshanhu Road 1988, Nanchang, China*
[4]*Department of Mechanical Engineering, Ganzhou Institute of Technology, 88 Loutiling Rd, Ganzhou 341000, Jiangxi, China*

Correspondence should be addressed to Y. M. Chen; chenymao@mail.sysu.edu.cn

Academic Editor: Wen Bao

We propose an accurate approach, based on the precise integration method, to solve the aeroelastic system of an airfoil with a pitch hysteresis. A major procedure for achieving high precision is to design a predictor-corrector algorithm. This algorithm enables accurate determination of switching points resulting from the hysteresis. Numerical examples show that the results obtained by the presented method are in excellent agreement with exact solutions. In addition, the high accuracy can be maintained as the time step increases in a reasonable range. It is also found that the Runge-Kutta method may sometimes provide quite different and even fallacious results, though the step length is much less than that adopted in the presented method. With such high computational accuracy, the presented method could be applicable in dynamical systems with hysteresis nonlinearities.

1. Introduction

As more and more approaches have been developed successfully, numerical solution techniques play a pivotal role in nonlinear dynamic research. It has still been a tough job, however, to obtain a reliable numerical solution in a long-term duration for nonlinear dynamical systems [1], especially when nonsmooth nonlinearities are included [2, 3]. Hysteresis is one typical kind of complex nonsmooth nonlinearities, which can result from smart materials and composite structures [4–6]. This study aims at proposing an effective approach for dynamical systems, illustrated by the aeroelastic systems of airfoils with hysteresis.

Nonlinear aeroelastic problems have received a large amount of research interests. Well known, Dowell et al. have done many significant contributions to nonlinear aeroelastics [7–9]. Xu et al. have proposed a control strategy appropriate for the suppression of a nonlinear aeroelastic system [10]. The nonlinearities of interests to aeroelasticians consist of two categories, that is, the structural and aerodynamics nonlinearities [11]. Specifically, the concentrated structural nonlinearities usually refer to cubic stiffnesses, freeplay, and hysteresis. And the freeplay and hysteresis are the typical nonsmooth nonlinearities arising from aeroelastic systems. They can result in complex nonlinear dynamic responses, such as limit cycle oscillations (LCOs), bifurcations, and chaos [12–16].

It is tougher and even impossible to find an analytic solution of a nonlinear aeroelastic system. Alternatively, researchers pay their attention to numerical and/or semianalytical solution approaches. The nonlinear aeroelastic responses are usually generated by numerical solution techniques such as the Runge-Kutta (RK) method and the finite-difference method [9]. Semi-analytical solvers are usually based on harmonic balancing [17–19], such as the perturbation-based method [12] and the homotopy analysis method [20] to mention a few.

To the best of our knowledge, additional theoretical and/or computational obstacles will usually be confronted when using semianalytical approaches to solving aeroelastic systems with nonsmooth nonlinearities. For example, Liu and Dowell [21] introduced FFT technique to handle the freeplay nonlinearities when applying the harmonic balance method to solve an airfoil aeroelastic system with a control surface. Liu et al. [22] proposed a new technique to transform time domain solutions into frequency ones so that the incremental harmonic balance method is capable of analyzing the LCOs of an airfoil aeroelastic system with a hysteresis.

Though in principle numerical solution approaches can be directly implemented to nonsmooth aeroelastic systems, troublesome problems regarding reliability and accuracy of the results still have been observed. For instance, Conner et al. [23] found that switching points in piecewise linear systems can affect the computational accuracy significantly and applied Hénon's method to locate the switching points. More recently, Cui et al. [24] employed the precise integration method (PIM) initiated by Zhong [25] to propose a promising solution technique for aeroelastic systems with freeplay. In this method, the switching points are accurately and effectively determined by a predictor-corrector algorithm. Furthermore, Chen and Liu [26] implemented this method to study LCOs and bifurcation of an airfoil with both an external store and a pitch freeplay.

Hysteresis phenomena have been observed in aeroelastic systems of airfoil for decades [11, 27]. To some extent, hysteresis can be considered as a composite of two directional freeplays, that is, the value of the nonlinearity depends not only on the vibration displacement but also on the velocity. Therefore, it is much tougher to solve hysteresis systems. Maybe for this reason, it is rare to find effective solution approaches to do this job [12, 22, 28]. It is necessary and worthwhile to extending our newly proposed method [24, 26] to study the airfoil aeroelastic system with hysteresis. Due to the directional features of the hysteresis, we should split the nonlinear system into subsystems of much higher dimensions. Different from the routine procedures, moreover, a predictor-corrector algorithm has to be constructed to treat with the directional features when determining the switching points.

2. Areoelastic Model with Hysteresis

Figure 1 shows the physical model of a two-dimensional airfoil. It oscillates in pitch and plunge directions, denoted by α, positive with the nose up, and by h, positive in the downward direction, respectively. The symbol b denotes the semichord length. The elastic axis is located at a distance $a_h b$ from the midchord, and the mass center is located at a distance $x_\alpha b$ from the elastic axis. Both distances are positive when measured towards the trailing edge of the airfoil.

The coupled motions of the airfoil in incompressible unsteady flow can be modeled as follows [17]:

$$\begin{aligned}
&c_0\ddot{\xi} + c_1\ddot{\alpha} + \left(c_2 + 2\zeta_\xi \frac{\overline{\omega}}{U^*}\right)\dot{\xi} + c_3\dot{\alpha} + c_4\xi + c_5\alpha \\
&\quad + c_6 w_1 + c_7 w_2 + c_8 w_3 + c_9 w_4 + \left(\frac{\overline{\omega}}{U^*}\right)^2 G(\xi) = f(t), \\
&d_0\ddot{\xi} + d_1\ddot{\alpha} + d_2\dot{\xi} + \left(d_3 + 2\zeta_\alpha \frac{1}{U^*}\right)\dot{\alpha} + d_4\xi + d_5\alpha + d_6 w_1 \\
&\quad + d_7 w_2 + d_8 w_3 + d_9 w_4 + \left(\frac{1}{U^*}\right)^2 M(\alpha) = g(t),
\end{aligned} \tag{1}$$

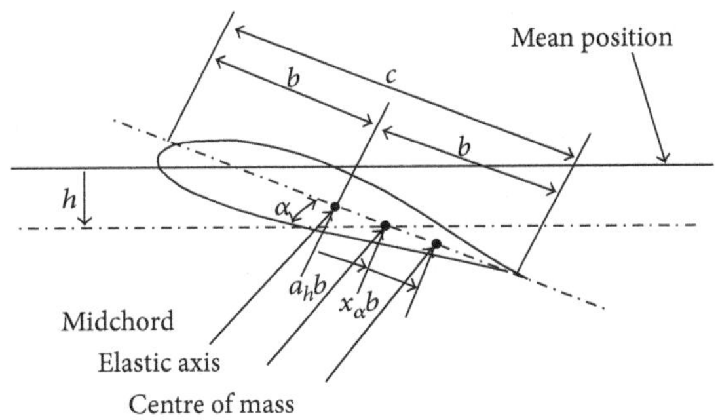

FIGURE 1: Sketch of an airfoil oscillating in pitch (α) direction with respect to the elastic axis and in plunge (h) direction measured from the mean position [14].

where $\xi = h/b$ is the nondimensional plunge displacement and the dot denotes the differentiation with respect to nondimensional time $\tau = Vt_1/b$ with t_1 as the real time and V as the flow speed. And $\overline{\omega}$ is given by $\overline{\omega} = \omega_\xi/\omega_\alpha$, with the symbols ω_ξ and ω_α as the natural frequencies of the uncoupled plunging and pitching modes, respectively. The symbol U is a nondimensional flow velocity defined as $U = V/(b\omega_\alpha)$. The damping coefficients are denoted as ζ_α and ζ_ξ corresponding with the pitch and plunge degrees of freedom, respectively; $G(\xi)$ and $M(\alpha)$ represent the nonlinear plunge and pitch stiffnesses, respectively. The other variables, w_i's, are given as

$$\begin{aligned}
w_1 &= \int_0^t e^{-\varepsilon_1(t-\sigma)}\alpha(\sigma)d\sigma, \\
w_2 &= \int_0^t e^{-\varepsilon_2(t-\sigma)}\alpha(\sigma)d\sigma, \\
w_3 &= \int_0^t e^{-\varepsilon_1(t-\sigma)}\xi(\sigma)d\sigma, \\
w_4 &= \int_0^t e^{-\varepsilon_2(t-\sigma)}\xi(\sigma)d\sigma,
\end{aligned} \tag{2}$$

with the constants $\varepsilon_1 = 0.0455$ and $\varepsilon_2 = 0.3$. The coefficients $c_0, \ldots, c_9$ and $d_0, \ldots, d_9$ are given in Liu and Dowell [17]. In this paper, we consider the hysteresis nonlinearity in pitch direction and a linear spring in plunge direction. Introducing a variable vector $X = (x_1, x_2, \ldots, x_8)^T$ with $x_1 = \alpha$, $x_2 = \dot{\alpha}$, $x_3 = \xi$, $x_4 = \dot{\xi}$, $x_5 = w_1$, $x_6 = w_2$, $x_7 = w_3$, and $x_8 = w_4$, the coupled state space system can be described by a set of eight first-order ordinary differential equations as follows:

$$\dot{X} = AX + B(X). \tag{3}$$

The expressions for A and $B(X)$ are given in the Appendix. Given any initial conditions, the state space vector X can be generated by time-stepping integration methods such as the RK method and Newmark method.

A freeplay nonlinearity usually results from backlash in loose and/or worn control surface hinges. When friction and backlash occur simultaneously, the nonlinearity usually exhibits a hysteresis feature. Actually, the considered model shown in Figure 2 is a simplified one, which was

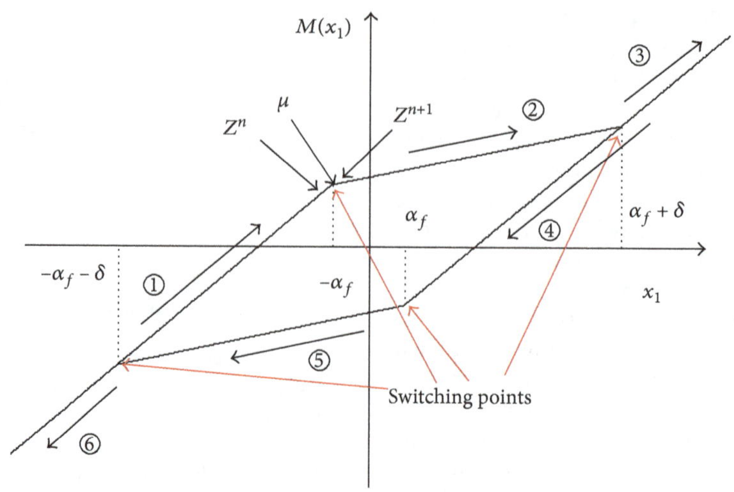

FIGURE 2: A hysteresis nonlinearity and illustration of the predictor-corrector algorithm for finding exact switching points.

widely used in aeroelastic analysis for airfoil [11, 18, 21, 26, 27]. As mentioned above, mathematically, the hysteresis can be treated as a directional composite of two freeplay nonlinearities [11]. Figure 2 is the general sketch of a hysteresis stiffness denoted by $M(x_1)$ with $x_1 = \alpha$ as

$$M(x_1) = \begin{cases} x_1 - \alpha_f + M_0, & x_1 < \alpha_f, \dot{x}_1 > 0, \\ x_1 + \alpha_f - M_0, & x_1 > -\alpha_f, \dot{x}_1 < 0, \\ M_0 + M_f(x_1 - \alpha_f), & \alpha_f \le x_1 \le \alpha_f + \delta, \dot{x}_1 > 0, \\ M_f(x_1 + \alpha_f) - M_0, & -\alpha_f - \delta \le x_1 \le -\alpha_f, \dot{x}_1 < 0, \\ x_1 + M_0 - \alpha_f - \delta(1 - M_f), & x_1 > \alpha_f + \delta, \dot{x}_1 > 0, \\ x_1 - M_0 + \alpha_f + \delta(1 - M_f), & x_1 < -\alpha_f - \delta, \dot{x}_1 < 0, \end{cases} \tag{4}$$

where the preload M_0, the freeplay δ, the beginning of the freeplay α_f, and M_f are all constants. Graphically, the hysteresis is divided into six portions shown in Figure 2. From portions (1) to (3), the direction coincides with the positive angular velocity in pitch direction, and vice versa for portions (4) to (6). The switching points, separating different portions as subsystems, depend not only on the angular displacement but also on the velocity.

3. PIM with Predictor-Corrector Algorithm

The PIM is chosen because it can provide highly precise numerical results for linear dynamical systems given by ordinary differential equations [25]. The nonsmooth system (3) can be rewritten as a series of linear subsystems.

$$\dot{X} = A_i X + B_i, \tag{5}$$

where A_i and B_i $(i = 1, \ldots, 6)$ are the respective coefficient matrices for subsystems. Note that the inhomogeneous terms B_i in (5) is different from $B(X)$ in (3). In fact, B_i is a constant vector for each linear subsystem. The function $B(X)$ in (3) is a matrix associated with the nonlinear force in pitch direction denoted as $M(x_1)$. $M(x_1)$ is divided into six portions, and in each portion, the nonlinear function is linear and only related to x_1. The linear parts related to x_1 in $B(X)$ is put to coefficient matrix A. As a result, the nonsmooth system (3) can be rewritten as a series of linear subsystems, that is, (5). A simple-yet-efficient dimension expansion method can be employed to avoid computing the reverse of the coefficient matrix. With $Y = B_i$ and $Z = [X, Y]^T$, (5) is further transformed into

$$\dot{Z} = C_i Z \tag{6}$$

in which

$$C_i = \begin{bmatrix} A_i & I \\ 0 & 0 \end{bmatrix}, \tag{7}$$

where I is an identity matrix.

For each subsystem, the analytical solution of (6) can be expressed as

$$Z = \exp(C_i \tau) Z(0), \tag{8}$$

with $Z(0)$ as the initial condition. A time step length (γ) is chosen repeatedly to generate solutions for a given time series, denoted as $[0, \gamma, 2\gamma, \ldots, N\gamma]$. Calculating the exponential matrix $T = \exp(C_i \gamma)$ according to the procedures introduced by Zhong [25], we can obtain the state vectors by the following recurrence scheme

$$Z^{n+1} = T Z^n, \tag{9}$$

where Z^n represents the solution at the nth time point, that is, $\tau = n\gamma$.

If two consecutive vibration states, Z^n and Z^{n+1}, are located in the same subsystem, Z^{n+1} can be obtained straightforwardly by implementing (9). As the recurrence computation proceeds, the vibration state sometimes inevitably switches from one subsystem to another, as illustrated in Figure 2. The latter state is in essence not a true solution unless it locates exactly at a switching point. Instead, it can only be considered as an approximation. Though the latter state can still provide better approximation as long as the time step (γ) is refined, the computation cost will dramatically increase. As we know, the most outstanding merit for PIM lies in its high accuracy and efficiency. In order to maintain this merit, a predictor-corrector method has to be proposed to accurately search the vibration response happening exactly at the switching point.

Intuitively, the switches between subsystems can be judged by the fact that the vibration displacement increases (decreases) beyond the corresponding parametric region to one subsystem. For instance, the state vector switches from portion (1) to (2), or (2) to (3), etc., as shown in Figure 2, provided that the vibration velocity is unidirectional. That means that Z^{n+1} passes one of the switching points such as $(\alpha_f, \alpha_f+\delta, -\alpha_f, -\alpha_f-\delta)$. Introduce μ to represent one of switching values $(\alpha_f, \alpha_f+\delta, -\alpha_f, -\alpha_f-\delta)$, which is located between the two consecutive states ($Z^n$and Z^{n+1}). We define a coefficient as

$$\lambda = \frac{\mu - \alpha^n}{\alpha^{n+1} - \alpha^n} \tag{10}$$

to reduce the time as $\lambda\gamma$ for Z^n reaching the solution at the switching point. Here, α^n corresponds to the counterpart in Z^n.

Different from the case with a freeplay nonlinearity [24, 26], the considered aeroelastic system with a hysteresis exhibits directional features. Implicitly, a switch may happen as the direction changes, even though the displacement does not pass any one of the abovementioned switching points. This kind of switch depends on the fact that the angular velocity passes 0 rather than the displacement, for example, from portion (2) to (4) and vice versa. In this case, we define

$$\lambda = \frac{-(\alpha)^n}{\left[(\alpha)^{n+1} - (\alpha)^n\right]}, \tag{11}$$

where $(\alpha)^n$ denotes the angular velocity at Z^n. According to (11), the coefficient converges to 1 as $(\alpha)^{n+1}$ approaches 0, which denotes implicitly a switch point.

Once an explicit or implicit switching is revealed, an approximation of the solution at the switching point is given as

$$Z^{n+1} = T^{\lambda} Z^n. \tag{12}$$

In Figure 3, we can see sometimes that the vibration state switches from one subsystem to another. Once the vibration state switches from one subsystem to another, we defined a ratio as $\lambda = (\mu - \alpha^n)/(\alpha^{n+1} - \alpha^n)$ being located between 0 and 1. The vibration state (α^{n+1}) will be the real one appearing at the switching point as long as $\lambda = 1$. This ratio is in fact the proportional mean of the switching value (μ) between the vibration states (α^n, α^{n+1}). With this coefficient, α^{n+1} (as the first entry of Z^{n+1}) will be updated because the solution Z^{n+1} is corrected as $Z^{n+1} = T^{\lambda}Z^n$. Generally, $Z^{n+1} = T^{\lambda}Z^n$ will approach the real vibration state at the switching point more

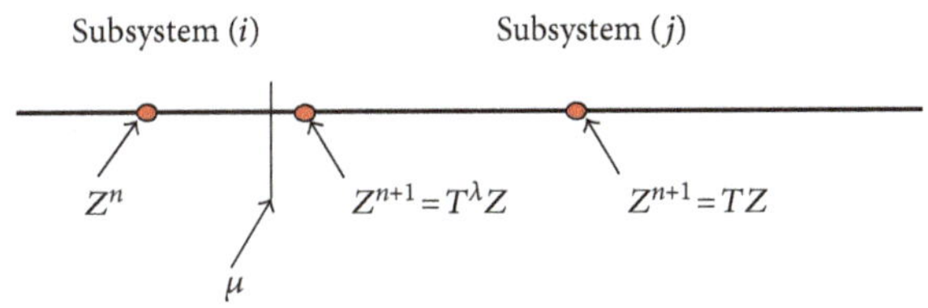

Figure 3: Sketch of the predictor-corrector algorithm.

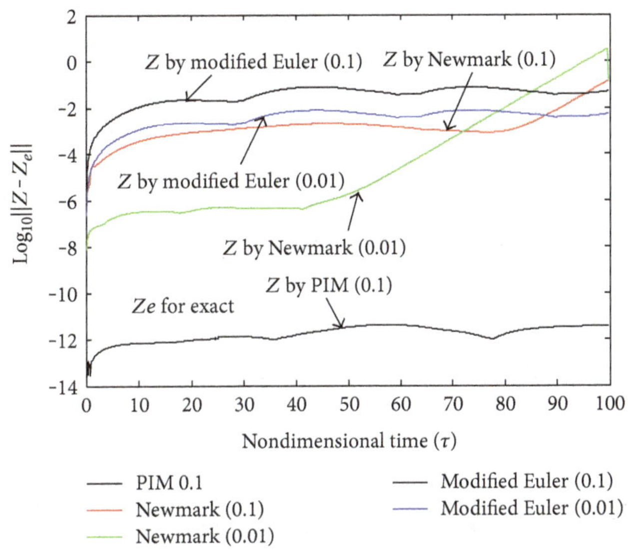

Figure 4: Logarithmic errors of modified Euler, Newmark, or PIM results with respect to the exact solutions of the aeroelastic system at $U/U_0 = 0.9$ in pitch direction.

closely than $Z^{n+1} = TZ^n$. Therefore, the updated solution $Z^{n+1} = T^{\lambda}Z^n$ presents a better approximation for the real vibration state at the switching point. The approximation precision can be improved by further updating the ratio λ according to the updated solution Z^{n+1}, and vice visa. The predictor-corrector algorithm is performed by repeating (10) (or (11)) and (12) until λ approaches 1 close enough. In the following calculation, the tolerance error between λ and 1 is set to be 10^{-10}.

4. Numerical Examples

The system parameters are chosen as $\mu = 100$, $a_h = -0.5$, $x_\alpha = 0.25$, $r_\alpha = 0.5$, $\zeta_\xi = \zeta_\alpha = 0$, $\overline{\omega} = 0.2$, $M_0 = M_f = 0.5°$, $\delta = 0.1°$, and $\alpha_f = M_0 - (\delta/2)(1 - M_f)$.

The results provided by the presented method will be compared with those by the modified Euler method, Newmark method, widely-used fourth-order RK method, and the exact results, respectively. Note that the exact solutions are obtained by solving the subsystems one by one using the theories for ordinary differential equations. The procedures are tedious due to the switches of the vibration state, because the solution at the switching points has to be determined by trial-and-error calculations.

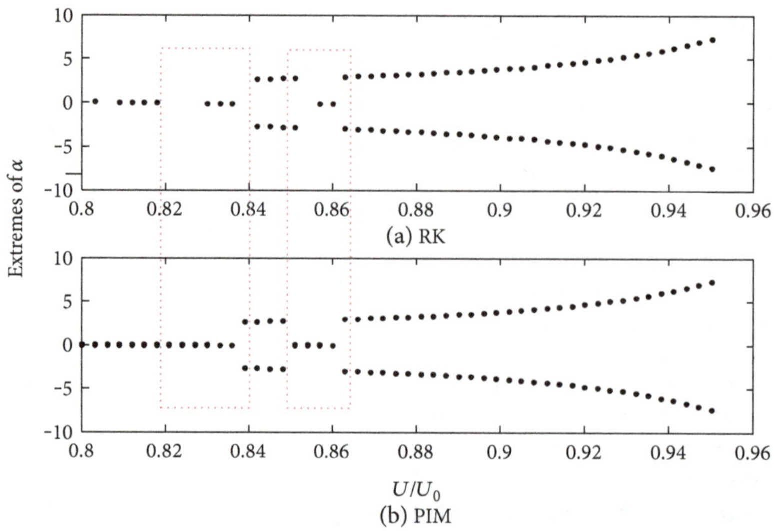

FIGURE 5: Bifurcation charts of the aeroelastic system obtained by RK (a) and PIM (b) with $\Delta t = 0.1$, respectively. Major differences are boxed by red dashed lines.

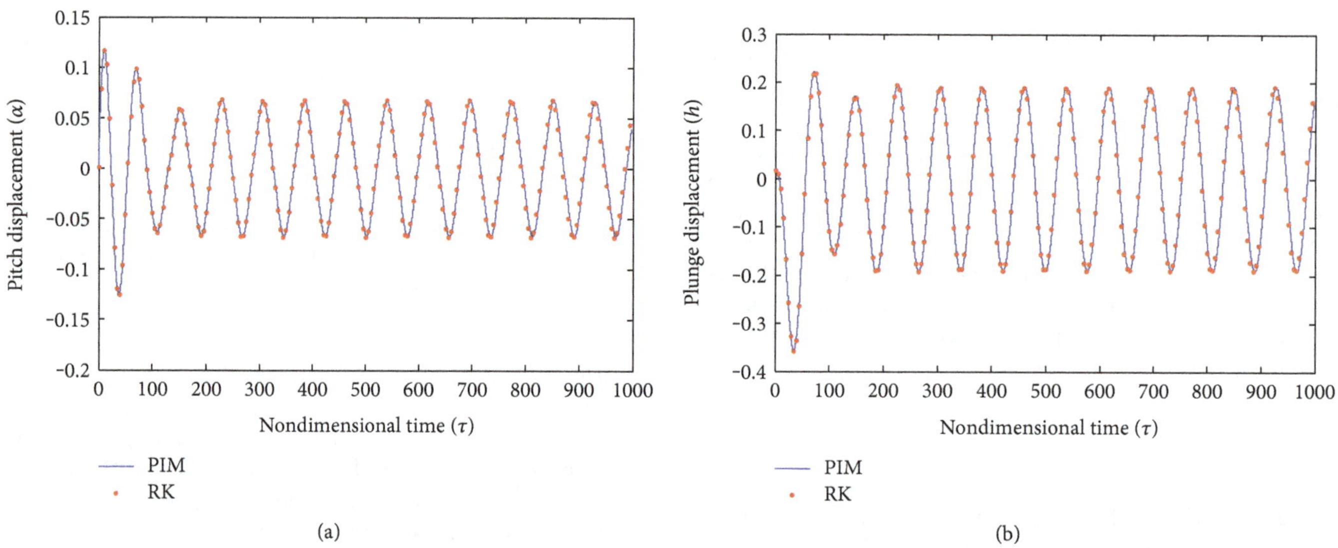

FIGURE 6: Time histories of the aeroelastic system at $U/U_0 = 0.9$ for (a) pitch and (b) plunge directions, respectively.

The critical flutter speed is obtained as $U_0 = 6.2851$ by linear flutter analysis method. When the speed is beyond the critical flutter speed, the system will lose its stability and all dynamic responses will increase infinitely. The initial condition is chosen as $Z(0) = [0.1, 1, 1, 0, 0, 0, 0, 0]^T$ as an illustrative example in the following discussions.

In Figure 4, we compare the numerical results provided by the PIM, modified Euler method, and Newmark method with the exact ones. We can see that when the time step chosen is much smaller, the modified Euler method improves the accuracy about one order of magnitude, while the Newmark method improves about three orders of magnitude at the beginning and produces a big error in the back. We find that the algorithm PIM with a predictor-corrector has high computational accuracy. The RK method that is widely used has better accuracy and stability than the modified Euler method and Newmark method. So in the following examples, we use the RK method compared with the proposed approach to illustrate the effectiveness.

The bifurcation charts shown in Figure 5 are obtained by the PIM and RK results, respectively, with the same time step $\Delta t = 0.1$ and $0.8 < U/U_0 < 0.95$. As a whole, when $U/U_0 < 0.84$ and $0.85 < U/U_0 < 0.86$, the pitch motions tends to the zero equilibrium even as the flow speed increases. Sudden jumping for the pitch motion is observed at $U/U_0 = 0.84$ or 0.862. The amplitudes increase as the flow speed increases beyond $U/U_0 > 0.86$. Interestingly, there are two remarkable discrepancies between the two results as $0.82 < U/U_0 < 0.84$ and $0.85 < U/U_0 < 0.86$ or so, respectively, boxed by red dashed lines in Figure 5. When $U/U_0 > 0.86$, both methods provide

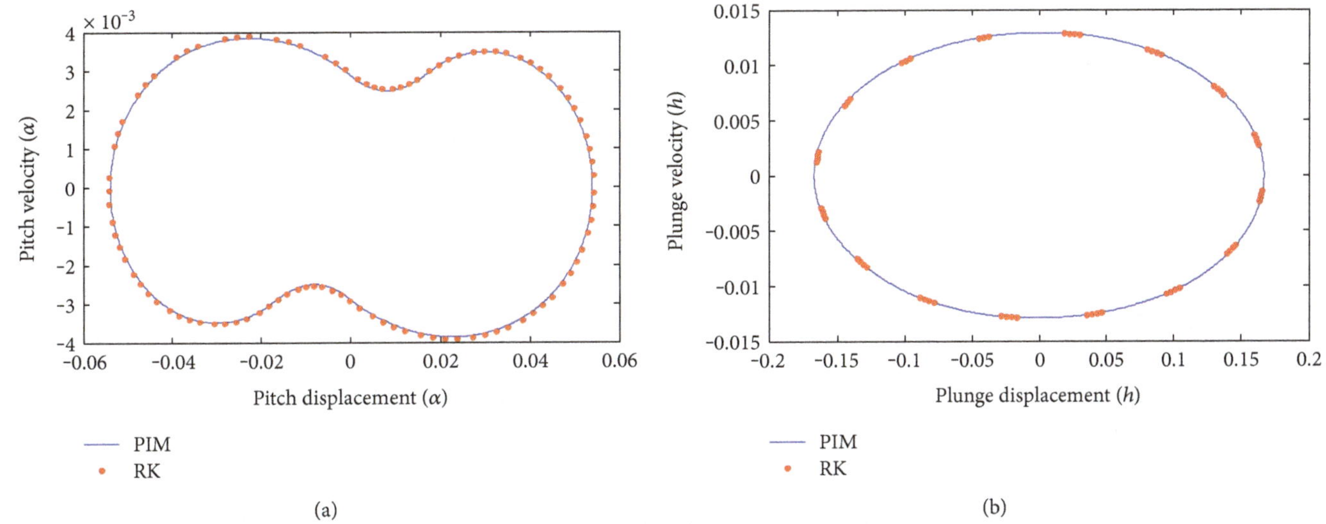

FIGURE 7: Phase planes of LCOs for (a) pitch and (b) plunge directions, respectively, at $U/U_0 = 0.87$.

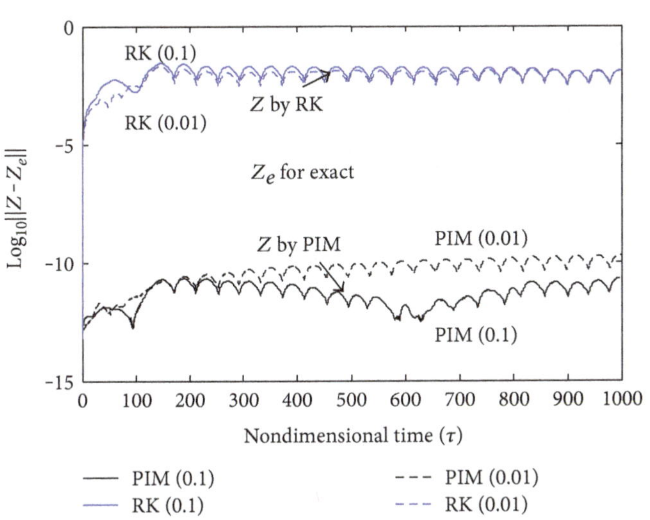

FIGURE 8: Logarithmic errors of RK or PIM result with respect to the exact solutions of the aeroelastic system at $U/U_0 = 0.87$ in pitch direction.

nearly the same solutions. To further investigate the differences, more refined simulations will be presented in the following discussion as the flow speed near the jump positions.

Figure 6 shows the time histories in pitch and plunge with $U/U_0 = 0.9$, respectively. Figure 7 shows phase planes for LCOs with $U/U_0 = 0.87$. Both the time histories and phase planes indicate that the PIM results are in nice agreement with the RK solutions. Note that the time steps are both chosen as 0.1.

In order to check the accuracy of PIM and RK method, we compare the numerical results provided by the PIM and RK with the exact ones. Figure 8 shows the logarithmic error of PIM as well as RK results compared with the exact ones. The PIM solution is much more accurate than the RK result by about eight orders of magnitude with the same time step as 0.01 (or 0.1). Importantly, the high precision can be maintained as the step length is chosen much longer for PIM. When RK is used, the accuracy improves at the earlier stage of simulation, but it remains at the same order in long-term duration.

More importantly, it should be pointed out the RK solutions are sometimes not reliable. As marked by red dashed lines as $U/U_0 < 0.86$, in some cases the PIM provides LCOs, while the RK solutions converges to equilibriums. When $U/U_0 < 0.84$, both the PIM and RK results imply that the system will not give rise to LCOs. As $U/U_0 = 0.8$, Figure 9 shows that the solutions in pitch direction obtained by PIM are well consistent with those by the RK method. However, observable discrepancies appear in the plunge direction. Figure 10 shows the time histories in pitch at $U/U_0 = 0.85$ obtained by the PIM, RK, and exact, respectively. We can see that the PIM solutions are in excellent agreement with the exact ones, whereas the discrepancy for the RK result grows to an observable magnitude during a relative long-term duration. Compared with PIM, possibly, the error for RK method is caused by its relatively lower accuracy and the approximate determining of the switching points.

Last but not least, the PIM can maintain a high precision as long as the time step is chosen at a reasonable range [25], whereas the RK method could possibly provide fallacious results even though the time step is refined significantly. Figure 11 shows the time histories presented by PIM, RK, and the exact solutions at $U/U_0 = 0.1$, respectively. The solutions obtained by PIM with the time step 0.1 agrees well with the exact ones. The RK solutions are very different from the PIM and exact solutions, even as the step is refined from 0.1 to 0.01 and 0.001.

5. Conclusion and Remarks

We have employed the PIM to propose an accurate algorithm for solving dynamical systems with hysteresis nonlinearities. The procedures are illustrated by the aeroelastic system of a two-dimensional airfoil. The solutions obtained by PIM are in excellent agreement with those by exact methods. Importantly, the accuracy can remain to be a high level as long as

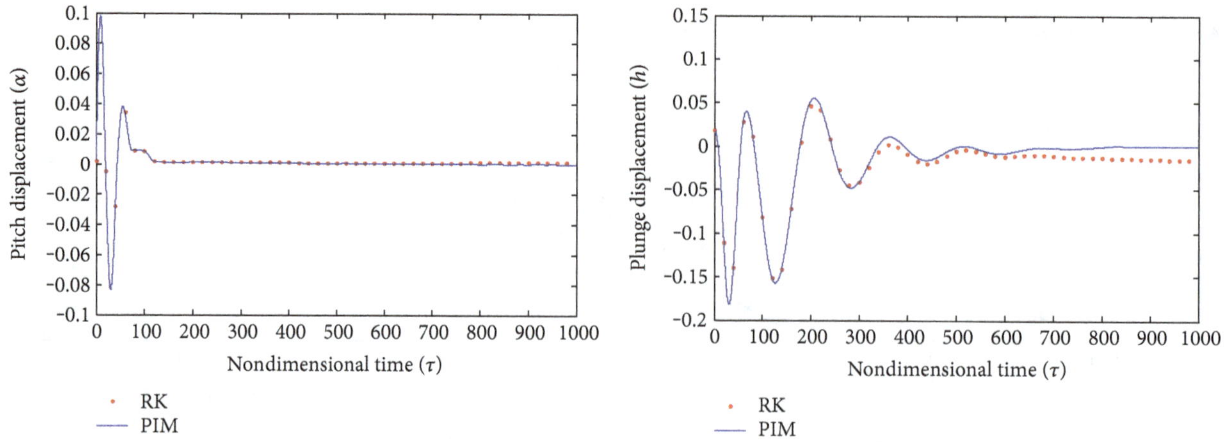

(a) (b)

FIGURE 9: Time histories of the aeroelastic system at $U/U_0 = 0.8$ (a) for pitch and (b) for plunge direction, respectively.

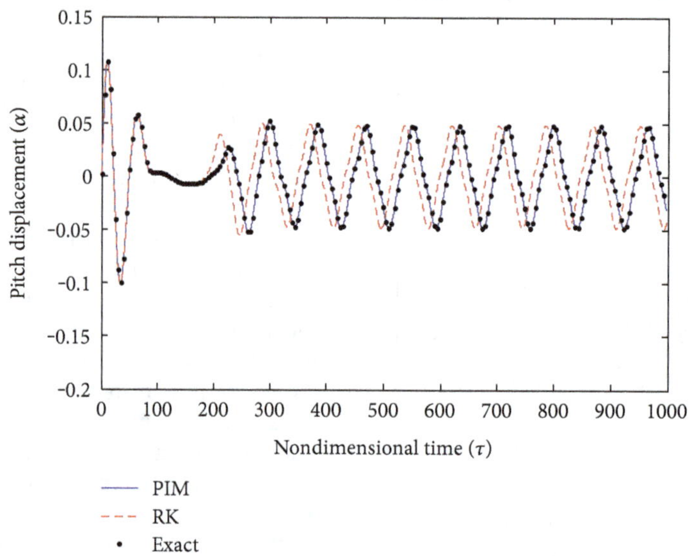

FIGURE 10: Time histories of the aeroelastic system at $U/U_0 = 0.85$ in pitch direction provided by PIM, RK, and exact solutions, respectively.

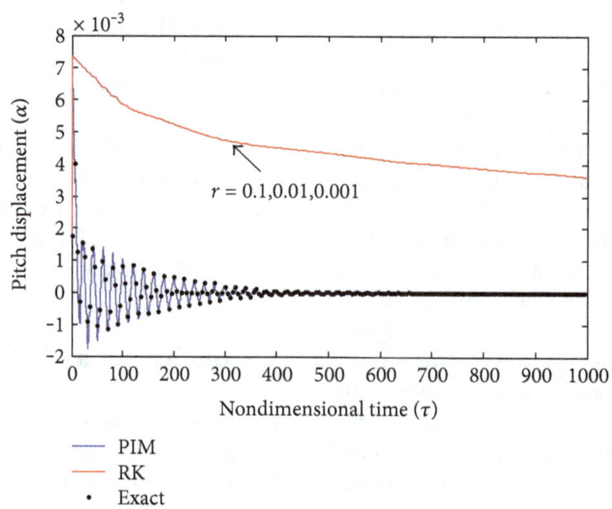

FIGURE 11: Time histories of the aeroelastic system at $U/U_0 = 0.1$ obtained by the exact solutions, PIM with the time step as 0.1, and RK with the time step as 0.1, 0.01, or 0.001, respectively.

the step length locates at a reasonable range. In principle, the accuracy can reach any desired magnitude.

Special attention has been paid on the RK method, as it has been implemented widely even considered as a benchmark for comparison. It is found that, however, the RK may sometimes provide erroneous results even if the step is very refined. With such high accuracy, the presented method could be applicable to more aeroelastic systems with nonsmooth nonlinearities, especially those with hysteresis. Moreover, it has the potential to become a benchmark for comparison in verifying some other numerical solution techniques for such systems.

It has been reported a lot on the complexity in numerical simulation of the control problems raising in nonlinear dynamical systems, especially when nonsmooth nonlinearities are included [29, 30]. For instance, Li et al. [31] employed Hénon's method to locate the switching points in the numerical integration procedures, for closed-loop controlled systems of airfoil with a freeplay. In the near future, we will try to investigate the feasibility and effectiveness of using the presented method in such problems.

Appendix

$$A=\begin{bmatrix} 0 & 1 & 0 & 0 & 0 & 0 & 0 & 0 \\ a_{21} & a_{22}+2jc_0\zeta_\alpha\frac{1}{U^*} & a_{23}-jd_0\left(\frac{\overline{\omega}}{U^*}\right)^2 & a_{24}-2jd_0\zeta_\xi\frac{\overline{\omega}}{U^*} & a_{25} & a_{26} & a_{27} & a_{28} \\ 0 & 0 & 0 & 1 & 0 & 0 & 0 & 0 \\ a_{41} & a_{42}-2jc_1\zeta_\alpha\frac{1}{U^*} & a_{43}+jd_1\left(\frac{\overline{\omega}}{U^*}\right)^2 & a_{44}+2jd_1\zeta_\xi\frac{\overline{\omega}}{U^*} & a_{45} & a_{46} & a_{47} & a_{48} \\ 1 & 0 & 0 & 0 & -\varepsilon_1 & 0 & 0 & 0 \\ 1 & 0 & 0 & 0 & 0 & -\varepsilon_2 & 0 & 0 \\ 0 & 0 & 1 & 0 & 0 & 0 & -\varepsilon_1 & 0 \\ 0 & 0 & 1 & 0 & 0 & 0 & 0 & -\varepsilon_2 \end{bmatrix}, \tag{A.1}$$

$$B=\left[0 \;\; -jc_0\left(\frac{1}{U^*}\right)^2 M(x_1) \;\; 0 \;\; jc_1\left(\frac{1}{U^*}\right)^2 M(x_1) \;\; 0 \;\; 0 \;\; 0 \;\; 0\right]^T.$$

The parameters $j, a_{21}, \ldots, a_{28}, a_{41}, \ldots, a_{48}$ are given in Liu and Dowell [17].

Conflicts of Interest

The authors declare that they have no conflicts of interest.

Acknowledgments

This work was supported by the National Natural Science Foundation of China (11572356, 11672337).

References

[1] S. J. Liao and P. F. Wang, "On the mathematically reliable long-term simulation of chaotic solutions of Lorenz equation in the interval [0, 10000]," *Science China Physics, Mechanics and Astronomy*, vol. 57, no. 2, pp. 330–335, 2014.

[2] I. Sgura, "A finite difference approach for the numerical solution of non-smooth problems for boundary value ODEs," *Mathematics and Computers in Simulation*, vol. 95, pp. 146–162, 2014.

[3] F. Angulo, G. Olivar, G. A. Osorio, C. M. Escobar, J. D. Ferreira, and J. M. Redondo, "Bifurcations of non-smooth systems," *Communications in Nonlinear Science and Numerical Simulation*, vol. 17, no. 12, pp. 4683–4689, 2012.

[4] E. H. Dowell, J. W. Edwards, and T. Strganac, "Nonlinear aeroelasticity," *Journal of Aircraft*, vol. 40, no. 5, pp. 857–874, 2003.

[5] Y. H. Tan and L. Deng, "Modeling the dynamic sandwich system with hysteresis using ARMAX model," *Mathematics and Computers in Simulation*, vol. 97, pp. 162–188, 2014.

[6] X. Q. Fan and Y. Tao, "Investigation of flow control for the hypersonic inlets via counter flow," *International Journal of Aerospace Engineering*, vol. 2015, Article ID 956317, 8 pages, 2015.

[7] E. H. Dowell, J. P. Thomas, and K. C. Hall, "Nonlinear inviscid aerodynamic effects on transonic divergence, flutter, and limit-cycle oscillations," *American Institute of Aeronautics and Astronautics Journal*, vol. 40, pp. 638–646, 2012.

[8] E. H. Dowell and K. C. Hall, "Modeling of fluid-structure interaction," *Annual Review of Fluid Mechanics*, vol. 33, pp. 445–490, 2001.

[9] E. H. Dowell and D. Tang, "Nonlinear aeroelasticity and unsteady aerodynamics," *American Institute of Aeronautics and Astronautics Journal*, vol. 40, pp. 1697–1707, 2002.

[10] X. Z. Xu, Y. K. Gao, and W. G. Zhang, "Iterative learning control of a nonlinear aeroelastic system despite gust load," *International Journal of Aerospace Engineering*, vol. 2015, Article ID 237804, 13 pages, 2015.

[11] B. H. K. Lee, S. J. Price, and Y. S. Wong, "Nonlinear aeroelastic analysis of airfoils: bifurcation and chaos," *Progress in Aerospace Science*, vol. 35, pp. 205–334, 1999.

[12] K. Chung, Y. He, and B. Lee, "Bifurcation analysis of a two-degree-of-freedom aeroelastic system with hysteresis structural nonlinearity by a perturbation-incremental method," *Journal of Sound and Vibration*, vol. 320, no. 1-2, pp. 163–183, 2009.

[13] L. Liu, E. Dowell, and J. Thomas, "A high dimensional harmonic balance approach for an aeroelastic airfoil with cubic restoring forces," *Journal of Fluids and Structures*, vol. 23, no. 3, pp. 351–363, 2007.

[14] C. Popescu, Y. Wong, and B. Lee, "An expert system for predicting nonlinear aeroelastic behavior of an airfoil," *Journal of Sound and Vibration*, vol. 319, no. 3–5, pp. 1312–1329, 2009.

[15] D. Li, S. Guo, and J. Xiang, "Study of the conditions that cause chaotic motion in a two-dimensional airfoil with structural nonlinearities in subsonic flow," *Journal of Fluids and Structures*, vol. 33, pp. 109–126, 2012.

[16] C. C. Cui, J. K. Liu, and Y. M. Chen, "Quantification of limit cycle oscillations in nonlinear aeroelastic systems with stochastic parameters," *International Journal of Aerospace Engineering*, vol. 2016, Article ID 3049163, 9 pages, 2016.

[17] L. Liu and E. H. Dowell, "The secondary bifurcation of an aeroelastic airfoil motion: effect of high harmonics," *Nonlinear Dynamics*, vol. 37, no. 1, pp. 31–49, 2004.

[18] F. X. Chen, Y. M. Chen, and J. K. Liu, "Calculation of the bounds on limit cycle oscillations for nonlinear aeroelastic systems based on equivalent linearization," *Science China Technological Sciences*, vol. 57, pp. 1249–1256, 2014.

[19] R. Xiong, X. D. Liu, and J. Geng, "IHB method for response computation of a class of second-order hysteretic nonlinear control systems under a sinusoidal excitation," *Journal of Vibration and Shock*, vol. 32, pp. 99–105, 2013.

[20] Y. M. Chen and J. K. Liu, "Homotopy analysis method for limit cycle oscillations of an airfoil with cubic nonlinearities," *Journal of Vibration and Control*, vol. 16, pp. 163–179, 2010.

[21] L. Liu and E. H. Dowell, "Harmonic balance approach for an airfoil with a freeplay control surface," *American Institute of Aeronautics and Astronautics Journal*, vol. 43, pp. 802–815, 2005.

[22] J. K. Liu, F. X. Chen, and Y. M. Chen, "Bifurcation analysis of aeroealstic systems with hysteresis by incremental harmonic balance method," *Applied Mathematics and Computation*, vol. 219, no. 5, pp. 2398–2411, 2012.

[23] M. D. Conner, D. M. Tang, E. H. Dowell, and L. N. Virgin, "Nonlinear behavior of a typical airfoil section with control surface freeplay: a numerical and experimental study," *Journal of Fluids and Structures*, vol. 11, no. 1, pp. 89–109, 1997.

[24] C. C. Cui, J. K. Liu, and Y. M. Chen, "Simulating nonlinear aeroelastic responses of an airfoil with freeplay based on precise integration method," *Communications in Nonlinear Science and Numerical Simulation*, vol. 22, no. 1–3, pp. 933–942, 2015.

[25] W. X. Zhong, "On precise integration method," *Journal of Computational and Applied Mathematics*, vol. 163, no. 1, pp. 59–78, 2004.

[26] Y. M. Chen and J. K. Liu, "Nonlinear aeroelastic analysis of an airfoil-store system with a freeplay by precise integration method," *Journal of Fluids and Structures*, vol. 46, pp. 149–164, 2014.

[27] S. Mittal and P. Saxena, "Prediction of hysteresis associated with the static stall of an airfoil," *American Institute of Aeronautics and Astronautics Journal*, vol. 38, pp. 933–935, 2000.

[28] L. Liu, Y. S. Wong, and B. H. K. Lee, "Non-linear aeroelastic analysis using the point transformation method, part 2: hysteresis model," *Journal of Sound and Vibration*, vol. 253, no. 2, pp. 471–483, 2002.

[29] L. Fortuna and G. Muscato, "A roll stabilization system for a monohull ship: modeling, identification, and adaptive control," *IEEE Transactions on Control Systems Technology*, vol. 4, no. 1, pp. 18–28, 1996.

[30] P. Yu, Z. Chen, L. Librescu, and P. Marzocca, "Implications of time-delayed feedback control on limit cycle oscillation of a two-dimensional supersonic lifting surface," *Journal of Sound and Vibration*, vol. 304, no. 3–5, pp. 974–986, 2007.

[31] D. C. Li, S. J. Guo, and J. W. Xiang, "Aeroelastic dynamic response and control of an airfoil section with control surface nonlinearities," *Journal of Sound and Vibration*, vol. 329, no. 22, pp. 4756–4771, 2010.

7

Adaptive Fuzzy Sliding Mode Guidance Law considering Available Acceleration and Autopilot Dynamics

Yulin Wang[1] **Shengjing Tang**[1] **Wei Shang,**[2] **and Jie Guo**[1]

[1]*School of Aerospace Engineering, Beijing Institute of Technology, Beijing 100081, China*
[2]*System Design Institute of Hubei Aerospace Technology Academy, Hubei, 430040, China*

Correspondence should be addressed to Jie Guo; guojie1981@bit.edu.cn

Academic Editor: Vaios Lappas

Terminal guidance law for missiles intercepting high maneuvering targets considering the limited available acceleration and autopilot dynamics of interceptor is investigated. Conventional guidance laws based on adaptive sliding mode control theory were designed to intercept a maneuvering target. However, they demand a large acceleration for interceptor at the end of the terminal guidance, which may have acceleration saturation especially when the target acceleration is close to the available acceleration of interceptor. In this paper, a terminal guidance law considering the available acceleration and autopilot dynamics of interceptor is proposed. Then, a fuzzy system is utilized to approximate and replace the variable structure term, which can handle the unknown target acceleration. And an adaptive neural network system is adopted to compensate the effects caused by the designed overlarge acceleration of interceptor such that the interceptor with small available acceleration can intercept the high maneuvering target. Simulation results show that the guidance law with available acceleration and autopilot dynamics (AAADG) is highly effective for reducing the acceleration command and achieving a small final miss distance.

1. Introduction

In recent years, the hypersonic flight vehicles with high maneuverabilities have been widely developed [1]. It is necessary to design an appropriate guidance law to intercept the hypersonic flight vehicles for its high strike ability. However, the widely used PNG law [2, 3] and its variants [4–8] would fail to satisfy the requirement of high accuracy for the task of intercepting the hypersonic flight vehicles with high maneuverabilities [9]. An effective approach to intercept maneuverable targets is applying robust control methods, such as sliding mode control (SMC) [10], adaptive control [11], nonlinear $H\infty$ control [12], and optimal control [13].

The guidance law based on SMC has become an active research area because of its strong robustness and easy implementation [10, 14, 15]. In the guidance law design, the zero-effort miss distance is usually used to define sliding mode surface. For example, Shima et al. [16] proposed a sliding mode guidance law for integrated missile autopilot with the zero-effort miss distance as sliding mode surface. The sliding mode guidance laws are also designed by guiding the line-of-sight (LOS) angular rate converge to zero. Then, Zhang et al. [17] developed a novel guidance law based on integral sliding mode control and employed a nonlinear disturbance observer to estimate the target acceleration. By estimating the target acceleration via an extended state observer, Zhu et al.[18] designed a guidance law without requiring the information on the target acceleration. However, a compromise should be made between the tracking accuracy and control input since the high tracking accuracy will lead to a very large control input. Thus, the proper values for the control parameters, such as the amplitude of variable structure term, are hard to choose. Considering finite-time convergence, Zhou et al. [19] proposed a variable structure finite-time-convergent guidance law and proved that the LOS angular rate would converge to zero before the ending of the terminal guidance. To attack the target with a predefined impact angle, the guidance laws with impact angle constraints were proposed in [20–23]. Lee et al. [24] developed a guidance law using the high-performance sliding mode control theory to intercept a stationary or slowly moving target. Kumar et al. [20] presented a guidance law to intercept the targets from any initial heading angle without exhibiting any singularity. Furthermore, some guidance laws take autopilot dynamics into consideration [25, 26]. Sun et al. [27] studied a second order sliding mode guidance law with autopilot dynamics. Li et al. [28] took both the autopilot dynamics and uncertainties into consideration and proposed a finite-time-convergent guidance law to intercept a maneuvering target.

To the best of our knowledge, the existing sliding mode guidance laws may require a greater guidance command than the available acceleration of interceptor when the maximum acceleration of target is equal to or larger than the available acceleration of interceptor. To solve this problem, an acceleration limiter is adopted to reduce the saturated acceleration demand [29, 30]. Xiong et al. [31] proposed a strategy making the coefficients of the sliding mode guidance law vary according to the fuzzy rule to reduce the acceleration demand at the beginning of the terminal guidance. However, it cannot guarantee the stability of the guidance system when the designed acceleration is significantly reduced at the end of the homing guidance. In this paper, a neural network system is employed to compensate the effects of the designed overlarge acceleration and the stability of the guidance system is proved. However, to guarantee a good performance of the guidance system when the missile intercepts the hypersonic flight vehicle with a high maneuverability, the gain of switching term needs to be chosen larger than the upper bound of target acceleration, which is hard to choose in advance and will cause the chattering problem. Then, with the universal approximation property of fuzzy systems, a continuous adaptive fuzzy system is used to approximate and replace the variable structure term, which can handle the unknown target maneuver and avoid the chattering phenomenon. The input information of the adaptive fuzzy system is the relative distance and the LOS angular rate which can be measured by the interceptor seeker easily. Besides, before designing a guidance law, it is necessary to consider the autopilot dynamics because the autopilot lag of a homing missile will decrease the precision of guidance. Then, a sliding mode guidance law considering the available acceleration and autopilot dynamics of missile for intercepting a high maneuvering target is proposed.

The main contribution of this paper is that, by adopting neural network system to reduce the acceleration command, so that the interceptor is able to intercept the hypersonic flight vehicles whose maximum acceleration is equal to or larger than the available acceleration of interceptor. In addition, the information needed to estimate the upper bound of target acceleration is limited and easy to obtain, so the proposed guidance law is easy to be implemented in practice. The rest of this paper is organized as follows. Section 2 describes the formulations of the guidance system considering the available acceleration and autopilot dynamics of interceptor. In Section 3, the control methods for the proposed guidance law are introduced. In Section 4, the stability of the guidance law is verified. At last, compared with proportional navigation guidance (PNG) and finite-time convergence guidance (FTCG), simulation results demonstrate the effectiveness of AAADG.

2. The Dynamics of the Relative Motion between the Interceptor and Target

This section presents the mathematic model of the guidance system for intercepting. The planar relative motion of the interceptor and target is shown in Figure 1. The interceptor and the target are denoted by the subscripts M and T, respectively. In order to simplify the guidance law design, the interceptor and target are assumed to have two mass points. The corresponding equations are given by as follows:

$$\dot{r} = -v_{\mathrm{M}}\cos\sigma_{\mathrm{M}} + v_{\mathrm{T}}\cos\sigma_{\mathrm{T}}, \tag{1}$$

$$r\dot{q} = v_{\mathrm{M}}\sin\sigma_{\mathrm{M}} - v_{\mathrm{T}}\sin\sigma_{\mathrm{T}}, \tag{2}$$

$$\sigma_{\mathrm{M}} = q - \theta_{\mathrm{M}}, \tag{3}$$

$$\sigma_{\mathrm{T}} = q - \theta_{\mathrm{T}}, \tag{4}$$

$$\dot{\theta}_{\mathrm{M}} = \frac{a_{\mathrm{M}}}{v_{\mathrm{M}}}, \tag{5}$$

$$\dot{\theta}_{\mathrm{T}} = \frac{a_{\mathrm{T}}}{v_{\mathrm{T}}}, \tag{6}$$

where r is the relative distance and $\dot{r}$ is the relative velocity between the interceptor and the target. v_{M} is the velocity of the interceptor. a_{M} is the normal acceleration of the interceptor. θ_{M} is the flight-path angle of the interceptor. v_{T} is the target velocity. a_{T} is the normal acceleration of the target. θ_{T} is the flight-path angle of the target. q and $\dot{q}$ are the light-of-sight (LOS) angle and LOS angular rate between the interceptor and target, respectively.

Assumption 1: [31]. In the terminal guidance, the magnitudes of the speeds of the interceptor and target are assumed as two constants such that $\dot{v}_{\mathrm{M}} = 0$ and $\dot{v}_{\mathrm{T}} = 0$.

In practice, because of the physical limitation of the interceptor actuator, the interceptor cannot achieve the required acceleration when the target acceleration is equal to or larger than the available acceleration of the interceptor. It may lead to a large final miss distance. Therefore, the required acceleration of the interceptor should be limited to a certain level. From (5), it can be seen that the acceleration of interceptor varies with the change of $\dot{\theta}_{\mathrm{M}}$. So, we can limit $\dot{\theta}_{\mathrm{M}}$ as follows:

$$\bar{\dot{\theta}}_{\mathrm{M}} = \mathrm{sat}\left(\dot{\theta}_{\mathrm{M}}\right) = \begin{cases} -\dot{\theta}_{\mathrm{M\,max}} & \dot{\theta}_{\mathrm{M}} > -\dot{\theta}_{\mathrm{M\,max}} \\ \dot{\theta}_{\mathrm{M}} & -\dot{\theta}_{\mathrm{M\,max}} \le \dot{\theta}_{\mathrm{M}} \le \dot{\theta}_{\mathrm{M\,max}} \\ \dot{\theta}_{\mathrm{M\,max}} & \dot{\theta}_{\mathrm{M}} > \dot{\theta}_{\mathrm{M\,max}} \end{cases}, \tag{7}$$

$$\bar{\dot{\theta}}_{\mathrm{M}} - \dot{\theta}_{\mathrm{M}} = \delta, \tag{8}$$

where $\dot{\theta}_{\mathrm{M\,max}} = \mathrm{const} > 0$ is the maximum flight-path angular rate of the interceptor, which is influenced by the flight environment of the interceptor, and $\bar{\dot{\theta}}_{\mathrm{M}}$ is the actual flight-path angular rate. δ is a nonlinear function caused by saturation.

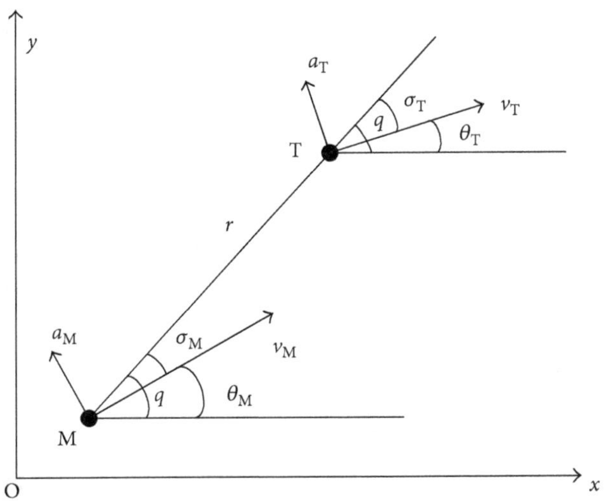

FIGURE 1: Relative motion between the interceptor and target.

The normal acceleration command generated by the guidance law is controlled by the autopilot of the interceptor, which tracks the expected acceleration command by modulating fins or thrusters. However, there is a lag between the expected acceleration and the achieved acceleration. Therefore, the performance of the interceptor is expected to be improved by using a guidance law considering the autopilot dynamics. The autopilot dynamics can be well modeled and approximated as the following first-order system:

$$v_M \ddot{\theta}_M = -\frac{1}{T_1} v_M \dot{\theta}_M + \frac{1}{T_1} v_M \dot{\theta}_C, \tag{9}$$

where $\dot{\theta}_C$ is the expected flight-path angular rate command of the interceptor. $\dot{\theta}_M$ is the achieved flight-path angular rate. T_1 is the time constant of the autopilot, which is easy to obtain in advance.

The designing purpose of the guidance law is adjusting a_M so that the LOS angular rate will converge to zero. Based on this idea, we define three state variables as follows:

$$\begin{aligned} x_1 &= q, \\ x_2 &= \dot{q}, \\ x_3 &= \dot{\theta}_M. \end{aligned} \tag{10}$$

Then, we obtain the integrated dynamics equations for the guidance system with the autopilot dynamics of the interceptor as follows:

$$\begin{aligned} \dot{x}_1 &= x_2, \\ \dot{x}_2 &= -\frac{2\dot{r}}{r} x_2 - \frac{v_M \cos \sigma_M}{r} (x_3 + \delta) + \frac{1}{r} \omega, \\ \dot{x}_3 &= -\frac{1}{T_1} x_3 + \frac{1}{T_1} u, \end{aligned} \tag{11}$$

where $\omega = v_T \cos \sigma_T \dot{\theta}_T$ is the target motion information. u is the actual control input.

3. Guidance Law Design considering Available Acceleration and Autopilot Dynamics of the Interceptor

In this section, the guidance law considering the available acceleration and autopilot dynamics of the interceptor is designed. It can be seen that the guidance system described by (11) is in a strict feedback form. However, the guidance law designed using backstepping method contains the high-order time derivatives of LOS angle and interceptor-to-target distance. To avoid this problem, the dynamic surface control method is proposed by introducing a first-order low-pass filter. The first-order low-pass filter allows the design where the model is not differentiable. For the guidance system described by (11), the goal of the guidance law is to make the LOS angular rate converge to zero. The design of the guidance law is given as follows:

$$\begin{aligned} s_1 &= x_2, \\ x_{3d} &= \frac{(k_1 + 2)|\dot{r}| x_2 + \varepsilon \operatorname{sgn} s_1 - \dot{v}_M \sin \sigma_M}{v_M \cos \sigma_M} - \hat{\delta}, \\ \frac{x_{3c}}{x_{3d}} &= \frac{1}{T_2 s + 1}, x_3(0) = x_{3c}(0), \\ s_2 &= x_3 - x_{3c}, \\ u &= x_3 + \frac{T_1}{T_2}(x_{3d} - x_{3c}) - k_2 T_1 s_2, \end{aligned} \tag{12}$$

where s_1 and s_2 are designed as the dynamic surfaces. x_{3d} is the virtual control obtained from the first step. k_1 and k_2 are the dynamic surface gains. x_{3c} is the command input of the second step, which is obtained by x_{3d} passing through the low-pass filter. T_2 is the time constant of the filter. ε sgn s_1 is the variable structure term to enhance the robustness of the system against the unknown target movement, where the switching term amplitude ε satisfies $\varepsilon > |\omega|$. $\hat{\delta}$ is the adaptive acceleration compensator, which will be designed later.

Remark 1. The choice of the switching term amplitude ε is connected to the maximum acceleration of the target. Because of the uncertain maneuver of the target, the bound of target acceleration cannot be estimated in advance. It is difficult to determine the value of switching term amplitude, since the small value of ε makes the $\dot{q}$ diverge and cannot converge to zero, and the large value of ε will cause heavy chattering. Moreover, as ε sgn s_1 is not a continuous function, the interceptor actuator cannot response quickly to realize ε sgn s_1 in practice, which will cause large miss distance.

In this paper, the adaptive fuzzy function $\hat{f}(\dot{q}, r \mid \hat{\eta})$ is chosen to approximate the switching term ε sgn s_1. The inputs of adaptive fuzzy function are the relative distance r between the interceptor and target and LOS angular rate $\dot{q}$. The switching term is replaced by the continuous function which is the output of adaptive fuzzy system, so that the improper value of ε cannot influence the precision of the guidance law and the chattering phenomenon is avoided. The adaptive fuzzy system is designed as follows:

Define $b_1 = \dot{q}$ and $b_2 = r/r_0$, where r_0 is the initial relative distance between the interceptor and target. The knowledge based on adaptive fuzzy systems comprises a collection of fuzzy if-then rules as the following form:

(i) R^1: if b_1 is F_1^1 and b_2 is F_2^1, then y is G^1.

(ii) R^2: if b_2 is F_1^2 and b_2 is F_2^2, then y is G^2.

(iii) R^3: if b_3 is F_1^3 and b_2 is F_2^3, then y is G^3.

(iv) R^4: if b_4 is F_1^4 and b_2 is F_2^4, then y is G^4.

(v) R^5: if b_5 is F_1^5 and b_2 is F_2^5, then y is G^5.

In order to approximate the switching term $\varepsilon \operatorname{sgn} s_1$ properly, the fuzzy rules should be symmetric and uniform and cover all the potential regions of b_1 and b_2. Thus, the adaptive fuzzy membership functions of the fuzzy rules in this paper are chosen as follows:

$$\begin{aligned} u^1(b_1, b_2) &= \frac{1}{1 + e^{(25b_1+1)}} \times \frac{1}{1 + e^{0.5b_2}}, \\ u^2(b_1, b_2) &= e^{-(5b_1+0.2)^2} \times e^{-(0.1b_2+0.2)^2}, \\ u^3(b_1, b_2) &= e^{-(5b_1)^2} \times e^{-(0.1b_2)^2}, \\ u^4(b_1, b_2) &= e^{-(5b_1-0.2)^2} \times e^{-(0.1b_2-0.2)^2}, \\ u^5(b_1, b_2) &= \frac{1}{1 + e^{(-25b_1+1)}} \times \frac{1}{1 + e^{-0.5b_2}}. \end{aligned} \tag{13}$$

Define the fuzzy basis functions as follows:

$$\xi_m(\dot{q}, r) = \frac{u^m(b_1, b_2)}{\sum_{i=1}^{5} u^i(b_1, b_2)}, \quad m = 1, 2, \ldots, 5. \tag{14}$$

The switching term can be approximated and replaced by the adaptive fuzzy system. Using singleton function, center average defuzzification, and product inference, the adaptive fuzzy logic system can be expressed as follows:

$$\hat{f}(\dot{q}, r|\hat{\boldsymbol{\eta}}) = \hat{\boldsymbol{\eta}}^T \xi(\dot{q}, r), \tag{15}$$

where $\hat{\boldsymbol{\eta}}$ is the adaptive parameter vector of the fuzzy system.

If the designed acceleration exceeds the available acceleration of interceptor too much during the terminal guidance, the control law may cause the system instable and lead to a large final miss distance. In order to solve this problem, the adaptive radial basis function neural network $\hat{\delta}$ is designed as the acceleration compensator which can compensate the influence of designed overlarge acceleration. Using the universal approximation property of neural network function, there exists an ideal neural network that approximates the nonlinear function δ in (8) closely, such that

$$\delta = \mathbf{W}^T \mathbf{h}(x_3) + \varsigma, \tag{16}$$

where ς is the approximation error which is very small and can be ignored. $\mathbf{W}$ is the ideal constant weight vector of neural network. And the adaptive radial basis function neural network is as follows:

$$\hat{\delta} = \widehat{\mathbf{W}}^T \mathbf{h}(x_3), \tag{17}$$

where $\mathbf{h}(x_3)$ is the basis vector and $\mathbf{h}_j(x_3) = \exp(-(x_3 - c_j)^2/2b_j^2)$, $j = 1, 2, \ldots, 5$. $\mathbf{c} = [c_1, c_2, \ldots, c_5]$ is the center of the receptive field and $\mathbf{b} = [b_1, b_2, \ldots, b_5]$ is the width of the Gaussian function. $\widehat{\mathbf{W}}$ is the adaptive weight vector of neural network.

Neural network system is designed to compensate the influence caused by acceleration saturation of the interceptor. Therefore, the system can eliminate the limit of the available acceleration.

4. Stability Analysis

Lemma 1: [32] *Consider the system*

$$\dot{x} = g(x, t). \tag{18}$$

Suppose that there is a continuously differentiable Lyapunov function $V(x, t)$ defined in a neighborhood of the origin. For the equation $\dot{V}(x, t) \le -\alpha V(x, t) + C$ and conditions $V(0) = p$, $\alpha > C/p$. Then, the zero solution of the system described by (18) is stable.

Theorem 1. *Employing the guidance law described in (12), whose adaptive control parameters for the fuzzy system in (15) and neural network system in (17) are designed as $\dot{\hat{\boldsymbol{\eta}}} = (T_\eta r_0/r)\xi s_1$ and $\dot{\widehat{\mathbf{W}}} = -(T_W v_M \cos\sigma_M/r)\mathbf{h}s_1$, respectively, the guidance system described by (11) considering the available acceleration and autopilot dynamics of the interceptor is stable.*

Proof 1. Define the state error vector of the low-pass filter as follows:

$$\tilde{x}_3 = x_{3c} - x_{3d}. \tag{19}$$

The approximation error of the radial basis function neural network weights is defined as follows:

$$\widetilde{\mathbf{W}} = \mathbf{W} - \widehat{\mathbf{W}}. \tag{20}$$

Furthermore, define the optimal parameter vector $\boldsymbol{\eta}^*$ for the fuzzy system in (15) as follow:

$$\boldsymbol{\eta}^* = \arg \min_{\eta \in \Omega} \left[\sup_{\dot{q}, r \in \mathbb{R}} \left(\boldsymbol{\eta}^T \xi(\dot{q}, r) - \varepsilon \operatorname{sgn} s_1 \right) \right]. \tag{21}$$

Then, the estimated error of the adaptive parameter vector for the fuzzy system is defined as follows:

$$\tilde{\boldsymbol{\eta}} = \boldsymbol{\eta}^* - \hat{\boldsymbol{\eta}}. \tag{22}$$

With the guidance law designed in (12), the time derivative of the state error vector in (19) is calculated as follows:

$$\dot{\tilde{x}}_3 = \dot{x}_{3c} - \dot{x}_{3d} = \frac{1}{T_2} x_{3d} - \frac{1}{T_2} x_{3c} - \dot{x}_{3d}. \tag{23}$$

As the optimal weight vector $\boldsymbol{\eta}^*$ of the fuzzy system in (21) and the ideal weight vector $\mathbf{W}$ of the neural network system in (16) are two constant vectors, the time derivative of the adaptive parameter error vectors in (20) and (22) are calculated as follows:

$$\dot{\tilde{\mathbf{W}}} = -\dot{\hat{\mathbf{W}}} = \frac{T_W v_M \cos\sigma_M}{r}\mathbf{h}s_1, \tag{24}$$

$$\dot{\tilde{\boldsymbol{\eta}}} = -\dot{\hat{\boldsymbol{\eta}}} = -\frac{T_\eta r_0}{r}\xi s_1. \tag{25}$$

With the guidance law designed in (12), the time derivative of s_1 is as follows:

$$\begin{aligned}\dot{s}_1 = &-\frac{k_1|\dot{r}|}{r}s_1 - \frac{\hat{\eta}^T\xi}{r} - \frac{v_M\cos\sigma_M}{r}\left(\delta-\hat{\delta}\right)\\ &+\frac{1}{r}\omega - \frac{v_M\cos\sigma_M}{r}(\tilde{x}_3+s_2).\end{aligned} \tag{26}$$

Similar to the process of (23), with the guidance law in (12), the time derivative of s_2 is calculated as follows:

$$\dot{s}_2 = -k_2 s_2. \tag{27}$$

Define the Lyapunov function candidate as follows:

$$V = \frac{1}{2}s_1^2 + \frac{1}{2}s_2^2 + \frac{1}{2}\tilde{x}_3^2 + \frac{1}{2T_\eta r_0}\tilde{\boldsymbol{\eta}}^T\tilde{\boldsymbol{\eta}} + \frac{1}{2T_W}\tilde{\mathbf{W}}^T\tilde{\mathbf{W}}, \tag{28}$$

where T_η and T_W are two positive constant. The time derivative of the Lyapunov function is computed as follows:

$$\dot{V} = s_1\dot{s}_1 + s_2\dot{s}_2 + \tilde{x}_3\dot{\tilde{x}}_3 + \frac{1}{T_\eta r_0}\tilde{\boldsymbol{\eta}}^T\dot{\tilde{\boldsymbol{\eta}}} + \frac{1}{T_W}\tilde{\mathbf{W}}^T\dot{\tilde{\mathbf{W}}}. \tag{29}$$

With the results calculated in (23), (26), and (27), it holds that

$$\begin{aligned}s_1\dot{s}_1 = &-\frac{k_1|\dot{r}|}{r}s_1^2 - \frac{\hat{\eta}\theta}{r}s_1 - \frac{v_M\cos\sigma_M}{r}\left(\delta-\hat{\delta}\right)s_1\\ &-\frac{v_M\cos\sigma_M}{r}(\tilde{x}_3+s_2)s_1 + \frac{1}{r}\omega s_1,\end{aligned} \tag{30}$$

$$s_2\dot{s}_2 = -k_2 s_2^2, \tag{31}$$

$$\begin{aligned}\tilde{x}_3\dot{\tilde{x}}_3 = (x_{3c}-x_{3d})\left(\frac{1}{T_2}x_{3d} - \frac{1}{T_2}x_{3c} - \dot{x}_{3d}\right) \le -\frac{1}{T_2}(x_{3c}-x_{3d})^2\\ +\frac{1}{2}\dot{x}_{3d}^2(x_{3d}-x_{3c})^2 + \frac{1}{2} = -\left(\frac{1}{T_2} - \frac{1}{2}\dot{x}_{3d}^2\right)\tilde{x}_3^2 + \frac{1}{2}.\end{aligned} \tag{32}$$

Substituting the results of (30), (31), and (32) into (29), the time derivative of the Lyapunov function can be reformulated as follows:

$$\begin{aligned}\dot{V} \le &-\frac{k_1|\dot{r}|}{r}s_1^2 - \frac{\hat{\boldsymbol{\eta}}\xi}{r}s_1 - \frac{v_M\cos\sigma_M}{r}\left(\delta-\hat{\delta}\right)s_1\\ &-\frac{v_M\cos\sigma_M}{r}(\tilde{x}_3+s_2)s_1 + \frac{s_1}{r}\omega - k_2 s_2^2 - \left(\frac{1}{T_2} - \frac{1}{2}\dot{x}_{3d}^2\right)\tilde{x}_3^2\\ &+\frac{1}{2} + \frac{1}{T_\eta r_0}\tilde{\boldsymbol{\eta}}^T\dot{\tilde{\boldsymbol{\eta}}} + \frac{1}{T_W}\tilde{\mathbf{W}}^T\dot{\tilde{\mathbf{W}}} \le -\frac{k_1|\dot{r}|}{r}s_1^2 - k_2 s_2^2\\ &-\left(\frac{1}{T_2} - \frac{1}{2}\dot{x}_{3d}^2\right)\tilde{x}_3^2 + \frac{s_1}{r}\omega - \frac{\boldsymbol{\eta}*^T\xi}{r}s_1\\ &+\frac{v_M\cos\sigma_M}{2r}\left(\tilde{x}_3^2 + 2s_1^2 + s_2^2\right) + \tilde{\boldsymbol{\eta}}^T\left(\frac{1}{r}\xi s_1 - \frac{1}{T_\eta r_0}\dot{\hat{\boldsymbol{\eta}}}\right)\\ &-\tilde{\mathbf{W}}^T\left(\frac{v_M\cos\sigma_M}{r}\mathbf{h}s_1 + \frac{1}{T_W}\dot{\hat{\mathbf{W}}}\right) + \frac{1}{2}.\end{aligned} \tag{33}$$

With the time derivatives of the estimated parameter error vectors in (24) and (25), and considering the universal approximation property of the fuzzy system as shown in (21), (33) can be rewritten as follows:

$$\begin{aligned}\dot{V} \le &-\left(\frac{k_1|\dot{r}|}{r} - \frac{v_M\cos\sigma_M}{r}\right)s_1^2 - \left(k_2 - \frac{v_M\cos\sigma_M}{2r}\right)s_2^2\\ &-\left(\frac{1}{T_2} - \frac{1}{2}\dot{x}_{3d}^2 - \frac{v_M\cos\sigma_M}{2r}\right)\tilde{x}_3^2 + \frac{1}{2} + \frac{s_1}{r}\omega\\ &-\frac{\boldsymbol{\eta}*^T\xi}{r}s_1 \le -cV + M,\end{aligned} \tag{34}$$

where c and M are two positive constants which satisfy the following:

$$\begin{aligned}c \le \min\Bigg(&2\left(\frac{k_1|\dot{r}|}{r} - \frac{v_M\cos\sigma_M}{r}\right), 2k_2 - \frac{v_M\cos\sigma_M}{r},\\ &\cdot 2\left(\frac{1}{T_2} - \frac{1}{2}\dot{x}_{3d}^2 - \frac{v_M\cos\sigma_M}{2r}\right)\Bigg),\end{aligned} \tag{35}$$

$$M \ge \max\left(\frac{1}{2T_\eta r_0 c}\tilde{\boldsymbol{\eta}}^T\tilde{\eta}\right) + \max\left(\frac{1}{2T_W c}\tilde{\mathbf{W}}^T\tilde{\mathbf{W}}\right) + \frac{1}{2}. \tag{36}$$

With Lemma 1, it can be concluded from (34) that the system is stable. The proof is finished.

Remark 1. From the guidance law designed in (12), we can get that the main parameters affecting the precisions of guidance are k_1, k_2, and T_2. (k_1+2) is similar to the navigation ratio of the traditional PNG law, which is usually created from 2 to 6. And considering the choice of the constant c in (35), k_1 can take a value from 1 to 4. According to some empirical tests, k_2 can be chosen from 1 to 5. T_2 takes a value from 0.01 to 0.5. The adaptive parameters $\hat{\mathbf{W}}$ and $\hat{\boldsymbol{\eta}}$ can be adjusted on line using the states of the interceptor and target, and their initial values are chosen as zero.

Remark 2. Although AAADG needs to know more information than PNG guidance law, it is not difficult to realize the guidance law. In the proposed guidance law in (12), all the parameters needed can be measured by a target seeker easily.

Remark 3. Compared with the existing guidance laws, AAADG neither needs to estimate the exact maneuver information of the target nor simply treats the unknown target acceleration as zero. Meanwhile, this guidance law ensures small final miss distance under the challenging situation, where the acceleration of target is equal to or larger than the available acceleration of the interceptor.

5. Simulation Results

In this section, the effectiveness of AAADG is verified by numerical simulations. For the comparison purpose, we apply the proposed guidance law, the PNG and the FTCG designed in [19] to each simulation case. The proportional parameter of PNG is chosen as $k_p = 2$. The parameters of FTCG are chosen as $N = 3$, $\eta = 0.5$, $\beta = 2$, and $f = 100$. In FTCG, the sign function $\operatorname{sgn} x$ is replaced with a saturation function $\operatorname{sat}(x)$ which is expressed as follows:

$$\operatorname{sat}(x) = \begin{cases} 1, & x > \vartheta, \\ \dfrac{x}{\vartheta}, & |x| \le \vartheta, \\ -1, & x < -\vartheta, \end{cases} \qquad (37)$$

where ϑ is the small positive constant, which is chosen as 0.001 in simulation.

Case 1. Intercepting the target with periodic maneuver and the available acceleration of interceptor is less than the maximum acceleration of the target.

In this case, we use the proposed guidance law to intercept a target with constant velocity and sinusoidal normal acceleration. The initial position of the target is $x_{T0} = 2300$ m and $y_{T0} = 3000$ m. The initial flight-path angle is $\theta_{T0} = 0$ deg. The velocity of the target is $v_T = 1000$ m/s. And the target maneuvers with a normal acceleration is $a_T = 200 \cos t$ m/s^2.

The initial position of the interceptor is set as $x_{M0} = 0$ m and $y_{M0} = 0$ m, and the initial flight-path angle is $\theta_{M0} = 30$ deg. The velocity of the interceptor is $v_M = 1200$ m/s. The initial relative distance between the interceptor and target is calculated as $r_0 = 3780$ m. The initial light-of-sight angle is calculated as $q_0 = 52.5$ deg. And the available acceleration of the interceptor is 160 m/s^2, which is smaller than the maximum acceleration of the target. The parameters of the AAADG are chosen as $k_1 = 2$, $k_2 = 4$, $T_1 = 0.05$ s, $T_2 = 0.06$ s, $T_W = 1$, $b = [1\,1\,1\,1\,1]$, $c = [-2, -1, 0, 1, 2]$, and $T_\eta = 100$.

For Case 1, the final miss distances and impact times under three guidance laws are listed in Table 1. The accelerations of the interceptor and the LOS angle rates under three guidance laws are plotted in Figures 2 and 3, respectively. The trajectories of the interceptor and target under three guidance laws are plotted in Figure 4. The relative distances between the interceptor and target under three guidance laws are plotted in Figure 5.

Table 1: Final miss distance and impact time.

Guidance law	PNG	FTCG	AAADG
Final miss distance (m)	35.2539	8.7406	0.6663
Impact time (s)	13.17	12.97	12.97

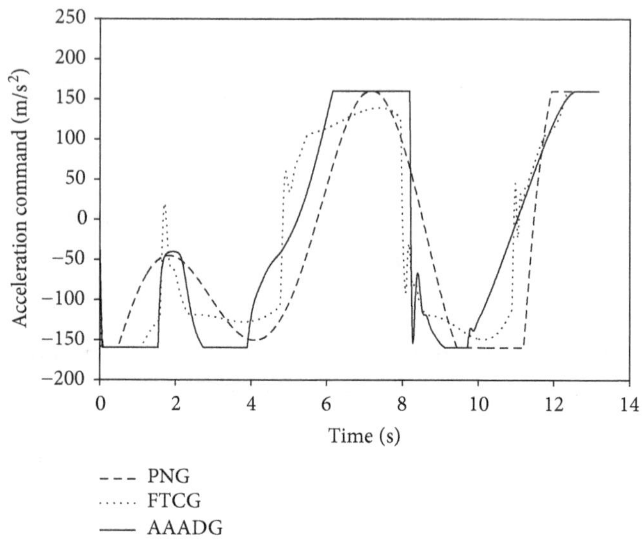

Figure 2: Accelerations under three guidance laws.

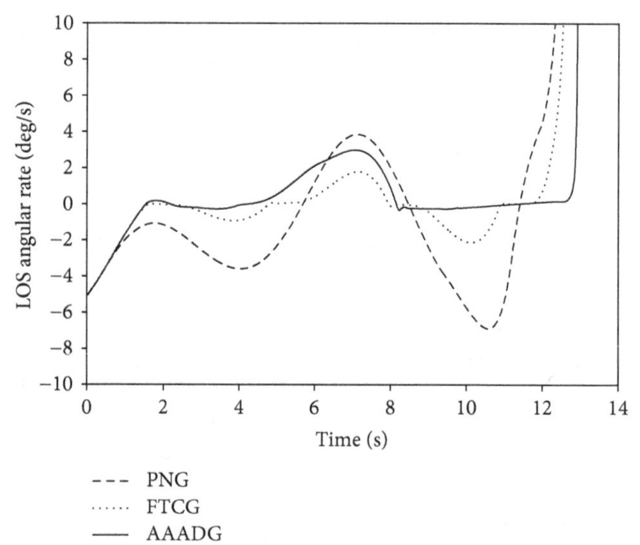

Figure 3: LOS angular rates under three guidance laws.

Figure 2 shows that the accelerations under three guidance laws are limited to 160 m/s^2. And at the end of the terminal guidance, the acceleration command under PNG reaches to 160 m/s^2 earlier than those under other two guidance laws. Moreover, from Figure 2, the acceleration commands under FTCG and AAADG reached to 160 m/s^2 at

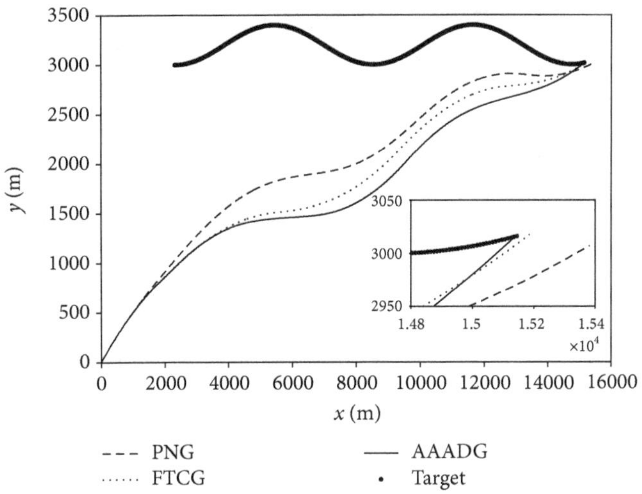

FIGURE 4: Trajectories of the interceptor and target under three guidance laws.

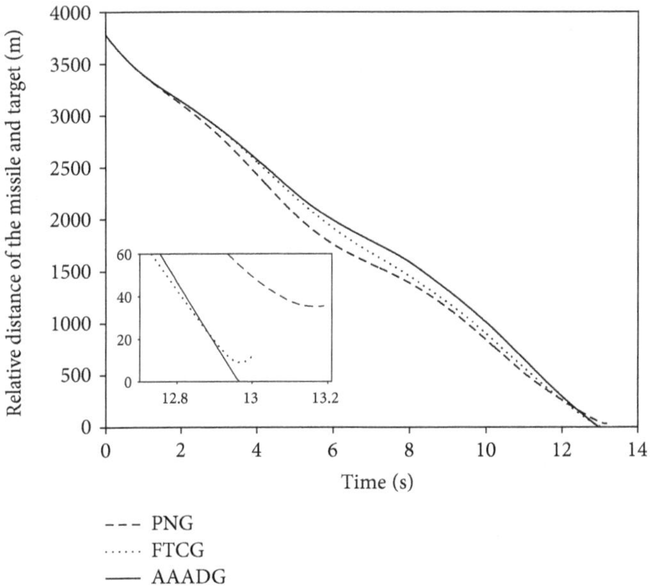

FIGURE 5: Relative distances between the interceptor and target under three guidance laws.

the end of the terminal guidance, but the acceleration command under FTCG has the chattering phenomenon at 11 s.

Figure 3 shows that the LOS angular rate at the beginning of the guidance is extremely large because of the initial state errors between the interceptor and target. The LOS angular rate under PNG becomes divergent too early and does not converge to zero in the terminal guidance, so the final miss distance under PNG is very large. The LOS angular rates under FTCG and AAADG converge to zero in 2 s. However, the LOS angular rate under FTCG diverges earlier than that under AAADG, such that the final miss distance under PNG is larger than that under AAADG.

Figure 4 shows the trajectories of the interceptor and target under three guidance laws. Figure 5 shows the relative

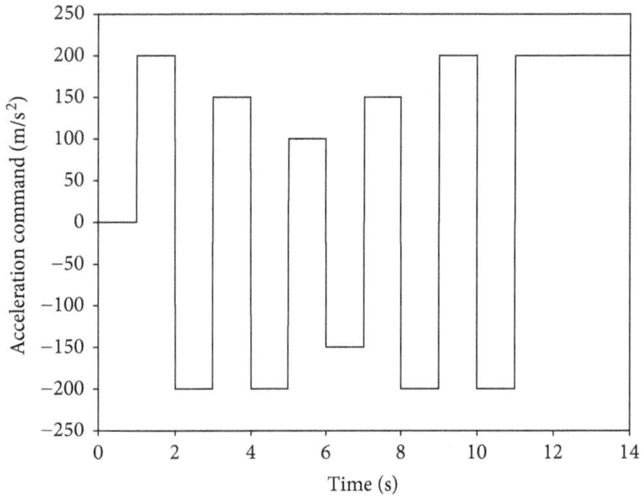

FIGURE 6: Target acceleration.

distance between the interceptor and target as well as the impact time. The impact time under AAADG is earlier than that under other two guidance laws, which can be seen from Figures 4 and 5 and Table 1. And the final miss distance under AAADG is the smallest.

Case 2. Intercepting a target with nonperiodic maneuver and the available acceleration of interceptor is equal to the maximum acceleration of the target.

In this case, the initial positions and velocities of the target and interceptor are the same as those in Case 1. The nonperiodic acceleration of the target is shown in Figure 6. The maximum acceleration of the target is 200 m/s^2. And the available acceleration of the interceptor is 200 m/s^2 which is equal to the maximum acceleration of the target.

For Case 2, the final miss distances and impact time under three guidance laws are listed in Table 2. The accelerations of the interceptor and the LOS angular rates under three guidance laws are plotted in Figures 7 and 8, respectively. The trajectories of the interceptor and target under three guidance laws are plotted in Figure 9. The relative distances between the interceptor and target under three guidance laws are plotted in Figure 10.

Figure 7 shows that the accelerations under PNG and FTCG reached to 200 m/s^2 too early and the acceleration under AAADG dis not reach to 200 m/s^2 at the end of the terminal guidance. Moreover, from Figure 8, the LOS angular rates under PNG and FTCG are divergent at the end of the terminal guidance. However, the LOS angular rate under AAADG will converge to zero at the end of the terminal guidance. Figures 9 and 10 and Table 2 show that the final miss distance under AAADG is smaller than that under other two guidance laws.

Consequently, considering the available acceleration and autopilot dynamics of the interceptor, AAADG is able to intercept the periodic and nonperiodic high maneuvering targets with small final miss distance.

Table 2: Final miss distance and impact time.

Guidance law	PNG	FTCG	AAADG
Final miss distance (m)	51.5315	42.9289	0.8256
Impact time (s)	13.52	13.44	13.31

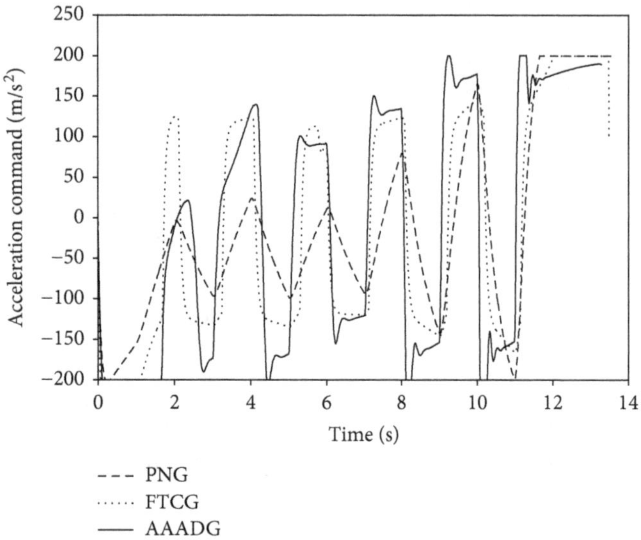

Figure 7: Accelerations under three guidance laws.

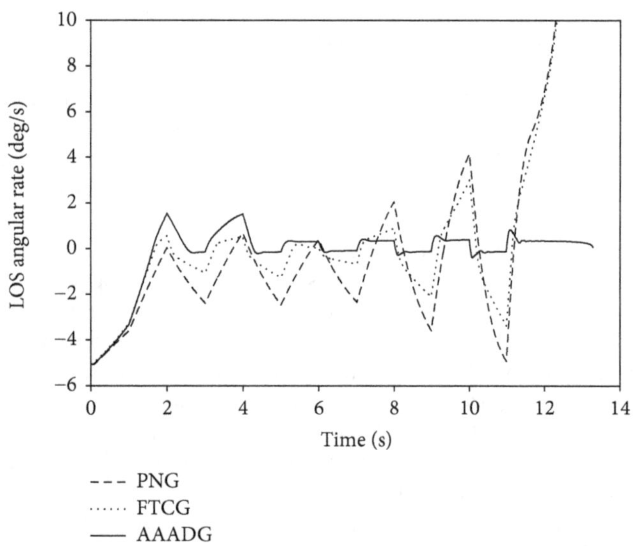

Figure 8: LOS angular rates under three guidance laws.

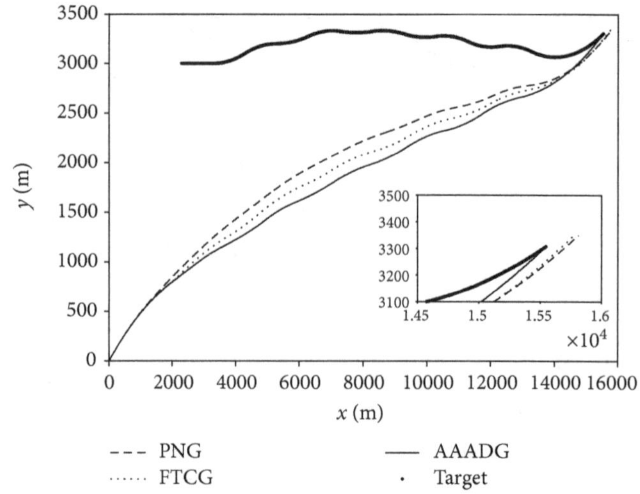

Figure 9: Trajectories of the interceptor and target under three guidance laws.

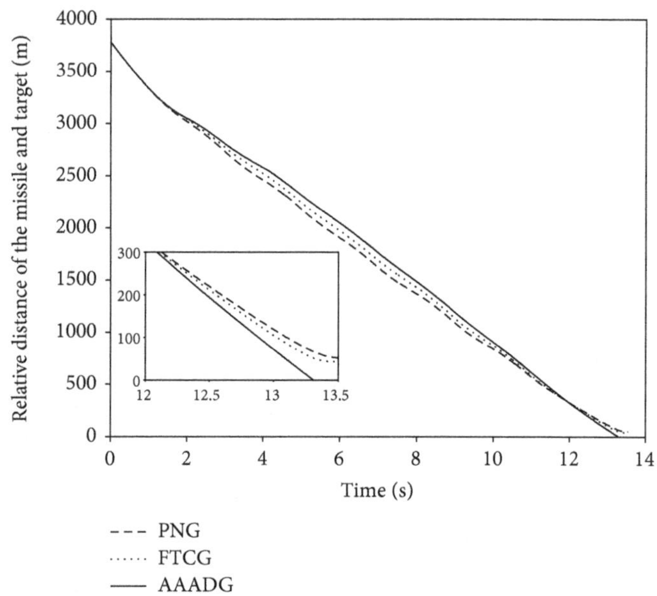

Figure 10: Relative distances between the interceptor and target under three guidance laws.

6. Conclusion

In this paper, the terminal guidance law considering the available acceleration and autopilot dynamics of the interceptor has been designed to intercept a strong maneuverable target. In the proposed guidance law, the adaptive fuzzy system is employed to approximate and replace the variable structure term, which can handle the unknown target maneuver and eliminate the chattering phenomenon. Moreover, the adaptive radial basis function neural network system is designed to compensate the influence caused by the acceleration saturation of the interceptor. Therefore, the interceptor under AAADG can intercept the hypersonic flight vehicle whose maximum acceleration is greater than or equal to the available acceleration of the interceptor. Moreover, the target information needed for AAADG is easy to obtain, so that the proposed AAADG is highly practical.

Conflicts of Interest

The authors declare that there is no conflict of interests regarding the publication of this paper.

Acknowledgments

This paper was sponsored by the National Natural Science Foundation of China 11202024.

References

[1] E. N. Johnson, A. J. Calise, M. D. Curry, K. D. Mease, and J. E. Corban, "Adaptive guidance and control for autonomous hypersonic vehicles," *Journal of Guidance, Control, and Dynamics*, vol. 29, no. 3, pp. 725–737, 2006.

[2] P. J. Yuan and J. S. Chern, "Solutions of true proportional navigation for maneuvering and nonmaneuvering targets," *Journal of Guidance, Control, and Dynamics*, vol. 15, no. 1, pp. 268–271, 1992.

[3] C.-D. Yang, F.-B. Hsiao, and F.-B. Yeh, "Generalized guidance law for homing missiles," *IEEE Transactions on Aerospace and Electronic Systems*, vol. 25, no. 2, pp. 197–212, 1989.

[4] M. Guelman and J. Shinar, "Optimal guidance law in the plane," *Journal of Guidance, Control, and Dynamics*, vol. 7, no. 4, pp. 471–476, 1984.

[5] S. E. Talole and R. N. Banavar, "Proportional navigation through predictive control," *Journal of Guidance, Control, and Dynamics*, vol. 21, no. 6, pp. 1004–1006, 1998.

[6] T.-H. Kim, B.-G. Park, and M.-J. Tahk, "Bias-shaping method for biased proportional navigation with terminal-angle constraint," *Journal of Guidance, Control, and Dynamics*, vol. 36, no. 6, pp. 1810–1816, 2013.

[7] Y. A. Zhang, G. X. Ma, and H. L. Wu, "A biased proportional navigation guidance law with large impact angle constraint and the time-to-go estimation," *Proceedings of the Institution of Mechanical Engineers, Part G: Journal of Aerospace Engineering*, vol. 228, no. 10, pp. 1725–1734, 2014.

[8] B. S. Kim, J. G. Lee, and H. S. Han, "Biased PNG law for impact with angular constraint," *IEEE Transactions on Aerospace and Electronic Systems*, vol. 34, no. 1, pp. 277–288, 1998.

[9] C.-D. Yang and C.-C. Yang, "A unified approach to proportional navigation," *IEEE Transactions on Aerospace and Electronic Systems*, vol. 33, no. 2, pp. 557–567, 1997.

[10] D. Zhou, C. Mu, and W. Xu, "Adaptive sliding-mode guidance of a homing missile," *Journal of Guidance, Control, and Dynamics*, vol. 22, no. 4, pp. 589–594, 1999.

[11] D. Chwa and J. Y. Choi, "Adaptive nonlinear guidance law considering control loop dynamics," *IEEE Transactions on Aerospace and Electronic Systems*, vol. 39, no. 4, pp. 1134–1143, 2003.

[12] C. D. Yang and H. Y. Chen, "Nonlinear H robust guidance law for homing missiles," *Journal of Guidance, Control, and Dynamics*, vol. 21, no. 6, pp. 882–890, 1998.

[13] B.-G. Park, T.-H. Kim, and M.-J. Tahk, "Optimal impact angle control guidance law considering the seeker's field-of-view limits," *Proceedings of the Institution of Mechanical Engineers, Part G: Journal of Aerospace Engineering*, vol. 227, no. 8, pp. 1347–1364, 2013.

[14] J. Moon, K. Kim, and Y. Kim, "Design of missile guidance law via variable structure control," *Journal of Guidance, Control, and Dynamics*, vol. 24, no. 4, pp. 659–664, 2001.

[15] S. D. Brierley and R. Longchamp, "Application of sliding-mode control to air-air interception problem," *IEEE Transactions on Aerospace and Electronic Systems*, vol. 26, no. 2, pp. 306–325, 1990.

[16] T. Shima, M. Idan, and O. M. Golan, "Sliding-mode control for integrated missile autopilot guidance," *Journal of Guidance, Control, and Dynamics*, vol. 29, no. 2, pp. 250–260, 2006.

[17] Z. Zhang, S. Li, and S. Luo, "Terminal guidance laws of missile based on ISMC and NDOB with impact angle constraint," *Aerospace Science and Technology*, vol. 31, no. 1, pp. 30–41, 2013.

[18] Z. Zhu, D. Xu, J. Liu, and Y. Xia, "Missile guidance law based on extended state observer," *IEEE Transactions on Industrial Electronics*, vol. 60, no. 12, pp. 5882–5891, 2013.

[19] D. Zhou, S. Sun, and K. L. Teo, "Guidance laws with finite time convergence," *Journal of Guidance, Control, and Dynamics*, vol. 32, no. 6, pp. 1838–1846, 2009.

[20] S. R. Kumar, S. Rao, and D. Ghose, "Nonsingular terminal sliding mode guidance with impact angle constraints," *Journal of Guidance, Control, and Dynamics*, vol. 37, no. 4, pp. 1114–1130, 2014.

[21] Y. Zhao, Y. Sheng, and X. Liu, "Sliding mode control based guidance law with impact angle constraint," *Chinese Journal of Aeronautics*, vol. 27, no. 1, pp. 145–152, 2014.

[22] Y. Zhang, M. Sun, and Z. Chen, "Finite-time convergent guidance law with impact angle constraint based on sliding-mode control," *Nonlinear Dynamics*, vol. 70, no. 1, pp. 619–625, 2012.

[23] Z. Zhang, S. Li, and S. Luo, "Composite guidance laws based on sliding mode control with impact angle constraint and autopilot lag," *Transactions of the Institute of Measurement and Control*, vol. 35, no. 6, pp. 764–776, 2013.

[24] C.-H. Lee, T.-H. Kim, and M.-J. Tahk, "Design of impact angle control guidance laws via high-performance sliding mode control," *Proceedings of the Institution of Mechanical Engineers, Part G: Journal of Aerospace Engineering*, vol. 227, no. 2, pp. 235–253, 2013.

[25] P. Qu and D. Zhou, "A dimension reduction observer-based guidance law accounting for dynamics of missile autopilot," *Proceedings of the Institution of Mechanical Engineers, Part G: Journal of Aerospace Engineering*, vol. 227, no. 7, pp. 1114–1121, 2013.

[26] S. S. Moosapour, G. Alizadeh, S. Khanmohammadi, and H. Moosapour, "A novel robust proportional navigation guidance law design for missile considering autopilot dynamic," *Transactions of the Institute of Measurement and Control*, vol. 35, no. 5, pp. 703–710, 2013.

[27] S. Sun, D. Zhou, and W. Hou, "A guidance law with finite time convergence accounting for autopilot lag," *Aerospace Science and Technology*, vol. 25, no. 1, pp. 132–137, 2013.

[28] G.-L. Li, H. Yan, and H.-B. Ji, "A guidance law with finite time convergence considering autopilot dynamics and uncertainties," *International Journal of Control Automation and Systems*, vol. 12, no. 5, pp. 1011–1017, 2014.

[29] W. Wang, S. Xiong, X. Liu, S. Wang, and L. Ma, "Adaptive nonsingular terminal sliding mode guidance law against maneuvering targets with impact angle constraint," *Proceedings of the Institution of Mechanical Engineers, Part G: Journal of Aerospace Engineering*, vol. 229, no. 5, pp. 867–890, 2014.

[30] C.-M. Lin and C.-F. Hsu, "Guidance law design by adaptive fuzzy sliding-mode control," *Journal of Guidance, Control, and Dynamics*, vol. 25, no. 2, pp. 248–256, 2002.

[31] J. H. Xiong, S. J. Tang, J. Guo, and T. N. Wang, "Improved sliding mode guidance law based on fuzzy variable coefficients strategy," *The Aeronautical Journal*, vol. 118, no. 1202, pp. 435–451, 2014.

[32] M. Hou, X. Liang, and G. Duan, "Adaptive block dynamic surface control for integrated missile guidance and autopilot," *Chinese Journal of Aeronautics*, vol. 26, no. 3, pp. 741–750, 2013.

8

Squeeze Film Dampers Executing Small Amplitude Circular-Centered Orbits in High-Speed Turbomachinery

Sina Hamzehlouia and Kamran Behdinan

Department of Mechanical and Industrial Engineering, University of Toronto, Toronto, ON, Canada

Correspondence should be addressed to Sina Hamzehlouia; shamzehl@mie.utoronto.ca

Academic Editor: Linda L. Vahala

This work represents a pressure distribution model for finite length squeeze film dampers (SFDs) executing small amplitude circular-centered orbits (CCOs) with application in high-speed turbomachinery design. The proposed pressure distribution model only accounts for unsteady (temporal) inertia terms, since based on order of magnitude analysis, for small amplitude motions of the journal center, the effect of convective inertia is negligible relative to unsteady (temporal) inertia. In this work, the continuity equation and the momentum transport equations for incompressible lubricants are reduced by assuming that the shapes of the fluid velocity profiles are not strongly influenced by the inertia forces, obtaining an extended form of Reynolds equation for the hydrodynamic pressure distribution that accounts for fluid inertia effects. Furthermore, a numerical procedure is represented to discretize the model equations by applying finite difference approximation (FDA) and to numerically determine the pressure distribution and fluid film reaction forces in SFDs with significant accuracy. Finally, the proposed model is incorporated into a simulation model and the results are compared against existing SFD models. Based on the simulation results, the pressure distribution and fluid film reaction forces are significantly influenced by fluid inertia effects even at small and moderate Reynolds numbers.

1. Introduction

According to the classic lubrication theory, the pressure distribution in the annular region of a squeeze film damper is determined by using Reynolds equation, where it is assumed that the inertial forces are negligible relative to viscous forces (i.e., $Re \approx 0$) [1]. In recent years, increase in velocity and size of turbomachinery and application of low-viscosity lubricants requires the fluid inertia effect to be included in design and analysis of SFDs.

Generally, SFDs incorporate an oil film to provide load support with a suitable damping action. SFDs are typically implemented in rotor systems to attenuate vibration amplitudes and to isolate structural components in high-speed turbomachinery applications, including aircraft turbine engines, high performance compressors, and automotive turbochargers. The main advantages of SFDs are their capacity in reducing the effects of unbalance-induced rotor vibrations at resonance frequencies, reducing the level of forces transmitted at supports, and preventing nonsynchronous instabilities. Figure 1 demonstrates the geometry of a conventional SFD. A typical SFD consists of a stationary outer bearing and an inner journal with approximately identical diameters. The journal is assembled on the outer surface of a rolling element and is prevented from rotation by using an antirotation pin. The annular region between the bearing and the housing is generally smaller than 0.25 mm and is filled with a lubricant. The precession motion of the journal is induced by residual unbalance of the rotor and generates a hydrodynamic squeeze film pressure distribution that exerts reaction forces over the journal, providing the damping force to attenuate the transmitted forces and in turn reducing the rotor vibration. The dynamic force response of SFDs is determined by the damper geometry, operating speed, and lubricant properties. Furthermore, SFDs do not produce direct stiffness; a suitable degree of stiffness is introduced by using retaining springs parallel to the squeeze film or by using a pair of elastomer O-rings in radial disposition.

The pioneering work by Cooper [2] demonstrated the benefits conferred on the rotor operation from coupling of

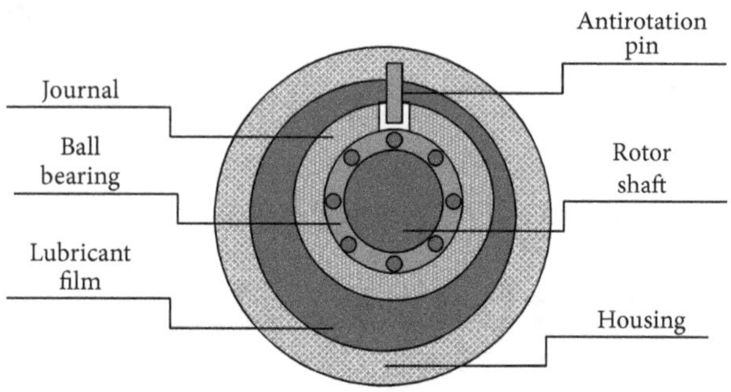

Figure 1: Schematic diagram of a squeeze film damper.

a damping element, represented by an oil squeeze film, to an elastic element placed within the rotor support. Ever since, research works have focused on providing greater insight into different features of SFDs, including the effect of fluid inertia. The effect of fluid inertia on the performance of SFDs has been the subject of several studies in the past few decades [3, 4]. Smith [5] used simplified journal bearing geometries, including the short and long bearing models, to determine the effect of fluid inertia on the dynamic characteristics of journal bearings. He concluded that the effect of fluid inertia introduces an added mass to the rotor system, which significantly influences the dynamic characteristics of short stiff rotors. Several investigations [6, 7] have used the limiting damper geometries (i.e., short bearing and long bearing approximations) to determine the effect of fluid inertia on bearing force characteristics. El-Shafei and Crandall [8] have used an energy approximation method by applying Lagrange's equation to determine the force coefficients in SFDs. They have shown that the energy approximation provides an accurate estimation of the inertia forces in SFDs. Moreover, they have determined the force coefficients for long and short SFDs. San Andres et al. [9–12] have determined SFD force coefficients for short and long bearing geometries and extended the application of these force coefficients to finite length bearings by applying a leakage coefficient. Hashimoto [13] has developed analytical pressure expressions for infinitely long journal bearings with turbulence and inertia effects. El-Shafei [14–16] has determined the SFD force coefficients including inertia effects for both short and long dampers and incorporated the results into rotordynamic models. Although the analysis of limiting bearing geometries provides enhanced understanding of the flow dynamics for SFDs, they demonstrate boundary properties of the dampers, while, in practice, dampers have a finite length and application of the above approximations leads to significant calculation error for arbitrary finite length SFD geometries and operating conditions.

Reinhardt and Lund [17] introduced a first-order perturbation technique for small Reynolds numbers to determine the force coefficients for hydrodynamic journal bearings. They suggested that the inertial corrections to the damping and stiffness coefficients are small, while a large inertia coefficient is introduced that significantly affects the bearing dynamics. However, application of the first-order perturbation method is limited to small and moderate inertia effects.

An alternative approach is to introduce a bulk-flow model [18, 19] for the SFD. Subsequently, the bulk-flow model is solved by using finite element method (FEM). While the bulk-flow modeling and the numerical solution significantly improve the accuracy of the calculation for SFD parameters, it is generally computationally very inefficient, especially for integration into rotordynamic models, where the SFD parameters are calculated over several iterations.

Andrés and Delgado [20] have represented a modified Reynolds equation to describe the lubricant pressure distribution in SFDs and proposed a solution based on FEM. San Andrés and Vance [21] have proposed an analytical closed-form expression for the lubricant pressure in SFDs including fluid inertia and turbulence.

Circular-centered motions of the journal center are a very common type of journal motion in industrial applications of squeeze film dampers, including vertical rotors mounted on SFDs, horizontal rotors mounted on SFDs with centralizing springs, and rotors operating close to critical velocities and for rotor response to large unbalance forces [9]. This CCO condition is typically assumed for SFDs with centering elements. Furthermore, the assumption of small amplitude motions of the journal center is a very common practice in the turbomachinery industry, which further reduces the SFD flow equations by eliminating the nonlinear convective inertia terms. This model reduction facilitates the investigation of unsteady inertia terms. Tichy [22] has studied the effect of small amplitude CCOs for viscoelastic fluids, showing the significant effect of fluid inertia on the fluid film reaction forces, even at moderate Reynolds numbers. San Andres and Vance [9, 11] have further emphasized the significant effect of fluid inertia on dynamic performance of squeeze film dampers executing small amplitude CCOs by representing the effect of fluid inertia on SFD force coefficients for both central and off-centered motions of the journal center.

In most studies, the dynamic characteristics of SFDs are investigated for an isolated bearing system and the models are constructed to satisfy computational requirements for a single SFD system; however, in practice, SFDs are integrated into a rotor system. The rotordynamic analysis of the rotor system, including the dynamic interaction between the SFDs and the rotor, requires the development of an SFD model that is designed for rotordynamic applications, including significant accuracy and maintaining an acceptable computational efficiency. The complexity of most current SFD dynamic models makes them computationally inefficient for rotordynamic analysis. Furthermore, the application of approximate SFD geometries (i.e., long and short bearing models) for the finite length bearings results in considerable degradation of the calculation accuracy.

This work introduces a dynamic model for the calculation of fluid film reaction forces for finite length squeeze film dampers executing small amplitude CCOs. The proposed model numerically determines the SFD fluid film reaction force components under different operating conditions. By using an efficient numerical technique, the direct calculation of SFD fluid forces makes the proposed model very ideal for rotordynamic applications. The resulting extended Reynolds equation is used to numerically determine the pressure distribution in the lubricant. Finally, the pressure distribution is numerically integrated over the journal surface to calculate the fluid film reaction force components.

2. Governing Equations

The dynamic behavior of a viscous Newtonian fluid within boundaries is generally characterized by using the 3-dimensional Navier-Stokes equations as follows [1]:

$$\frac{\partial \rho}{\partial t} + \frac{\partial (\rho u)}{\partial x} + \frac{\partial (\rho v)}{\partial y} + \frac{\partial (\rho w)}{\partial z} = 0, \tag{1}$$

$$\rho \left\{ \frac{\partial u}{\partial t} + u\frac{\partial u}{\partial x} + v\frac{\partial u}{\partial y} + w\frac{\partial u}{\partial z} \right\} = -\frac{\partial P}{\partial x} + \mu \left\{ \frac{\partial^2 u}{\partial x^2} + \frac{\partial^2 u}{\partial y^2} + \frac{\partial^2 u}{\partial z^2} \right\}, \tag{2}$$

$$\rho \left\{ \frac{\partial v}{\partial t} + u\frac{\partial v}{\partial x} + v\frac{\partial v}{\partial y} + w\frac{\partial v}{\partial z} \right\} = -\frac{\partial P}{\partial y} + \mu \left\{ \frac{\partial^2 v}{\partial x^2} + \frac{\partial^2 v}{\partial y^2} + \frac{\partial^2 v}{\partial z^2} \right\}, \tag{3}$$

$$\rho \left\{ \frac{\partial w}{\partial t} + u\frac{\partial w}{\partial x} + v\frac{\partial w}{\partial y} + w\frac{\partial w}{\partial z} \right\} = -\frac{\partial P}{\partial z} + \mu \left\{ \frac{\partial^2 w}{\partial x^2} + \frac{\partial^2 w}{\partial y^2} + \frac{\partial^2 w}{\partial z^2} \right\}, \tag{4}$$

where (1) is the continuity equation corresponding to the conservation of mass within the fluid boundaries and (2) to (4) correspond to the conservation of momentum within the fluid boundaries.

According to the hydrodynamic lubrication theory, the thin film assumption in hydrodynamic bearings is characterized by the small ratio of film thickness to the bearing's other physical dimensions; that is, $c \ll R$. Consequently, we have the following:

(1) The effect of the curvature of the film is negligible; hence a linear coordinate system is used to describe the lubricant dynamics.

(2) The variation of the pressure across the film is negligible (i.e., $\partial P/\partial y = 0$).

The SFD configuration in this work is a symmetric damper about its midplane with open ends (i.e., no seal). The geometry of the system is represented in Figure 2. According to the thin film theory, which facilitates the application of Cartesian coordinates, an orthogonal Cartesian coordinate system $\{x, y, z\}$ is fixed in the plane of the lubricant, where the z-axis is perpendicular to the plane of motion. Furthermore, an orthogonal Cartesian system $\{x', y', z'\}$ translating with angular velocity $R\omega$ is introduced, where the x'-axis is perpendicular to the line connecting the centers of the inner and outer cylinders, and the y'-axis is in the direction of the minimum thickness. The angle θ' starts from the origin of the fixed Cartesian system and the angle θ is measured from the maximum film thickness in the direction of the whirling motion. Finally, a fixed inertial coordinate system $\{X, Y\}$ is defined at the center of the bearing. Based on the thin film theory, for an incompressible and isoviscous lubricant, the flow equations in the SFD are reduced to the following:

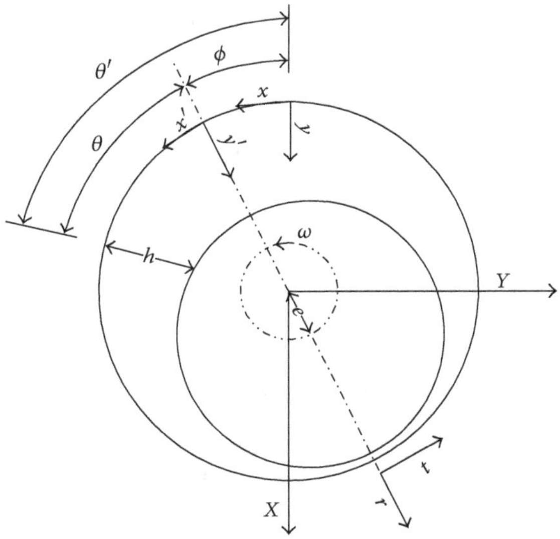

FIGURE 2: Squeeze film damper geometry and coordinate systems.

$$\frac{\partial u}{\partial x} + \frac{\partial v}{\partial y} + \frac{\partial w}{\partial z} = 0, \tag{5}$$

$$\rho \left\{ \frac{\partial u}{\partial t} + u\frac{\partial u}{\partial x} + v\frac{\partial u}{\partial y} + w\frac{\partial u}{\partial z} \right\} = -\frac{\partial P}{\partial x} + \mu\frac{\partial^2 u}{\partial y^2}, \tag{6}$$

$$\frac{\partial P}{\partial y} = 0, \tag{7}$$

$$\rho \left\{ \frac{\partial w}{\partial t} + u\frac{\partial w}{\partial x} + v\frac{\partial w}{\partial y} + w\frac{\partial w}{\partial z} \right\} = -\frac{\partial P}{\partial z} + \mu\frac{\partial^2 w}{\partial y^2}. \tag{8}$$

Subsequently, the flow equations are normalized by means of dimensionless quantities to demonstrate the effect of Reynolds number on the flow characteristics:

$$\theta = \theta' - \phi = \frac{x}{R} - \omega t,\ \eta = \frac{y}{c},\ \xi = \frac{z}{R},\ \tau = \omega t,$$
$$\bar{u} = \frac{u}{R\omega},\ \bar{v} = \frac{v}{c\omega},\ \bar{w} = \frac{w}{R\omega}, \quad (9)$$
$$\bar{P} = \frac{Pc^2}{R^2\omega\mu},\ H = \frac{h}{c} = 1 + \varepsilon\cos\theta,\ \mathrm{Re} = \frac{\rho\omega c^2}{\mu}.$$

Furthermore, according to order of magnitude calculations, for small amplitude motions, convective inertia terms in (6) and (8) are negligible relative to unsteady (temporal) inertia terms [23]. Therefore, the dimensionless flow equations are introduced as follows:

$$\frac{\partial\bar{u}}{\partial\theta} + \frac{\partial\bar{v}}{\partial\eta} + \frac{\partial\bar{w}}{\partial\xi} = 0,$$
$$\mathrm{Re}\frac{\partial\bar{u}}{\partial\tau} = -\frac{\partial\bar{P}}{\partial\theta} + \frac{\partial^2\bar{u}}{\partial\eta^2}, \quad (10)$$
$$\mathrm{Re}\frac{\partial\bar{w}}{\partial\tau} = -\frac{\partial\bar{P}}{\partial\xi} + \frac{\partial^2\bar{w}}{\partial\eta^2}.$$

The boundary conditions for the fluid velocities in SFDs are defined as follows:

$$\bar{u} = 0,\ \bar{v} = 0,\ \bar{w} = 0, \quad \eta = 0,$$
$$\bar{u} = 0,\ \bar{v} = \frac{\partial H}{\partial\tau},\ \bar{w} = 0, \quad \eta = H. \quad (11)$$

Additionally, the boundary conditions for the hydrodynamic pressure in an open ended SFD are as follows.

(1) The pressure is periodic and continuous in the circumferential direction (θ); that is,

$$\bar{P}(\theta,\xi) = \bar{P}(\theta + 2\pi,\xi). \quad (12)$$

(2) The lubricant pressure equals atmospheric pressure at the axial ends of the bearing; that is,

$$\bar{P}\left(\theta,\frac{L}{D}\right) = \bar{P}\left(\theta,-\frac{L}{D}\right) = 0. \quad (13)$$

(3) The hydrodynamic pressure must be above the liquid cavitation pressure; that is,

$$\bar{P} \geq \bar{P}_{\mathrm{cav}} \quad 0 \leq \theta \leq 2\pi,\ -\frac{L}{D} \leq z \leq \frac{L}{D}, \quad (14)$$

where P_{cav} is the saturation pressure of the lubricant or the saturation pressure for the release of entrapped gases in the lubricant, typically ambient pressure. Subsequently, (10) are integrated along the film thickness and the velocity boundary conditions are applied:

$$\frac{\partial}{\partial\theta}\int_0^H \bar{u}\,d\eta + \frac{\partial}{\partial\xi}\int_0^H \bar{w}\,d\eta + \frac{\partial H}{\partial\tau} = 0,$$
$$\mathrm{Re}\frac{\partial}{\partial\tau}\int_0^H \bar{u}\,d\eta = -H\frac{\partial\bar{P}}{\partial\theta} + \Delta\bar{\tau}_{\theta\eta}, \quad (15)$$
$$\mathrm{Re}\frac{\partial}{\partial\tau}\int_0^H \bar{w}\,d\eta = -H\frac{\partial\bar{P}}{\partial\xi} + \Delta\bar{\tau}_{\xi\eta}.$$

The dimensionless lubricant average flows are defined by the following:

$$\bar{q_i} = \int_0^H \bar{u}_i d\eta. \quad (16)$$

Substituting (16) into (15) gives the following:

$$\frac{\partial\bar{q_\theta}}{\partial\theta} + \frac{\partial\bar{q_\xi}}{\partial\xi} + \frac{\partial H}{\partial\tau} = 0,$$
$$\mathrm{Re}\frac{\partial\bar{q_\theta}}{\partial\tau} = -H\frac{\partial\bar{P}}{\partial\theta} + \Delta\bar{\tau}_{\theta\eta}, \quad (17)$$
$$\mathrm{Re}\frac{\partial\bar{q_\xi}}{\partial\tau} = -H\frac{\partial\bar{P}}{\partial\xi} + \Delta\bar{\tau}_{\xi\eta}.$$

Assuming that the shapes of the velocity profiles are not strongly influenced by the inertia forces [13], the wall shear stress differences can be expressed by the following:

$$\Delta\bar{\tau}_{\theta\eta} = -\frac{12}{H^2}\bar{q_\theta},$$
$$\Delta\bar{\tau}_{\xi\eta} = -\frac{12}{H^2}\bar{q_\xi}; \quad (18)$$

hence,

$$\frac{\partial\bar{q_\theta}}{\partial\theta} + \frac{\partial\bar{q_\xi}}{\partial\xi} + \frac{\partial H}{\partial\tau} = 0, \quad (19)$$

$$\mathrm{Re}\frac{\partial\bar{q_\theta}}{\partial\tau} = -H\frac{\partial\bar{P}}{\partial\theta} - \frac{12}{H^2}\bar{q_\theta}, \quad (20)$$

$$\mathrm{Re}\frac{\partial\bar{q_\xi}}{\partial\tau} = -H\frac{\partial\bar{P}}{\partial\xi} - \frac{12}{H^2}\bar{q_\xi}. \quad (21)$$

Furthermore, assuming that the temporal inertia terms in the above equations are approximated by using inertialess and dimensionless average flows [24], the equations become

$$\mathrm{Re}\frac{\partial \overline{q_{\theta_0}}}{\partial \tau} = -H\frac{\partial \overline{P}}{\partial \theta} - \frac{12}{H^2}\overline{q_\theta},$$
$$\mathrm{Re}\frac{\partial \overline{q_{\xi_0}}}{\partial \tau} = -H\frac{\partial \overline{P}}{\partial \xi} - \frac{12}{H^2}\overline{q_\xi}, \quad (22)$$

where based on Reynolds equation

$$\overline{q_{\theta_0}} = -\frac{H^3}{12}\frac{\partial \overline{P_0}}{\partial \theta},$$
$$\overline{q_{\xi_0}} = -\frac{H^3}{12}\frac{\partial \overline{P_0}}{\partial \xi}. \quad (23)$$

Substituting (23) into (20) and (21) gives the following:

$$\mathrm{Re}\frac{\partial}{\partial \tau}\left(-\frac{H^3}{12}\frac{\partial \overline{P_0}}{\partial \theta}\right) = -H\frac{\partial \overline{P}}{\partial \theta} - \frac{12}{H^2}\overline{q_\theta},$$
$$\mathrm{Re}\frac{\partial}{\partial \tau}\left(-\frac{H^3}{12}\frac{\partial \overline{P_0}}{\partial \xi}\right) = -H\frac{\partial \overline{P}}{\partial \xi} - \frac{12}{H^2}\overline{q_\xi}. \quad (24)$$

Equations (24) are rearranged to solve for the flows:

$$\overline{q_\theta} = -\frac{H^3}{12}\frac{\partial \overline{P}}{\partial \theta} - \frac{H^2}{12}\mathrm{Re}\frac{\partial}{\partial \tau}\left(-\frac{H^3}{12}\frac{\partial \overline{P_0}}{\partial \theta}\right),$$
$$\overline{q_\xi} = -\frac{H^3}{12}\frac{\partial \overline{P}}{\partial \xi} - \frac{H^2}{12}\mathrm{Re}\frac{\partial}{\partial \tau}\left(-\frac{H^3}{12}\frac{\partial \overline{P_0}}{\partial \xi}\right). \quad (25)$$

Substituting (25) into the continuity equation (19) gives

$$\frac{\partial}{\partial \theta}\left(H^3\frac{\partial \overline{P}}{\partial \theta}\right) + \frac{\partial}{\partial \xi}\left(H^3\frac{\partial \overline{P}}{\partial \xi}\right) = 12\frac{\partial H}{\partial \tau} + G(\theta,\xi), \quad (26)$$

where

$$G(\theta,\xi) = \frac{\partial}{\partial \theta}\left[H^2\mathrm{Re}\frac{\partial}{\partial \tau}\left(\frac{H^3}{12}\frac{\partial \overline{P_0}}{\partial \theta}\right)\right] + \frac{\partial}{\partial \xi}\left[H^2\mathrm{Re}\frac{\partial}{\partial \tau}\left(\frac{H^3}{12}\frac{\partial \overline{P_0}}{\partial \xi}\right)\right]. \quad (27)$$

The above equation is expanded and rearranged as follows:

$$G(\theta,\xi) = \frac{\partial}{\partial \theta}\left(H^2\mathrm{Re}\right)\left[\frac{\partial}{\partial \tau}\left(\frac{H^3}{12}\frac{\partial \overline{P_0}}{\partial \theta}\right)\right] + H^2\mathrm{Re}\frac{\partial}{\partial \tau}\left\{\frac{\partial}{\partial \theta}\left(\frac{H^3}{12}\frac{\partial \overline{P_0}}{\partial \theta}\right) + \frac{\partial}{\partial \xi}\left(\frac{H^3}{12}\frac{\partial \overline{P_0}}{\partial \xi}\right)\right\}. \quad (28)$$

According to classical Reynolds equation, the inertialess (zeroth-order) pressure distribution in SFDs is described by the following:

$$\frac{\partial}{\partial \theta}\left(\frac{H^3}{12}\frac{\partial \overline{P_0}}{\partial \theta}\right) + \frac{\partial}{\partial \xi}\left(\frac{H^3}{12}\frac{\partial \overline{P_0}}{\partial \xi}\right) = \frac{\partial H}{\partial \tau}. \quad (29)$$

Substituting (29) into (28) gives the following:

$$G(\theta,\xi) = \frac{\partial}{\partial \theta}\left(H^2\mathrm{Re}\right)\left[\frac{\partial}{\partial \tau}\left(\frac{H^3}{12}\frac{\partial \overline{P_0}}{\partial \theta}\right)\right] + H^2\mathrm{Re}\frac{\partial^2 H}{\partial \tau^2}; \quad (30)$$

hence:

$$\frac{\partial}{\partial \theta}\left(H^3\frac{\partial \overline{P}}{\partial \theta}\right) + \frac{\partial}{\partial \xi}\left(H^3\frac{\partial \overline{P}}{\partial \xi}\right) = 12\frac{\partial H}{\partial \tau} + H^2\mathrm{Re}\frac{\partial^2 H}{\partial \tau^2} + \frac{\partial}{\partial \theta}\left(H^2\mathrm{Re}\right)\left[\frac{\partial}{\partial \tau}\left(\frac{H^3}{12}\frac{\partial \overline{P_0}}{\partial \theta}\right)\right]. \quad (31)$$

Consequently, the zeroth-order pressure distribution is calculated by using Reynolds equation and the solution is substituted into (31) to determine the total pressure distribution in SFDs. Finally, the fluid film reaction force components are determined by integrating the total pressure distribution over the journal surface:

$$\begin{bmatrix}\overline{F_r}\\ \overline{F_t}\end{bmatrix} = \int_{-L/D}^{L/D}\int_{\theta_1}^{\theta_2}\overline{P}(\theta,\xi)\begin{bmatrix}\cos(\theta)\\ \sin(\theta)\end{bmatrix}d\theta\, d\xi, \quad (32)$$

where θ_1 to θ_2 is the positive range of the pressure distribution. In this work, the simplistic π-film cavitation model is applied where the film cavitation region is developed in half the damper circumference. The following section represents the numerical solution for the pressure distribution and fluid film reaction forces for SFDs.

3. Numerical Solution

This section describes the numerical technique that is applied to (31) to determine the pressure distribution and the fluid film forces in it. In order to determine the numerical solution, firstly, a solution domain is defined for the problem. Subsequently, the partial differential equations are discretized over the solution domain. Finally, an iterative numerical algorithm is developed to calculate the pointwise pressure distribution in the lubricant domain.

3.1. Zeroth-Order Pressure Solution. In order to determine the total fluid film pressure distribution, the zeroth-order pressure distribution must be first calculated. The zeroth-order pressure distribution is characterized by (29). This equation is firstly expanded to facilitate the discretization of the equation terms:

$$\left[3H^2\frac{\partial H}{\partial \theta}\frac{\partial \overline{P_0}}{\partial \theta}+H^3\frac{\partial^2 \overline{P_0}}{\partial \theta^2}\right]+H^3\frac{\partial^2 \overline{P_0}}{\partial \xi^2}=12\frac{\partial H}{\partial \tau}. \quad (33)$$

The partial derivatives of the dimensionless fluid film thickness (H) are given as follows:

$$\begin{aligned}\frac{\partial H}{\partial \theta}&=-\varepsilon\sin(\theta),\\ \frac{\partial H}{\partial \tau}&=\dot{\varepsilon}\cos(\theta)+\varepsilon\sin(\theta).\end{aligned} \quad (34)$$

Assuming that the SFD executes CCOs, the radial velocity and acceleration of the journal center become zero and

$$\begin{aligned}\frac{\partial H}{\partial \tau}&=\varepsilon\sin(\theta),\\ \frac{\partial^2 H}{\partial \tau^2}&=-\varepsilon\cos(\theta).\end{aligned} \quad (35)$$

Substituting (35) into (33) gives

$$\begin{aligned}&\left[-3\varepsilon H^2\sin(\theta)\frac{\partial \overline{P_0}}{\partial \theta}+H^3\frac{\partial^2 \overline{P_0}}{\partial \theta^2}\right]+H^3\frac{\partial^2 \overline{P_0}}{\partial \xi^2}\\ &=12\varepsilon\sin(\theta).\end{aligned} \quad (36)$$

Subsequently, (36) is discretized based on a finite difference approximation (FDA) technique that uses a backward difference approximation for the first derivative terms and a central difference approximation for the second derivative terms, where

$$\begin{aligned}\frac{\partial \overline{P}}{\partial \theta}&=\frac{\overline{P}(\theta,\xi)-\overline{P}(\theta-\Delta\theta,\xi)}{\Delta\theta}=\frac{\overline{P}_{i,j}-\overline{P}_{i-1,j}}{\Delta\theta},\\ \frac{\partial \overline{P}}{\partial \xi}&=\frac{\overline{P}(\theta,\xi)-\overline{P}(\theta,\xi-\Delta\xi)}{\Delta\xi}=\frac{\overline{P}_{i,j}-\overline{P}_{i,j-1}}{\Delta\xi},\\ \frac{\partial^2 \overline{P}}{\partial \theta^2}&=\frac{\overline{P}(\theta+\Delta\theta,\xi)-2\overline{P}(\theta,\xi)+\overline{P}(\theta-\Delta\theta,\xi)}{\Delta\theta^2}\\ &=\frac{\overline{P}_{i+1,j}-2\overline{P}_{i,j}+\overline{P}_{i-1,j}}{\Delta\theta^2},\\ \frac{\partial^2 \overline{P}}{\partial \xi^2}&=\frac{\overline{P}(\theta,\xi+\Delta\xi)-2\overline{P}(\theta,\xi)+\overline{P}(\theta,\xi-\Delta\xi)}{\Delta\xi^2}\\ &=\frac{\overline{P}_{i,j+1}-2\overline{P}_{i,j}+\overline{P}_{i,j-1}}{\Delta\xi^2}.\end{aligned} \quad (37)$$

Substituting (37) into (36) gives the discretized Reynolds equation:

$$\begin{aligned}&\left[-3\varepsilon H^2\sin(\theta)\frac{\overline{P}_{0i,j}-\overline{P}_{0i-1,j}}{\Delta\theta}\right.\\ &\left.+H^3\frac{\overline{P}_{0i+1,j}-2\overline{P}_{0i,j}+\overline{P}_{0i-1,j}}{\Delta\theta^2}\right]+H^3\\ &\cdot\frac{\overline{P}_{0i,j+1}-2\overline{P}_{0i,j}+\overline{P}_{0i,j-1}}{\Delta\xi^2}=12\varepsilon\sin(\theta).\end{aligned} \quad (38)$$

In order to determine the pointwise pressure distribution, (38) is rearranged as follows:

$$P_{0i,j}=\frac{[12\varepsilon\sin\theta_i]-\left[3H_i^2(\varepsilon\sin\theta_i)/\Delta\theta+H_i^3/\Delta\theta^2\right]\overline{P}_{0i-1,j}-\left[H_i^3/\Delta\theta^2\right]\overline{P}_{0i+1,j}-\left[H_i^3/\Delta\xi^2\right]\left(\overline{P}_{0i,j+1}+\overline{P}_{0i,j-1}\right)}{\left[-3H_i^2(\varepsilon\sin\theta_i)/\Delta\theta-2H_i^3/\Delta\theta^2-2H_i^3/\Delta\xi^2\right]}. \quad (39)$$

Reynolds equation is classified as an elliptical PDE. Assuming that the lubricant is incompressible and isoviscous and the journal center executes CCO whirls, the following numerical procedure is used to determine the fluid film pressure distribution for a specified SFD eccentricity ratio.

(1) The boundary points are initialized to their prescribed values and the interior points are adjusted to zero.

(2) Equation (39) is iteratively solved for the interior points.

(3) The iteration is only interrupted when the solution error reaches a convergence criterion.

(4) Finally, an overrelaxation technique is applied to accelerate the convergence of the solution:

$$P_{0_{i,j}}^{(k)}=P_{0_{i,j}}^{(k-1)}+\lambda\left(P_{0_{i,j}}^{(k)}-P_{0_{i,j}}^{(k-1)}\right). \quad (40)$$

3.2. Total Pressure Solution. The numerical procedure that is employed to determine the total pressure distribution is very similar to the one for the zeroth-order pressure. The total pressure distribution in the fluid film is characterized by (31). This equation is first expanded to facilitate the discretization of the differential terms:

$$\begin{aligned}&H^3\frac{\partial^2 \overline{P}}{\partial \theta^2}+3H^2\frac{\partial H}{\partial \theta}\frac{\partial \overline{P}}{\partial \theta}+H^3\frac{\partial^2 \overline{P}}{\partial \xi^2}=12\frac{\partial H}{\partial \tau}+G(\theta,\xi),\\ &G(\theta,\xi)\\ &=H^2\,\mathrm{Re}\frac{\partial^2 H}{\partial \tau^2}\\ &+\frac{\mathrm{Re}\,H}{6}\frac{\partial H}{\partial \theta}\left[H^3\frac{\partial^2 \overline{P_0}}{\partial \tau\partial \theta}+3H^2\frac{\partial H}{\partial \tau}\frac{\partial \overline{P_0}}{\partial \theta}\right],\end{aligned} \quad (41)$$

where

$$\frac{\partial^2 \overline{P_0}}{\partial \tau \partial \theta} = \frac{\partial}{\partial \tau}\left(\frac{\partial \overline{P_0}}{\partial \theta}\right) = \frac{\partial}{\partial \theta}\left(\frac{\partial \overline{P_0}}{\partial \theta}\right)\frac{\partial \theta}{\partial \tau} = -\frac{\partial^2 \overline{P_0}}{\partial \theta^2}. \quad (42)$$

Equations (41) are discretized by using (37) and the partial derivatives of the fluid film thickness for CCO in (35) are substituted into the following equations:

$$H^3 \frac{\overline{P}_{i+1,j} - 2\overline{P}_{i,j} + \overline{P}_{i-1,j}}{\Delta\theta^2} - 3H^2 \varepsilon \sin(\theta) \frac{\overline{P}_{i,j} - \overline{P}_{i-1,j}}{\Delta\theta} + H^3 \frac{\overline{P}_{i,j+1} - 2\overline{P}_{i,j} + \overline{P}_{i,j-1}}{\Delta\xi^2} = 12\varepsilon \sin(\theta) + G_{i,j}, \quad (43)$$

$$G_{i,j} = -H^2 \operatorname{Re} \varepsilon \cos(\theta) - \frac{\operatorname{Re} H}{6} \varepsilon \sin(\theta) \left[3H^2 \varepsilon \sin(\theta) \frac{\overline{P}_{0i,j} - \overline{P}_{0i-1,j}}{\Delta\theta} - H^3 \frac{\overline{P}_{0i,j+1} - 2\overline{P}_{0i,j} + \overline{P}_{0i,j-1}}{\Delta\xi^2}\right], \quad (44)$$

$$P_{i,j} = \frac{\left[12\varepsilon \sin\theta_i + G_{i,j}\right] - \left[3H_i^2 (\varepsilon \sin\theta_i)/\Delta\theta + H_i^3/\Delta\theta^2\right]\overline{P}_{i-1,j} - \left[H_i^3/\Delta\theta^2\right]\overline{P}_{i+1,j} - \left[H_i^3/\Delta\xi^2\right]\left(\overline{P}_{i,j+1} + \overline{P}_{i,j-1}\right)}{\left[-3H_i^2 (\varepsilon \sin\theta_i)/\Delta\theta - 2H_i^3/\Delta\theta^2 - 2H_i^3/\Delta\xi^2\right]}. \quad (45)$$

Subsequently, (45) is iteratively solved by using the numerical procedure that is described in the previous section. Finally, the fluid film reaction forces are determined by numerically integrating the pressure field over the journal surface:

$$\begin{bmatrix} \overline{F}_r \\ \overline{F}_t \end{bmatrix} = \sum_{i=1}^{N} \sum_{j=1}^{M} \overline{P}_{i,j} \begin{bmatrix} \cos(\theta_i) \\ \sin(\theta_i) \end{bmatrix} \Delta\theta \Delta\xi. \quad (46)$$

4. Results and Discussion

This section represents the simulation results for the pressure distribution and the fluid film reaction forces for the proposed hydrodynamic SFD model, under different operating conditions, including journal eccentricity ratios, journal slenderness ratios, and Reynolds numbers (i.e., inertia effects). The numerical algorithm that was developed in the previous section is incorporated into Matlab to evaluate the effect of SFD operating parameters on the lubricant pressure distribution and the fluid film reaction forces. In order to validate the proposed model, the results are compared against the existing SFD models. San Andres and Vance [9] have proposed a linearized force coefficient model for SFDs executing small amplitude circular-centered orbits. They proposed a closed-form analytical model for the SFD pressure distribution as follows:

$$\overline{P} = \varepsilon \operatorname{Re} Z(\xi) \left[k_1 \cos\theta - k_2 \sin\theta\right], \quad (47)$$

where

$$Z(\xi) = 1 - \frac{\cosh(\xi)}{\cosh(L/D)}, \quad (48)$$

and k_1 and k_2 are the real and imaginary parts of k, which is defined as follows:

$$k = \frac{\zeta \sinh(\zeta)}{\left[\zeta \sinh(\zeta) + 2 - 2\cosh(\zeta)\right]}, \quad (49)$$

where

$$\zeta = (1 - i)\sqrt{\frac{\operatorname{Re}}{2}}. \quad (50)$$

Figure 3 represents the effect of fluid inertia on the dimensionless fluid pressure distribution at the bearing axial midplane ($\xi = 0$) at small eccentricity ratios for the numerical solution based on (45), the analytical pressure expression based on (47), and Reynolds equation (i.e., no inertia effect) in (39). In general, the fluid inertia effect causes a significant increase in the pressure amplitude, a change in the shape of the pressure profile, and a phase shift of the pressure peak in the direction of the journal precession. Furthermore, at small Reynolds numbers, the influence of the viscous forces makes the pressure profile closer to a sinusoid; however, at moderate and large Reynolds numbers the pressure is in phase with the gap acceleration and transforms into a cosine wave shape.

Figure 4 illustrates the effect of journal eccentricity ratio on the dimensionless midplane pressure distribution at moderate inertia effects (i.e., moderate Reynolds number). According to the results, the pressure amplitude considerably increases with the journal eccentricity ratio. Furthermore, the shape and the phase of the pressure profile are not considerably influenced by the eccentricity ratio.

Figure 5 demonstrates the effect of journal slenderness ratio on the dimensionless midplane pressure profile at small eccentricity ratios and moderate Reynolds numbers. Increasing the slenderness ratio significantly increases the pressure amplitude; however, the shape and the phase of the pressure profile are unaffected.

Furthermore, Vance [25] has developed the force coefficients for the short-length SFD with open ends by using the simplistic π-film assumption (i.e., film cavitation region is developed in half the damper circumference). The force coefficients are represented as follows:

$$C_{tt} = \frac{\pi \mu D}{4\left(1 - \varepsilon^2\right)^{3/2}} \left(\frac{L}{c}\right)^3, \quad (51)$$

$$C_{rt} = \frac{\mu \varepsilon D}{\left(1 - \varepsilon^2\right)^2} \left(\frac{L}{c}\right)^3, \quad (52)$$

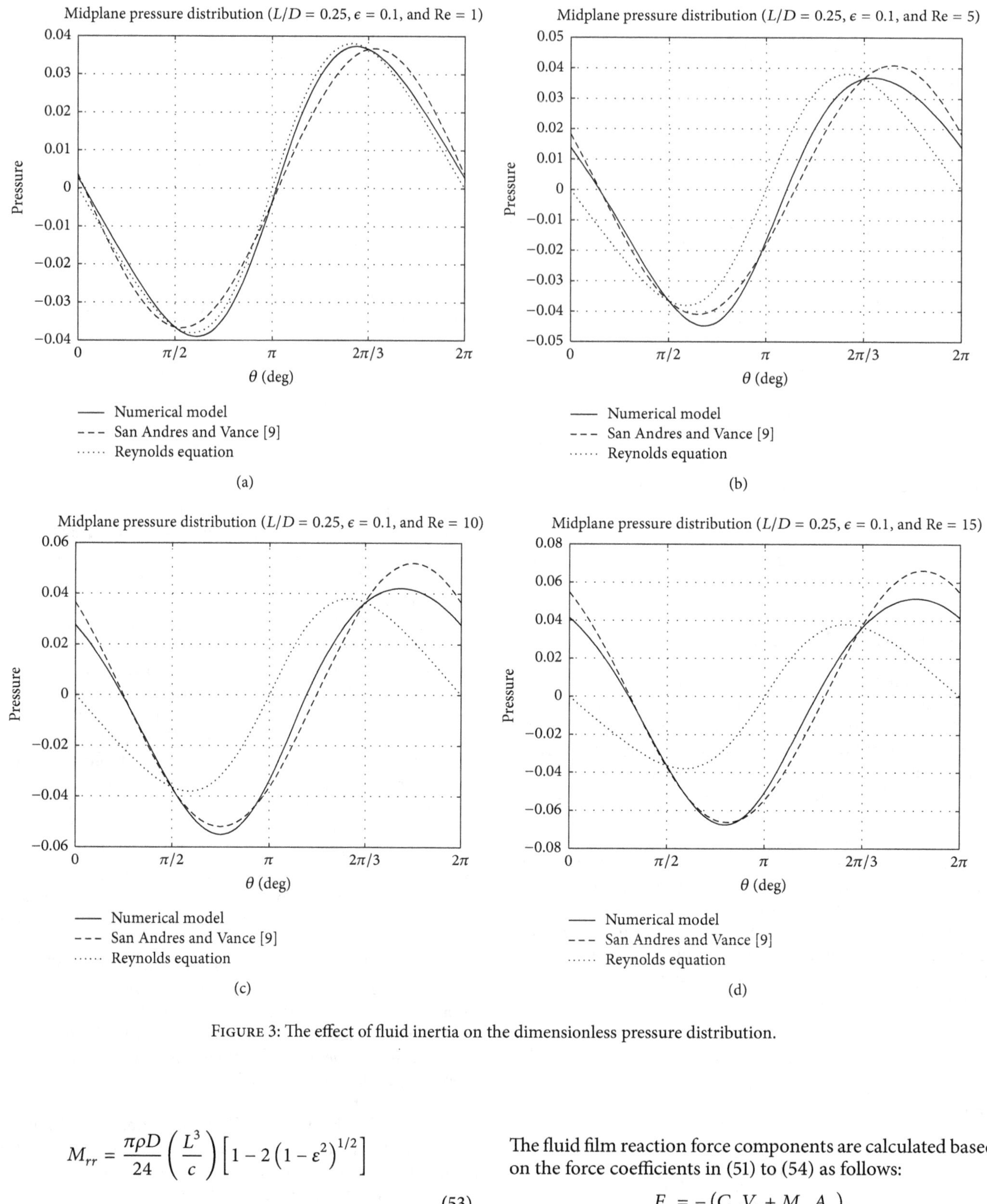

FIGURE 3: The effect of fluid inertia on the dimensionless pressure distribution.

$$M_{rr} = \frac{\pi \rho D}{24}\left(\frac{L^3}{c}\right)\left[1 - 2\left(1 - \varepsilon^2\right)^{1/2}\right] \cdot \left\{\frac{\left(1-\varepsilon^2\right)^{1/2} - 1}{\varepsilon^2\left(1-\varepsilon^2\right)^{1/2}}\right\}, \tag{53}$$

$$M_{tr} = -\frac{27\rho D}{140\varepsilon}\left(\frac{L^3}{c}\right)\left[2 + \frac{1}{\varepsilon}\ln\left(\frac{1-\varepsilon}{1+\varepsilon}\right)\right]. \tag{54}$$

The fluid film reaction force components are calculated based on the force coefficients in (51) to (54) as follows:

$$\begin{aligned} F_r &= -\left(C_{rt}V_t + M_{rr}A_r\right), \\ F_r &= -\left(C_{tt}V_t + M_{tr}A_r\right), \end{aligned} \tag{55}$$

where

$$\begin{aligned} V_t &= \varepsilon c \omega, \\ A_r &= -\varepsilon c \omega^2, \end{aligned} \tag{56}$$

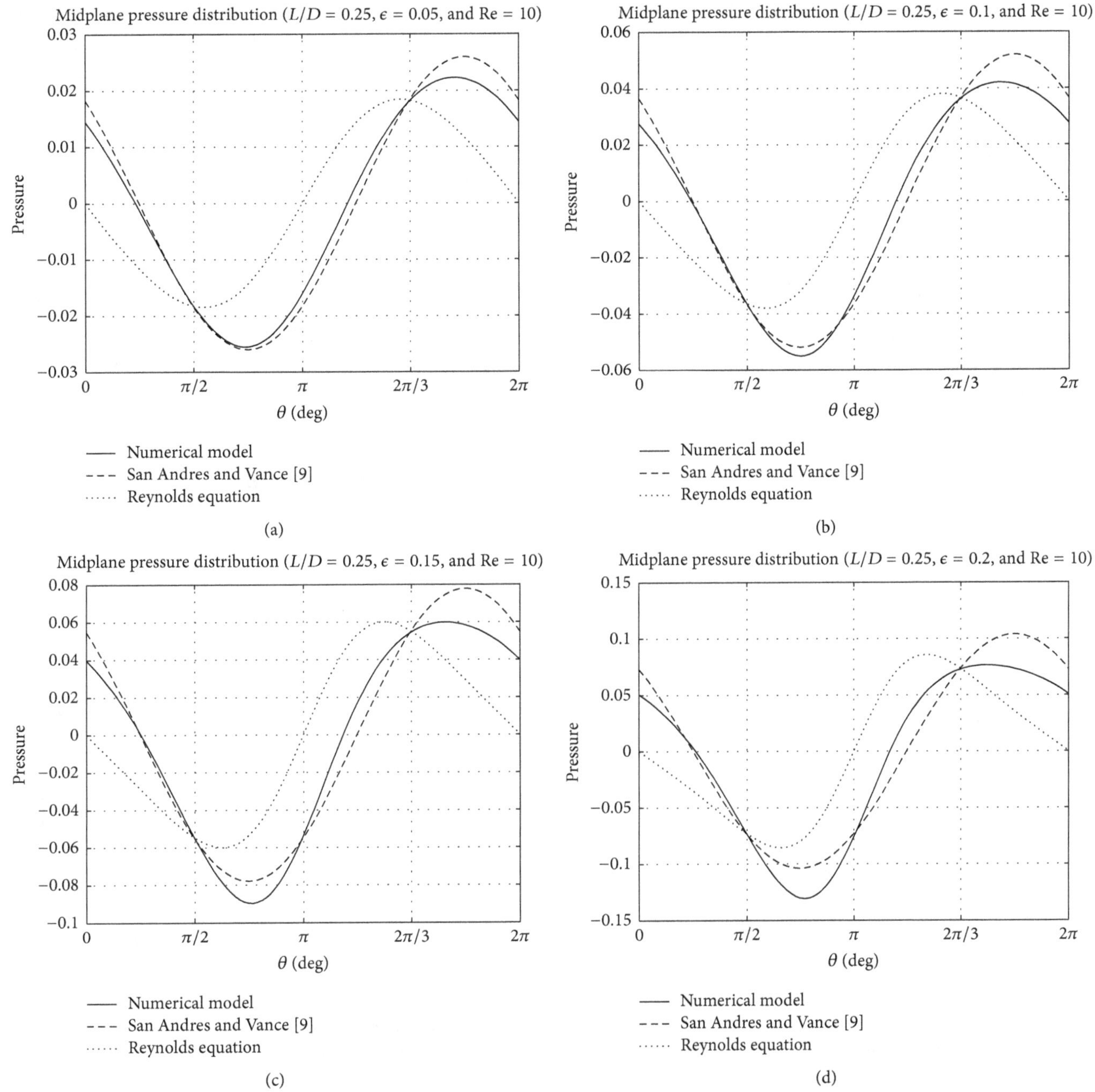

FIGURE 4: The effect of SFD journal eccentricity ratio on the dimensionless pressure distribution.

$$\overline{F_r} = \frac{C_f}{F_r},$$
$$\overline{F_t} = \frac{C_f}{F_t}, \quad (57)$$

where

$$C_f = \frac{\mu\omega R^4}{c^2}. \quad (58)$$

Subsequently, the fluid film reaction forces for the proposed model are compared against the results for San Andres and Vance [9] and Vance [25]. Figure 6 represents the effect of fluid inertia on the SFD fluid film reaction forces at small eccentricity ratios. The fluid forces are calculated by numerically integrating the lubricant hydrodynamic pressure distribution over the journal surface. The magnitude and the direction of the radial fluid film forces change with the fluid inertia effects. For SFDs executing small amplitude CCOs, the effect of fluid inertia is characterized by using a temporal inertia component. At no inertia effects (i.e., Re = 0) the effect of temporal inertia is neglected and the fluid film is characterized by Reynolds equation. In this case, the fluid film radial forces are negative, meaning that the direction of the

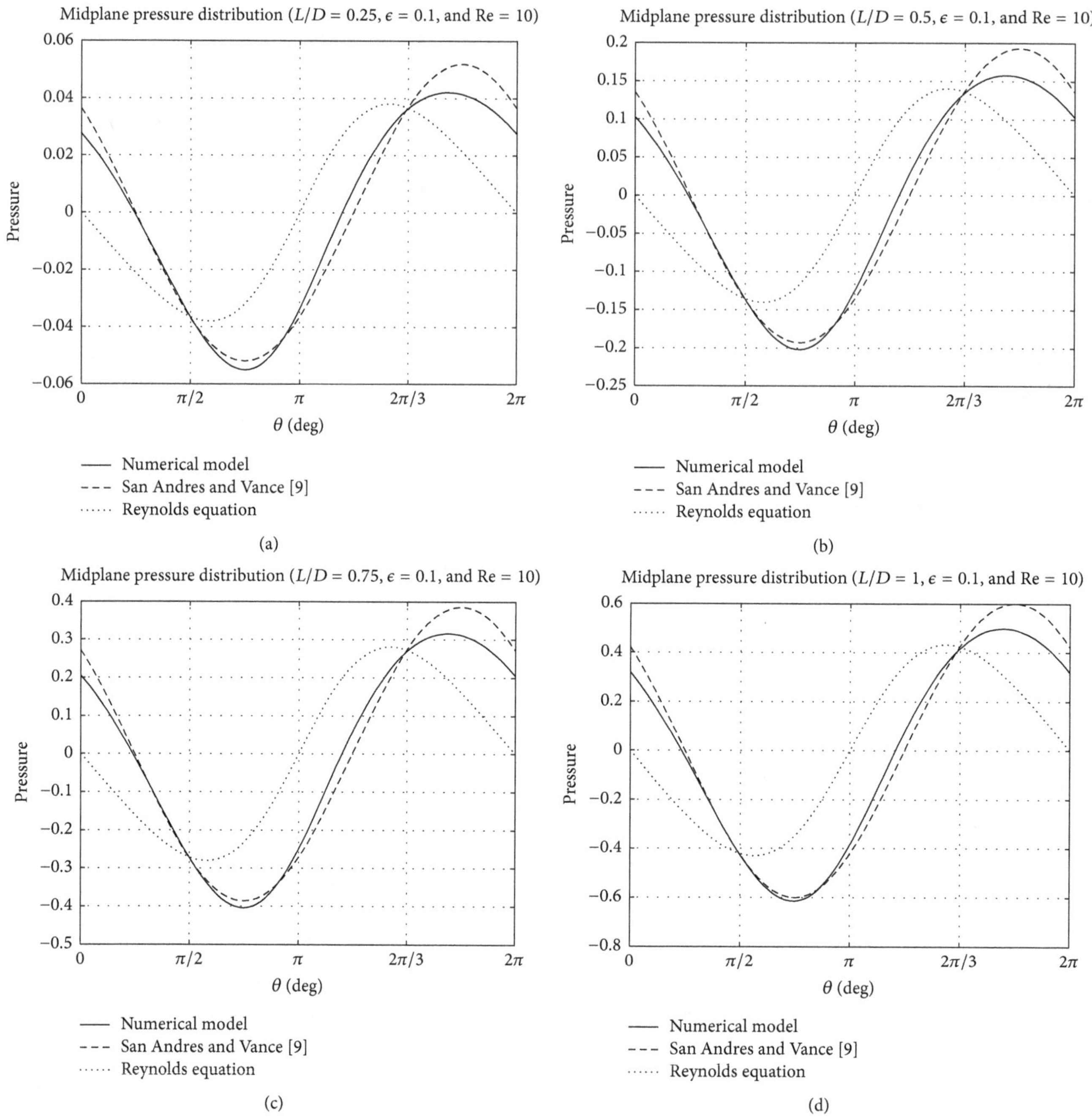

FIGURE 5: The effect of SFD journal slenderness ration on the dimensionless pressure distribution.

forces is towards the bearing center. At small Reynolds numbers (Figure 6(a)) the radial forces are positive at very small eccentricity ratios, meaning that the direction of the forces is outwards; however, as the eccentricity ratio grows slightly, the direction of the forces switches back to inwards. At moderate to large Reynolds numbers, the radial forces are positive over the range of the small eccentricity ratios and the direction of the forces stays outwards. Furthermore, at small eccentricity ratios, the magnitude of the radial forces is larger relative to the inertialess forces, meaning that the positive inertial forces dominate the negative viscous forces. Additionally, the tangential fluid film forces are not considerably influenced by fluid inertia at small Reynolds numbers. However, at larger Reynolds numbers, the magnitude of the tangential forces is slightly smaller relative to the inertialess forces. Moreover, the comparison between the numerical model and the analytical model in [9] firstly shows that the analytical model is incapable of predicting the fluid film radial forces at small Reynolds numbers, since the radial forces linearly increase with both Re and ε. Additionally, the disagreement between the results is smaller at very small eccentricity ratios and this error significantly increases for larger eccentricity

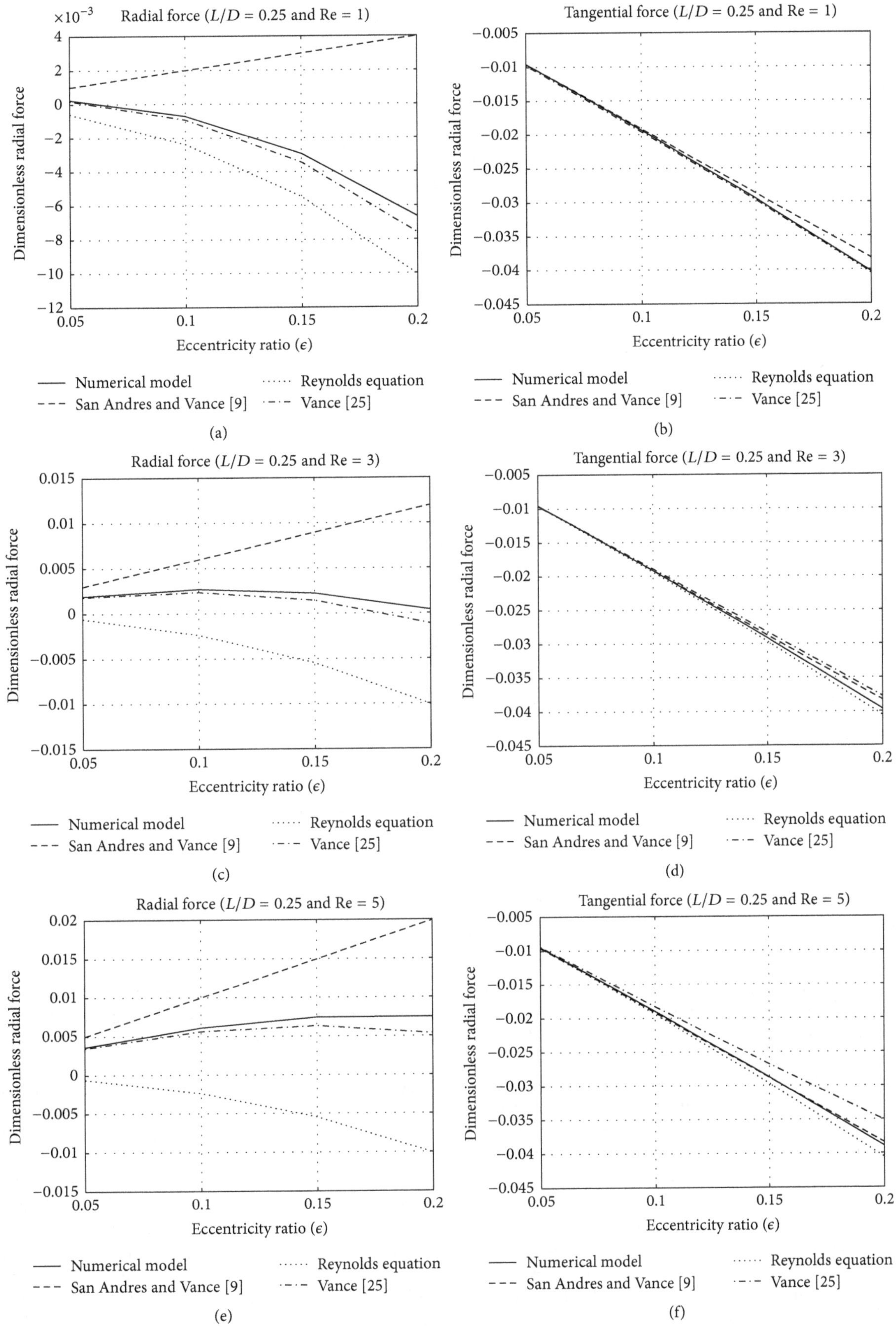

Figure 6: Continued.

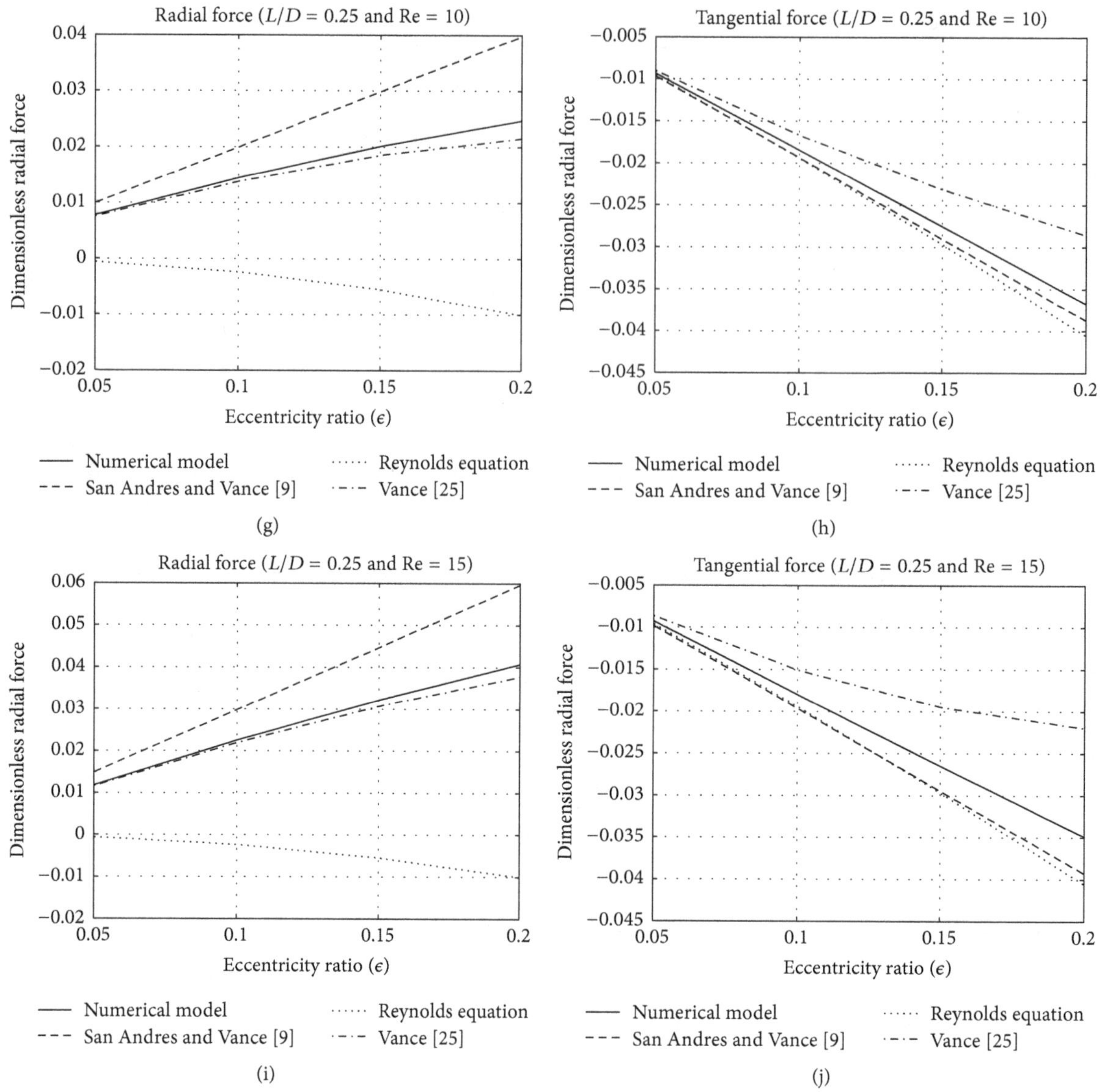

FIGURE 6: The effect of fluid inertia on the fluid film force components for small SFD journal eccentricity ratios.

ratios, showing that [9] is valid only for very small eccentricity ratios and moderate Reynolds numbers. Furthermore, the predicted tangential forces proposed by the two models are in good agreement. Finally, the radial forces based on the numerical model demonstrate very close agreement with the analytical model in [25], especially for smaller eccentricity ratios. The tangential forces are in very close agreement for small eccentricity ratios; however, the results diverge at larger eccentricity ratio values since the analytical model is driven for short SFDs.

Figure 7 demonstrates the evolution of the numerical calculation error for the proposed numerical procedure at different eccentricity ratios. In the simulations, the solution convergence criterion is fixed at 10^{-6} and the iteration is interrupted once the error drops below this criterion. For all the represented calculation examples, the solution converges rapidly and the iteration is stable. Similar convergence results are expected for arbitrary SFD parameter values, including eccentricity ratios, slenderness ratios, and Reynolds numbers.

Additionally, it is typically assumed that the higher order inertia terms in the SFD pressure calculations are negligible relative to the first-order and second-order inertia terms. An important contribution of this work is including a third-order nonlinear term in the fluid film pressure model in (31). Including this third-order expression increases the accuracy of the pressure calculations and ultimately the fluid film forces. Figure 8 demonstrates the contribution of this term to the radial and tangential fluid film forces. The fluid film forces are calculated for two cases where (1) the third-order term is neglected and (2) the third-order expression is included, and the relative error for the reaction forces is calculated at different eccentricity ratios and inertia effects. The results of the calculations demonstrate the significant contribution of the third-order term on the accuracy of the fluid film

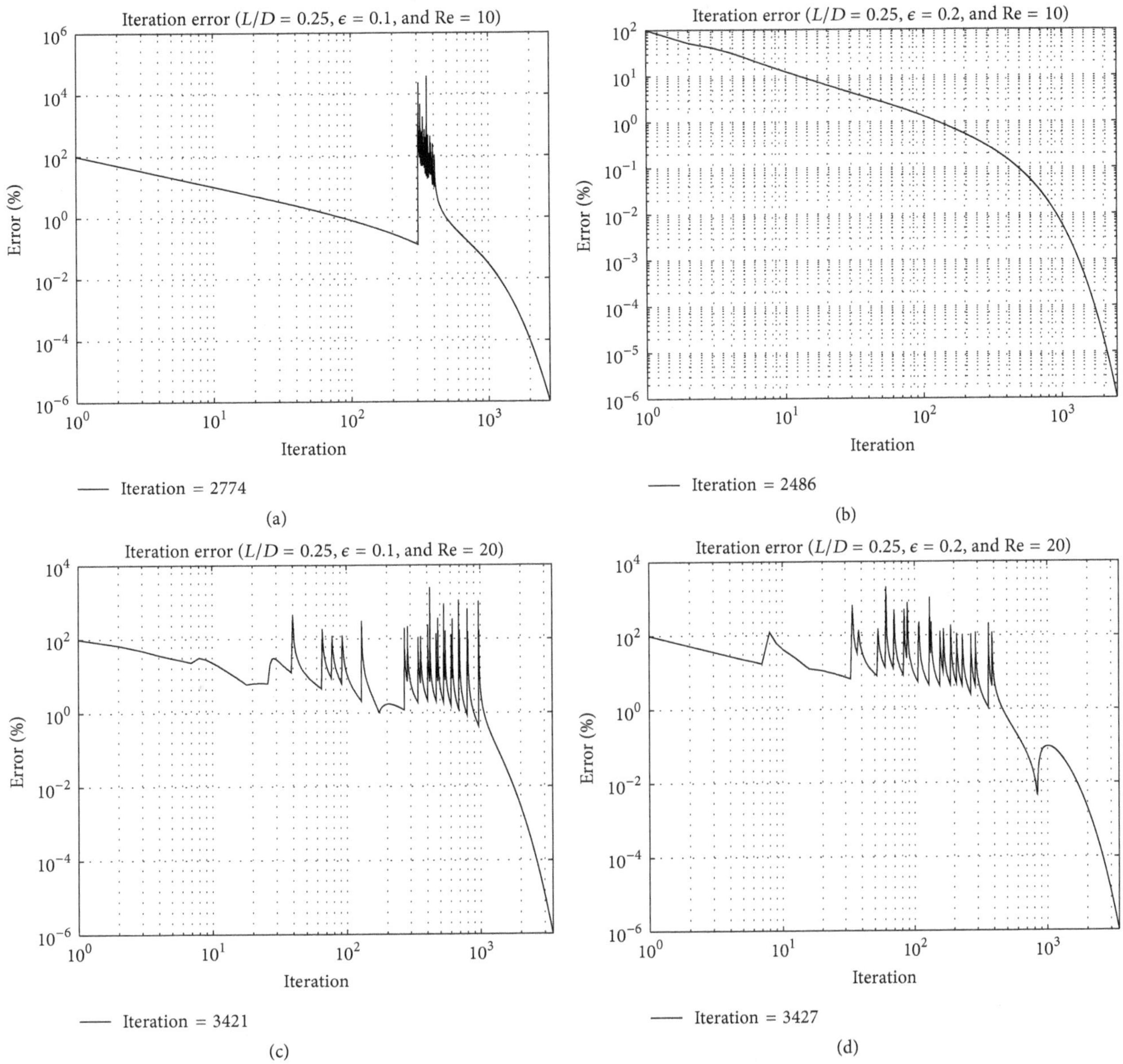

Figure 7: Evolution of the calculation error for several SFD operating conditions.

forces. In general, the calculation error for the radial forces declines by increasing the Reynolds number. Furthermore, the calculation error for the radial forces increases with the journal eccentricity ratio. Nevertheless, at small inertia effects (i.e., Re = 1 to Re = 4) the calculation error for the radial forces demonstrates irregular nonlinear behavior. The inertia effects introduce an inertial force component for the radial forces that is in the opposite direction to the viscous forces. For moderate and large Reynolds numbers the inertial forces dominate the viscous forces at small eccentricity ratios and the resultant radial forces are positive (Figures 6(e), 6(g), and 6(i)). However, at very small Reynolds numbers, the inertial forces dominate the viscous forces only over a very short range of eccentricity ratios and the radial forces quickly drop to negative (Figures 6(a) and 6(c)). This sudden variation in the radial forces results in the significant fluctuation in the calculation error. Finally, the calculation error for the tangential force increases with both Reynolds number and journal eccentricity ratio.

5. Conclusion

This work represented a hydrodynamic pressure distribution model for squeeze film dampers incorporated into high-speed turbomachinery. It is assumed that the journal center executes small amplitude CCO whirls, which is a very common practice in the turbomachinery industry. The proposed pressure model provides superior accuracy compared to the existing pressure models, including limiting geometry models. Moreover, a numerical procedure is proposed to solve the model for pointwise pressure distribution, which is significantly more computationally efficient relative to the

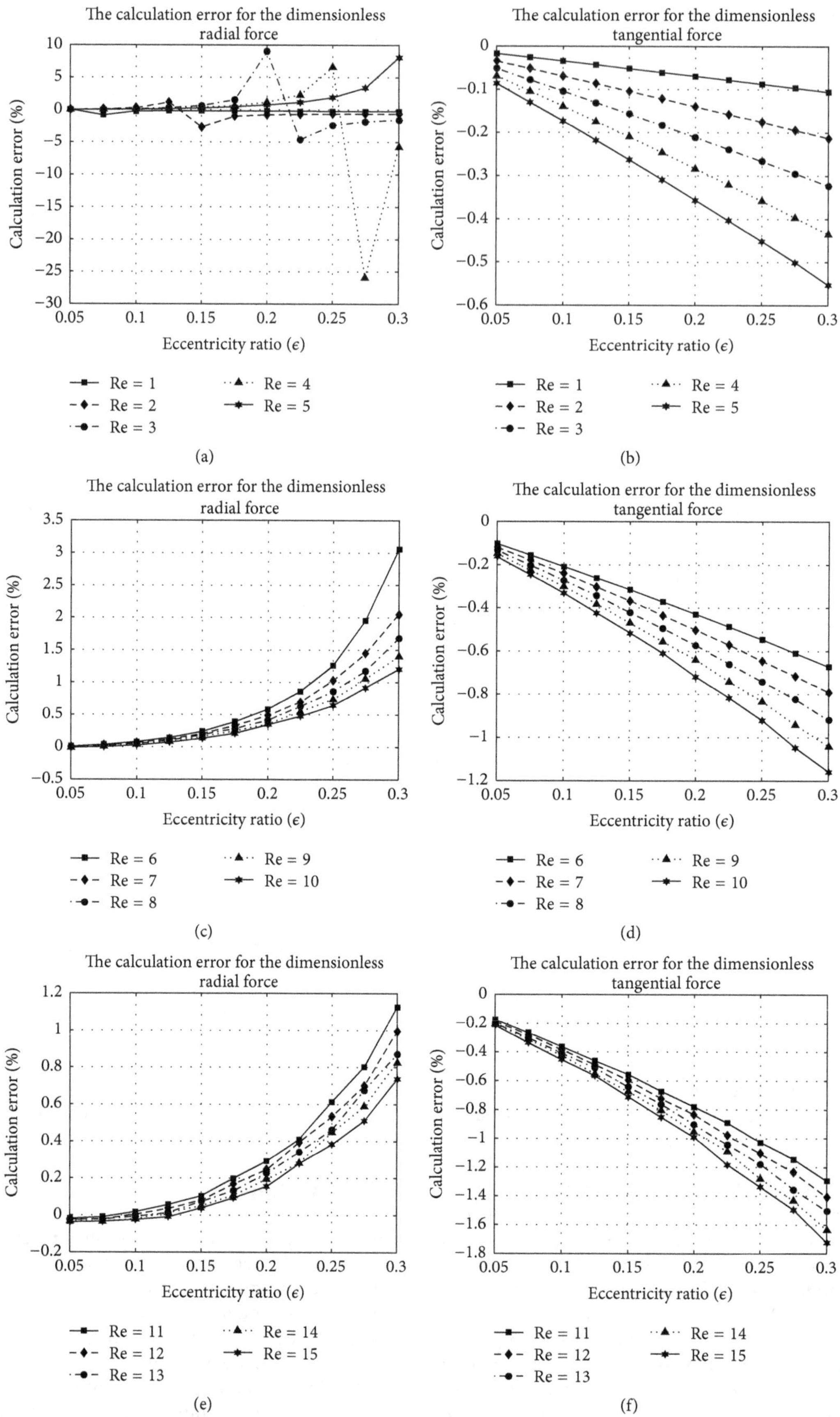

FIGURE 8: Contribution of the third-order nonlinear term in the pressure model to the dimensionless force components.

bulk-flow technique. Finally, the lubricant pressure field is numerically integrated over the journal surface to determine the fluid film reaction forces.

The analysis of the pressure model confirms the significant contribution of the fluid inertia to the dynamics of the lubricant. According to the results, we have the following.

(1) Including the inertia effects increases the pressure magnitude, shifts the position of the pressure peak in the direction of the journal precession, and changes the shape of the pressure profile. The shape of the pressure is sinusoidal at small Reynolds numbers and it changes to cosine-like at high Reynolds numbers.

(2) The amplitudes of the pressure profile significantly change with eccentricity ratio.

(3) The slenderness ratio only influences the magnitude of the pressure profile.

(4) The direction and the magnitude of the radial forces change with inertia effects.

(5) The tangential forces are not significantly influenced by fluid inertia effects at small eccentricity ratios.

The results of this work are especially valuable to the high-speed turbomachinery industry (i.e., jet engines and gas turbines) since it provides a method to quickly and accurately review damper designs and provides results as inputs to the system engineering team during the conceptual design phase of an engine/turbine. This could lead to significant development cost reduction by decreasing the number of system design iterations. Furthermore, the damper model is effectively integrated into the rotordynamic model, providing a very powerful simulation tool to accurately predict the system vibrations during the development phase.

Nomenclature

c: Radial clearance (m)
C_f: Force conversion coefficient
C_{ij}: SFD damping coefficient (N·s/m)
e: Eccentricity (m)
F_r, F_t: Radial and tangential fluid film reaction forces (N)
$h = c + e\cos\theta$: Fluid film thickness (m)
$H = 1 + \varepsilon\cos\theta$: Dimensionless fluid film thickness
L: Journal length (m)
M_{ij}: SFD inertia coefficient (kg)
P: Fluid pressure (Pa)
R: Journal radius (m)
$\mathrm{Re} = \rho\omega c^2/\mu$: Squeeze Reynolds number
t: Time (s)
u, v, w: Fluid velocity components in the x, y, z directions (m/s)
$\overline{F} = Fc^2/R^4\omega\mu$: Dimensionless fluid force
$\overline{P} = Pc^2/R^2\omega\mu$: Dimensionless fluid pressure
$\overline{P_0}$: Dimensionless zeroth-order pressure
$\overline{u} = u/R\omega$: Dimensionless circumferential velocity component
$\overline{v} = v/c\omega$: Dimensionless radial velocity component
$\overline{w} = w/R\omega$: Dimensionless axial velocity component
$\varepsilon = e/c$: Eccentricity ratio
$\eta = y/c$: Dimensionless radial component
$\theta = \theta' - \phi = x/R - \omega t$: Dimensionless circumferential component (rad)
λ: Relaxation parameter
μ: Fluid dynamic viscosity (N·s/m^2)
$\xi = z/R$: Dimensionless axial component
ρ: Fluid density (kg/m^3)
$\tau = \omega t$: Dimensionless temporal component
ω: Whirl velocity (rad/s).

Competing Interests

The authors declare that they have no competing interests.

Acknowledgments

This work was supported by grants from Natural Science and Engineering Research Council (NSERC) and Pratt and Whitney Canada.

References

[1] A. Z. Szeri, *Fluid Film Lubrication: Theory and Design*, Cambridge University Press, Cambridge, UK, 2005.

[2] S. Cooper, *Preliminary Investigation of Oil Films for the Control of Vibration*, Institution of Mechanical Engineers, 1963.

[3] L. Della Pietra and G. Adiletta, "The squeeze film damper over four decades of investigations—part I: characteristics and operating features," *Shock and Vibration Digest*, vol. 34, no. 1, pp. 3–26, 2002.

[4] L. Della Pietra and G. Adiletta, "The squeeze film damper over four decades of investigations. Part II: rotordynamic analyses with rigid and flexible rotors," *Shock and Vibration Digest*, vol. 34, no. 2, pp. 97–126, 2002.

[5] D. M. Smith, "Journal bearing dynamic characteristics-effect of inertia of lubricant," *Proceedings of the Institution of Mechanical Engineers, Conference Proceedings*, vol. 179, no. 10, pp. 37–44, 1964.

[6] T. Qingchang, L. Wei, and Z. Jun, "Fluid forces in short squeeze-film damper bearings," *Tribology International*, vol. 30, no. 10, pp. 733–738, 1997.

[7] S. Dousti, J. Cao, A. Younan, P. Allaire, and T. Dimond, "Temporal and convective inertia effects in plain journal bearings with eccentricity, velocity and acceleration," *Journal of Tribology*, vol. 134, no. 3, Article ID 031704, 2012.

[8] A. El-Shafei and S. Crandall, "Fluid inertia forces in squeeze film dampers," *Rotating Machinery and Vehicle Dynamics*, vol. 35, pp. 219–228, 1991.

[9] L. A. San Andres and J. M. Vance, "Effect of fluid inertia on squeeze-film damper forces for small-amplitude circular-centered motions," *ASLE Transactions*, vol. 30, no. 1, pp. 63–68, 1987.

[10] L. A. San Andrés, "Effect of fluid inertia on force coefficients for the long squeeze film damper," *Tribology Transactions*, vol. 31, no. 3, pp. 370–375, 1988.

[11] L. A. San Andres and J. M. Vance, "Force coefficients for open-ended squeeze-film dampers executing small-amplitude motions about an off-center equilibrium position," *ASLE Transactions*, vol. 30, no. 1, pp. 69–76, 1987.

[12] L. A. San Andres and J. M. Vance, "Effects of fluid inertia on finite-length squeeze-film dampers," *ASLE Transactions*, vol. 30, no. 3, pp. 384–393, 1987.

[13] H. Hashimoto, "Boundary conditions for the calculation of the dynamic characteristics of infinitely long journal bearings with turbulence and inertia effects," *Wear*, vol. 96, no. 1, pp. 1–16, 1984.

[14] A. El-Shafei, "Unbalance response of a Jeffcott Rotor incorporating short squeeze film dampers," *Journal of Engineering for Gas Turbines and Power*, vol. 112, no. 4, pp. 445–453, 1990.

[15] A. El-Shafei, "Modeling fluid inertia forces of short journal bearings for rotordynamic applications," *Journal of Vibration and Acoustics*, vol. 117, no. 4, pp. 462–469, 1995.

[16] A. El-Shafei, "Unbalance response of a Jeffcott Rotor incorporating long squeeze film dampers," *Journal of Vibration and Acoustics*, vol. 113, no. 1, pp. 85–94, 1991.

[17] E. Reinhardt and J. W. Lund, "The influence of fluid inertia on the dynamic properties of journal bearings," *Journal of Lubrication Technology*, vol. 97, no. 2, pp. 159–167, 1975.

[18] W. Duan, F. Chu, C.-H. Kim, and Y.-B. Lee, "A bulk-flow analysis of static and dynamic characteristics of floating ring seals," *Tribology International*, vol. 40, no. 3, pp. 470–478, 2007.

[19] J. Gehannin, M. Arghir, and O. Bonneau, "Complete squeeze-film damper analysis based on the bulk flow equations," *Tribology Transactions*, vol. 53, no. 1, pp. 84–96, 2009.

[20] L. S. Andrés and A. Delgado, "A novel bulk-flow model for improved predictions of force coefficients in grooved oil seals operating eccentrically," *Journal of Engineering for Gas Turbines and Power*, vol. 134, no. 5, Article ID 052509, 2012.

[21] L. San Andrés and J. M. Vance, "Effects of fluid inertia and turbulence on the force coefficients for squeeze film dampers," *Journal of Engineering for Gas Turbines and Power*, vol. 108, no. 2, pp. 332–339, 1986.

[22] J. A. Tichy, "Effects of fluid inertia and viscoelasticity on the one-dimensional squeeze-film bearing," *ASLE Transactions*, vol. 27, no. 2, pp. 164–167, 1984.

[23] M. F. Modest and J. A. Tichy, "Squeeze film flow in arbitrarily shaped journal bearings subject to oscillations," *Journal of Lubrication Technology*, vol. 100, no. 2, pp. 323–329, 1978.

[24] C. H. T. Pan, "Calculation of pressure, shear, and flow in lubricating films for high speed bearings," *Journal of Lubrication Technology*, vol. 96, no. 1, pp. 80–94, 1974.

[25] J. M. Vance, *Rotordynamics of Turbomachinery*, vol. 9, John Wiley & Sons, New York, NY, USA, 1988.

9

Triple-Frequency GPS Precise Point Positioning Ambiguity Resolution using Dual-Frequency based IGS Precise Clock Products

Fei Liu and Yang Gao

Department of Geomatics Engineering, University of Calgary, Calgary, AB, Canada

Correspondence should be addressed to Fei Liu; liu19@ucalgary.ca

Academic Editor: Salvatore Gaglione

With the availability of the third civil signal in the Global Positioning System, triple-frequency Precise Point Positioning ambiguity resolution methods have drawn increasing attention due to significantly reduced convergence time. However, the corresponding triple-frequency based precise clock products are not widely available and adopted by applications. Currently, most precise products are generated based on ionosphere-free combination of dual-frequency L1/L2 signals, which however are not consistent with the triple-frequency ionosphere-free carrier-phase measurements, resulting in inaccurate positioning and unstable float ambiguities. In this study, a GPS triple-frequency PPP ambiguity resolution method is developed using the widely used dual-frequency based clock products. In this method, the interfrequency clock biases between the triple-frequency and dual-frequency ionosphere-free carrier-phase measurements are first estimated and then applied to triple-frequency ionosphere-free carrier-phase measurements to obtain stable float ambiguities. After this, the wide-lane L2/L5 and wide-lane L1/L2 integer property of ambiguities are recovered by estimating the satellite fractional cycle biases. A test using a sparse network is conducted to verify the effectiveness of the method. The results show that the ambiguity resolution can be achieved in minutes even tens of seconds and the positioning accuracy is in decimeter level.

1. Introduction

With precise satellite orbit and clock products, Precise Point Positioning (PPP) using ionosphere-free (IF) code and carrier-phase observations can achieve centimeter-level accuracy if advanced error calibration models are applied [1]. The main disadvantage of PPP is that it needs significant time to reach convergence. Fast ambiguity resolution (AR) is requested to reduce this convergence time. Ambiguity fixed solutions can also further improve the PPP accuracy. Several PPP integer ambiguity resolution methods have been explored and developed in recent years [2–4]. However, it still takes a few tens of minutes to obtain reliable ambiguity resolution (AR) if only with dual-frequency observations. This is because large noise of code measurements leads to long time smoothing with Melbourne-Wübbena (MW) measurement combination [5, 6]. Moreover, the narrow-lane (NL) ambiguities need more than ten minutes to be fixed due to short wavelength [7].

At present, the third Global Positioning System (GPS) civil signal L5 is available with the launch of the latest Block IIF satellites, which enables more flexible ambiguity resolution strategies. Triple-frequency PPP AR can be achieved faster with longer wavelength, which has been studied by researchers [4, 8–10]. Gu et al. [9] verified the effectiveness of triple-frequency PPP using BeiDou datasets. For GPS triple-frequency PPP, Geng and Bock [8] fixed the ambiguities in each frequency and proved higher efficiency using simulated GPS datasets; Wang [10] only fixed the extra-wide-lane (EWL) and wide-lane (WL) ambiguities with a windowed phase smoothing phase technique to reduce the large noise of triple-frequency IF measurements. However, both GPS triple-frequency PPP AR researches mentioned above used simulated datasets, which would not suffer the inconsistency between the dual-frequency based precise clock products and the triple-frequency IF carrier-phase measurements. Very few researchers implemented the GPS triple-frequency PPP with measured datasets. Laurichesse [11] implemented the GPS triple-frequency PPP AR with integer clock products applicable for each single frequency carrier-phase measurements, which is however not widely adopted by applications yet. If widely applied dual-frequency based precise clock products are used in GPS triple-frequency PPP with observed data, there is obvious inconsistency between the dual-frequency based precise clock products and the triple-frequency IF measurements. This is because unlike the interfrequency clock biases (IFCBs) between dual-frequency IF code and phase measurement, which is stable over certain time [3], the IFCB between dual-frequency IF code and triple-frequency IF phase measurement varies up to meters over time [12, 13]. In this paper, GPS triple-frequency PPP AR method using widely used dual-frequency based precise clock products and measured datasets is proposed. In this method, the IFCBs between the dual-frequency and triple-frequency IF carrier-phase measurements are first estimated and then applied to triple-frequency IF carrier-phase measurements before using triple-frequency IF carrier-phase measurements in PPP. In this way, the widely applied dual-frequency precise clock products can be used in triple-frequency PPP. After applying the estimated IFCBs, the satellite fractional cycle biases (FCBs) can be estimated and used to recover the integer property of ambiguities. In other words, compared to dual-frequency PPP, in addition to FCBs, IFCBs between dual-frequency and triple-frequency IF carrier-phase measurements are also broadcasted to users to implement the triple-frequency GPS PPP AR.

Montenbruck et al. [13] and Li et al. [12] provided solutions for estimating L1/L2 and L1/L5 IFCBs, which can be also applied for estimating L1/L2 and triple-frequency IF carrier-phase IFCBs. Then, the triple-frequency GPS PPP AR using dual-frequency based IGS precise clock products and observed datasets can be conducted. Specifically, single-difference between-satellite PPP AR fixing only EWL and WL ambiguities is implemented. Due to limited number of GPS satellites with triple-frequency observables, dual-frequency and triple-frequency IF carrier-phase measurements are applied together to obtain PPP solutions. Datasets of one sparse network with eight reference stations in Europe are used to first estimate the IFCBs. The extra-wide-lane (EWL) and wide-lane (WL) FCBs can be generated by the reference stations in the networks after applying the estimated IFCBs. Both IFCB and FCB products have to be sent to test user stations. At the test user stations, the IFCBs are used to obtain stable float solutions. Then, after applying the estimated EWL and WL FCBs, the EWL ambiguities can be fixed instantaneously with the Melbourne-Wübbena combination while the WL ambiguities can be determined by the LAMBDA method [14]. Different from dual-frequency PPP AR which has to resolve the narrow-lane (NL) ambiguity, L1/L2 wide-lane ambiguity with valid wavelength of 3.4 m needs to be fixed in triple-frequency PPP AR, which means that fast AR can be achieved with longer wavelength. The paper is organized as follows. In Section 2, the procedures of dual-frequency single-difference PPP ambiguity resolution using dual-frequency based IGS precise products are presented. Section 3 introduces the problem of implementing triple-frequency PPP ambiguity resolution using the method in Section 2 and provides the solution. The test results and conclusions are shown in Sections 4 and 5, respectively.

2. Dual-Frequency PPP AR Using Dual-Frequency Based Precise Products

The dual-frequency undifferenced ionosphere-free (IF) combination of code and phase measurements after applying each error model (e.g., Sagnac effect, relative effect) can be expressed as

$$\begin{aligned}
P_{\text{dual}} &= \frac{f_1^2}{f_1^2 - f_2^2} P_1 - \frac{f_2^2}{f_1^2 - f_2^2} P_2 \\
&= \rho + \left(cdt^r + b^r_{P_{\text{dual}}}\right) - \left(cdt^s + b^s_{P_{\text{dual}}}\right) + T \\
&\quad + \varepsilon\left(P_{\text{dual}}\right) \\
&= \rho + cdt^r_{P_{\text{dual}}} - cdt^s_{P_{\text{dual}}} + T + \varepsilon\left(P_{\text{dual}}\right) \\
\Phi_{\text{dual}} &= \frac{f_1^2}{f_1^2 - f_2^2} \Phi_1 - \frac{f_2^2}{f_1^2 - f_2^2} \Phi_2 \\
&= \rho + \left(cdt^r + b^r_{\Phi_{\text{dual}}}\right) - \left(cdt^s + b^s_{\Phi_{\text{dual}}}\right) + T \\
&\quad + \lambda_{\text{dual}} N_{\text{dual}} + \varepsilon\left(\Phi_{\text{dual}}\right) \\
&= \rho + cdt^r_{\Phi_{\text{dual}}} - cdt^s_{\Phi_{\text{dual}}} + T + \lambda_{\text{dual}} N_{\text{dual}} \\
&\quad + \varepsilon\left(\Phi_{\text{dual}}\right) \\
cdt^r_{P_{\text{dual}}} &= cdt^r + b^r_{P_{\text{dual}}} \\
cdt^s_{P_{\text{dual}}} &= cdt^s + b^s_{P_{\text{dual}}} \\
cdt^r_{\Phi_{\text{dual}}} &= cdt^r + b^r_{\Phi_{\text{dual}}} \\
cdt^s_{\Phi_{\text{dual}}} &= cdt^s + b^s_{\Phi_{\text{dual}}},
\end{aligned} \tag{1}$$

where P_{dual} and Φ_{dual} represent the undifferenced dual-frequency ionosphere-free (IF) code and carrier-phase measurements (m), f_1 and f_2 represent the L1 and L2 frequencies, P_1, P_2 and Φ_1, Φ_2 are L1 and L2 code and phase measurements in unit of meter, respectively, ρ is the geometric range between receiver and satellite, c is the speed of light in vacuum, $cdt^s_{P_{\text{dual}}}$ (satellite code clock error), $cdt^s_{\Phi_{\text{dual}}}$ (satellite phase clock error), $cdt^r_{P_{\text{dual}}}$ (receiver code clock error), and $cdt^r_{\Phi_{\text{dual}}}$ (receiver phase clock error) are a function of the actual satellite clock error dt^s, receiver clock error dt^r, satellite dual-frequency IF code and phase biases, $b^s_{P_{\text{dual}}}$, $b^s_{\Phi_{\text{dual}}}$, and receiver dual-frequency IF code and phase biases $b^r_{P_{\text{dual}}}$, $b^r_{\Phi_{\text{dual}}}$. T is the tropospheric delay, $\varepsilon(P_{\text{dual}})$ and $\varepsilon(\Phi_{\text{dual}})$ are the noise including multipath of dual-frequency IF code and carrier-phase measurements, and λ_{dual} and N_{dual} represent the wavelength and ambiguity of dual-frequency IF combination, which is given as

$$\begin{aligned}\lambda_{\text{dual}} N_{\text{dual}} &= \frac{f_1^2 \lambda_1 N_1 - f_2^2 \lambda_2 N_2}{f_1^2 - f_2^2} \\ &= \lambda_{\text{NL}} N_1 + \frac{f_2}{f_1 + f_2} \lambda_{\text{WL}} N_{\text{WL}},\end{aligned} \tag{2}$$

where N_1 and N_2 are the integer ambiguities of L1 and L2 frequencies, N_{WL} and N_{NL} are the wide-lane and narrow-lane integer ambiguities formed by N_1 and N_2, and λ_{WL} and λ_{NL} represent the wavelength of wide-lane and narrow-lane combination, respectively. In float PPP solution, the satellite code clock error $cdt^s_{P_{\text{dual}}}$ can be provided by IGS precise clock products. The position, tropospheric delay, and receiver code clock error $cdt^r_{P_{\text{dual}}}$, together with the float ambiguities are going to be estimated.

In this study, PPP AR is implemented using the strategy proposed by Ge et al. [3] where single-difference (SD) FCBs are applied to recover the single-difference ambiguity integer property. The single-difference eliminates the receiver clock and bias. Here the dual-frequency PPP AR is introduced first to better illustrate the triple-frequency PPP AR. The single-difference dual-frequency observation model after applying various error correction models can be rewritten as

$$\begin{aligned}\nabla P_{\text{dual}} &= \frac{f_1^2}{f_1^2 - f_2^2} \nabla P_1 - \frac{f_2^2}{f_1^2 - f_2^2} \nabla P_2 \\ &= \nabla\rho - \nabla cdt^s_{P_{\text{dual}}} + \nabla T + \varepsilon(\nabla P_{\text{dual}}) \\ \nabla\Phi_{\text{dual}} &= \frac{f_1^2}{f_1^2 - f_2^2} \nabla\Phi_1 - \frac{f_2^2}{f_1^2 - f_2^2} \nabla\Phi_2 \\ &= \nabla\rho - c\nabla dt^s_{\Phi_{\text{dual}}} + \nabla T + \lambda_{\text{dual}} \nabla N_{\text{dual}} \\ &\quad + \varepsilon(\nabla\Phi_{\text{dual}}) \\ &= \nabla\rho - c\nabla dt^s_{P_{\text{dual}}} + \nabla T + \nabla A_{\text{dual}} \\ \nabla A_{\text{dual}} &= c\nabla dt^s_{P_{\text{dual}}} - c\nabla dt^s_{\Phi_{\text{dual}}} + \lambda_{\text{dual}} \nabla N_{\text{dual}} \\ &\quad + \varepsilon(\nabla\Phi_{\text{dual}}) \\ &= \nabla b^s_{P_{\text{dual}}} - \nabla b^s_{\Phi_{\text{dual}}} + \lambda_{\text{NL}} \nabla N_1 \\ &\quad + \frac{f_2}{f_1 + f_2} \lambda_{\text{WL}} \nabla N_{\text{WL}} + \varepsilon(\nabla\Phi_{\text{dual}}),\end{aligned} \tag{3}$$

where ∇ is the single-difference between-satellite operator. ∇A_{dual} is the single-difference float ambiguity. ∇N_{WL} and ∇N_{NL} are the single-difference wide-lane and narrow-lane integer ambiguities. In this section, because signals in only two frequencies are involved, the wide-lane and narrow-lane represent the corresponding combination of L1 and L2. In the next section, the wide-lane combinations will be specified. It can be seen that the IF ambiguities can be resolved by fixing the wide-lane and L1 ambiguities sequentially, which is illustrated as follows.

The single-difference between-satellite Melbourne-Wübbena (MW) combination can be formed as

$$\begin{aligned}\nabla A_{\text{MW}} &= \nabla\Phi_{\text{WL}} - \nabla P_{\text{NL}} \\ &= \lambda_{\text{WL}} \nabla N_{\text{WL}} - \nabla b^s_{\text{MW}} + \varepsilon(\nabla A_{\text{MW}}) \\ \nabla b^s_{\text{MW}} &= (\alpha_{\text{WL}} \nabla b^s_{\Phi 1} + \beta_{\text{WL}} \nabla b^s_{\Phi 2}) \\ &\quad - (\alpha_{\text{NL}} \nabla b^s_{P1} + \beta_{\text{NL}} \nabla b^s_{P2}) \\ \alpha_{\text{WL}} &= \frac{f_1}{f_1 - f_2} \\ \beta_{\text{WL}} &= -\frac{f_2}{f_1 - f_2} \\ \alpha_{\text{NL}} &= \frac{f_1}{f_1 + f_2} \\ \beta_{\text{NL}} &= \frac{f_2}{f_1 + f_2},\end{aligned} \tag{4}$$

where ∇A_{MW} is the single-difference MW combination, WL represents wide-lane combination while NL means narrow-lane combination, b^s_{MW} is the satellite MW bias, and $b^s_{\Phi 1}$, b^s_{P1}, $b^s_{\Phi 2}$, and b^s_{P2} represent the satellite phase and code biases of L1 and L2 frequencies. α_{WL}, β_{WL}, α_{NL}, and β_{NL} are the corresponding wide-lane and narrow-lane coefficients. The single-difference satellite wide-lane FCB at a receiver in unit of meter can be achieved by

$$\begin{aligned}\nabla N_{\text{WL}} &= \left\langle \frac{\nabla A_{\text{MW}}}{\lambda_{\text{WL}}} \right\rangle \\ \nabla b^s_{\text{MW}} &= \nabla A_{\text{MW}} - \lambda_{\text{WL}} \nabla N_{\text{WL}},\end{aligned} \tag{5}$$

where $\langle * \rangle$ denotes rounding of the real value to the nearest integer value. The wide-lane FCB correction with high precision can be calculated by averaging the corrections obtained in a network. For users, after applying the averaged wide-lane FCBs, the wide-lane ambiguities can be calculated by rounding of the WL ambiguity real value to the nearest integer value.

After solving the wide-lane ambiguities, if the receiver coordinate is already known, (3) can be rewritten with known parameters on the left side and unknowns on the right side as

$$\begin{aligned}&\nabla\Phi_{\text{dual}} - \frac{f_2}{f_1 + f_2} \lambda_{\text{WL}} \nabla N_{\text{WL}} - \nabla\rho + c\nabla dt^s_{P_{\text{dual}}} \\ &= \nabla T + \lambda_{\text{NL}} \nabla N_1 + \nabla b^s_{P_{\text{dual}}} - \nabla b^s_{\Phi_{\text{dual}}} + \varepsilon(\nabla\Phi_{\text{dual}}) \\ &= \nabla T + \nabla A_1 \\ &\nabla A_1 = \lambda_{\text{NL}} \nabla N_1 + \nabla b^s_{P_{\text{dual}}} - \nabla b^s_{\Phi_{\text{dual}}} + \varepsilon(\nabla\Phi_{\text{dual}}),\end{aligned} \tag{6}$$

where ∇A_1 is the float ambiguity of N_1. Similar to calculation of wide-lane FCB corrections, N_1 FCB corrections at a receiver in a network after convergence can be calculated by

$$\nabla N_1 = \left\langle \frac{\nabla A_1}{\lambda_{\mathrm{NL}}} \right\rangle$$
$$\nabla b^s_{N_1} = \nabla b^s_{P_{\mathrm{dual}}} - \nabla b^s_{\Phi_{\mathrm{dual}}} = \nabla A_1 - \lambda_{\mathrm{NL}} \nabla N_1. \quad (7)$$

By averaging the calculated N_1 FCB corrections from multiple receivers in a network, a precise value can be achieved. When users receive and apply the N_1 FCBs in the carrier-phase measurements, the integer property of N_1 ambiguities can be recovered. Then, N_1 can be resolved by applying LAMBDA method.

3. Triple-Frequency PPP AR Using Dual-Frequency Based Precise Products

According to Wang [10] and Feng [15], undifferenced triple-frequency IF carrier-phase observation model after applying error models can be written as

$$\begin{aligned}\Phi_{\mathrm{triple}} &= \frac{f_1}{f_1 - f_5}\Phi_{\mathrm{WL}_{\mathrm{L1L2}}} - \frac{f_5}{f_1 - f_5}\Phi_{\mathrm{WL}_{\mathrm{L2L5}}} \\ &= \rho + \left(cdt^r + b^r_{\Phi_{\mathrm{triple}}}\right) - \left(cdt^s + b^s_{\Phi_{\mathrm{triple}}}\right) + T \\ &\quad + \lambda_{\mathrm{triple}} N_{\mathrm{triple}} + \varepsilon\left(\Phi_{\mathrm{triple}}\right) \\ &= \rho + cdt^r_{\Phi_{\mathrm{triple}}} - cdt^s_{\Phi_{\mathrm{triple}}} + T + \lambda_{\mathrm{triple}} N_{\mathrm{triple}} \\ &\quad + \varepsilon\left(\Phi_{\mathrm{triple}}\right),\end{aligned} \quad (8)$$

where Φ_{triple} is the undifferenced triple-frequency IF carrier-phase measurement, $\Phi_{\mathrm{WL}_{\mathrm{L1L2}}}$ is the wide-lane L1 and L2 carrier-phase measurement, $\Phi_{\mathrm{WL}_{\mathrm{L2L5}}}$ is the wide-lane L2 and L5 carrier-phase measurement, $cdt^s_{\Phi_{\mathrm{triple}}}$ (satellite phase clock error) and $cdt^r_{\Phi_{\mathrm{triple}}}$ (receiver phase clock error) are a function of the actual satellite clock error dt^s, receiver clock error dt^r, and satellite and receiver triple-frequency IF phase biases $b^s_{\Phi_{\mathrm{triple}}}$, $b^r_{\Phi_{\mathrm{triple}}}$. $\lambda_{\mathrm{triple}}$ and N_{triple} represent the wavelength and ambiguity of triple-frequency IF combination, which is given as

$$\begin{aligned}\lambda_{\mathrm{triple}} N_{\mathrm{triple}} &= \frac{f_1}{f_1 - f_5}\lambda_{\mathrm{WL}_{\mathrm{L1L2}}} N_{\mathrm{WL}_{\mathrm{L1L2}}} \\ &\quad - \frac{f_5}{f_1 - f_5}\lambda_{\mathrm{WL}_{\mathrm{L2L5}}} N_{\mathrm{WL}_{\mathrm{L2L5}}}.\end{aligned} \quad (9)$$

Similar to (3), after single-difference between satellites, the observation equation can be rewritten as

$$\begin{aligned}\nabla\Phi_{\mathrm{triple}} &= \frac{f_1}{f_1 - f_5}\nabla\Phi_{\mathrm{WL}_{\mathrm{L1L2}}} - \frac{f_5}{f_1 - f_5}\nabla\Phi_{\mathrm{WL}_{\mathrm{L2L5}}} \\ &= \nabla\rho - c\nabla dt^s_{\Phi_{\mathrm{triple}}} + \nabla T + \lambda_{\mathrm{triple}}\nabla N_{\mathrm{triple}} \\ &\quad + \varepsilon\left(\nabla\Phi_{\mathrm{triple}}\right) \\ &= \nabla\rho - c\nabla dt^s_{P_{\mathrm{dual}}} + c\nabla dt^s_{P_{\mathrm{dual}}} - c\nabla dt^s_{\Phi_{\mathrm{triple}}} + \nabla T \\ &\quad + \lambda_{\mathrm{triple}}\nabla N_{\mathrm{triple}} + \varepsilon\left(\nabla\Phi_{\mathrm{triple}}\right) \\ &= \nabla\rho - c\nabla dt^s_{P_{\mathrm{dual}}} + \nabla T + \nabla A_{\mathrm{triple}} \\ \nabla A_{\mathrm{triple}} &= c\nabla dt^s_{P_{\mathrm{dual}}} - c\nabla dt^s_{\Phi_{\mathrm{triple}}} + \lambda_{\mathrm{triple}}\nabla N_{\mathrm{triple}} \\ &\quad + \varepsilon\left(\nabla\Phi_{\mathrm{triple}}\right) \\ &= \nabla b^s_{P_{\mathrm{dual}}} - \nabla b^s_{\Phi_{\mathrm{triple}}} + \frac{f_1}{f_1 - f_5}\lambda_{\mathrm{WL}_{\mathrm{L1L2}}} N_{\mathrm{WL}_{\mathrm{L1L2}}} \\ &\quad - \frac{f_5}{f_1 - f_5}\lambda_{\mathrm{WL}_{\mathrm{L2L5}}} N_{\mathrm{WL}_{\mathrm{L2L5}}} + \varepsilon\left(\nabla\Phi_{\mathrm{triple}}\right),\end{aligned} \quad (10)$$

where ∇ is the single-difference between-satellite operator. $cdt^s_{P_{\mathrm{dual}}}$ can be provided by IGS precise clock product and $\nabla A_{\mathrm{triple}}$ is the single-difference triple-frequency float ambiguity.

One issue in triple-frequency PPP AR is that, unlike the dual-frequency PPP AR in which $\nabla b^s_{P_{\mathrm{dual}}} - \nabla b^s_{\Phi_{\mathrm{dual}}}$ is relatively stable over certain time [3], triple-frequency PPP suffers a totally different situation where the variation of the single-difference ambiguities using dual-frequency based precise products can be up to meters [12, 13]. This is caused by the instability between satellite dual-frequency IF code bias and satellite triple-frequency IF phase bias, namely, $\nabla b^s_{P_{\mathrm{dual}}} - \nabla b^s_{\Phi_{\mathrm{triple}}}$. To implement PPP AR, one prerequisite is to obtain relatively stable float ambiguities. Then, after recovering the integer property of ambiguities, integer ambiguity resolution can be achieved. Since the stability of $\nabla b^s_{P_{\mathrm{dual}}} - \nabla b^s_{\Phi_{\mathrm{dual}}}$ has been proved, to obtain relative stable triple-frequency IF float ambiguities, one straightforward way is to recover $\nabla b^s_{\Phi_{\mathrm{dual}}}$ after correcting the difference between $\nabla b^s_{\Phi_{\mathrm{dual}}}$ and $\nabla b^s_{\Phi_{\mathrm{triple}}}$, namely, transforming $\nabla b^s_{P_{\mathrm{dual}}} - \nabla b^s_{\Phi_{\mathrm{triple}}}$ to $\nabla b^s_{P_{\mathrm{dual}}} - \nabla b^s_{\Phi_{\mathrm{dual}}}$.

3.1. Interfrequency Clock Bias (IFCB) Estimation. As mentioned above, the IFCB $\nabla b^s_{P_{\mathrm{dual}}} - \nabla b^s_{\Phi_{\mathrm{triple}}}$ is unstable, which is absorbed into float ambiguities, resulting in incorrect position estimation and unstable float ambiguity. In order to transform $\nabla b^s_{P_{\mathrm{dual}}} - \nabla b^s_{\Phi_{\mathrm{triple}}}$ to $\nabla b^s_{P_{\mathrm{dual}}} - \nabla b^s_{\Phi_{\mathrm{dual}}}$, the difference between $\nabla b^s_{\Phi_{\mathrm{triple}}}$ and $\nabla b^s_{\Phi_{\mathrm{dual}}}$ has to be estimated. According to Montenbruck et al. [13] and Li et al. [12], the IFCBs can be obtained based on the difference between the two ionosphere-free phase combinations (L1/L2/L5-minus-L1/L2) since the ionosphere impact is greatly reduced and other nondispersive errors contained in the observations can be eliminated. One thing needs to be mentioned is that the L2 phase center offset (PCO) and phase center variation (PCV) is used to correct for L5 due to similar frequency and unavailability of L5 PCO and PCV. To improve the computation efficiency, Li et al. [12] adopted two strategies namely epoch-differenced (ED), and satellite-differenced and epoch-differenced (SDED) method to calculate the IFCB. Difference between epochs removes the ambiguities while difference between satellites eliminates the contribution of receiver IFCB. The SDED method is applied to calculate the IFCB between dual-frequency and triple-frequency IF phase measurements in this work. IFCB in this section refers to the single-difference between-satellite IFCB. In SDED method, IFCB can be obtained by adding all the IFCB difference between two consecutive epochs from the beginning and the IFCB value at the initial epoch. The specific procedures are presented as follows.

To calculate the IFCB, the difference between the single-difference between-satellite dual-frequency and triple-frequency IF carrier-phase measurements using (3) and (10) can be formed as

$$\begin{aligned}\nabla\Phi_{\text{dual}} - \nabla\Phi_{\text{triple}} &= \nabla A_{\text{dual}} - \nabla A_{\text{triple}} \\ &= \nabla b^s_{\Phi_{\text{triple}}} - \nabla b^s_{\Phi_{\text{dual}}} + \lambda_{\text{dual}}\nabla N_{\text{dual}} \\ &\quad - \lambda_{\text{triple}}\nabla N_{\text{triple}},\end{aligned} \tag{11}$$

where $\nabla b^s_{\Phi_{\text{triple}}} - \nabla b^s_{\Phi_{\text{dual}}}$ is the IFCB between triple-frequency and dual-frequency IF phase measurements. As can be seen from (11), the variation of IFCB is reflected on the variation of difference between triple-frequency and dual-frequency IF phase measurements if no cycle slip occurs. To calculate the IFCB, the constant $\lambda_{\text{dual}}\nabla N_{\text{dual}} - \lambda_{\text{triple}}\nabla N_{\text{triple}}$ needs to be eliminated, which can be achieved by calculating the difference between two consecutive epochs, given by

$$\begin{aligned}d &= \nabla\Phi_{\text{dual}} - \nabla\Phi_{\text{triple}} \\ \Delta d &= d_k - d_{k-1} \\ &= \left(\nabla b^s_{\Phi_{\text{triple}}} - \nabla b^s_{\Phi_{\text{dual}}}\right)_k - \left(\nabla b^s_{\Phi_{\text{triple}}} - \nabla b^s_{\Phi_{\text{dual}}}\right)_{k-1},\end{aligned} \tag{12}$$

where d is the difference between the single-difference dual-frequency and triple-frequency IF phase measurements, d_k is the d at epoch k, and Δd is the difference of d at two consecutive epochs. It can be seen from (12) that Δd is the change of IFCB between two consecutive epochs. In order to improve the redundancy to calculate Δd, datasets of a network are usually utilized to calculate the elevation-angle-dependent weighted average Δd as

$$\begin{aligned}\Delta d &= \sum \frac{p_i \Delta d_i}{\sum p_i}, \\ p_i &= \frac{\sin\theta_i \sin\theta_{ri}}{\sqrt{\sin^2\theta_i + \sin^2\theta_{ri}}},\end{aligned} \tag{13}$$

where Δd_i is Δd obtained at i station in the network, θ_i is the corresponding satellite elevation angle at i station, θ_{ri} represents the reference satellite elevation angle at i station, and p_i represents the weight of Δd at i station.

The IFCB at any epoch can be calculated by adding Δd from the beginning and the IFCB at the initial epoch, expressed as

$$\delta_k = \delta_0 + \sum \Delta d, \tag{14}$$

where δ_k is the IFCB at epoch k and δ_0 is the IFCB at the initial epoch, which can be set as arbitrary value.

So far, $\nabla b^s_{\Phi_{\text{triple}}} - \nabla b^s_{\Phi_{\text{dual}}}$ can be obtained, which can be applied as another correction to triple-frequency IF carrier-phase measurements as $\nabla\Phi_{\text{triple}} + \delta_k$. One thing needs to be mentioned is that there is a constant bias between the IFCB calculated and the actual value shown as below

$$a = \left(\nabla b^s_{\Phi_{\text{triple}}} - \nabla b^s_{\Phi_{\text{dual}}}\right)_k - \delta_k. \tag{15}$$

This is because the IFCB at the initial epoch is assumed to be arbitrary value. However, this would not affect integer ambiguity resolution because the majority part of the constant bias only changes the integer values of WL ambiguities while the less than one cycle part will be treated as part of the FCB corrections.

3.2. Ambiguity Resolution. After obtaining the IFCBs, relative stable float ambiguities can be achieved. The next step is to recover the integer property of ambiguities, which is discussed in this part. From (10), it can be seen that the triple-frequency IF ambiguity resolution can be achieved by resolving the wide-lane L2/L5 and wide-lane L1/L2 integer ambiguities sequentially. Since the wavelength of WL L2/L5 is as long as 5.86 m, it will be called extra-wide-lane (EWL) and WL refers to wide-lane of L1 and L2 from now on. Similar to (4) and (5), EWL FCBs at a receiver can be determined by forming MW combination of L2 and L5 measurements as

$$\begin{aligned}\nabla A_{\text{MW}_{\text{L2L5}}} &= \nabla\Phi_{\text{WL}_{\text{L2L5}}} - \nabla P_{\text{NL}_{\text{L2L5}}} \\ &= \lambda_{\text{WL}_{\text{L2L5}}}\nabla N_{\text{WL}_{\text{L2L5}}} - \nabla b^s_{\text{MW}_{\text{L2L5}}} \\ &\quad + \varepsilon\left(\nabla A_{\text{MW}_{\text{L2L5}}}\right) \\ \nabla b^s_{\text{MW}_{\text{L2L5}}} &= \left(\alpha_{\text{WL}}\nabla b^s_{\Phi 2} + \beta_{\text{WL}}\nabla b^s_{\Phi 5}\right) \\ &\quad - \left(\alpha_{\text{NL}}\nabla b^s_{P2} + \beta_{\text{NL}}\nabla b^s_{P5}\right) \\ \alpha_{\text{WL}} &= \frac{f_1}{f_1 - f_2} \\ \beta_{\text{WL}} &= -\frac{f_2}{f_1 - f_2} \\ \alpha_{\text{NL}} &= \frac{f_1}{f_1 + f_2} \\ \beta_{\text{NL}} &= \frac{f_2}{f_1 + f_2}\end{aligned}$$

$$\nabla N_{\mathrm{MW_{L2L5}}} = \left\langle \frac{\nabla A_{\mathrm{MW_{L2L5}}}}{\lambda_{\mathrm{WL_{L2L5}}}} \right\rangle$$
$$\nabla b^s_{\mathrm{MW_{L2L5}}} = \nabla A_{\mathrm{MW_{L2L5}}} - \lambda_{\mathrm{WL_{L2L5}}} \nabla N_{\mathrm{WL_{L2L5}}}. \tag{16}$$

By averaging the satellite EWL FCBs obtained by the receivers in the network, precise EWL FCBs can be achieved and broadcasted to users. After applying the EWL FCBs, the EWL integer ambiguity can be obtained by rounding the real-valued EWL ambiguity to its nearest integer value.

After solving EWL ambiguities, similar to (6), the WL FCBs can be obtained by

$$\begin{aligned}
&\nabla\Phi_{\mathrm{triple}} - \nabla\rho + \frac{f_5}{f_1 - f_5}\lambda_{\mathrm{WL_{L2L5}}} N_{\mathrm{WL_{L2L5}}} + c\nabla dt^s_{P_{\mathrm{dual}}} + \delta \\
&\quad = \nabla T + \nabla b^s_{P_{\mathrm{dual}}} - \nabla b^s_{\Phi_{\mathrm{dual}}} + b_a \\
&\quad + \frac{f_1}{f_1 - f_5}\lambda_{\mathrm{WL_{L1L2}}} \widetilde{N}_{\mathrm{WL_{L1L2}}} + \varepsilon\left(\nabla\Phi_{\mathrm{dual}}\right) \\
&\quad = \nabla T + \nabla A_{\mathrm{WL}}
\end{aligned} \tag{17}$$
$$\begin{aligned}
&\nabla A_{\mathrm{WL}} \\
&\quad = \nabla b^s_{P_{\mathrm{dual}}} - \nabla b^s_{\Phi_{\mathrm{dual}}} + b_a + \frac{f_1}{f_1 - f_5}\lambda_{\mathrm{WL_{L1L2}}} \widetilde{N}_{\mathrm{WL_{L1L2}}} \\
&\qquad + \varepsilon\left(\nabla\Phi_{\mathrm{dual}}\right),
\end{aligned}$$

where ∇A_{WL} is the float WL ambiguity, δ is the IFCB obtained by (14), b_a is the less than one cycle part of the bias a in (15), and $\widetilde{N}_{\mathrm{WL_{L1L2}}}$ is the pseudo WL ambiguities after absorbing the majority part of bias a in (15). Due to application of IFCB, the dual-frequency IF satellite phase bias $\nabla b^s_{\Phi_{\mathrm{dual}}}$ is on the right side of the equation above instead of the triple-frequency one in (10), which means stable float ambiguities can be expected. After convergence, the float WL ambiguities can be obtained. Similar to (8), the WL FCB corrections at one receiver can be obtained as

$$\begin{aligned}
\widetilde{N}_{\mathrm{WL_{L1L2}}} &= \left\langle \frac{\nabla A_{\mathrm{WL}}}{\left(f_1/\left(f_1 - f_5\right)\right)\lambda_{\mathrm{WL_{L1L2}}}} \right\rangle \\
\nabla b^s_{N_{\mathrm{WL}}} &= \nabla b^s_{P_{\mathrm{dual}}} - \nabla b^s_{\Phi_{\mathrm{dual}}} + b_a \\
&= \nabla A_{\mathrm{WL}} - \frac{f_1}{f_1 - f_5}\lambda_{\mathrm{WL_{L1L2}}} \widetilde{N}_{\mathrm{WL_{L1L2}}}.
\end{aligned} \tag{18}$$

The WL FCBs broadcasted to users is the average of WL FCBs in a network. With the WL FCB corrections, the integer property of WL ambiguities can be recovered. The integer WL ambiguity will be searched by LAMBDA method. It can be seen that the valid wavelength of the WL ambiguity $f_1/(f_1-f_5)\lambda_{\mathrm{WL_{L1L2}}}$ is 3.40 m, which makes it much easier to obtain the WL ambiguities.

According to Geng and Bock [8] and Teunissen [16], the search area of ambiguity search space can be represented as

$$S = \pi\chi^2\sqrt{|Q|} \approx \pi\chi^2\frac{\sigma_L\delta_P}{k\lambda_1\lambda_2}, \tag{19}$$

FIGURE 1: Distribution of observation stations.

where S is the area of ambiguity search space, χ^2 is the defined threshold, $|Q|$ is the determinant of float ambiguity variance-covariance matrix, and λ_1 and λ_2 are the carrier-phase wavelength. It indicates that, with longer wavelength and smaller measurement noise, the ambiguities would be easier to be fixed. Although, for the triple-frequency IF carrier-phase, the noise is enlarged around 100 times, the valid wavelength for LAMBDA to be fixed is 3.40 m, which makes it much easier to be fixed, compared to the situation in dual-frequency PPP where NL ambiguities need to be fixed.

4. Tests and Results

To test the validity of the proposed triple-frequency ambiguity resolution method, one sparse network with 10 stations in Europe where more than four GPS Block IIF satellites (transmitting the L5 signal) can be observed for more than 60 minutes is used. Among the 10 stations, 8 stations are used as reference stations to generate the single-difference IFCBs and FCBs while the other two stations are used as test user stations. PPP AR is implemented at the test user stations. All the datasets used in this work can be downloaded from the IGS Multi-GNSS Experiment (MGEX) website. Trimble R9 receiver is used in every station to receive triple-frequency signals. The distribution of 10 stations used is shown in Figure 1. The eight stations to generate IFCBs and FCBs are cebr, vill, dlf1, gop7, metg, kiru, dyng, and masl, which are denoted in blue. The red stations redu and tlse are used to implement PPP AR. Dual-frequency PPP AR is first implemented in Section 4.1 using the same datasets as triple-frequency PPP AR in Section 4.2 to make a comparison. The same datasets are applied to fix the ambiguities of the same Block IIF satellites in both dual-frequency and triple-frequency PPP AR tests.

The observation span for redu and tlse is chosen on UTC May 9, 2015, from 02:35 am to 03:35 am when four Block IIF satellites (PRN 01, 03, 06, and 09) are in view simultaneously

TABLE 1: Details of test (unit for distance is kilometer).

Name	Interval (s)	Average distance	Min distance	Max distance	Number of Ref stations	Inter-Ref station distance
redu	1	1538	227	3006	8	35-4494
tlse	1	1620	568	2913	8	35-4494

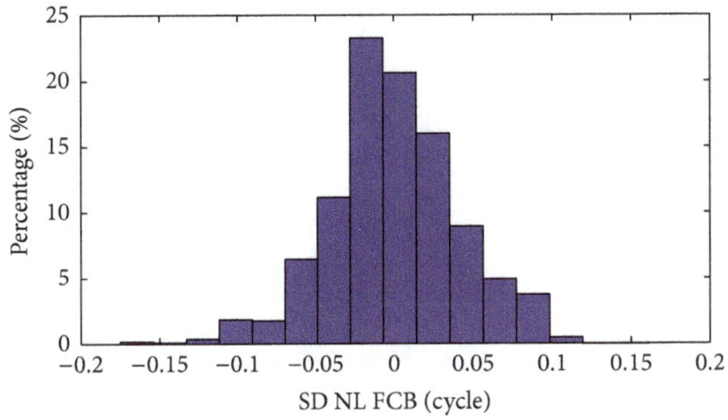

FIGURE 2: Histogram of the fractions of SD NL ambiguities after applying FCBs.

for all the stations mentioned above. The further details of the network can be found in Table 1. In both dual-frequency and triple-frequency processing, IGS final precise orbit and clock products are applied by both reference stations and test user stations. Single-difference is applied to eliminate the receiver contribution to biases in estimating the IFCBs, FCBs, and PPP AR implementation. In the tests, PRN 03 is set as the reference satellite due to high elevation. The datasets of reference stations are used to generate the FCBs to recover the integer property of ambiguities in dual-frequency processing. In triple-frequency processing, apart from the WL FCBs, IFCBs are first estimated to compensate the inconsistency in triple-frequency IF carrier-phase observations.

4.1. Dual-Frequency PPP AR. As illustrated in previous researches [2–4], the dual-frequency ionosphere-free ambiguities can be decomposed as WL and NL ambiguities. To recover the integer property of ambiguities, the WL and NL FCBs are first estimated by the reference stations. The WL ambiguities can be determined by MW combinations easily with a wavelength of 0.86 m. With known coordinates of reference stations, the float ambiguities can be converged quickly. Therefore, the NL FCBs can be determined quickly as well. Shown in Figure 2 is the histogram of NL ambiguities after applying estimated NL FCBs. It can be seen that over 95% fraction parts of NL ambiguities fall into the −0.1 to 0.1 cycle after applying the estimated NL FCBs.

The one-hour observation at user test stations redu and tlse is divided into 2 30-minute sessions at each station. Due to short wavelength of narrow-lane combination, it takes tens of minutes for the test stations to fix the ambiguities. The positioning results for the four sessions at the test user stations are shown in Figure 2. It can be seen that, after successfully fixing the correct integer ambiguities, the positioning results are improved immediately. The time to first fix (TTFF) in these four sessions is 15, 21, 19, and 29 minutes, respectively. The average TTFF is 21 minutes. For the two sessions at station redu, the RMS of horizontal positioning errors is reduced from 2.30 cm and 8.12 cm to 1.45 cm and 0.36 cm after fixing the ambiguities. The vertical RMS is reduced from 2.06 cm and 12.04 cm to 0.59 cm and 6.83 cm. The positioning results for the two sessions at tlse station are improved as well. The horizontal errors are reduced from 5.47 cm and 3.05 cm to 0.76 cm and 0.53 cm. As for the vertical direction, the error is reduced from 13.57 cm to 4.63 cm in the first session while the second session suffered slight degradation from 3.79 cm to 3.98 cm.

The results of the dual-frequency PPP AR of GPS Block IIF satellites show that the dual-frequency PPP AR can achieve centimeter-level accuracy, but it takes long time to fix the ambiguities.

4.2. Triple-Frequency PPP AR. In triple-frequency PPP processing, dual-frequency IF code measurements of Block IIF satellites are applied instead of triple-frequency IF code measurements. This is because the triple-frequency IF combination would enlarge the noise about 100 times, which degrades the positioning results. Moreover, there is inconsistency between the triple-frequency IF code measurements and the dual-frequency based precise products. The bias between the dual-frequency and triple-frequency IF code measurements is shown in Figure 4.

It can be seen from Figure 3 that the biases between dual-frequency and triple-frequency IF code measurements fluctuate greatly. Therefore, it is more appropriate to apply dual-frequency IF code measurements. Unlike dual-frequency PPP, the IFCBs between triple-frequency and dual-frequency IF phase measurements need to be estimated to compensate the inconsistency between the triple-frequency IF carrier-phase measurements and the IGS dual-frequency based precise orbit and clock precise products. The weighted IFCBs based on (12)–(14) are first calculated with the datasets of reference stations in the network. Shown in Figure 5 are the estimated weighted IFCBs with reference station datasets for PRN 01, PRN 06, and PRN 09 with PRN 03 as reference satellite and the first common epoch of each satellite as reference epoch. The initial single-difference IFCB (δ_0) is set as 0. From Figure 5, we can see that the variation of IFCB can be up to almost 3 meters over 2 hours which accounts for more than half a valid cycle (3.4 m) of WL L1/L2 in triple-frequency PPP AR. The aim of IFCB estimation is to obtain converged stable float ambiguities, which is the prerequisite to fix the ambiguities. Otherwise, the large variation IFCBs would contaminate the ambiguities, resulting in unstable float ambiguities which cannot be fixed.

As introduced before, the triple-frequency IF ambiguities can be decomposed as EWL ambiguities and WL ambiguities.

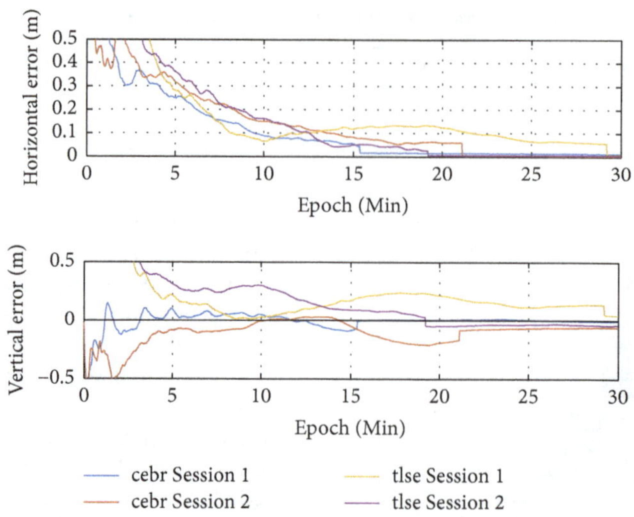

FIGURE 3: Dual-frequency positioning results of 4 30-minute sessions.

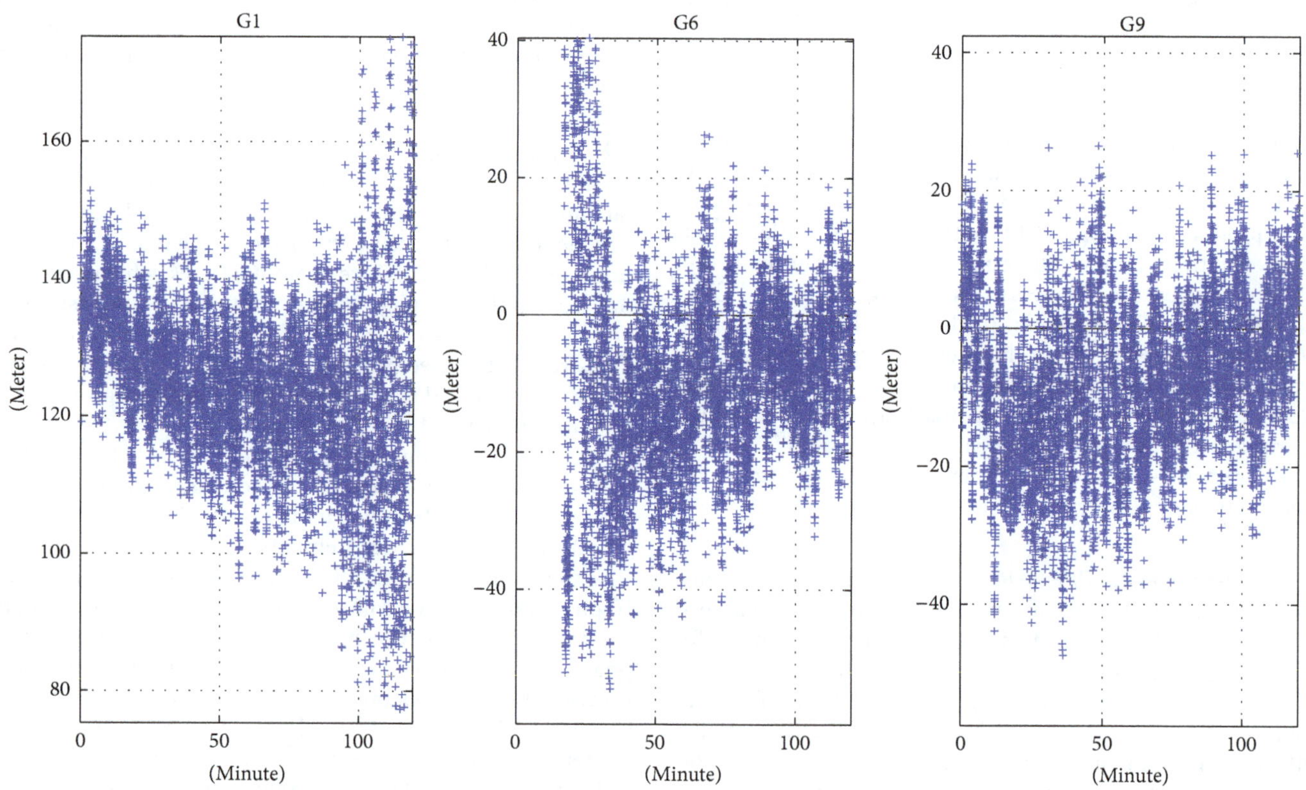

FIGURE 4: Bias between dual-frequency and triple-frequency IF code measurements.

After applying the correction of single-difference IFCBs and achieving stable single-difference float ambiguities, WL FCBs can be obtained after fixing the EWL ambiguities at reference stations with known coordinates. Single-difference EWL FCBs at reference station can be obtained by rounding the WL L2/L5 SD MW combinations directly due to large valid wavelength (5.86 m). Similarly, after the float ambiguities get converged and EWL ambiguities are fixed at reference stations, the WL FCBs can be obtained by rounding the converged WL ambiguities. Figure 6 shows that over 95% fractional parts of WL ambiguities are within 0.1 cycles after applying the WL FCBs.

With estimated IFCBs, EWL FCBs, and WL FCBs, the triple-frequency PPP AR can be implemented at the test user stations redu and tlse. The cut-off angle is set as 15° and an elevation-dependent weighted strategy was applied to measurements to reduce the impact of poor quality of measurements in low elevation. For EWL ambiguities, bias rounding

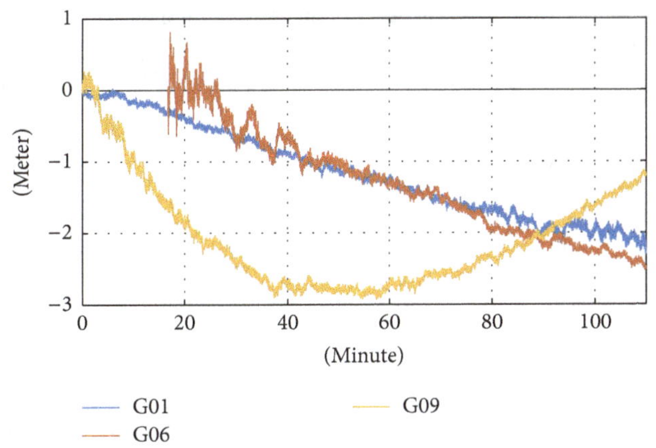

FIGURE 5: Reference epoch based SD IFCB.

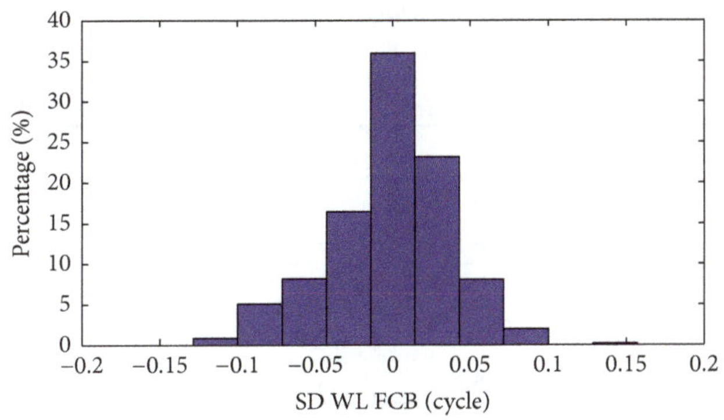

FIGURE 6: Histogram of the fractions of SD WL ambiguities after applying FCBs.

is directly applied due to the large wavelength of EWL. As for WL, LAMBDA method is applied to search for integer ambiguities and the ratio of the second minimum to the minimum quadratic form of residuals (a threshold of 3) is calculated to validate an ambiguity resolution. Because of the large noise of triple-frequency IF measurements, the weight of triple-frequency IF measurements is lower, compared to dual-frequency IF measurements. The weight ratio of dual-frequency IF phase measurements to triple-frequency IF phase measurements is about 33. Because of large noise of triple-frequency IF carrier-phase measurements and limited number of Block IIF satellites, the ambiguity resolution results can be expected that it could improve the results, but in limited extent.

As noted in the previous sections, the valid WL wavelength is about 3.4 meters, so ambiguity resolution is expected to be accomplished faster than conventional dual-frequency GPS PPP. With collected datasets at test user stations, one-hour observation with four Block IIF satellites is utilized for triple-frequency PPP AR. SD EWL and SD WL FCBs can be obtained at reference stations to recover the integer property of ambiguities. In the first test, the one-hour observation at each station is divided into 10 sessions with 6 minutes in one session. In order to ensure the success of fixing ambiguities, ambiguities are attempted to be resolved after 3 minutes. Figures 7 and 8 are the positioning errors in horizontal and vertical directions for the 10 sessions at redu and tlse, respectively. Each color represents positioning error of one session. Incorrect ambiguities are fixed for the last session at test user station tlse. It can be seen that, in most cases, the ambiguity resolution would have immediate impact on the positioning results. However, the positioning results still keep at decimeter level with slight improvement for some sessions. This is because only WL ambiguities are fixed in this work and the large noise of triple-frequency IF carrier-phase measurements and limited number of Block IIF satellites would contribute little to improving positioning results. To see the improvement of positioning results, the RMS of positioning errors after fixing ambiguities is listed in Table 2. The fixed solution results are compared with the float solution in the same period. It can be seen that correct triple-frequency WL ambiguities can be fixed in 19 sessions out of the 20 test sessions, except for the last session at station tlse. For the WL ambiguities fixed sessions at tlse, the positioning accuracy is slightly improved. The positioning accuracy after ambiguities fixed still keeps at decimeter level. For station redu, positioning accuracy degrades after WL ambiguity resolution in 4 sessions. The positioning accuracy keeps at decimeter level as well after WL ambiguity resolution. To further see the fast ambiguity resolution using triple-frequency measurements, the one-hour dataset is divided into 30 sessions with 2-minute observation in each session. The ambiguities are attempted to be fixed after 30 seconds of processing. The positioning results are shown in Table 3. Similarly, the positioning errors of fixed solution and float solution in the same period are compared. This time all the WL ambiguities can be fixed for all the 60 sessions at two test user stations. For station redu, the positioning accuracy degrades slightly after fixing the WL ambiguities in 40% sessions (12/30), while 2 sessions suffer positioning accuracy degradation at tlse. The fixed ambiguities of all sessions are correctly fixed, regardless of improvement of positioning or not. Therefore, the reason that leads to positioning degrading is not incorrect fixed ambiguities. There is great chance to obtain worse positioning results with limited number of WL ambiguities fixed partly due to the large noise of triple-frequency IF carrier-phase measurements. Besides, the quality of SD IFCBs and SD FCBs also has great impact on the positioning results.

5. Conclusions

In this paper, triple-frequency GPS PPP AR is investigated based on observed datasets. A general data processing method is proposed and verified. The collected datasets of triple-frequency IF carrier-phase measurements indicate an inconsistency with the IGS dual-frequency based precise clock product obtained by L1/L2 ionosphere-free combination. The interfrequency biases are estimated and applied to triple-frequency IF carrier-phase measurements in order to obtain relatively stable float ambiguities. Limited by the available triple-frequency datasets, one network with ten stations is utilized. In the network, the PPP AR is implemented at the two test user stations with the IFCBs and FCBs generated by eight reference stations. Compared with dual-frequency PPP

TABLE 2: RMS of positioning results with fixed ambiguities and float ambiguities during the same period (after fixing ambiguities) for 10 6-minute sessions at redu and tlse. In the table, "float" means RMS of positioning error with float ambiguities, "fixed" means RMS of positioning error with fixed WL ambiguities, "—" means there is no corresponding float solution RMS for comparison, and "fail" means failure of fixing correct WL ambiguities during the session. The unit in this table is centimeter.

Station	1	2	3	4	5	6	7	8	9	10
redu										
Float	33.5	41.4	43.1	43.1	56.9	49.1	62.1	69.8	60.9	54.5
Fixed	30.8	41.1	50.4	38.6	72.2	49.3	54.3	56.7	70.9	49.1
tlse										
Float	45.8	63.1	97.6	81.3	89.0	125.8	106.0	61.8	98.0	—
Fixed	44.1	69.5	66.3	67.3	85.6	84.9	64.0	57.5	45.8	Fail

TABLE 3: RMS of positioning results with fixed ambiguities and float ambiguities during the same period (after fixing ambiguities) for 30 2-minute sessions at redu and tlse. In the table, "float" means RMS of positioning error with float ambiguities and "fixed" means RMS of positioning error with fixed WL ambiguities.

Station	1	2	3	4	5	6	7	8	9	10
redu										
Float	58.2	68.6	68.6	59.2	96.7	86.7	94.3	67.6	76.2	51.6
Fixed	58.1	41.3	61.9	48.3	53.5	68.7	76.8	70.3	51.9	44.8
tlse										
Float	132.8	54.1	88.8	132.7	91.6	80.1	118.8	173.0	146.7	97.1
Fixed	89.1	44.1	80.7	123.8	88.3	59.5	55.2	113.9	92.2	73.5
Station	11	12	13	14	15	16	17	18	19	20
redu										
Float	92.6	84.0	48.3	65.2	92.7	66.8	63.8	67.8	80.5	41.6
Fixed	41.7	44.3	55.1	104.7	100.6	47.3	51.1	80.6	61.2	46.8
tlse										
Float	86.4	142.0	141.0	85.1	86.9	186.6	142.3	122.1	140.6	121.0
Fixed	53.7	134.3	99.2	96.3	74.9	113.5	93.5	67.7	72.2	68.5
Station	21	22	23	24	25	26	27	28	29	30
redu										
Float	60.8	78.5	87.9	55.5	64.9	57.9	76.7	58.3	120.6	51.8
Fixed	74.8	75.9	96.9	59.4	62.0	80.5	103.4	61.9	62.2	85.8
tlse										
Float	102.2	80.4	75.0	159.9	123.9	161.0	100.1	163.8	160.9	162.4
Fixed	57.8	78.2	39.3	99.6	44.7	51.2	52.8	59.8	249.8	108.8

AR, triple-frequency ambiguities can be fixed much faster. However, the fixed triple-frequency solution still keeps at decimeter level. In some cases, the accuracy even degrades after correctly fixing the ambiguities, because only a limited number of WL ambiguities are fixed and the triple-frequency IF combination enlarges the measurement noise. The test results show that most positioning results become better after fixing ambiguities, which confirms the effectiveness of the proposed fast triple-frequency PPP AR method. In this study, the satellite L5 PCO and PCV are assumed as the same value of satellite L1 and L2. The receiver L5 PCO and PCV are assumed to be the same value of L2 due to similar frequency. Actually, due to the large coefficients in triple-frequency IF combination, the effect of PCO and PCV could have a decimeter-level impact on the measurements in some situations, which may cause inaccurate estimation of IFCBs. The estimation of IFCBs can be omitted if triple-frequency precise clock product is available in the future, and the results can be better if precise information (e.g., satellite and receiver PCO and PCV) for the third frequency signal can be available.

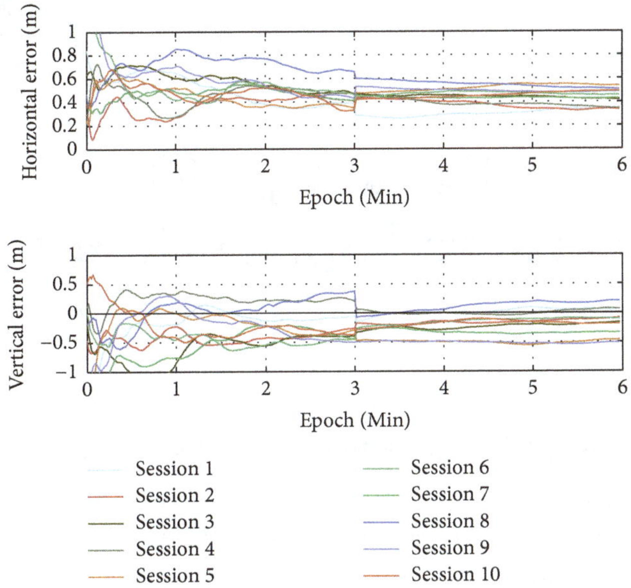

FIGURE 7: Positioning results of 10 6-minute sessions at redu.

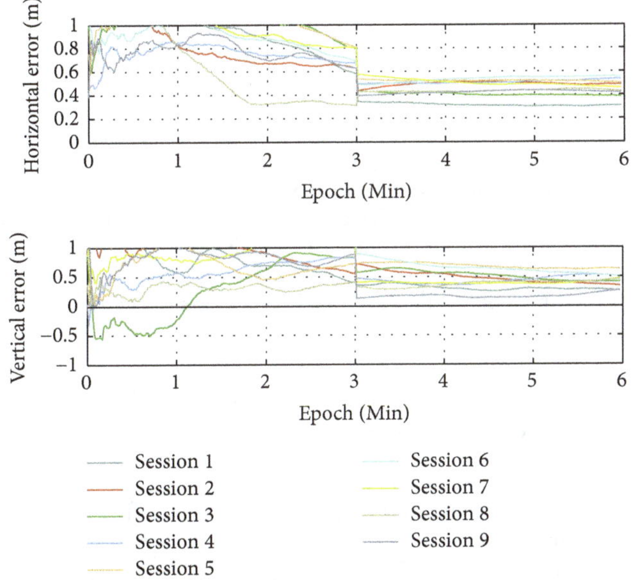

FIGURE 8: Positioning results of 10 6-minute sessions at tlse.

Competing Interests

The authors declare that there is no conflict of interests regarding the publication of this paper.

References

[1] J. F. Zumberge, M. B. Heflin, D. C. Jefferson, M. M. Watkins, and F. H. Webb, "Precise point positioning for the efficient and robust analysis of GPS data from large networks," *Journal of Geophysical Research B: Solid Earth*, vol. 102, no. 3, pp. 5005–5017, 1997.

[2] P. Collins, "Isolating and estimating undifferenced GPS integer ambiguities," in *Proceedings of the Institute of Navigation National Technical Meeting (NTM '08)*, pp. 720–732, San Diego, Calif, USA, January 2008.

[3] M. Ge, G. Gendt, M. Rothacher, C. Shi, and J. Liu, "Resolution of GPS carrier-phase ambiguities in Precise Point Positioning (PPP) with daily observations," *Journal of Geodesy*, vol. 82, no. 7, pp. 389–399, 2008.

[4] D. Laurichesse, F. Mercier, J.-P. Berthias, P. Broca, and L. Cerri, "Integer ambiguity resolution on undifferenced GPS phase measurements and its application to PPP and satellite precise orbit determination," *Navigation, Journal of the Institute of Navigation*, vol. 56, no. 2, pp. 135–149, 2009.

[5] W. G. Melbourne, "The case for ranging in GPS-based geodetic systems," in *Proceedings of the 1st International Symposium on Precise Positioning with the Global Positioning System*, vol. 1519, pp. 373–386, Rockville, Md, USA, April 1985.

[6] G. Wübbena, "Software developments for geodetic positioning with GPS using TI-4100 code and carrier measurements," in *Proceedings of the 1st International Symposium on Precise Positioning with the Global Positioning System*, vol. 19, Rockville, Md, USA, April 1985.

[7] J. Geng, F. N. Teferle, X. Meng, and A. H. Dodson, "Towards PPP-RTK: ambiguity resolution in real-time precise point positioning," *Advances in Space Research*, vol. 47, no. 10, pp. 1664–1673, 2011.

[8] J. Geng and Y. Bock, "Triple-frequency GPS precise point positioning with rapid ambiguity resolution," *Journal of Geodesy*, vol. 87, no. 5, pp. 449–460, 2013.

[9] S. Gu, Y. Lou, C. Shi, and J. Liu, "BeiDou phase bias estimation and its application in precise point positioning with triple-frequency observable," *Journal of Geodesy*, vol. 89, no. 10, pp. 979–992, 2015.

[10] M. Wang, *Ambiguity resolution with precise point positioning [Ph.D. thesis]*, Department of Geomatics Engineering, University of Calgary, 2014.

[11] D. Laurichesse, *Handling the Biases for Improved Triple-Frequency PPP Convergence*, GPS World, 2015.

[12] H. Li, X. Zhou, B. Wu, and J. Wang, "Estimation of the inter-frequency clock bias for the satellites of PRN25 and PRN01," *Science China: Physics, Mechanics and Astronomy*, vol. 55, no. 11, pp. 2186–2193, 2012.

[13] O. Montenbruck, U. Hugentobler, R. Dach, P. Steigenberger, and A. Hauschild, "Apparent clock variations of the Block IIF-1 (SVN62) GPS satellite," *GPS Solutions*, vol. 16, no. 3, pp. 303–313, 2012.

[14] P. J. G. Teunissen, "The least-squares ambiguity decorrelation adjustment: a method for fast GPS integer ambiguity estimation," *Journal of Geodesy*, vol. 70, no. 1-2, pp. 65–82, 1995.

[15] Y. Feng, "GNSS three carrier ambiguity resolution using ionosphere-reduced virtual signals," *Journal of Geodesy*, vol. 82, no. 12, pp. 847–862, 2008.

[16] P. J. G. Teunissen, "On the sensitivity of the location, size and shape of the GPS ambiguity search space to certain changes in the stochastic model," *Journal of Geodesy*, vol. 71, no. 9, pp. 541–551, 1997.

10

Autonomous Rendezvous and Docking with Nonfull Field of View for Tethered Space Robot

Panfeng Huang,[1,2] Lu Chen,[1,2] Bin Zhang,[1,2] Zhongjie Meng,[1,2] and Zhengxiong Liu[1,2]

[1]*National Key Laboratory of Aerospace Flight Dynamics, Northwestern Polytechnical University, Xi'an, Shaanxi 710072, China*
[2]*Research Center for Intelligent Robotics, School of Astronautics, Northwestern Polytechnical University, Xi'an, Shaanxi 710072, China*

Correspondence should be addressed to Panfeng Huang; pfhuang@nwpu.edu.cn

Academic Editor: Christian Circi

In the ultra-close approaching phase of tethered space robot, a highly stable self-attitude control is essential. However, due to the field of view limitation of cameras, typical point features are difficult to extract, where commonly adopted position-based visual servoing cannot be valid anymore. To provide robot's relative position and attitude with the target, we propose a monocular visual servoing control method using only the edge lines of satellite brackets. Firstly, real time detection of edge lines is achieved based on image gradient and region growing. Then, we build an edge line based model to estimate the relative position and attitude between the robot and the target. Finally, we design a visual servoing controller combined with PD controller. Experimental results demonstrate that our algorithm can extract edge lines stably and adjust the robot's attitude to satisfy the grasping requirements.

1. Introduction

The increasing number of space debris has been a serious threat for the safety of space activities. Although many debris removal methods have already been proposed [1–3], such as the electrodynamic tether [4] and foam method [5], the majority of them are limitedly examined on ground and far from being practical. In practice, the traditional "space platform + multi-DOF manipulator arm" system has been the preferred solution; however, it has several disadvantages, such as small operating range, complicated control operation, and being only applicable for cooperative target capturing, whose attitude and movement are stable. Therefore, developing on-orbit service technologies, especially for noncooperative targets, is critical. In order to overcome the problems caused by manipulator, Huang et al. [6, 7] proposed the tethered space robot (TSR) system for noncooperative target capturing and detumbling [8, 9]. TSR is a new type of space robot, which comprises a platform, space tether, and an operational robot, as shown in Figure 1. The operational robot is released by the platform through the space tether to approach the target and accomplish an on-orbit task. Compared with the manipulator arm, TSR has two main merits: a large operational radius (as long as 1,000 m) and the enhanced flexibility.

The flow of providing an on-orbit service can be divided into three sequential stages [10]:

(a) The platform distinguishes the target from space environment and flies towards it gradually from a far distance.

(b) The platform flies around the target and detects corresponding regions to identify a suitable position to release the operational robot. Then, the operational robot will be launched out and freely fly towards the frontal target. Meanwhile, it keeps detecting the suitable grasping region of the target to guide end-effector in precise manipulation.

(c) The operational robot grasps the target once reaching the appointed position (usually less than 0.15 m) and eliminates the tumbling of the robot-target combination with its own propellants. Then it provides on-orbit services, such as injecting propellants and dragging the target into a graveyard orbit.

In the third stage, the operational robot needs to capture the target spacecraft through the docking system (such as the docking ring) in order to provide on-orbit services. But the docking system usually works for cooperative targets

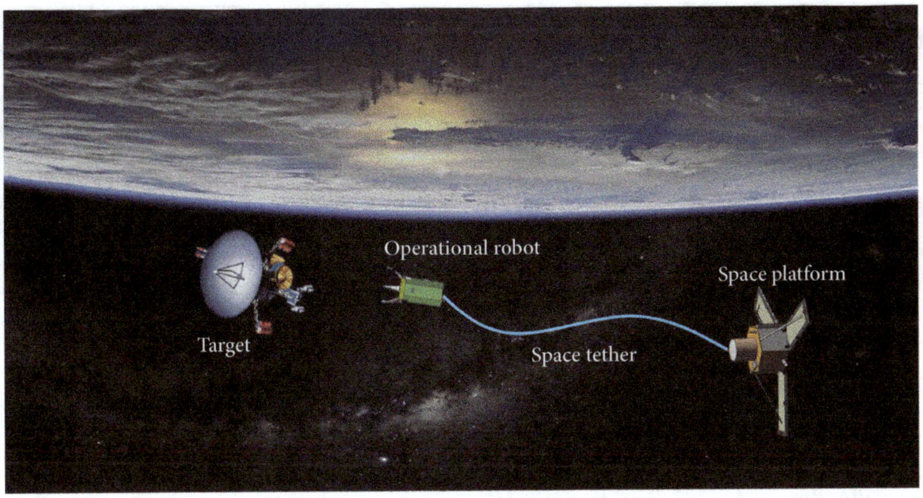

FIGURE 1: Diagram of tethered space robot.

only. When it comes to noncooperative targets, any device or structure mounted on them is unknown in advance. To facilitate the determination of the grasping position, we analyze the common spacecraft structures and select satellite bracket as a suitable location for grasp, which is usually structurally robust and widely present in spacecraft [11].

During the approaching process, it is necessary for TSR to recognize the grasping position and measure their relative positions and attitudes. The calculated information will be fed into the control system to adjust TSR's self-attitude and guide its approaching path. Though microwave radar and laser radar have been used in relative navigation, we select charge-coupled device (CCD) cameras mounted on the operational robot to provide vision-based measurements considering their low mass and rich view information [12].

Providing the on-orbit service involves many key techniques and one of them is the control of the relative position and attitude between TSR and the target.

Yuga et al. [13] proposed a coordinated control method using tension and thruster. Wang et al. [14, 15] proposed that, through altering the position of the connecting point between the tether and the robot, the tether tension force was able to achieve the orbit and attitude control simultaneously. Mori and Matunaga [16] designed tether tension/rope length controller and realized the attitude control. Experiments showed that this controller could save propellants. But these methods all require that the relative position between the space platform, tethered robot, and the target is known. In order to measure the relative position and orientation between them, visual servoing control problem has been widely studied. Xu et al. [17] proposed a stereo vision-based method to measure the relative position and orientation. The basic idea is to identify the feature points of the target in 3D space. Thienel et al. [18] presented a nonlinear method to estimate the attitude of the target spacecraft and then realized the tracking to it. But it assumed that the target attitude was already estimated using vision method. Hafez et al. [19] proposed a visual servoing control method of the dual-arm space robot. In [20], Dong and Zhu used the extended Kalman filter to estimate pose and motion of noncooperative target in real time. Cai et al. [21] proposed a monocular visual servoing control system for the tethered space robot. It realized the real time tracking of noncooperative target satellite in complex scenes. For these position and orientation measurement methods, the key technology is the detection and recognition of the characteristics of the target spacecraft. Considering the wide existence of circle-shape device, Hough Transform methods [22] are generally adopted to detect space target in autonomous rendezvous and docking. Casonato and Palmerini [23] used Hough Transform to identify basic geometric shapes, such as circles and straight lines, in rendezvous monitoring. The work [24] also used Hough Transform to convert the image from spatial space to parameter space, facilitating the detection of launch bolt holes. In [25], two marks, an external circle and an internal circle, are laid on different planes of the target to indicate the relative attitude and position. They can be easily detected by Hough Transform methods too. In addition, Ramirez et al. [26] proposed a quadrilateral recognition method, which could be used to recognize the spacecraft's solar panels. Grompone Von Gioi et al. [27] proposed a line segment detection algorithm based on image gradients. It could achieve rapid detection of line segments and had been applied to detect the edge lines of the spacecraft's solar panels. Wang et al. [28, 29] proposed a new approach to dynamic eye-in-hand visual tracking for the robot manipulator in constrained environments with an adaptive controller.

The above methods generally aim at detecting the characteristics of the target from a relatively far distance, where the point features, linear features, or geometric features of the target could be fully detected [30], as shown in Figures 2(a) and 2(b). But when in extremely close distance (usually less than 1 m), these features are generally difficult to extract and are incomplete due to the limited FOV of cameras. The brackets supporting the solar panels are the element present in the majority of spacecraft, excluding the small cube satellites which often accommodate body-mounted panels. These elements can be easily recognized by cameras using

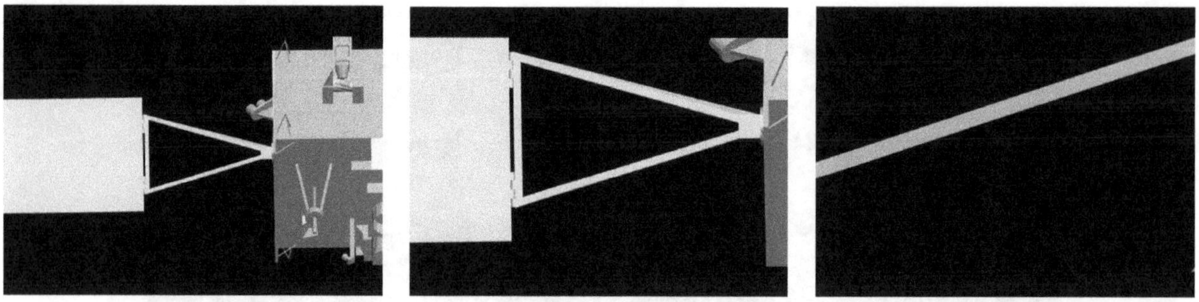

(a) Far distance (b) Close distance (c) Extremely close distance

FIGURE 2: Images captured in different approaching distances.

their edge lines (Figure 2(c)). But for the above-mentioned methods, simply using the detected edge line features is inadequate and they tend to fail [31].

In this paper, we propose a visual servoing control method based on the detection of margin line on the satellite brackets. Firstly, a gradient based edge line detection method is proposed to extract edges of the satellite brackets. Then, we construct the model to acquire the relative position and attitude between TSR and the target. Lastly, we integrate the PD controller to design a new visual servoing controller. It is able to control TSR's attitude to satisfy the grasping requirements. This method is appropriate for addressing the relative position and attitude problem in extremely close distance when TSR approaches the target satellite.

The remainder of this paper is organized as follows. In Section 2, we introduce the project model of edge lines. The detailed description of the line detection algorithm and servoing control method is presented in Sections 3 and 4. Section 5 illustrates and analyzes the experiments and we conclude our work in Section 6.

2. Projection Model of Edge Lines

2.1. Coordinates Definition. Observing system is composed of the operational robot and the satellite bracket. In extremely close stage, TSR moves towards the grasping position and only edge information of the brackets is available. In order to facilitate the relative measurement and future expression, the coordinate systems used in our paper are defined as follows:

(1) The coordinate system of operating robot body $O_GX_GY_GZ_G$: the origin O_G is located in the centroid of the robot; O_GX_G axis points to the direction of flight; O_GZ_G axis points to the geocenter; O_GY_G axis direction is determined by the right-hand rule.

(2) The coordinate system of the capture point $O_ZX_ZY_ZZ_Z$: the origin O_Z is located at the capture point of the target panel bracket; directions of O_ZZ_Z axis and O_CZ_C axis are same; the direction of axis O_ZX_Z is along the grasping pole of panel bracket; axis O_ZY_Z is determined by the right-hand rule.

(3) Camera coordinate system $O_CX_CY_CZ_C$: origin O_C is at the optical center of the camera, O_CZ_C axis is the optical axis, and O_CX_C axis and O_CY_C axis are horizontal axis and vertical axis which are parallel to the image plane.

(4) Image coordinate system $O_DX_DY_D$: coordinate plane of image coordinate system $X_DO_DY_D$ is on the camera's CCD imaging plane, the origin O_D is the intersection of optical axis and imaging plane, and O_DX_D axis and O_DY_D axis are parallel to the O_CX_C and the O_CY_C axis.

(5) Computer image coordinate system UV: the origin is located in the upper left corner of the digital image, U and V denote the number of columns and rows of a digital image. The relationships of aforementioned coordinate systems are illustrated in Figure 3.

Besides, we choose the widely adopted Pinhole model to represent the camera imaging relationship [32]. Set the coordinate of a space point P at camera coordinate system $O_CX_CY_CZ_C$ as (x_c, y_c, z_c); then the imaging model of the point P is represented as

$$\begin{bmatrix} u \\ v \\ 1 \end{bmatrix} = \begin{bmatrix} \alpha_x & 0 & u_0 \\ 0 & \alpha_y & v_0 \\ 0 & 0 & 1 \end{bmatrix} \begin{bmatrix} \dfrac{x_c}{z_c} \\ \dfrac{y_c}{z_c} \\ 1 \end{bmatrix}. \tag{1}$$

(u, v) is the coordinate of point P in the coordinate system of computer image. (u_0, v_0) is the main point of the camera. $\alpha_x = f/dx$, $\alpha_y = f/dy$, f is the focal length of the camera, and dx and dy are a pixel's physical sizes in the horizontal and vertical direction. Generally, dx is equal to dy.

If two points, $A(u_1, v_1)$ and $B(u_2, v_2)$, which locate in the computer image coordinate system UV, are considered, then the line AB can be expressed as

$$u = kv + b, \quad k = \frac{u_1 - u_2}{v_1 - v_2}, \quad b = \frac{v_1u_2 - u_1v_2}{v_1 - v_2}. \tag{2}$$

2.2. Camera Imaging Model. Assume the rotation matrix and translation vector from grasping point coordinate system $O_ZX_ZY_ZZ_Z$ to camera coordinate system as R_{TC} and T,

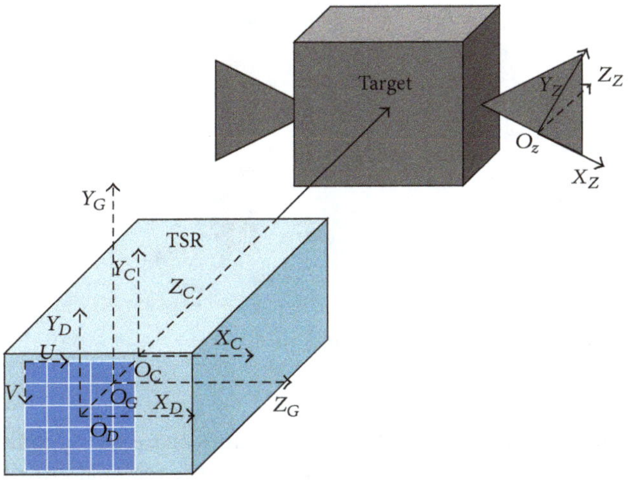

FIGURE 3: Definition of coordinate systems.

respectively [33]. The matrix transformed from grasping point coordinate system to camera coordinate system is

$$M_{TC} = \begin{bmatrix} R_{TC} & T \\ 0^T & 1 \end{bmatrix}. \quad (3)$$

Therefore, the transformational relationship of the space point P between homogeneous coordinates $(x_z, y_z, z_z, 1)^T$ in grasping point coordinates $O_Z X_Z Y_Z Z_Z$ and homogeneous coordinates $(x_c, y_c, z_c, 1)^T$ in the camera coordinate system $O_C X_C Y_C Z_C$ is

$$\begin{bmatrix} x_c \\ y_c \\ z_c \\ 1 \end{bmatrix} = M_{TC} \begin{bmatrix} x_z \\ y_z \\ z_z \\ 1 \end{bmatrix} = \begin{bmatrix} R_{TC} & T \\ 0^T & 1 \end{bmatrix} \begin{bmatrix} x_z \\ y_z \\ z_z \\ 1 \end{bmatrix},$$
$$z_c \begin{bmatrix} u \\ v \\ 1 \end{bmatrix} = \begin{bmatrix} \alpha_x & 0 & u_0 & 0 \\ 0 & \alpha_y & v_0 & 0 \\ 0 & 0 & 1 & 0 \end{bmatrix} \begin{bmatrix} R_{TC} & T \\ 0^T & 1 \end{bmatrix} \begin{bmatrix} x_z \\ y_z \\ z_z \\ 1 \end{bmatrix}. \quad (4)$$

3. Line Segment Detection

3.1. Image Gradients. Image gradients reflect the intensity changes between adjacent pixels in the image. A small gradient value corresponds to a region with smooth gray level in the image, and a large gradient value corresponds to a sharp edge region in the image. Pixels along the consistent gradient direction are likely to belong to an edge in the image. Thus, calculating the gradients of images is important for the edge structure detection. For image **I**, the gradients in pixel x are defined as

$$g_x(x, y) = \frac{I(x+1, y) + I(x+1, y+1) - I(x, y) - I(x, y+1)}{2},$$

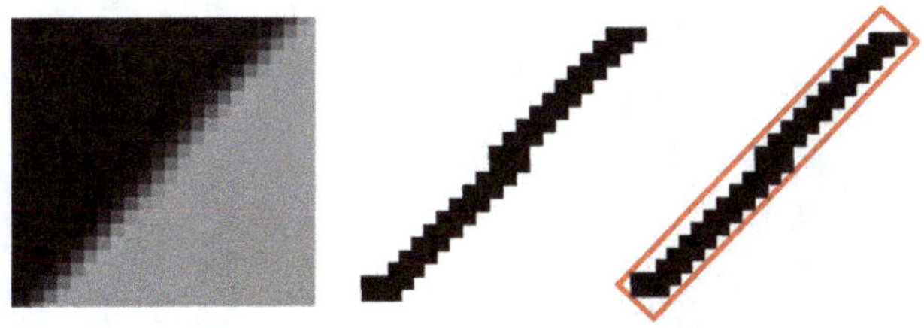

FIGURE 4: Line support areas.

$$g_y(x, y) = \frac{I(x, y+1) + I(x+1, y+1) - I(x, y) - I(x+1, y)}{2}, \quad (5)$$

where $I(x, y)$ represents the intensity value of pixel (x, y); $g_x(x, y)$ and $g_y(x, y)$ are horizontal and vertical gradients of pixel (x, y). Then, we calculate

$$G(x, y) = \sqrt{\left(g_x^2(x, y) + g_y^2(x, y)\right)},$$
$$\theta = \arctan\left(\frac{g_y(x, y)}{g_x(x, y)}\right), \quad (6)$$

where $G(x, y)$ and θ represent the gradient value and gradient orientation of pixel (x, y), respectively.

To speed up the edge structure detection and exclude the monotonous regions in image, we first preprocess the image by thresholding the gradient value of each pixel. Pixels whose gradient value is less than the predefined threshold are more likely to be from nonedge regions and will be removed directly. Beside, we define pixel consistency τ as the difference between adjacent pixel gradient orientations, which is $\tau = |\theta_1 - \theta_2|$. If $\tau < t$, it is considered that the orientations of adjacent pixels are consistent and these pixels should be in the same straight line, where t is an human-defined parameter. The determination of τ is described in Section 3.3.

3.2. Pseudo-Sorting Method. Line support area is composed of pixels with similar gradient directions in the image, as shown in Figure 4. In general, we use the external rectangle, for example, the red color one in Figure 4, to approximate the support area.

The support area of straight line segment is generated using region growing method [34, 35]. It first selects a small part of pixels as the seeds. Then, it iteratively integrates neighboring pixels with similar orientations and the seed region gradually grows. The detailed region growing strategy is introduced in Section 3.3. The performance of the region growing method is sensitive to the selection of seed pixels. One way to determine seeds is to sort image pixels according to their gradient magnitudes and selects pixels with larger values as the seed pixels. However, the computational efficiency of sorting these pixels is rather low due to the large amount of pixels contained in the image. Even though some fast sorting methods (e.g., bubble sort, quick sort) are used,

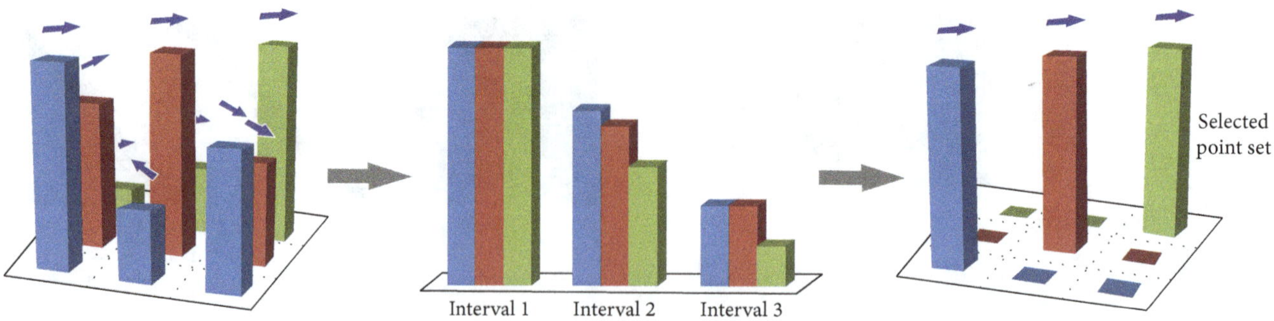

FIGURE 5: Illustration of pseudo-ordering method using a 9-pixel region.

the real time performance is still hard to obtain. Here, we adopt the pseudo-sorting method to sort the pixels (Figure 5).

Pseudo-ordering method is generally used to roughly arrange pixels in order according to their gradients. Firstly, a linear interval is generated according to the maximum and minimum values of the image gradients. Each interval represents a range of gradient values. Pixels are placed in the corresponding intervals according to their gradient values. Hence, as the increase of interval number, the pseudo-ordering results will become more accurate while the algorithmic efficiency becomes lower. Then, for pixels in the same interval, they have the same possibility of being selected as seeds.

To determine the number of intervals, we take several experiments and conclude that, for common 8-bit image, setting interval number to be 1024 could achieve a good tradeoff between performance and time complexity.

3.3. Region Growing Method. According to the results of the pseudo-sorting, the pixel point is selected as the seed point of the region growing according to the descending gradient value [36]. First, we define the direction of the line support area as follows:

$$\theta_{\text{region}} = \arctan\left(\frac{\sum_i \sin(\theta_i)}{\sum_i \cos(\theta_i)}\right), \tag{7}$$

where θ_{region} represents the angle of the support area and θ_i represents the linear direction angle of the pixel i. Set the seed point of region growing as (x, y); then the direction of the support area at the beginning is the direction of the seed point; namely, $\theta_{\text{region}} = \theta_{(x,y)}$.

Let $S_x = \cos(\theta_{\text{region}})$ and $S_y = \sin(\theta_{\text{region}})$. For each point $p(x, y)$ in the support area, we visit its 8 neighboring pixels, which are $p(x-1, y-1)$, $p(x, y-1)$, $p(x+1, y-1)$, $p(x-1, y)$, $p(x+1, y)$, $p(x-1, y+1)$, $p(x, y+1)$, and $p(x+1, y+1)$, and calculate the difference between their gradient orientations and θ_{region}. If the difference is less than τ, it is considered that they have the same direction, and the corresponding pixel will be added to the straight line support area. Then, we update the direction of the support area using

$$\theta_{\text{region}} = \arctan\left(\frac{S_y + \sin(\theta_{\overline{P}})}{S_x + \cos(\theta_{\overline{P}})}\right), \tag{8}$$

where $\overline{P}$ denotes the selected neighbor pixel of $p(x, y)$. The visited pixels will be removed from the candidate list and not be accessed again. Repeat above steps until no new pixel could be added to the linear support area, and the regional growth algorithm terminates.

During the region growing, the selection of the threshold value τ is key and will influence the final results of the algorithm. Intuitively, when a small τ is adopted, only pixels with almost identical orientations could be included in the line support area. The generated support area may be too short to provide the complete edge information. As τ increases, more nonedge pixels tend to be covered by the support area, which leads to its increase in length and width. This will cause a lot of false positive detection. In order to make a tradeoff between deficient edge information and high false positive rate, we evaluate the performance under different τ and select τ to be 22.5° by trial and error.

The region growing results under different τ are illustrated in Figure 6. Besides 22.5°, two typical thresholds, which are smaller and larger than 22.5°, are selected to provide the visual examples.

3.4. Straight Line Determination. According to the aforementioned region growing algorithm, the support area of straight line segment is generated. Then the next step is to produce the corresponding line segments from each support area. First, in order to reduce the computational burden, the support area with extremely few pixels is excluded, which is likely caused by the isolated noise. Excluding areas with few pixels can be regarded as a type of erosion operation in image processing. Both of them aim to eliminate the small regions in image. Specifically, we could use a hollow circle template with small radius and slide it over the image. For areas with few pixels, they will be totally enclosed by the template and hence will be discarded. For support region areas, they will intersect with the circle due to their length and will be retained.

In order to determine the straight line segment, the external rectangle is used to describe its support area. The support area is scored by summing up the gradient values of pixels contained in it. Then the support area centroid (c_x, c_y) can be calculated by the following formula:

$$\begin{aligned} c_x &= \frac{\sum_{j\in\text{Re}} G(j) * x(j)}{\sum_{j\in\text{Re}} G(j)}, \\ c_y &= \frac{\sum_{j\in\text{Re}} G(j) * y(j)}{\sum_{j\in\text{Re}} G(j)}, \end{aligned} \tag{9}$$

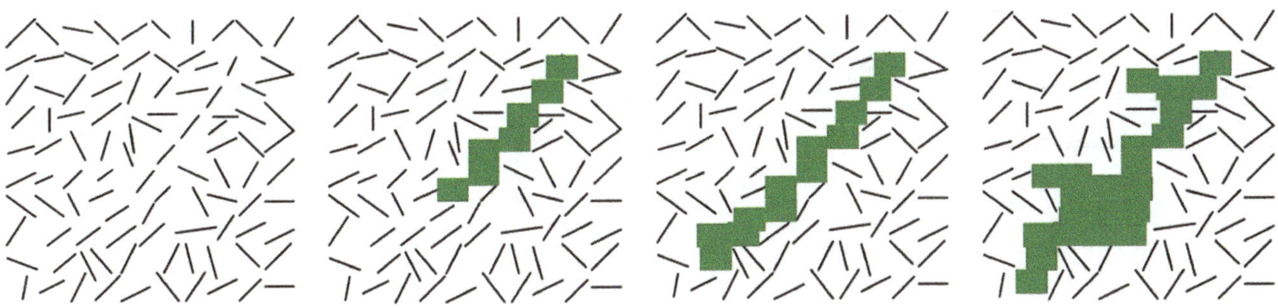

(a) Origin image (b) $\tau = 11.25^\circ$ (c) $\tau = 22.5^\circ$ (d) $\tau = 45^\circ$

FIGURE 6: Region growing results with different τ. (a) is the original gradient map, including the edge area with noise interference. (b), (c), and (d) are results corresponding to 11.25°, 22.5°, and 45°, respectively.

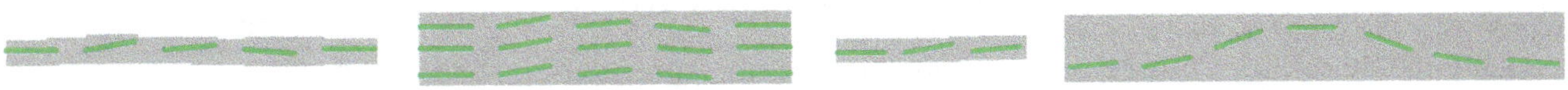

(a) (b) (c) (d)

FIGURE 7: Illustration of different support areas including (a) straight line, (b) monotonous region, (c) isolated noise, and (d) low-curvature curve. (b) and (c) could be discarded by criterion 1; (d) could be discarded by criterion 2.

where $G(j)$ is the gradient magnitude value of the pixel j in the support area Re; $x(j)$ and $y(j)$ are coordinates of pixel j. The center of the external rectangle is chosen as the center of the support area, and the long axis of the rectangle is chosen as the direction θ_{region} of the support area, with the short axis vertical to θ_{region}.

Since not all of the straight segment support areas correspond to a straight line segment model, they need to be further judged after obtaining the external rectangle of the support area. In this paper, we develop two criteria to achieve this.

(1) The ratio of long and short axis in the external rectangle should be larger than the set threshold (typically 4 to 7).

Due to the existence of noise, pixels in monotonous region may be wrongly selected as the seeds of region growing method, leading to the support areas shown in Figure 7(b). Besides, the noise is also likely to form some small line segments, which cannot be detected due to the pixel number it contained, as shown in Figure 7(c). It is obvious to see that by thresholding the ratio between long and short axis these support areas can both be eliminated.

(2) When the direction of the external rectangle is consistent with the direction of the long axis of rectangle (direction deviation is less than τ), the proportion of its pixel number to the total number of pixels contained in the rectangular should be larger than the predefined threshold (typically 50% to 70%).

Determining the straight lines with long-short axis ratio may include some low-curvature curves, such as the circular edge with a larger radius, as shown in Figure 7(d). It can be observed that although the curve shares a large long-short axis ratio, it tends to produce the increase of the support region area. Hence, by thresholding the proportion of the pixel number divided by the support region area, these regions can be eliminated.

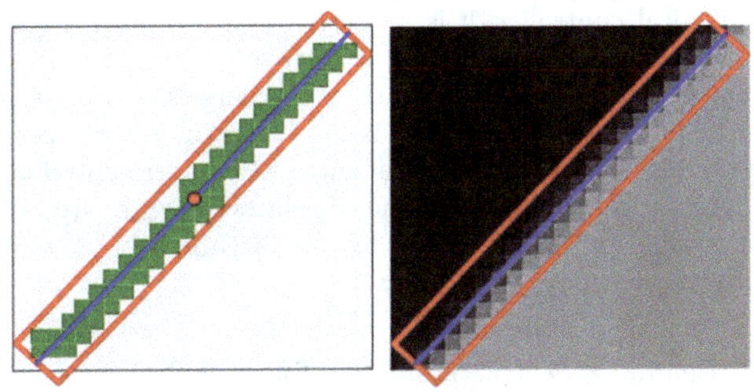

FIGURE 8: Schematic diagram of the line segment validation.

The support areas which satisfy these two criteria correspond to the straight line segment model, and the principal axis which comes through the center of the external rectangle is the straight line segment. As shown in Figure 8, the principal axis of the external rectangle, that is, the blue line segment, is the straight segment.

4. Servoing Controller Design

Through the above algorithm, we can detect the edge lines of the satellite brackets in the image. These operations are all based on the computer image coordinates. Hence, the detected edge lines can be expressed as

$$u = k_1 v + b_1,$$
$$u = k_2 v + b_2. \tag{10}$$

According to (10), we design a controller to achieve the control of relative attitude of the robot in the ultra-close approaching stage.

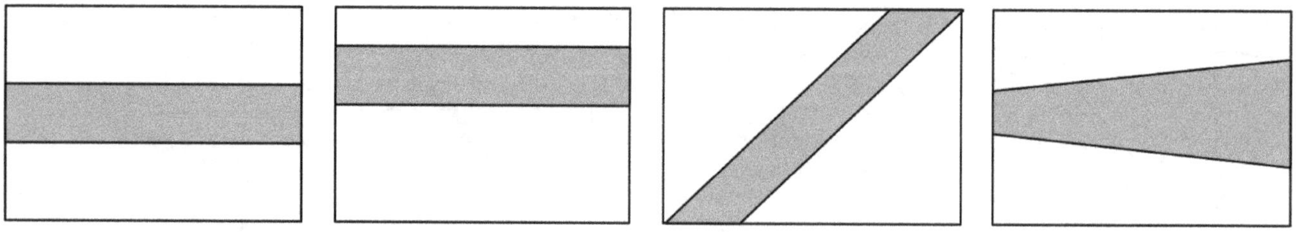

FIGURE 9: Relationships of two edge lines under different relative attitudes.

4.1. Edge Lines Analysis. The capturing device of TSR is usually installed in the front plate of the robot and close to the camera. Hence, we set that the capture coordinate system coincides with the camera coordinate system $O_CX_CY_CZ_C$, and assume the Euler angles between capture coordinate system $O_CX_CY_CZ_C$ and target coordinate system $O_ZX_ZY_ZZ_Z$ as α, β, γ, respectively [37]. It is easy to analyze that, for different relative attitude of the robot and the target, the imaging of the bracket edge lines in the camera image plane is shown in Figure 9.

The imaging result in Figure 9(a) shows the final state of an ideal controller. It is requested that the Euler angles β and γ should be zero, and the relative position deviation y along direction Y is zero. The Euler angles α of axis X can be any angle. Figure 9(b) is the case that y is not zero. Figure 9(c) shows that the Euler angle γ is not zero and there exists rotation along axis Z. The edge lines in Figure 9(d) are not parallel ($k_1 \neq k_2$) and it reveals that the robot rotates along axis Y.

4.2. Controller Design. From the above analysis, we can conclude that (1) the relative attitude angle γ can be measured by summing up the slopes of two edge lines; (2) the relative attitude angle β can be measured by subtracting the slopes of two edge lines; (3) the position deviation y along axis Y can be measured using the sum of b_1 and b_2. Therefore, we refer to the PD controller and design our controller as follows [38]:

$$\begin{aligned} M_z &= -p_1\left(k_1 + k_2\right) - p_2\left(\dot{k}_1 + \dot{k}_2\right), \\ M_y &= -p_3\left(k_1 - k_2\right) - p_4\left(\dot{k}_1 - \dot{k}_2\right), \\ F_y &= -p_5\left(b_1 + b_2 - 2u_0\right) - p_6\left(\dot{b}_1 + \dot{b}_2\right), \end{aligned} \tag{11}$$

where $p_1 \sim p_6$ are the controller parameters and should be adjusted in practice. M_z and M_y are the control torque of axes Z and Y. F_y is the control force of axis Y. The stability of the proposed controller could be proved by Lyapunov first method [39, 40].

We have to note that the attitude angle α is not controlled by the controller, because only the pole structure of the bracket is imaged in ultra-close distance (Figure 9(a)) and it is resistant to attitude change along axis X. But in practice α should be restricted during the approaching process using additional sensors, for example, the global camera, to keep TSR's self-attitude stable. The flowchart of our method is shown in Figure 10.

TABLE 1: The composition of visual perception subsystem.

Camera (VS-902H)	CCD: 1/3″
	Resolution: 752 (H) × 582 (V)
Lens (Computar M1614)	Focal length: 5 mm
	Field of view: 65.5° (H) × 51.4° (V)
Processor	Pentium Dual-Core E5200 with 2.50 GHz, 2G RAM
	Operating system: Windows XP SP3
	Programming environment: VS2005 IDE with OpenCV

5. Experiments and Results

5.1. Experiment Conditions. In order to verify the effectiveness of the proposed control algorithm, we set up a visual servoing control system, as shown in Figure 11. It mainly consists of visual perception subsystem, control subsystem, movement platform, and the target satellite model. The composition of visual perception subsystem is listed in Table 1. Since this paper mainly focuses on the visual servoing control in ultra-close distance, we adopt a single-pole model to substitute the satellite bracket model. In the experiment, we use the 6-DOF movement platform to simulate TSR's movement in space. The visual perception subsystem and target model are mounted on the movement platform to control their relative position and attitude.

5.2. Results Analysis. In order to evaluate the edge line detection performance in approaching process, we make a simulation and move TSR from initial position (about 3 m to the target) to grasping position (about 0.2 m to the target). Meanwhile, the single-pole model swings with an initial rate, about 10°/s. Figure 12 qualitatively demonstrates that despite the change of relative position and attitude between TSR and the target model our proposed algorithm is always able to detect the edge lines.

Figure 13 quantitatively reveals the measurement values of slope k and intercept b of the detected edge lines in every frame, respectively. In (a), the initial slope is −0.45, which means TSR's self-attitude needs to be further adjusted. As

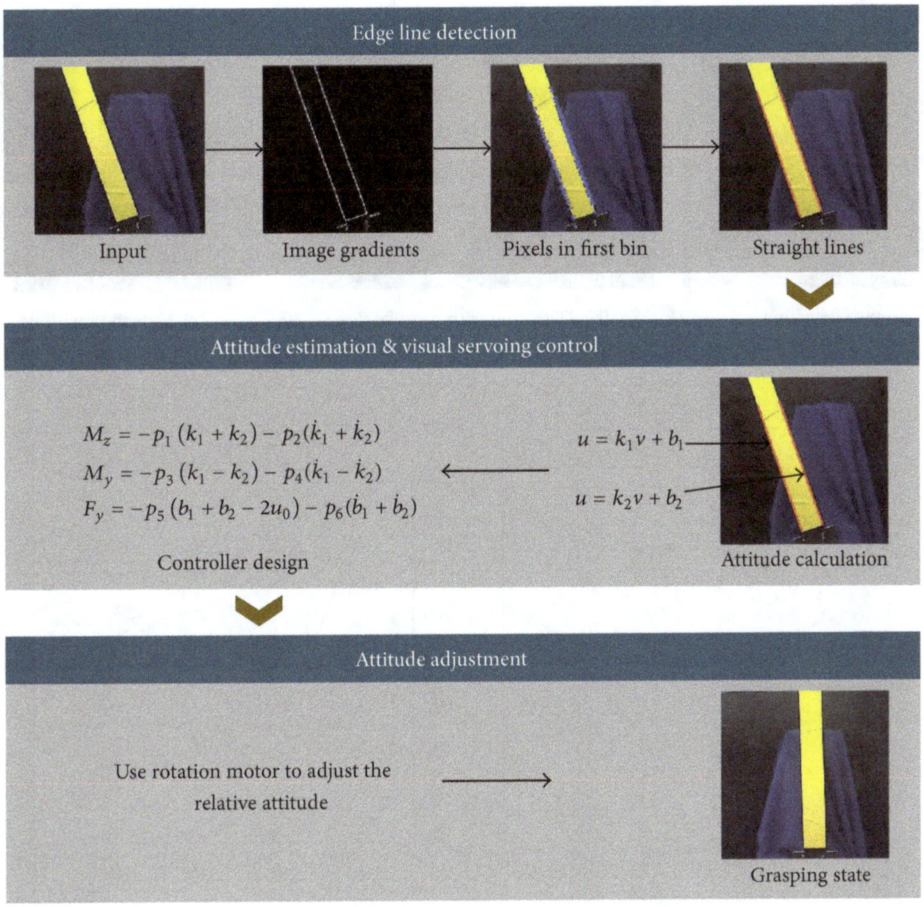

FIGURE 10: Flowchart of the proposed method.

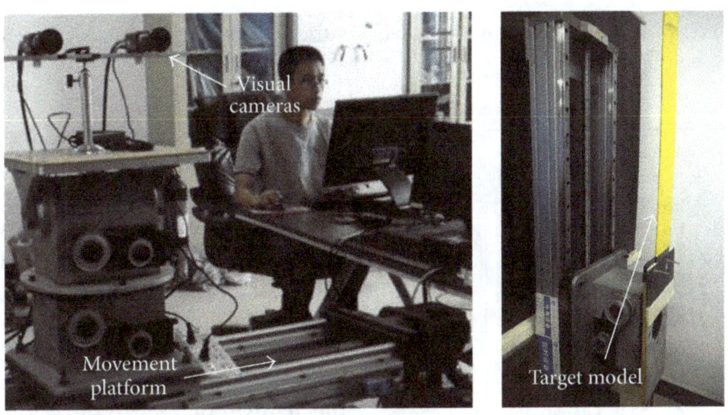

(a) Visual perception and movement subsystems (b) Single-pole satellite model

FIGURE 11: Visual servoing control system.

TSR approaches the target, slops of the two detected lines gradually converge to about 0, which means TSR reaches the ideal grasping attitude. Meanwhile, intercepts of the detected lines converge to about 400 and 300, respectively. It is easy to find that their sum is rather close to the image width $2u_0$. It indicates that the two edge lines are horizontally symmetric along the image center and TSR reaches the ideal grasping position.

In Figure 14, we give the time consumption of the proposed method. The average time of detecting one image is about 0.115 seconds. Considering the control period of the controller is 0.25 seconds, our method could satisfy the real time need.

5.3. Limitations. As mentioned in Section 4.2, TSR's rotation around axis x is not taken into consideration in this paper.

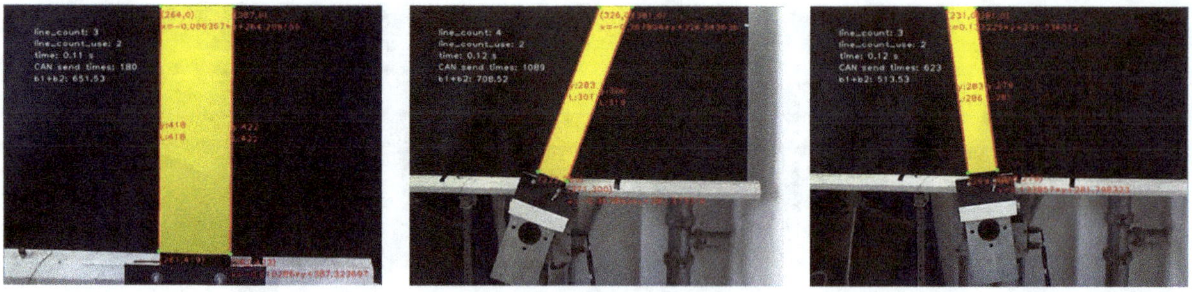

(a) Close-range detection (b) Detection with positive slope (c) Detection with negative slope

FIGURE 12: Edge line detection results under the different position and attitude.

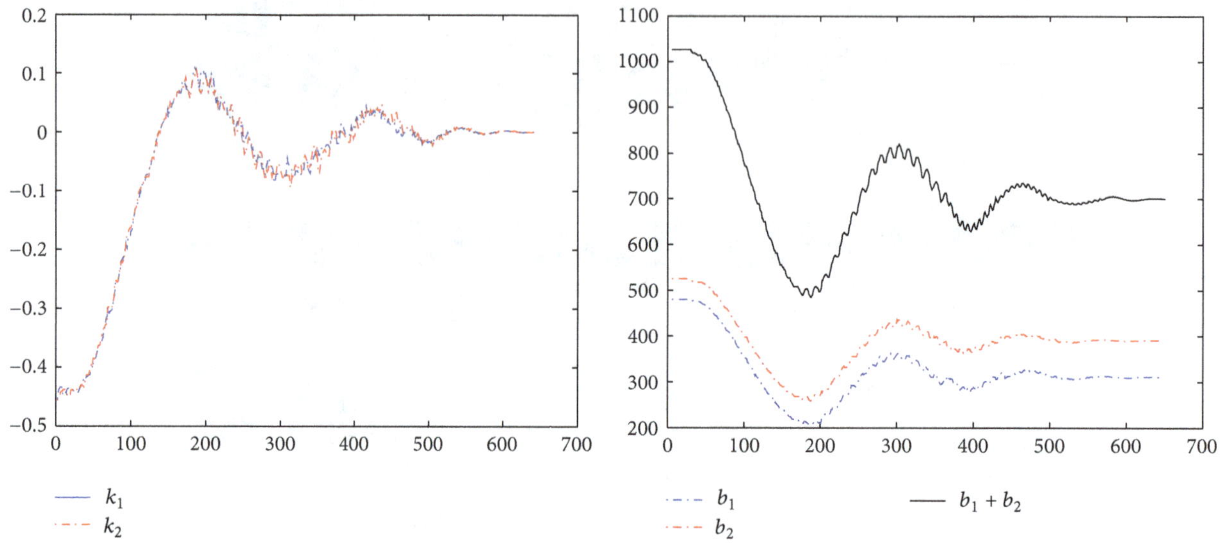

(a) (b)

FIGURE 13: Measurement value curves of slope k (a) and intercept b (b) of the detected edge lines. All curves converge at about 500th frame.

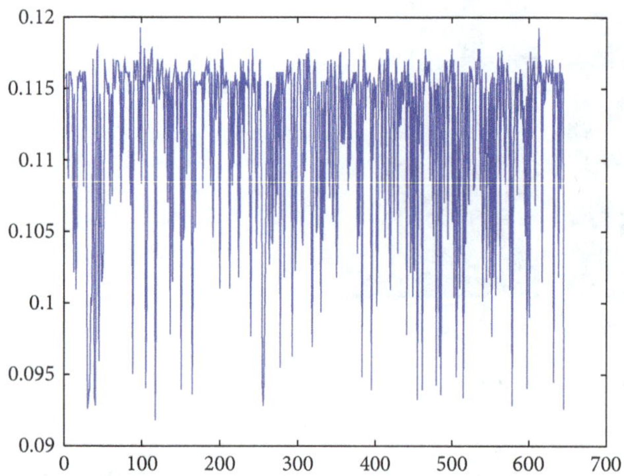

FIGURE 14: Comparison of time consumption per frame.

But in practice this rotation will lead to TSR's deviation from the approaching path and increase the possibility of collision with the target. Hence in future works the control to attitude angle α should be added to the controller.

Although our proposed method is able to work successfully, the number of frames it needs to offset the relative position and attitude is a bit high. It could be ascribed to the low frame frequency (4 Hz) of visual camera. The time span between adjacent frames leads to the inconsistency between actual measurement values and control parameters. In order to solve this, we consider (1) increasing the frame frequency of cameras as well as further reducing the complexity of our method and (2) introducing the motion prediction mechanism of TSR to offset the attitude change of target, in the future.

6. Conclusions

This paper presents a novel monocular visual servoing control method using only the edge lines. It is designed to measure the relative position and attitude between TSR and the target under nonfull field of view, which is intractable for traditional feature based visual servoing methods.

The novelty of this paper is constructed by the following three aspects: (1) We propose a novel edge line detection method. It is based on gradient growing and can stably extract lines in the image. Also, it meets the real time requirement of controller. (2) We build the model and analyze the relationship between edge line parameters and the relative attitude. (3) We design a PD-based visual servoing controller for ultra-close approaching process, which requires the two detected edge line parameters only.

Experiments on the semiphysical simulation system indicate that our method is invariant to rotation and scale change.

Competing Interests

The authors declare that they have no competing interests.

Acknowledgments

This research is sponsored by the National Natural Science Foundation of China (Grant nos. 11272256, 60805034) and the Fundamental Research Funds for the Central Universities (Grant no. 3102016BJJ03).

References

[1] M. Shan, J. Guo, and E. Gill, "Review and comparison of active space debris capturing and removal methods," *Progress in Aerospace Sciences*, vol. 80, pp. 18–32, 2016.

[2] P. Huang, F. Zhang, Z. Meng, and Z. Liu, "Adaptive control for space debris removal with uncertain kinematics, dynamics and states," *Acta Astronautica*, vol. 128, pp. 416–430, 2016.

[3] F. Zhang and P. Huang, "Releasing dynamics and stability control of maneuverable tethered space net," *IEEE/ASME Transactions on Mechatronics*, vol. PP, no. 99, pp. 1–1, 2016.

[4] P. Williams, "Optimal orbital transfer with electrodynamic tether," *Journal of Guidance, Control, and Dynamics*, vol. 28, no. 2, pp. 369–372, 2005.

[5] P. Pergola, A. Ruggiero, M. Andrenucci, and L. Summerer, "Low-thrust missions for expanding foam space debris removal," in *Proceedings of the 32nd International Electric Propulsion Conference*, Wiesbaden, Germany, September 2011.

[6] P. Huang, D. Wang, Z. Meng, F. Zhang, and Z. Liu, "Impact dynamic modeling and adaptive target capturing control for tethered space robots with uncertainties," *IEEE/ASME Transactions on Mechatronics*, vol. 21, no. 5, pp. 2260–2271, 2016.

[7] P. Huang, D. Wang, Z. Meng, F. Zhang, and J. Guo, "Adaptive postcapture backstepping control for tumbling tethered space robot–target combination," *Journal of Guidance, Control, and Dynamics*, vol. 39, no. 1, pp. 150–156, 2016.

[8] P. Huang, M. Wang, Z. Meng, F. Zhang, Z. Liu, and H. Chang, "Reconfigurable spacecraft attitude takeover control in post-capture of target by space manipulators," *Journal of the Franklin Institute*, vol. 353, no. 9, pp. 1985–2008, 2016.

[9] P. Huang, D. Wang, F. Zhang, Z. Meng, and Z. Liu, "Postcapture robust nonlinear control for tethered space robot with constraints on actuator and velocity of space tether," *International Journal of Robust and Nonlinear Control*, 2016.

[10] L. Chen, P. Huang, J. Cai, Z. Meng, and Z. Liu, "A non-cooperative target grasping position prediction model for tethered space robot," *Aerospace Science and Technology*, vol. 58, pp. 571–581, 2016.

[11] B. Bischof, L. Kerstein, J. Starke, H. Guenther, and W. Foth, "ROGER—robotic geostationary orbit restorer," *Science and Technology Series*, vol. 109, pp. 183–193, 2004.

[12] R. B. Friend, R. T. Howard, and P. Motaghedi, "Orbital express program summary and mission overview," in *SPIE Defense and Security Symposium*, Orlando, Fla, USA, April 2008.

[13] N. Yuga, S. Fumiki, and N. Shinichi, "Guidance and control of 'tethered retriever' with collaborative tension-thruster control for future on-orbit service missions," in *Proceedings of the 8th International Symposium on Artificial Intelligence: Robotics and Automation in Space (i-SAIRAS '05)*, Munich, Germany, September 2005.

[14] D. Wang, P. Huang, J. Cai, and Z. Meng, "Coordinated control of tethered space robot using mobile tether attachment point in approaching phase," *Advances in Space Research*, vol. 54, no. 6, pp. 1077–1091, 2014.

[15] D. Wang, P. Huang, and Z. Meng, "Coordinated stabilization of tumbling targets using tethered space manipulators," *IEEE Transactions on Aerospace and Electronic Systems*, vol. 51, no. 3, pp. 2420–2432, 2015.

[16] O. Mori and S. Matunaga, "Formation and attitude control for rotational tethered satellite clusters," *Journal of Spacecraft and Rockets*, vol. 44, no. 1, pp. 211–220, 2007.

[17] W. Xu, B. Liang, C. Li, and Y. Xu, "Autonomous rendezvous and robotic capturing of non-cooperative target in space," *Robotica*, vol. 28, no. 5, pp. 705–718, 2010.

[18] J. K. Thienel, J. M. VanEepoel, and R. M. Sanner, "Accurate state estimation and tracking of a non-cooperative target vehicle," in *Proceedings of the AIAA Guidance, Navigation, and Control Conference*, pp. 5511–5522, Keystone, Colo, USA, August 2006.

[19] A. H. A. Hafez, V. V. Anurag, S. V. Shah, K. M. Krishna, and C. V. Jawahar, "Reactionless visual servoing of a dual-arm space robot," in *Proceedings of the IEEE International Conference on Robotics and Automation (ICRA '14)*, pp. 4475–4480, June 2014.

[20] G. Dong and Z. H. Zhu, "Position-based visual servo control of autonomous robotic manipulators," *Acta Astronautica*, vol. 115, pp. 291–302, 2015.

[21] J. Cai, P. Huang, L. Chen, and B. Zhang, "An efficient circle detector not relying on edge detection," *Advanced in Space Research*, vol. 51, no. 11, pp. 2359–2375, 2014.

[22] H. Yuen, J. Princen, J. Illingworth, and J. Kittler, "Comparative study of hough transform methods for circle finding," *Image and Vision Computing*, vol. 8, no. 1, pp. 71–77, 1990.

[23] G. Casonato and G. B. Palmerini, "Visual techniques applied to the ATV/ISS rendez-vous monitoring," in *Proceedings of the IEEE Aerospace Conference Proceedings*, pp. 613–624, Big Sky, Mont, USA, March 2004.

[24] M. Aull, "Visual servoing for an autonomous rendezvous and capture system," *Intelligent Service Robotics*, vol. 2, no. 3, pp. 131–137, 2009.

[25] M. Sabatini, G. B. Palmerini, and P. Gasbarri, "A testbed for visual based navigation and control during space rendezvous operations," *Acta Astronautica*, vol. 117, pp. 184–196, 2015.

[26] V. A. Ramirez, S. A. M. Gutierrez, and R. E. S. Yanez, "Quadrilateral detection using genetic algorithms," *Computación y Sistemas*, vol. 15, no. 2, pp. 181–193, 2011.

[27] R. Grompone Von Gioi, J. Jakubowicz, J.-M. Morel, and G. Randall, "LSD: a fast line segment detector with a false detection control," *IEEE Transactions on Pattern Analysis and Machine Intelligence*, vol. 32, no. 4, pp. 722–732, 2010.

[28] H. Wang, Y.-H. Liu, W. Chen, and Z. Wang, "A new approach to dynamic eye-in-hand visual tracking using nonlinear observers," *IEEE/ASME Transactions on Mechatronics*, vol. 16, no. 2, pp. 387–394, 2011.

[29] H. Wang, B. Yang, Y. Liu, W. Chen, X. Liang, and R. Pfeifer, "Visual servoing of soft robot manipulator in constrained environments with an adaptive controller," *IEEE/ASME Transactions on Mechatronics*, 2016.

[30] A. Flores-Abad, O. Ma, K. Pham, and S. Ulrich, "A review of space robotics technologies for on-orbit servicing," *Progress in Aerospace Sciences*, vol. 68, pp. 1–26, 2014.

[31] J. Liu, N. Cui, F. Shen, and S. Rong, "Dynamics of RObotic GEostationary orbit restorer system during deorbiting," *IEEE Aerospace and Electronic Systems Magazine*, vol. 29, no. 11, pp. 36–42, 2014.

[32] A. Petit, E. Marchand, and K. Kanani, "Vision-based space autonomous rendezvous: a case study," in *Proceedings of the IEEE/RSJ International Conference on Intelligent Robots and Systems (IROS '11)*, pp. 619–624, IEEE, San Francisco, Calif, USA, September 2011.

[33] B. P. Larouche and Z. H. Zhu, "Autonomous robotic capture of non-cooperative target using visual servoing and motion predictive control," *Autonomous Robots*, vol. 37, no. 2, pp. 157–167, 2014.

[34] J.-P. Gambotto, "A new approach to combining region growing and edge detection," *Pattern Recognition Letters*, vol. 14, no. 11, pp. 869–875, 1993.

[35] S. A. Hojjatoleslami and J. Kittler, "Region growing: a new approach," *IEEE Transactions on Image Processing*, vol. 7, no. 7, pp. 1079–1084, 1998.

[36] K.-L. Chung, Y.-H. Huang, J.-P. Wang, T.-C. Chang, and H.-Y. Mark Liao, "Fast randomized algorithm for center-detection," *Pattern Recognition*, vol. 43, no. 8, pp. 2659–2665, 2010.

[37] E. Rosten, R. Porter, and T. Drummond, "Faster and better: a machine learning approach to corner detection," *IEEE Transactions on Pattern Analysis and Machine Intelligence*, vol. 32, no. 1, pp. 105–119, 2010.

[38] M. Nohmi, "Mission design of a tethered robot satellite "STARS" for orbital experiment," in *Proceedings of the IEEE International Conference on Control Applications (CCA '09)*, pp. 1075–1080, St. Petersburg, Russia, July 2009.

[39] O. Bourquardez, R. Mahony, N. Guenard, F. Chaumette, T. Hamel, and L. Eck, "Image-based visual servo control of the translation kinematics of a quadrotor aerial vehicle," *IEEE Transactions on Robotics*, vol. 25, no. 3, pp. 743–749, 2009.

[40] E. Malis, F. Chaumette, and S. Boudet, "2 1/2 D visual servoing," *IEEE Transactions on Robotics and Automation*, vol. 15, no. 2, pp. 238–250, 1999.

11

Experimental Investigation on the Feasibility of using a Fresnel Lens as a Solar-Energy Collection System for Enhancing On-Orbit Power Generation Performance

Tae-Yong Park,[1] Joo-Yong Jung,[2] and Hyun-Ung Oh[1]

[1] *Space Technology Synthesis Laboratory, Department of Aerospace Engineering, Chosun University, 375 Seosuk-dong, Dong-gu, Gwangju 501-759, Republic of Korea*
[2] *Spaceborne SAR R&D Lab, LIG Nex1, Yongin, Republic of Korea*

Correspondence should be addressed to Hyun-Ung Oh; ohu129@chosun.ac.kr

Academic Editor: Christopher J. Damaren

Cube satellites have a limitation for generating power because of their cubic structure and extremely small size. In addition, the incidence angle between the sun and the solar panels continuously varies owing to the revolution and rotation of the satellite according to the attitude control strategy. This angle is an important parameter for determining the power generation performance of the cube satellite. In this study, we performed an experimental feasibility study that uses a Fresnel lens as a solar-energy collection system for cube satellite applications, so that the power generation efficiency can be enhanced under the worst incidence angle condition between the sun and solar panels by concentrating and redirecting solar energy onto the solar panels with a commercial Fresnel lens. To verify the effectiveness of the proposed system, we conducted a power-measurement test using a solar simulator and Fresnel lenses at various angles to the light source. In addition, we predicted the on-orbit power-generation enhancement achieved by employing the solar-energy collection system with various attitude control strategies.

1. Introduction

A cube satellite is a parametric design specification for a pico-class satellite that is based on a volume of 1000 cm^3 and a required mass of 1.33 kg or less, as a standard size of 1 U (unit). Its low construction cost and short development period make it a cost-effective platform for various missions in space [1]. Universities around the world are actively participating in cube satellite development programs for academic and training purposes. These programs offer participants unique opportunities providing hands-on experience [2, 3]. Because of these advantages and the increasing adoption in the fields of scientific investigation, military, and technology demonstration, cube satellites are also attracting attention from industries and space-research organizations across the world [4–6].

Recent advances in microelectronics and component manufacturing technology for space applications have led to the miniaturization and reduction of payloads and cube satellite parts; thus, the mission capability of cube satellites has been continuously enhanced [7]. However, the missions of cube satellites are more challenging than those of regular satellites, and the overall required power of cube satellites also tends to be higher. Most satellites, including cube satellites, use photovoltaic solar cells, which are a well-known and reliable source of power. Therefore, a sufficient surface area for the solar cells is a very important system design parameter for satellites. However, cube satellites are considerably smaller than typical satellites; thus, it is far more difficult to provide a sufficiently large area for the installation of the solar cells. Additionally, the available area for solar cells must typically be shared with other appendages, such as the payload, communication antenna, camera, and access port. The number of panels that can be simultaneously oriented toward the sun is limited owing to the cube-shaped configuration. The variation in the incidence angles of sunlight with respect to the solar panels is also an important parameter for determining the power generation performance of satellites because this angle varies continuously according to the attitude control strategy and the revolution of the satellite. These are the main restrictions on the power generation performance of cube satellites. To overcome these performance limitations, cube satellite developers implement additional deployable solar panels to increase the power generation performance. Recently, many cube satellite vendors have produced deployable solar panels [8–10], most of which are based on passive deployment mechanisms using hinges and torsional springs. However, these deployable panels require an additional launch-locking mechanism to securely hold and safely release them during launch and on-orbit phases. The deployed solar panels can offer greatly improved power generation but require an increased system complexity and pose the risk of deployment failure, which might seriously affect the mission. In addition, while a deployable solar panel faces the sun and actively generates power, the power generation of the solar panels integrated on the satellite is limited because of the shadow effect induced by an existence of the on-board deployable panel.

To enhance the power generation capability of cube satellite, we focused on CPV (Concentrating PhotoVoltaic) systems, which are widely used for ground-system applications [11–19]. The CPV system for ground applications improves the power generation efficiency of solar cells by concentrating solar-energy onto a relatively small area of the solar cells by using optical systems such as lenses or curved-mirrors. A great benefit of the CPV system is its cost-effectiveness, which is achieved through the reduction of the solar cell area compared to the alternatives that reduce the solar cell area with standard non-CPV modules.

High-concentrating photovoltaic (HCPV) systems [11–13] use concentrating optics, such as dish reflectors or Fresnel lenses to focus sunlight with a concentration ratio ranging from 300 to over 1,000 suns and typically employ multijunction solar cells, which have a high efficiency and lower efficiency loss associated with increases in temperature or life degradation as compared to conventional Si solar cells. Rumyantsev et al. [13] proposed an HCPV module that can enhance the overall power generation efficiency by using a primary Fresnel lens to perform the main concentration and a convex lens as a secondary optical element to ensure uniform illumination and to reduce the accuracy requirements for alignment and tracking. Low-concentration photovoltaic (LCPV) systems [14–16] are less complex than HCPV systems and are widely used in ground-based system applications owing to their higher tracking tolerances, passive heat sinks, and lower construction costs. Uematsu et al. [16] proposed and developed a flat-plate static LCPV system, which can reduce the cost of photovoltaic modules and has a larger tolerance for the acceptance angle than the HCPV system. The flat-plate static LCPV module consists of a cover glass, solar cells, and a reflector sheet with submillimeter V-grooves. The solar energy reflected from the V-grooved reflector sheet is completely reflected at the cover-glass surface and illuminated on the solar cells. Thus, the system concentrates solar energy at a concentration ratio ranging from 1.5 to 2 suns.

Luminescent solar concentrators (LSC) [17–19] are an emerging type of solar concentration technology and are still in the research stage. A sheet of transparent material doped with fluorophores contains luminescent particles that absorb a fraction of the incident solar spectrum. The absorbed solar energy is directed to the edges of the sheet by total internal reflection and collected by the solar cells attached there. Thus, the LSC provides a high concentration ratio without a solar-tracker or an additional cooling system. Wang et al. [19] proposed an LSC with a silicon solar cell module that can provide a higher power generation efficiency than one using an organic dye with self-absorption by employing a rare-earth complex that can achieve zero self-absorption.

However, the CPV technologies have seen only limited consideration for cube satellite applications. To date, research into CPV systems for space applications has been mostly considered for deployable systems [20–22] or space-based solar power (SSP) systems [23, 24]. Habraken et al. [22] investigated the effectiveness of using CPV technologies for spacecraft with high power requirements. Two concepts of CPV technologies based on reflective and refractive concentrating elements were compared to conventional solar arrays in various aspects including environmental aging effects of optical systems, off-pointing sensitivity, and thermal management of the solar array. Mori et al. [24] investigated the SSP systems, which collect solar energy using an extra-large optical lenses or mirror systems in-orbit and transfer the power to an on-ground facility via microwaves or lasers.

Despite the great advantages of CPV systems regarding their power generation performance, applying the ground-based CPV systems to space missions leads to technical risks and increases the system complexity. Most current CPV technologies for ground-based systems require a relatively large optical system to efficiently focus the incident solar energy onto the solar cells. In addition, they require a solar tracking system to secure the acceptance angle of the CPV module with respect to the sun vector and an additional cooling system to avoid the power generation efficiency loss of the solar cell due to the increase in temperature caused by the solar-energy concentration. Therefore, the existing ground-based CPV technologies are impractical for space applications without design modification, especially for cube satellites, which have an extremely limited accommodation area.

To overcome the limitations of conventional ground-based CPV technologies, Oh and Park [25] proposed a CPV system for cube satellite applications. This system has the potential to increase the power generation efficiency using a commercial multiarray lens system installed on the edge of the solar panels even under the worst incidence angle conditions between the sun and the solar panels. The feasibility of the CPV system was verified through a power-measurement test using a solar simulator and a CPV system with a commercial multiarray lens system. However, in this study, the on-orbit power generation performance of a satellite with a CPV system according to the attitude control strategy was not investigated.

In this study, we propose a solar-energy collection system for cube satellite applications using a commercial PMMA (polymethyl methacrylate) Fresnel lens system, which provides an enhanced power generation efficiency compared to the multiarray lens system proposed by Oh and Park [25]. The effectiveness of the proposed solar-energy collection system was demonstrated by a power-measurement test using a solar simulator and Fresnel lens system under various incidence-angle conditions with a light source. In addition, we predicted the enhancement of the power generation efficiency of the proposed solar-energy collection system according to the various attitude control profiles of the satellite by an on-orbit power generation simulation using the commercial software STK (Systems Tool Kit) [26]. In the simulations, we confirmed that the proposed solar-energy collection system with the Fresnel lens system installed on the edge of the solar panel at an optimized installation angle is effective for providing a much higher power generation performance for the cube satellite as compared to the conventional configuration using a multiarray lens system [25].

2. Concept of Solar-Energy Collection Systems for Cube Satellite Applications

The solar-energy collection system proposed in this study is one of the main payloads of the STEP Cube Lab (Cube Laboratory for Space Technology Experimental Project), which is a 1 U standardized cube satellite developed by STSL (Space Technology Synthesis Laboratory) at Chosun University in Korea. Figure 1 shows the cube satellite configuration of the STEP Cube Lab, whose main mission objective is the on-orbit verification of core space technologies researched in domestic universities that are potential candidates for future space programs. To achieve the mission objectives of the STEP Cube Lab, five technologies were chosen: a microelectromechanical system-based solid propellant thruster, a novel nonexplosive holding and release mechanism triggered by nichrome burn wire heating, a variable-emittance radiator, a phase-change material, and the solar-energy collection system proposed in this study. The STEP Cube Lab mission will be performed in a sun-synchronous orbit at an altitude of 600 km for a duration of 1 year after the launch and is scheduled for 2017.

In the case of the STEP Cube Lab, the satellite employs a passive attitude-stabilization method [27] using a permanent magnet and hysteresis damper, which has been widely used for cube satellite. This method provides effective attitude-stabilization performance for satellites, especially cube satellites with an extremely small accommodation space, as the system has a considerably small mass of 50 g and does not consume power. A satellite using a permanent magnet aligns with the direction of the local earth magnetic field and performs two cycles of rotation per orbit. Therefore, the attitude control strategy is an important parameter for determining the power generation efficiency of the satellite, as the incidence angle β between the sun vector and the satellite varies continuously owing to the revolution and rotation of the satellite.

To overcome the aforementioned limitations of applying ground-based conventional CPV technologies to cube satellites, we propose a cube satellite with a solar-energy collection system by employing a Fresnel lens system. This system is a photovoltaic power generation system that can enhance the power generation efficiency by using Fresnel lenses. Figure 2 illustrates the basic principle of the proposed solar-energy collection system proposed in this study. The Fresnel lenses installed on the edge of the solar panel collect the solar energy and illuminate it onto the solar panel surface even under the worst condition of the angle β between the sun and the solar panel. By installing the Fresnel lenses on the edge of the solar panel at an optimized angle, the light intensity and duration of the light redirected onto the solar panel are increased. In addition, the Fresnel lenses installed on the edge of the solar panel can minimize the performance reduction occurring because of the presence of the lenses according to β angle variation, which is suitable for power generation. Thus, the system proposed in this study has the potential to enhance the power generation efficiency, even though the β angle between the sun and the solar panel varies continuously according to the attitude control strategy of the satellite. In the STEP mission, the Fresnel lenses are installed on the edge of the solar panels in the $+X$ and $-X$ directions, as shown in Figure 1, to verify the proposed solar-energy collection system by comparing the power generation performance between the solar panel with and without the Fresnel lenses on-orbit.

TABLE 1: Specification of a solar simulator.

Name	SolarTrainer Junior (IKS, Germany)
Irradiation power source	Halogen lamp
Illumination intensity	150 W/m^2
Solar cell	Polycrystalline Si solar cell
Cell efficiency	17%
Open-circuit voltage (V_{oc})	0.5 V (single cell)
Short-circuit current (I_{sc})	1.5 A (at 1000 W/m^2)
Cell size	5 cm × 10 cm (single cell, total 4 solar cells)

3. Power-Measurement Test Set-Up

To verify the effectiveness of the solar-energy collection system proposed in this study, a power-measurement test was performed using a solar simulator and a commercial Fresnel lens, as shown in Figure 3. Table 1 lists the specifications of the solar simulator (SolarTrainer Junior, IKS (Germany)) used in this study. The simulator consists of a polycrystalline Si solar cell with a conversion efficiency of 17%, an open-circuit voltage of 0.5 V, and a short-circuit current of 1.5 A. A halogen lamp with an illumination intensity of 150 W/m^2 is used as an irradiation source, and multimeters are used to measure the current and voltage output from the solar cells. The variation in the sun incidence angle with respect to the solar panel due to the rotation and revolution of the satellite was simulated by adjusting the angle of the light source and solar panel of the simulator. The angles α and β were defined as shown in

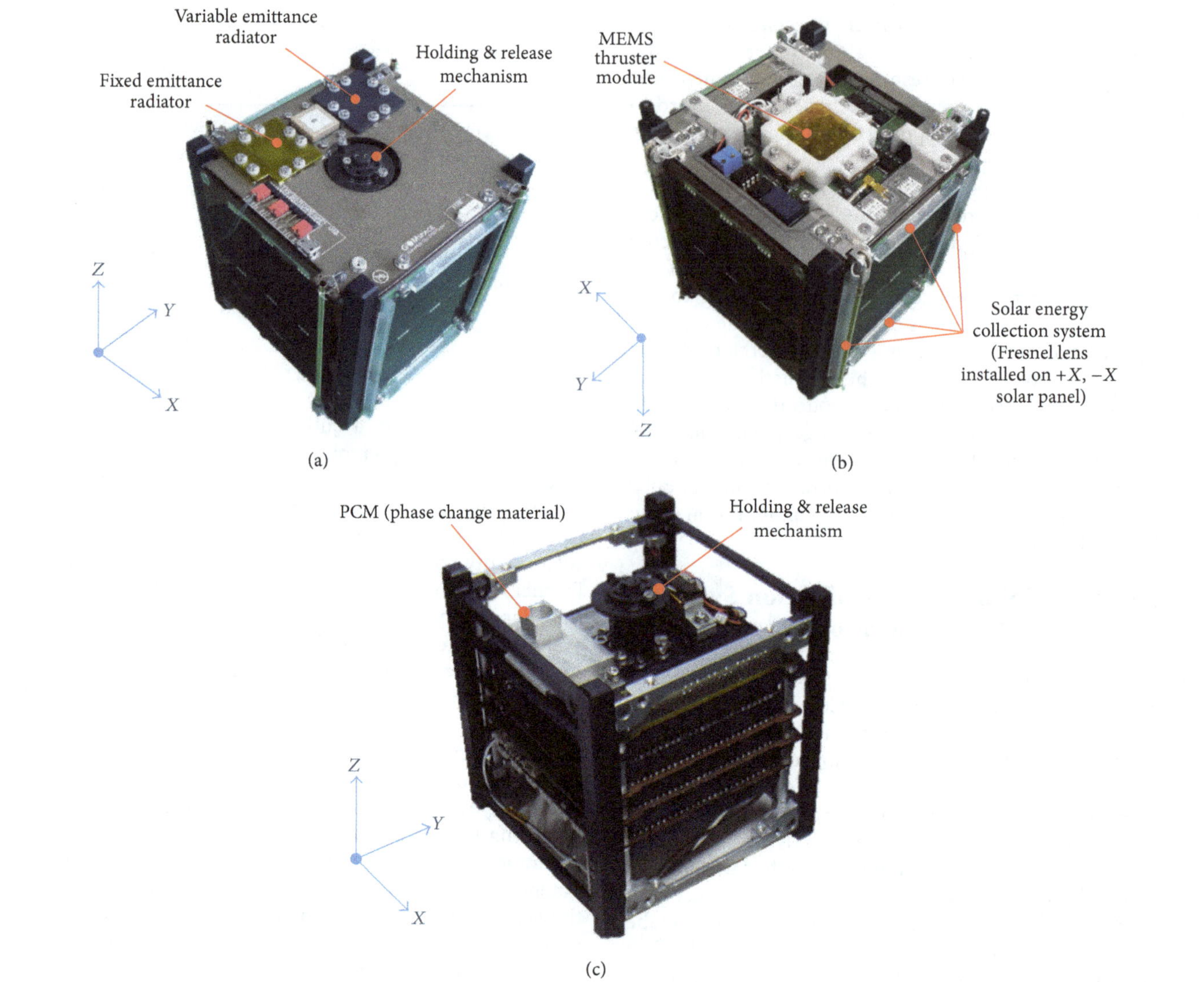

FIGURE 1: Configuration and payload accommodation of the STEP Cube Lab. ((a) external view from the $+Z$ axis, (b) external view from the $-Z$ Axis, and (c) internal view from the $+Z$ axis).

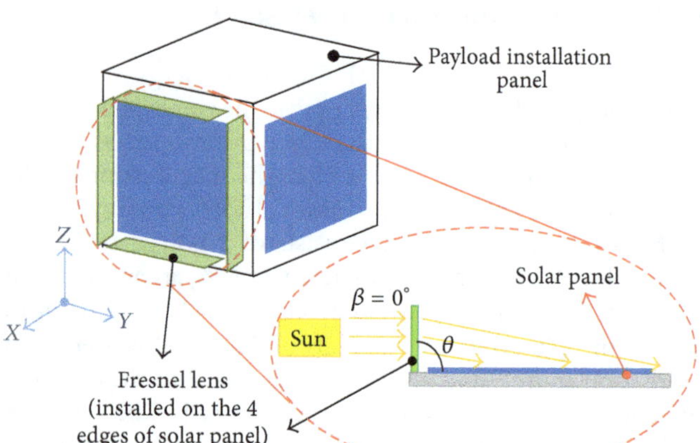

FIGURE 2: Schematic of the solar-energy collection system for cube satellite applications.

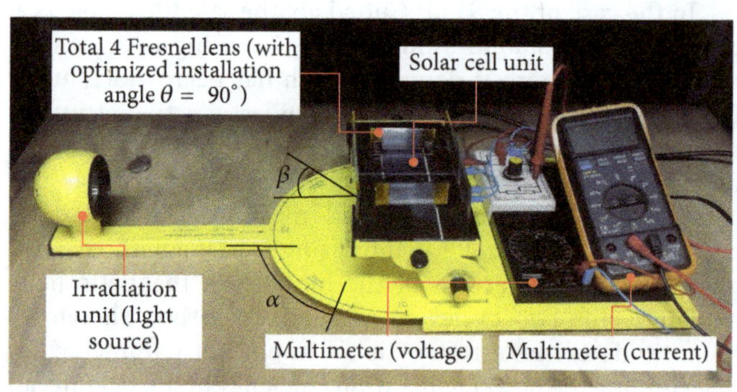

FIGURE 3: Power-measurement test set-up.

Figure 3. α is the horizontal rotation angle of the irradiation source, and β is the incidence angle between the irradiation

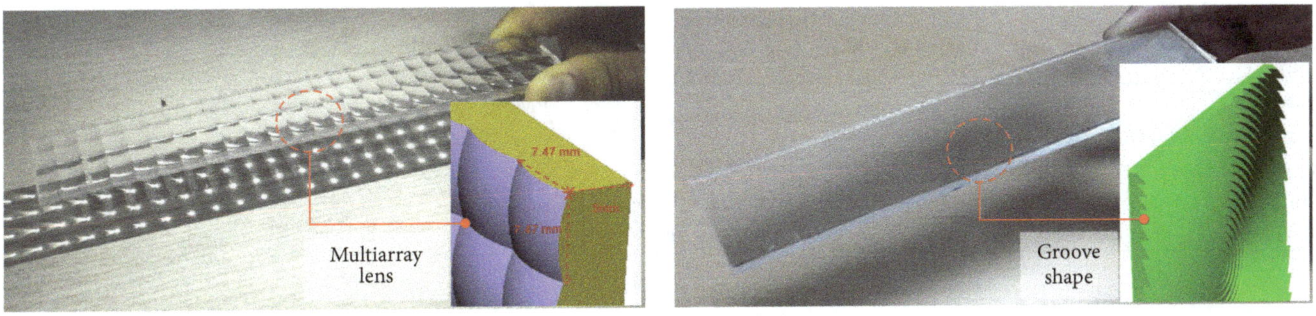

(a) (b)

FIGURE 4: Commercial PMMA convex lenses ((a) Fresnel lens, (b) multiarray lens).

source and the solar panel. The angle β represents the angle between the sun vector and the solar panel of the satellite in-orbit; the angle values of 0° and 90° correspond to the worst and best conditions, respectively, for generating power.

In the power-measurement test, a commercial convex Fresnel lens was used to verify the effectiveness of the proposed solar-energy collection system. For a comparison of the power generation performance, a commercial convex multiarray lens used in a previous study by Oh and Park [25] was applied. Figures 4(a) and 4(b) show the configurations of the commercial convex Fresnel and multiarray lenses, respectively. Table 2 presents the specifications of the convex lenses. The lenses were made of PMMA and had a transmission efficiency of 92% with wavelengths ranging from 400 to 1,100 nm. They were installed on the edge of the solar panel of the simulator, as shown in Figure 3, to enhance the power generation efficiency of the solar panel even under the worst angle condition of $\beta = 0°$, by effectively collecting the solar energy and redirecting it onto the solar panel. Through preliminary tests, the angle θ of the Fresnel lens system, which corresponds to the inclination angle between the Fresnel lens and the solar panel, was optimized to 90°, and that of the convex multiarray lens was optimized to 30°. The tests were conducted in a darkroom to prevent the influence of unintended light from external light sources on the solar cells. In this test, we measured the voltage and current from the solar panel, both with and without the PMMA lenses, under various incidence-angle conditions.

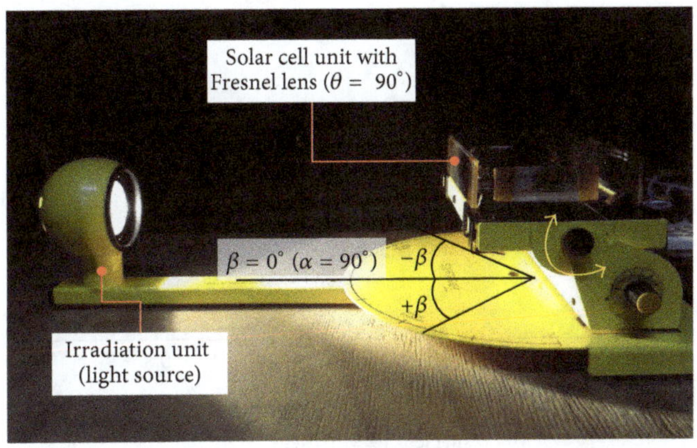

FIGURE 5: Configuration of the power-measurement test with $\beta = 0°$ and $\alpha = 90°$.

4. Power-Measurement Test Results

To verify the effectiveness of the proposed solar-energy collection system employing the Fresnel lens system under various conditions of the sun incidence angle, the current-voltage (I-V) characteristic curves of the solar panel were obtained both with and without the Fresnel lens system according to the variation in the incidence angle of the light source. In addition, power generation performance of the Fresnel lens system was compared to the results of the multiarray lens system.

Figure 5 shows the test configuration that simulates $\beta = 0°$ and $\alpha = 90°$, which is the worst sun incidence-angle condition on the solar panel. The I-V curves of the solar panels obtained with and without the Fresnel lens system at $\beta = 0°$ and $\alpha = 90°$ are shown in Figure 6. The I-V curves obtained for the multiarray lens system are also plotted in the figure in order to make a comparison with the Fresnel lens system. The test results of both lenses with an optimized angle of θ show a much higher power generation performance than those obtained without lenses because the lens can effectively redirect and illuminate the solar energy to the solar panels under the worst sun incidence-angle condition. The maximum power of the Fresnel lens system, calculated from the I-V curve, was 25.1 mW, which is 9.4 times greater than that of the multiarray lens system. This indicates that the Fresnel lens installed at the optimized inclination angle of $\theta = 90°$ is more effective for enhancing the power generation performance of the solar panel by efficiently collecting and redirecting a much larger amount of the solar energy than the multiarray lens installed with $\theta = 30°$.

Figure 7 shows the test results obtained when the β angle was varied from −30° to 90°, and when α was fixed as 90°, in order to evaluate the performance variation of the solar panel both with and without the Fresnel lens system over a wide range of sun incidence angles. The results for the multiarray lens system are also plotted here for comparison. The range of β from −30° to 0° is the worst condition with regard to the power generation of the solar panel because it cannot absorb

Table 2: Specification of commercial PMMA lenses.

Name	Convex Fresnel lens	Convex multiarray Lens
Material	PMMA (polymethyl methacrylate)	
Transmission efficiency	92% (at 400–1100 nm)	
Allowable service temperature	−70°C~80°C	
Dimension	30 mm × 189 mm × 3 mm	30 mm × 189 mm × 5 mm
Focal length	182 mm	29.88 mm

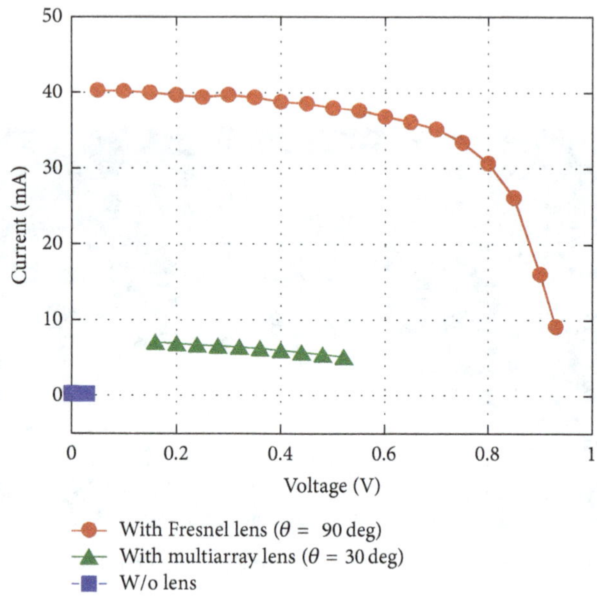

Figure 6: Results of the power-measurement test with $\beta = 0^\circ$ and $\alpha = 90^\circ$.

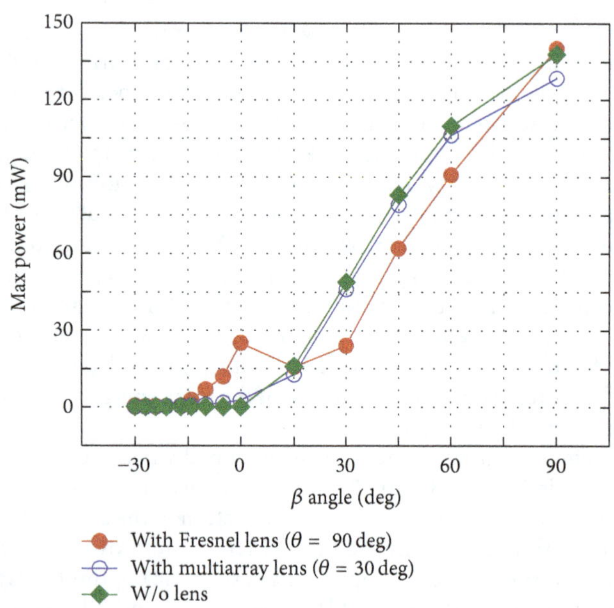

Figure 7: Results of power-measurement test with varying β and $\alpha = 90^\circ$.

the solar energy directly. However, the Fresnel lens system can effectively generate a much higher power than the other systems under the worst incidence-angle condition from −30° to 15°. In particular, the Fresnel lens system exhibited the highest performance at $\beta = 0^\circ$. These results indicate that the proposed Fresnel lens system might be effective for power generation in a three-axis stabilization attitude control satellite when the dedicated solar panel faces the sun. In contrast, as β increases beyond 15°, under a favorable incidence-angle condition for the solar panel, the performance obtained using the Fresnel lens system decreases although the performance obtained using the multiarray lens system is similar to that without the lens system. This is because of the shadowing effect of the lens system installed on the edge of the solar panel with a transmission efficiency of 92%. However, the test results for the Fresnel lens system with $\beta = 90^\circ$ under the best sun-incidence angle condition showed almost identical performance to the configuration without the lens system, and the multiarray lens system yielded a 5.5% decrease in the power generation performance. This is attributed to the minimization of the performance degradation by the transmission loss of the lens array at $\beta = 90^\circ$, which is the best condition for power generation on-orbit. These test results indicate that the power generation performance of the solar-energy collection system utilizing the PMMA lens system depends on the sun-incidence angle. Therefore, the effects induced by the attitude control strategies of the satellite, such as passive permanent-magnet stabilization and active three-axis attitude control, under sun incidence-angle variation were investigated through an on-orbit power simulation, as discussed in the last chapter of this paper.

Figure 8 shows the test configuration for validating the effectiveness of the proposed strategy with β fixed as 0° and α varying between 67.5°, 45°, 22.5°, and 0°. Figures 9(a) and 9(b) compare the test results, that is, the *I-V* curves between the Fresnel and multiarray lens systems. The test results obtained without the lens systems are also plotted in both figures. The solar-energy collection system utilizing the PMMA lens system exhibited a much higher performance than that without a lens system. In addition, as the α angle decreased, the performance of both lens systems tended to decrease until $\alpha = 22.5^\circ$, but the performances increased again at $\alpha = 0^\circ$. The calculated maximum power of the Fresnel lens system under the worst condition of $\alpha = 22.5^\circ$ was 1.77 mW. This is almost twice the power generation performance of the multiarray lens system. The test results indicate that the proposed system employing the Fresnel lens is much more

TABLE 3: Summary of the power-measurement test results.

Incidence angle α	β	P_{max} (mW) w/o PMMA lens	With Fresnel lens	With multiarray lens
67.5°	0°	0.0032	21	1.98
45°			1.79	1.41
22.5°			1.77	0.90
0°			7.63	1.01
90°	−30°	0.0003	0.42	0.014
	−25°	0.0003	0.35	0.08
	−20°	0.0003	0.35	0.16
	−15°	0.0003	2.86	0.48
	−10°	0.0003	7.02	0.88
	−5°	0.0003	14.02	1.66
	0°	0.0032	25.05	2.66
	15°	15.79	15.67	12.62
	30°	48.778	23.93	46
	45°	82.77	61.95	79.04
	60°	109.76	90.73	106.16
	90°	137.87	140.2	128.35

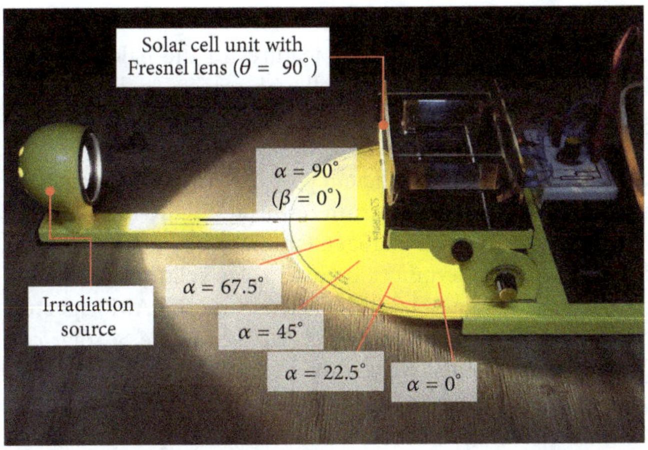

FIGURE 8: Configuration of the power-measurement test with $\beta = 0°$ and varying α.

effective for power generation than the multiarray lens system with a varying α angle.

Table 3 summarizes the calculated maximum power from the power-measurement test results of V_{max} and I_{max} for each test condition. The proposed solar-energy collection system using the Fresnel lens had a much higher power generation performance in almost all the test cases, except when the sun incidence angle β exceeded 15°. Particularly, the highest power generation performance was observed at $\beta = 0°$ and $\alpha = 90°$. The power generation performance of the Fresnel lens system was 7,828 times better than without a solar-energy collection system and 9.4 times greater than that of the multiarray lens system. However, when the incidence angle β exceeded 15°, the performance obtained with the Fresnel lens system was lower than that without the lens systems, and the multiarray lens system exhibited results similar to those obtained without the lens systems. These phenomena will be considered for the on-orbit prediction of the power generation efficiency of the solar-energy collection system with various attitude control strategies.

The halogen lamp used in this study is an uncollimated light source. However, the Fresnel lens used in this study basically redirects the collected light towards its focal point. By applying the Fresnel lens in a way that was proposed in this study, the solar energy collected to the Fresnel lens is spread over the area of solar panel without forming a focal point on it. In addition, the power measurement test using a xenon lamp has been performed by Oh and Park [25] to investigate additional spectral features of light source related to an air mass zero (AM0) solar irradiation because this more effectively simulates the solar energy in a wide solar wavelength spectrum compared to the halogen lamp. In the test, they used a multiarray lens, which has the same dimensions and material as the one used in this study. The test results were obtained under the incidence angle of $\beta = 0°$ and $\alpha = 90°$ condition. The results indicated that much higher power generation can be obtained using xenon lamp compared to the halogen lamp. The test results indicate that the PMMA Fresnel lens is effective over the solar wavelength spectrum because its principle to collect and illuminate the solar energy onto the solar panel is similar as that of the multiarray lens. These features indicate that the SEC system proposed in this study is feasible for efficient power generation on-orbit over the wavelength spectrum, although the test results were obtained under the condition using uncollimated light source. Therefore, we judged that the proposed SEC system will be feasible under a collimated light source or an actual sunlight in-orbit. In the near future, this

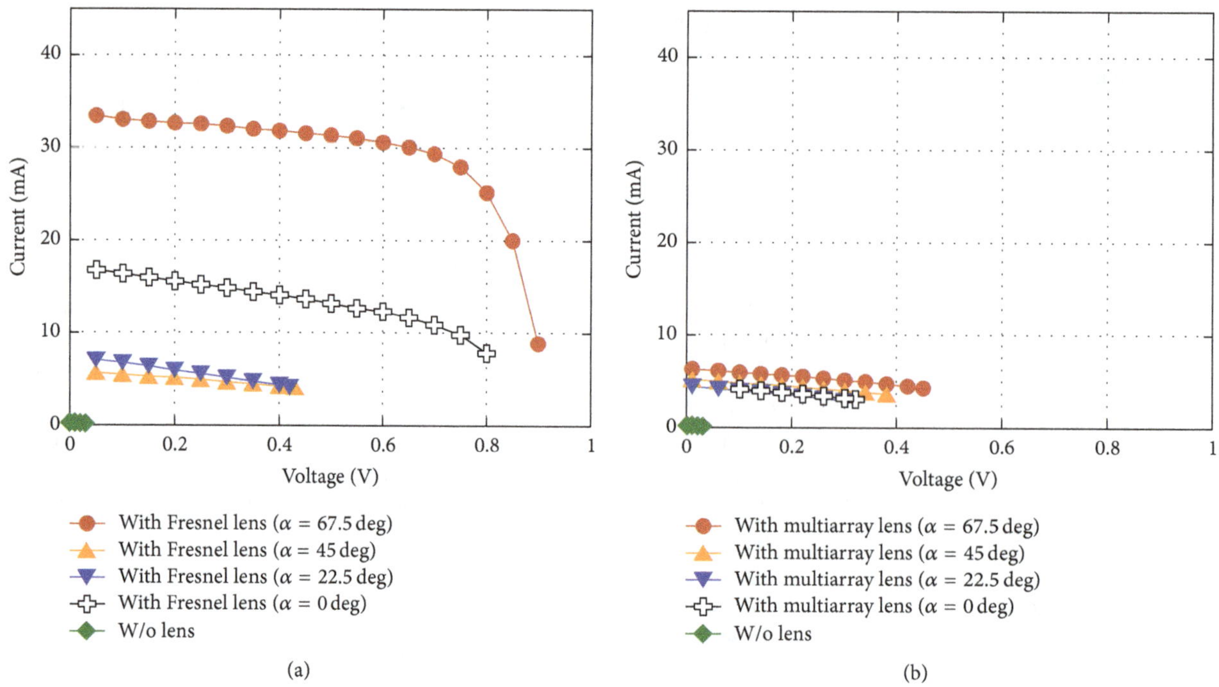

FIGURE 9: Results of the power-measurement test with $\beta = 0°$ and varying α ((a) Fresnel lens, (b) multiarray lens).

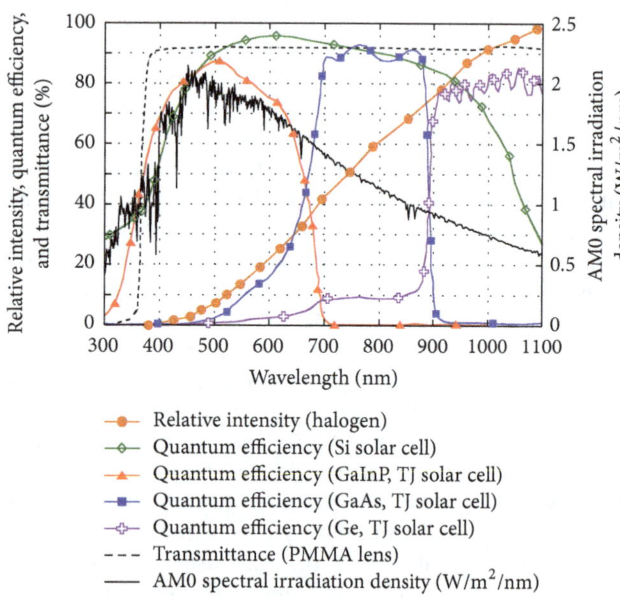

FIGURE 10: Spectral characteristics of the PMMA lens, solar cells, and halogen lamp with respect to AM0 solar irradiation.

will be verified by an on-orbit power measurement test of the SEC system using the cube satellite STEP Cube Lab.

5. Assessment of Applicability of a PMMA Lens for a Solar-Energy Collection System

To investigate the technical applicability of the PMMA lens for the solar-energy collection system, the transmittance of the PMMA lens, the spectral features of the halogen lamp light source, and the Si solar cell with respect to the air mass zero (AM0) environment used in the power-measurement test were analyzed, as these features are strongly related to the power generation efficiency of the proposed system in the solar wavelength spectrum on-orbit. Figure 10 shows the relative intensity of the halogen lamp, the quantum efficiency of the Si and triple-junction GaAs solar cells, the transmittance of the commercial PMMA lens, and the AM0 spectral irradiation density. As shown in the figure, the Si solar cell used in the test effectively simulates the spectral features of the triple-junction GaAs solar cell, which has been widely used for space programs. The halogen lamp used in this study has limitations for simulating solar-energy over a wide wavelength spectrum, especially at ultraviolet wavelengths. However, the PMMA lens is capable of transmitting solar-energy over a wide wavelength range, as shown in Figure 10. In addition, it covers the wide wavelength range of the triple-junction GaAs solar cell using multiple semiconductor materials GaInP, GaAs, and Ge. The test results also show a higher power generation performance, although the halogen lamp light source has limitations for simulating the spectral features of the solar cell and the AM0 spectral irradiation density. These assessments indicate that the solar-energy collection system using the PMMA lens is effective for efficient power generation on-orbit over the given wavelength spectrum.

The outgassing requirements of the material selection for space applications are a total mass loss (TML) less than 1.0% and a controlled volatile condensable material (CVCM) less than 0.1%, according to the ASTM E595 standard [28]. Therefore, we performed an outgassing-measurement test on

TABLE 4: Material and thermooptical properties of the PMMA lens.

Property	Parameter	Value
Thermooptical property	Absorptivity (a)	0.04
	Emissivity (ϵ)	0.04
Material property	Conductivity (W/m/k)	1.5
	Density (kg/m^3)	2210
	Specific heat (J/kg/k)	0.198

the PMMA lens. The test apparatus and procedures were in accordance with the ASTM E595 standard. In the TML measurement test, we used four samples of PMMA lens specimen with dimensions of 10 mm × 10 mm× 1 mm. The measured TML and CVCM values were 1.028% and 0.038%, respectively. The TML value was slightly over the requirement of 1.0%, but the difference was at most 0.078%. Therefore, we judge that this PMMA lens is applicable for the STEP cube mission.

The allowable service-temperature range of the PMMA lens given by the vendor is −70°C~80°C, as shown in Table 2. Therefore, to investigate the survivability of PMMA, the lens integrated on the solar panel, we performed an on-orbit thermal analysis under the worst hot and cold orbital conditions by using the commercial on-orbit thermal analysis software of Thermal Desktop [29] and SINDA/Fluint [30]. Table 4 presents the material and thermooptical properties of the PMMA lens. Table 5 shows the orbit and thermal parameters used as inputs for the analysis. In addition, we assumed that the satellite performs two cycles of rotation per orbit by using the passive permanent magnet attitude-stabilization method. The results of the on-orbit thermal analysis are summarized in Table 6. The maximum and minimum temperatures of the PMMA lens calculated in the on-orbit thermal analysis under the worst hot and cold conditions were 24.7°C and −38.4°C, respectively. The results indicate that the PMMA lens would operate safely within the allowable temperature conditions on-orbit and would have a sufficient temperature margin of 31.6°C with regard to the allowable temperature thresholds.

6. On-Orbit Prediction of Power Generation Efficiency

The power-measurement test results demonstrate that the power generation efficiency of the proposed solar-energy collection system using the PMMA lens depends on the sun incidence angle. This means that the attitude control strategy of the satellite is one of the most important parameters for determining the power generation efficiency of the satellite, as the incidence angle β between the sun vector and the satellite varies continuously owing to the revolution and rotation of the satellite. Therefore, we predicted the on-orbit power generation efficiency according to the attitude control strategies of a satellite with passive permanent-magnet stabilization, the active three-axis stabilization method, and the passive gravity-gradient attitude stabilization method. This prediction was performed using STK. The power-measurement test

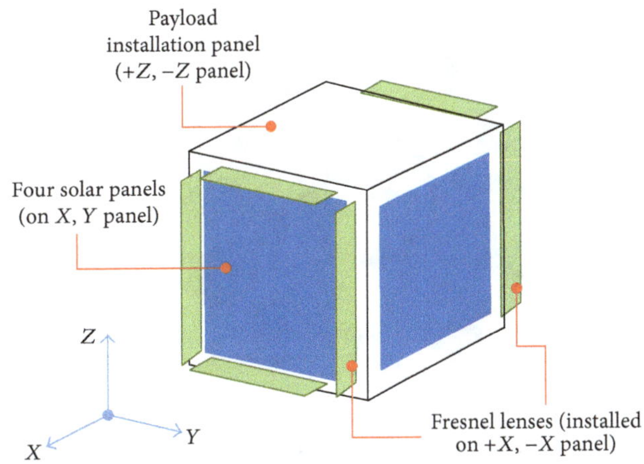

FIGURE 11: 1 U standardized cube satellite model employing the solar-energy collection system.

results described in this study were used as input data for the prediction. Table 7 summarizes the simulation parameters used for the prediction. In the simulation, we assumed the satellite to be operated in a sun-synchronous orbit at an altitude of 600 km. The orbital periods of sunlight and eclipse during the on-orbit operation were 3,682 and 2,119 seconds, respectively. The average incoming solar intensity was 1,358 W/m^2 and the conversion efficiency of the solar cell was 30%.

Figure 11 illustrates an example of the 1 U cube satellite with the proposed solar-energy collection system. The satellite employs a total of four solar panels in the X and Y side panels because it is assumed that the Z panels are used for other appendages such as the antenna or payloads. We also assumed that PMMA lens systems were installed on the $+X$ and $-X$ panels. Figure 12 shows the attitude profiles of the satellite with different attitude control strategies. Figure 12(a) illustrates the on-orbit profiles of the satellite with passive permanent-magnet stabilization, which involved two cycles per orbit whereby the satellite aligned itself with the local earth magnetic field. Therefore, the incidence angle between the sun vector and the solar panels varied continuously, as shown in Figure 12(a). In the case of active three-axis attitude control, the $+Y$ panel of the satellite maintained the attitude of the satellite toward the sun, as shown in Figure 12(b). In the case of the passive gravity-gradient attitude stabilization method, the $-Z$ panel of the satellite maintained the attitude of the satellite toward the direction of earth's center in-orbit, as shown in Figure 12(c). In the estimation, the incidence angle β between the sun vector and the solar panels and the power generated by each solar panel were calculated using STK. To estimate the power generation of the solar cells with the PMMA lens system, the power-measurement test data, as summarized in Table 3, were used for a power calculation while considering the variation of the sun incidence angle with respect to the solar cells with the PMMA lens. In the case of the 3-axis active attitude control method, we did not consider a counterbalancing effect due to the power consumption for maintaining attitude of the satellite in-orbit

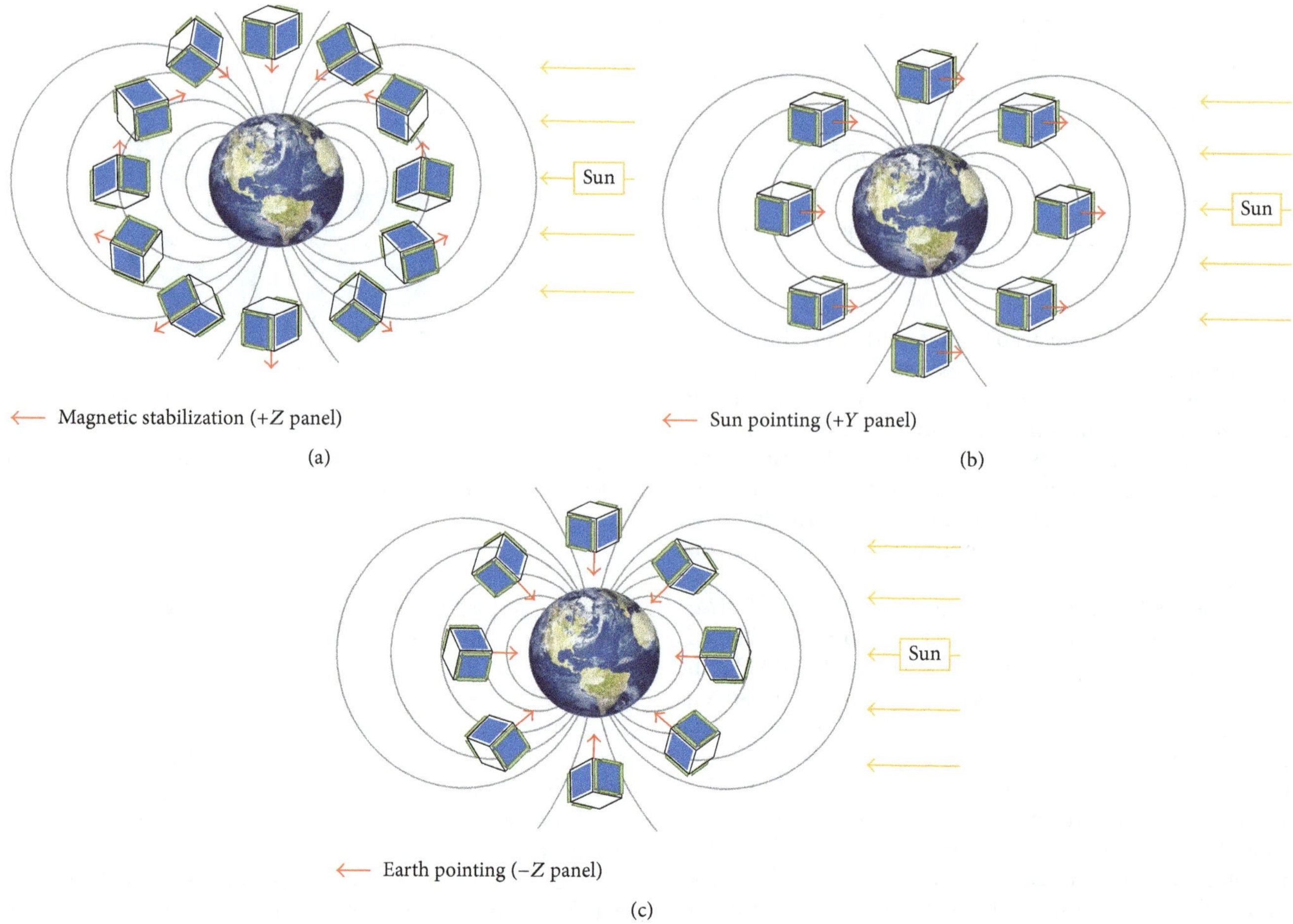

FIGURE 12: Attitude profile of the satellite ((a) passive attitude-stabilization method using permanent magnet and hysteresis rods, (b) active three-axis attitude control method, and (c) gravity-gradient stabilization method).

TABLE 5: Conditions for the on-orbit thermal analysis.

Orbit parameter	Worst hot case	Worst cold case
Orbit type	Sun-synchronous (Circular)	
Inclination (deg.)	97.78	
Altitude (km)	600	
Orbital period (sec)	5801.23	
LTAN (local time ascending node)	12:00 AM	
Solar flux (W/m^2)	1420	1287
Albedo	0.35	0.3
IR planet shine (W/m^2)	249	227
Solar term	Winter solstice	Summer solstice

because the objective of this study is to confirm the feasibility of the SEC system using Fresnel lens system by comparing its power generation performance of the cube satellite on-orbit with that using the multiarray lens.

Figure 13 shows the on-orbit time profiles of the power generation of the satellite with and without the Fresnel lens system for the permanent-magnet stabilization method. The on-orbit profile with the multiarray lens system is also plotted here for comparison. The results indicate that the power generation of the satellites with both PMMA lens systems depends on the variation of the sun incidence angle β with respect to the solar panel because of the revolution and rotation of the satellite according to the attitude control strategy, particularly for the Fresnel lens system. Consequently, the estimated total power generation of the satellite with the Fresnel lens system is 2.22 Wh, which is 10.13% and 9.8%

TABLE 6: Summary of the on-orbit thermal analysis results.

Component	Worst cold case		Worst hot case		Margin with respect to
	T_{min} (°C)	T_{max} (°C)	T_{min} (°C)	T_{max} (°C)	T_{min}/T_{max} (°C)
PMMA lens	−38.4	12.94	−33.2	24.7	−31.6/+55.3

TABLE 7: Conditions for the on-orbit power generation prediction.

Parameter	Condition
Orbit	Sun-synchronous, 600 km altitude
Orbital period	5801 seconds (sunlight/eclipse: 3682/2119 seconds)
Solar intensity	1358 W/m^2
Attitude control	(i) Passive attitude-stabilization using permanent magnet (ii) Active three-axis attitude control (iii) Passive gravity-gradient attitude stabilization
Solar cell	Efficiency: 30% Cell area: 60.36 cm^2 (per panel)
Solar-energy collection system	$+X$, $-X$ panel installation

TABLE 8: Results of the on-orbit power generation prediction.

Parameter	w/o PMMA lens	With Fresnel lens	With multiarray Lens
	Passive attitude-stabilization method using permanent magnet		
Generated power (Wh)	2.017	2.22	2.022
Solar cell efficiency (%)	30	33.04	30.08
	3-axis active attitude control method		
Generated power (Wh)	2.48	3.34	2.57
Solar cell efficiency (%)	30	40.36	31.1
	Passive gravity-gradient attitude stabilization method		
Generated power (Wh)	2.25	2.32	2.16
Solar cell efficiency (%)	30	30.82	28.72

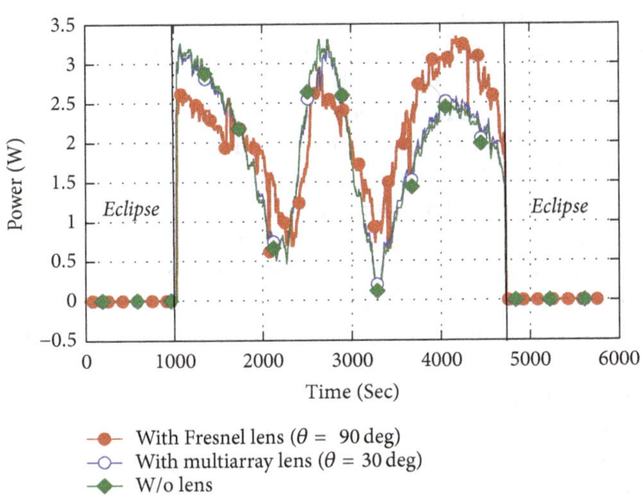

FIGURE 13: Results of the on-orbit power generation simulation of the satellite using the passive attitude-stabilization method with a permanent magnet and hysteresis rods.

higher than that of the satellites without lens systems and with the multiarray lens system, respectively. These results indicate that the solar-energy collection system using the Fresnel lens can effectively enhance the power generation performance, despite the power degradation due to the presence of Fresnel lens system in front of the solar panels as shown in Figure 7.

Table 8 summarizes the estimated power generation, that is, the results of the on-orbit power generation prediction for each case. Figure 14 shows the on-orbit time profiles of the power generation of the satellite with and without the Fresnel lens system under the active three-axis attitude control strategy. For comparison, the on-orbit profiles of the satellite with the multiarray lens system are also plotted here. In this case, we assumed that the $+Y$ panel of the satellite maintained a sun-pointing attitude, as shown in Figure 12(b). Therefore, the $+X$ and $-X$ panels of the satellite were located at the worst positions with regard to power generation because they cannot absorb the solar energy from the sun at a sun incidence angle of $\beta = 0°$.

The power estimation results indicate that both of the lens systems effectively enhance the power generation of the

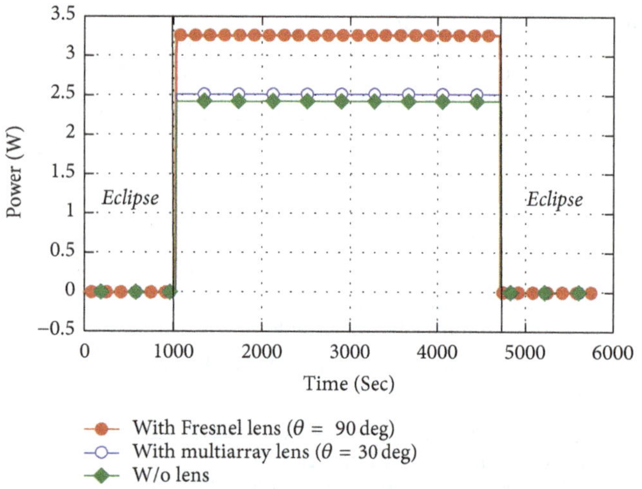

FIGURE 14: Results of the on-orbit power generation simulation of the satellite using the active three-axis attitude control method.

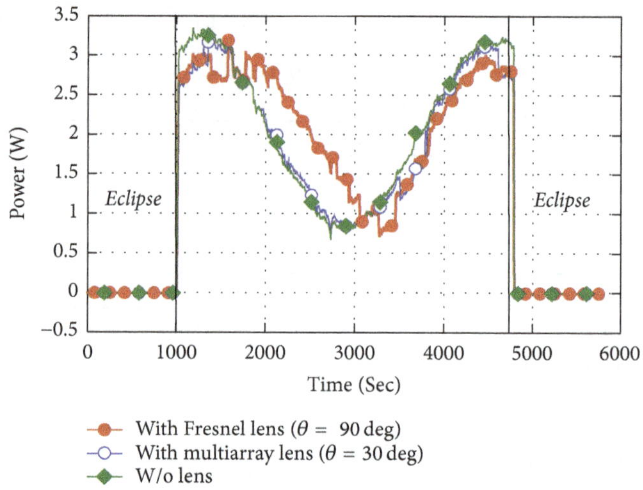

FIGURE 15: Results of the on-orbit power generation simulation of the satellite using the gravity-gradient attitude stabilization method.

satellite. The total generated power with the Fresnel lens system was 3.34 Wh, which is almost 34.68% higher than that without a solar-energy collection system and 30% higher than the multiarray lens system. In addition, when the three-axis attitude control strategy was applied to the SEC system, the power generation performance was 50% higher than the permanent-magnet stabilization method. If additional solar panels are installed on the $+Z$ and $-Z$ panels, a much higher power generation performance can be expected.

Figure 15 shows the on-orbit time profiles of the power generation of the satellite with and without the Fresnel lens system under the passive gravity-gradient attitude stabilization method. For comparison, this figure also shows the on-orbit time profile of the satellite with the multi-array lens system. The power estimation results indicate that both lens systems generate much lower power as compared to the other attitude control strategies. However, the Fresnel lens system showed much higher power generation than the multiarray lens system. The total generated power with the Fresnel lens system is 2.32 Wh, which is a 7.41% greater power generation than that of the satellite with the multiarray lens system. Therefore, we can say that the SEC system proposed in this study is also effective for enhancing power generation of cube satellite under the various attitude control strategies.

The primary goal of this study is to investigate the feasibility of using the Fresnel lens as a SEC system for cube satellite applications. Thus, the aging effects of the PMMA lens system in the space radiation environment were not investigated in this study. In the near future, this effect will be investigated through the STEP mission by obtaining power generation profiles using cube satellite in-orbit. In addition, since the SEC system proposed in this study was considered as a fixed configuration, the current size of the lens system shall be reduced to meet the dimensional constraints of the 1 U cube satellite standard. However, this also involves a power degradation of the SEC system in-orbit. Alternatively, the SEC system can be considered as a deployable one to avoid such power degradation. The most reasonable implementation method of the Fresnel lens system will be investigated in a future study considering dimension, mass, and power generation aspects.

7. Conclusions

This paper proposes a solar-energy collection system employing a Fresnel lens that is capable of providing an enhanced power generation efficiency for cube satellites with a limited area to install solar cells. Although the sun incidence angle with respect to the solar panel varies continuously because of the revolution and rotation of the satellite according to the attitude control strategy, the proposed system can effectively enhance the power generation efficiency of the satellite. The effectiveness of the proposed system was demonstrated by a power-measurement test using a solar simulator and a commercial PMMA Fresnel lens. The test results demonstrated that the Fresnel lens system installed on the edge of the solar panel at an optimized inclination angle is far more effective for power generation performance than the multiarray lens system even under the worst condition of the sun incidence angle with respect to the solar panel. Based on the test results, we predicted the on-orbit power generation efficiency of a cube satellite with the proposed solar-energy collection system using the commercial software STK. In the simulation, we confirmed an enhancement of 34.68% in the power generation performance of a cube satellite mploying a Fresnel lens with a three-axis attitude control strategy as compared to a satellite without the solar-energy collection system.

Competing Interests

The authors declare that they have no competing interests.

References

[1] R. A. Deepak and R. J. Twiggs, "Thinking out of the box: space science beyond the CubeSat," *Journal of Small Satellites*, vol. 1, no. 1, pp. 3–7, 2012.

[2] V. Lappas, N. Adeli, L. Visagie et al., "CubeSail: a low cost CubeSat based solar sail demonstration mission," *Advances in Space Research*, vol. 48, no. 11, pp. 1890–1901, 2011.

[3] X. Li, Q. Schiller, L. Blum et al., "First results from CSSWE CubeSat: characteristics of relativistic electrons in the near-earth environment during the October 2012 magnetic storms," *Journal of Geophysical Research: Space Physics*, vol. 118, no. 10, pp. 6489–6499, 2013.

[4] R. Hoyt, J. Slostad, and R. Twiggs, "The Multi-Application Survivable Tether (MAST) experiment," in *Proceedings of the 39th AIAA/ASME/SAE/ASEE Joint Propulsion Conference and Exhibit*, July 2003.

[5] S. Arnold, J. Armstrong, C. Person, and M. Tietz, "QbX—the CubeSat experiment," in *Proceedings of the 26th Annual AIAA/USU Conference on Small Satellites*, SSC12-XI-4, pp. 1–15, Logan, Utah, USA, August 2012.

[6] J. Kim and W. Kevin, "Sun tracking controller for UKube-1 using magnetic torquer only," *Automatic Control in Aerospace*, vol. 19, no. 1, pp. 541–546, 2013.

[7] N. F. de Rooij, S. Gautsch, D. Briand et al., "MEMS for space," in *Proceedings of the International Solid-State Sensors, Actuators and Microsystems Conference*, pp. 17–24, June 2009.

[8] http://www.cubesatkit.com/.

[9] http://www.gaussteam.com/.

[10] http://www.gomspace.com/.

[11] K. Ryu, J.-G. Rhee, K.-M. Park, and J. Kim, "Concept and design of modular Fresnel lenses for concentration solar PV system," *Solar Energy*, vol. 80, no. 12, pp. 1580–1587, 2006.

[12] P. Pérez-Higueras, E. Muñoz, G. Almonacid, and P. G. Vidal, "High concentrator photovoltaics efficiencies: present status and forecast," *Renewable and Sustainable Energy Reviews*, vol. 15, no. 4, pp. 1810–1815, 2011.

[13] V. D. Rumyantsev, N. Y. Davidyuk, E. A. Ionova et al., "HCPV modules with primary and secondary minilens panels," in *Proceedings of the 6th International Conference on Concentrating Photovoltaic (ICPV-6 '10)*, vol. 1277, pp. 97–100, Freiburg, Germany, April 2010.

[14] F. Morichetti, A. Righetti, G. Grasso, M. C. Ubaldi, and S. M. Pietralunga, "Gondola: trackless LCPV module for Si cells with low-aspect ratio and high energy rating," in *Proceedings of the 38th IEEE Photovoltaic Specialists Conference (PVSC '12)*, pp. 957–960, Austin, Tex, USA, June 2012.

[15] L. Fraas, L. Minkin, J. Avery, H. X. Huang, J. Fraas, and P. Uppal, "Portable concentrating solar power supplies," in *Proceedings of the 35th IEEE Photovoltaic Specialists Conference (PVSC '10)*, pp. 3025–3029, Honolulu, Hawaii, USA, June 2010.

[16] T. Uematsu, Y. Yazawa, T. Joge, and S. Kokunai, "Fabrication and characterization of a flat-plate static-concentrator photovoltaic module," *Solar Energy Materials & Solar Cells*, vol. 67, no. 1, pp. 425–434, 2001.

[17] L. Desmet, A. J. M. Ras, D. K. G. De Boer, and M. G. Debije, "Monocrystalline silicon photovoltaic luminescent solar concentrator with 4.2% power conversion efficiency," *Optics Letters*, vol. 37, no. 15, pp. 3087–3089, 2012.

[18] M. D. Hughes, S.-Y. Wang, D.-A. Borca-Tasciuc, and D. A. Kaminski, "Analysis of ultra-thin crystalline silicon solar cells coupled to a luminescent solar concentrator," *Solar Energy*, vol. 122, pp. 667–677, 2015.

[19] T. Wang, J. Zhang, W. Ma et al., "Luminescent solar concentrator employing rare earth complex with zero self-absorption loss," *Solar Energy*, vol. 85, no. 11, pp. 2571–2579, 2011.

[20] P. Jenkins, M. Krasowski, N. Prokop et al., "TACSAT-4 solar cell experiment: advanced solar cell technologies in a high radiation environment," in *Proceedings of the 34th IEEE Photovoltaic Specialists Conference (PVSC '09)*, Philadelphia, Pa, USA, June 2009.

[21] http://www.dss-space.com/index.html.

[22] S. Habraken, J.-M. Defise, J.-P. Collette, P. Rochus, P.-A. D'Odemont, and M. Hogge, "Space solar arrays and concentrators," *Acta Astronautica*, vol. 48, no. 5-12, pp. 421–429, 2001.

[23] W. Seboldt, M. Klimke, M. Leipold, and N. Hanowski, "European sail tower SPS concept," *Acta Astronautica*, vol. 48, no. 5-12, pp. 785–792, 2001.

[24] M. Mori, H. Kagawa, and Y. Saito, "Summary of studies on space solar power systems of Japan Aerospace Exploration Agency (JAXA)," *Acta Astronautica*, vol. 59, no. 1-5, pp. 132–138, 2006.

[25] H.-U. Oh and T. Park, "Experimental feasibility study of concentrating photovoltaic power system for cubesat applications," *IEEE Transactions on Aerospace and Electronic Systems*, vol. 51, no. 3, pp. 1942–1949, 2015.

[26] AGI, *STK User's Manual Version 8.0.1 for PCs*, Analytical Graphics Inc. (AGI), 2008.

[27] Y.-K. Chang, J.-H. Park, Y.-H. Kim, B.-Y. Moon, and M.-I. Min, "Design and development of HAUSAT- 1 picosatellite system (Cubesat)," in *Proceedings of the International Conference on Recent Advances in Space Technologies (RAST '03)*, pp. 47–54, IEEE, November 2003.

[28] ASTM E595-15, "Standard Test Method for Total Mass Loss and Collected Volatile Condensable Materials from Outgassing in a Vacuum Environment," ASTM International, West Conshohocken, Pa, USA, 2015, https://www.astm.org/.

[29] "Thermal Desktop User's Manual, Ver. 5.5," C&R Technologies, Boulder, Colo, USA, February 2012.

[30] C&R Technologies, *SINDA/FLUINT User's Manual, Version 5.5*, C&R Technologies, Boulder, Colo, USA, 2011.

12

Intelligent Situational Control of Small Turbojet Engines

Rudolf Andoga, Ladislav Főző, Jozef Judičák, Róbert Bréda, Stanislav Szabo, Róbert Rozenberg, and Milan Džunda

Faculty of Aeronautics, Technical University of Košice, Rampová 7, 041 21 Košice, Slovakia

Correspondence should be addressed to Rudolf Andoga; rudolf.andoga@tuke.sk

Academic Editor: Konstantinos Kyprianidis

Improvements in reliability, safety, and operational efficiency of aeroengines can be brought in a cost-effective way using advanced control concepts, thus requiring only software updates of their digital control systems. The article presents a comprehensive approach in modular control system design suitable for small gas turbine engines. The control system is based on the methodology of situational control; this means control of the engine under all operational situations including atypical ones, also integrating a diagnostic system, which is usually a separate module. The resulting concept has been evaluated in real-world laboratory conditions using a unique design of small turbojet engine iSTC-21v as well as a state-of-the-art small turbojet engine TJ-100. Our results show that such advanced control system can bring operational quality of an engine with old turbocompressor core iSTC-21v on par with state-of-the-art engines.

Dedicated to the memory of Ladislav Madarász

1. Introduction

From the cybernetic point of view, aeroengines are nonlinear systems with complex dynamics operating in stochastic environment in a broad range of conditions as principally described in traditional handbooks [1, 2]. Energetic efficiency, ecologic efficiency, and safety are today the key factors in their development. Small gas turbine aeroengines are specifically characterized by compact dimensions weighing from 0.5 to 50 kilograms and relatively high static thrust outputs in the range up to 1500 N, producing very high power to weight ratios as defined in [3]. Such engines represent a great potential for commercial applications in different areas of aviation for propulsion of drones and small UAVs described in [4], as well as small airplanes or helicopters [5]. Because of compact dimensions as well as cost-effective operation and production, these engines can serve as experimental testbeds for research purposes as can be seen in the work of Benini and Giacometti [3], where a small turbojet engine with static thrust of 200 N has been designed. The idea of utilizing small turbojet engines for rapid prototyping of algorithms and construction has also been pursued by other authors, Pecinka and Jilek designing a cost-effective test cell for small turbojet engines [6], application of small turbojet engines in education described in the works [7, 8]. Small turbojet engines have also been used as testbeds in research of application alternative fuels for gas turbine engines using synthetic as well as biofuels in the works of Badami et al. [9, 10] or optimization of jet fuel combustion processes described in [11]. Small turbojet engines can be also used for aeroengines' component design and optimization as illustrated in optimization of turbine blades [12] or centrifugal compressors [13]. Small gas turbine engines have also been utilized in general gas path optimization described in sources [14, 15]. This research review shows that small gas turbine engines are a rapidly developing technology, which is usable not only for specific propulsion of different aerial vehicles but also for usable testbeds for rapid prototyping of components for normal-sized gas turbine engine.

In order to improve operational efficiency of gas turbine engines, constructional design changes can be done, like compressor redesign [13], combustion chamber material and design improvements, fuel nozzle redesign [11], or turbine blade redesign [12]. The other approach is to increase efficiency through design and implementation of progressive control algorithms utilizing digital control systems with a self-tuning control system as designed by Adibhatla and Lewis [16]. This allows improving operation of the controlled engine only through software algorithm updates, which may be a cost-effective solution especially in the area of unmanned systems, where certification is not as complicated as in general aviation legislation. Based on the previous state-of-the-art review, it can be concluded that utilizing small gas turbine engines for prototyping of new control methodologies and approaches is a new area, which has not been thoroughly explored and has a great potential even for applications in gas turbine engines used in civil general aviation.

Gas turbine engines currently used in general aviation are mainly controlled using digital engine control systems with full control authority—FADEC (full authority digital engine control), utilizing certified standard algorithms for handling the engine's operational states as well as managing its auxiliary systems and thrust as is systematically described in the textbooks [17, 18].

Algorithms, which are certified and commonly used in control of gas turbine engines, are based on the proven methodology of PID control [17] as well as scheduling strategies for higher level controllers like power management [18]. On top of that, special strategies like limiters and auxiliary systems are used to protect the engine from departing its operational envelope [19]. Transient states of the engine are handled by acceleration/deceleration min/max schedulers [19], typical architecture of the baseline digital engine control using engine protection logic being shown in the recent work of Connolly et al. in [20]. Diagnostics and health management are usually running as a separate process influencing operation and control of the engine using limiters [18, 19]. Life-extending measures like employment of the N_dot control methodology can be taken as progressive [21].

Utilization of digital control systems having the engine's control algorithms and laws running as software allows improvement of its efficiency only by modification of those algorithms or updating the control laws. This has been pursued by many authors in the recent decades bringing advances in the field of control theory and computational intelligence into control systems of gas turbine engines in order to increase their efficiency and reliability without the need of structural optimizations of the engine's components as comprehensively described in a survey of intelligent technologies for application in gas turbine engines [21, 22].

An approach, which is quite often used in advanced concepts of gas turbine engine control, is the use of the methodology of adaptive control [23, 24], using specialized model-based algorithms to change or compute coefficients of PID controllers according to environmental or model parameters as presented for example in [16]; adaptive control methodology can also utilize engine performance and health computational models to adapt controllers as shown in the conceptual designs described in [25] and, on theoretical level, analyzed in [26]. These works are providing a link between engine diagnostic and control systems, while also using methodologies of computational intelligence (ARMAX model) as for example applied by Mu et al. in [27]. Other approaches which have been taken in advanced control of gas turbine engines were robust, LQ, and optimal control. These methodologies have been already found to be beneficial in solution of complex control problems as illustrated for example in [28, 29]. They were successfully applied in the envisioned engine control systems, the robust control algorithm being described in [30], and LQ control algorithms applied in gas turbine engine control [31, 32]. Even more advanced methodologies like model predictive control [33] or hybrid fuzzy-genetic algorithms have been proposed in adaptation of gas turbine engine controllers as described in [34].

The works from the area of adaptive control link control algorithms to diagnostic algorithms or engine health evaluation algorithms as described in [25, 26]. Digital diagnostic systems are directly connected to safety and reliability of any control system operation but are however often investigated as separate systems utilizing progressive methodologies and algorithms like support vector machines and decision trees as designed in [35] or application of neural networks in engine fault detection as described in [36] connecting the diagnostic system to the engine's control system. Fuzzy inference rule-based diagnostic combined with neural networks system has been also proposed for application in fault detection system of gas turbine engines in [37].

These works illustrate that by application of progressive control methodologies, efficiency, reliability, and safety of operation of gas turbine engines can be improved. However, most of them base their results and conclusions solely on simulations like [19, 27, 30, 34] or present only conceptual designs like [21, 22, 25]. Application of methodologies from the field of computational intelligence can also be seen together with interconnection of diagnostic and control systems [21, 22, 27], which can further enhance the control system's reliability and efficiency.

The aim of the research presented in this paper is to design a highly integrated control and diagnostic system suitable for control of gas turbine engines, which is modular and able to combine different progressive as well as classical control methodologies in a unified architecture, thus increasing efficiency of the complete system. Emphasis in this design is put on strong integration of the control system with diagnostics and the ability to control the engine also during atypical operational states using the most efficient controller in the current situation. The designed system is presented as a framework architecture using the methodology of situational control modified specifically for digital control of gas turbine engines.

Contrary to most results presented in research papers, the aim is to prove the operability of the framework control system's design in real-world laboratory conditions utilizing a small turbojet engine iSTC-21v with static thrust of 500 N based on an old turbostarter TS-21. The resulting prototype architecture of the control system is aimed to bring the efficiency of this old turbocompressor engine design to modern standards by application of the designed situational control framework. This framework is aimed to be specifically applied on small turbojet engines; however, it is expected to be also applicable on normal sized gas turbine engines, using the small engine as a demonstrator of the approach.

2. Situational Control Methodology Framework Design

It is very difficult to design a single controller, which will be able to control a complex dynamic nonlinear system operating in a stochastic environment with high quality in all its operational states. Progressive methodologies like robust control [30], linear quadratic control [31, 32], and model predictive and fuzzy control [33, 34] are able to produce controllers robust enough to cover a large spectrum of states and uncertainties; they are however often computationally too complex as shown in the references. Another approach, which has already been often used in solution of control problems, is to design simpler specific controllers for specific operational states of the investigated dynamic system for example in an application using different controllers for gas turbine generator and power system of aeropropulsion system as described in [38].

In aviation, this approach is widely used in flight control systems with a broad spectrum of applicable algorithms [39] combined in modern avionic digital control systems [40] as well as engine control system switching control algorithms for optimal dynamic characteristics [41]. In flight control, the control laws are different for different flight phases like ground, take-off, flight, and landing [40]. A similar pattern can be found in aeroengine control algorithms—the oldest approach is the application of acceleration schedulers, which are specialized controllers to handle acceleration as a special control strategy as shown in the traditional engine control concept [21]. Setpoint/trim controllers are then used to keep the engine at a stable operating speed or thrust [17, 21].

It is a logical step in solution of control problems being able to design simpler and more specialized controllers for certain operational states as defined in [21] having different control loops for limiting, governing, and acceleration/deceleration. These control approaches are however not integrated; individual controllers are not interacting and are switched only by the use of minimum or maximum selection logic [21, 22] or using energy-based switching as shown in [41]. On the other hand, situational control represents an interconnected controllers' framework usable in control of complex dynamic systems under all operational states, emphasizing control in the atypical ones as defined by Madarász et al. in [42], not just using limiters in order to protect the engine's envelope but allow active control during atypical operational state, like high angle of attack flight, engine overheat conditions, and compressor stalls.

The basic foundation design of a situational control system suitable for general application on turbojet engines has been developed in [42] and expanded in [43]; the resulting basic control concept is shown in Figure 1. Analyzers process the measured data describing the operational state of the system and are defined as input (ANX), output (ANY), state (ANZ), and control (ANR).

All operational states of the controlled complex system are then decomposed into n situational frames, which represent specific groups of operational states. In case of a gas turbine engine, these can be states like start-up, acceleration, and stable operation. The situational classifier is acting here as a decision-making element, selecting controllers in an intelligent manner to handle certain situational frames with the highest control efficiency. This high-level modular concept represents the design base for the situational control system, which can be further modified for application in gas turbine aeroengines with possible combination of any control algorithms nowadays used or proposed for gas turbine engines. The concept of situational control methodology designed by the authors also creates a platform for strong integration of the engine's diagnostic system into its control this integration being described in [43]. Development of the diagnostic system and its integration into the situational control system is further described in [44].

On the basis of these previous development iterations of the situational control system as presented in [45], a new complex general architecture has been developed, which is suitable for turbojet, turboshaft, and turboprop engines presented in Figure 2. The new core part of the system is the intelligent supervisory element, which largely expands functions of the situational classifier replacing the one designed in [45] and previous publications [42, 43]. The resulting design is original and based on practical experiments with the iSTC-21v engine in laboratory conditions. Intelligent supervisory element is integrating diagnostic module, selecting individual controllers, and computing command signals for individual situational controllers C_i. The diagnostic module generates disturbance indications, thus helping situational classification and validates the selected parameters of the engine and its environment; these data are then used by the individual controllers as well as the intelligent supervisory element. Description of the designed diagnostic module can be found in [44, 45].

The newly designed intelligent supervisory element has to solve the following tasks in the proposed expanded implementation shown in Figure 3:

(i) Power management—setting the optimal command values for speed, exhaust nozzle diameter, propeller pitch, or shaft load

(ii) Situational classifier—classification of the engine's operational state into a situational frame

(iii) Situational selector—an algorithm securing fluid switching between individual controllers and generating controller selection gating signals

(iv) Adaptive limiters—adaptive limits in order to keep the engine operating within changing operational envelope for increased safety

The main task of the power management is to calculate the optimal commands for the engine's main state parameters. These are commands for the controlled variables like the turbocompressor's setpoint defined as its shaft speed or the propeller's speed or defined by its angle of attack. The

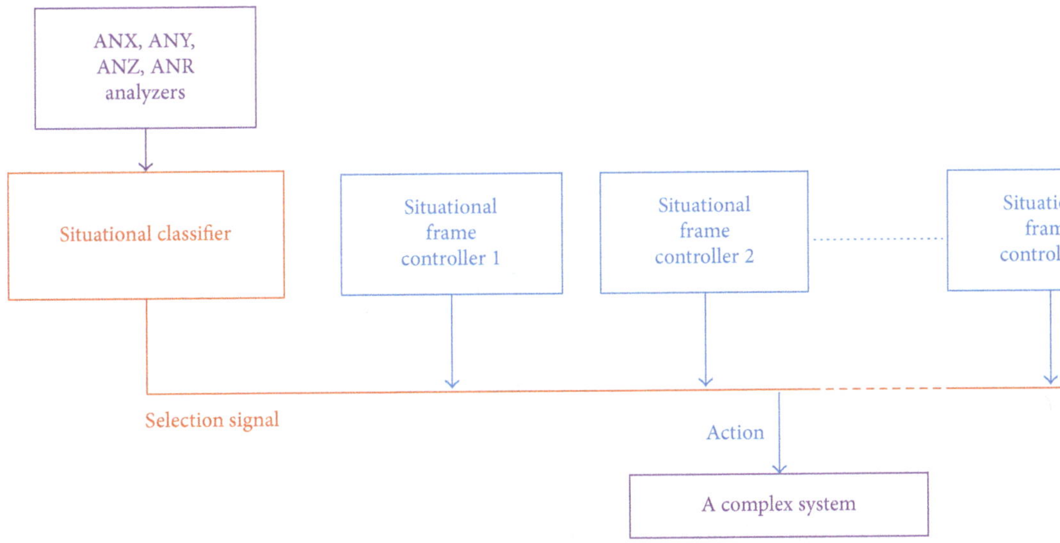

FIGURE 1: The basic concept of the situational control system [42, 43].

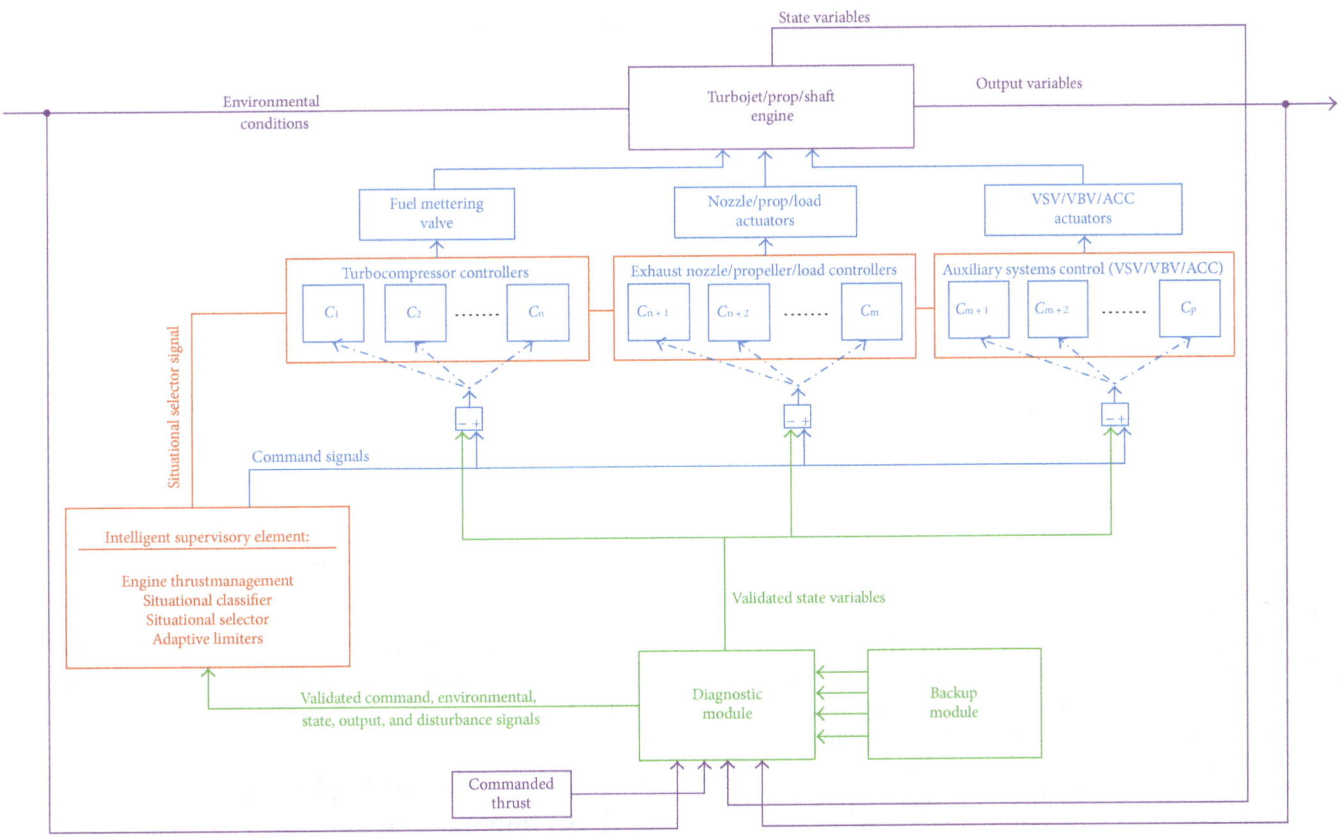

FIGURE 2: The designed situational control system for gas turbine aeroengines.

block also contains a situational classifier, which finds the situational frame (or mode) in which the engine currently operates. This indication is then sent to the situational selector, which secures fluent transitions between the switched controllers on the lower level and assigns the corresponding controller to the actual situational frame.

The composition of the system including the lower level controllers is shown in Figure 4.

Safety is maintained by operating the engine within its operational envelope using a set of adaptive limiters; the commanded parameter as computed by the lower level controllers is being compared to the output of the adaptive limiters, and a minimum or maximum is selected. Adaptivity of the limiters relies on the principle that the maximal shaft speed or exhaust gas temperature of the engine can be adjusted according to flight conditions or

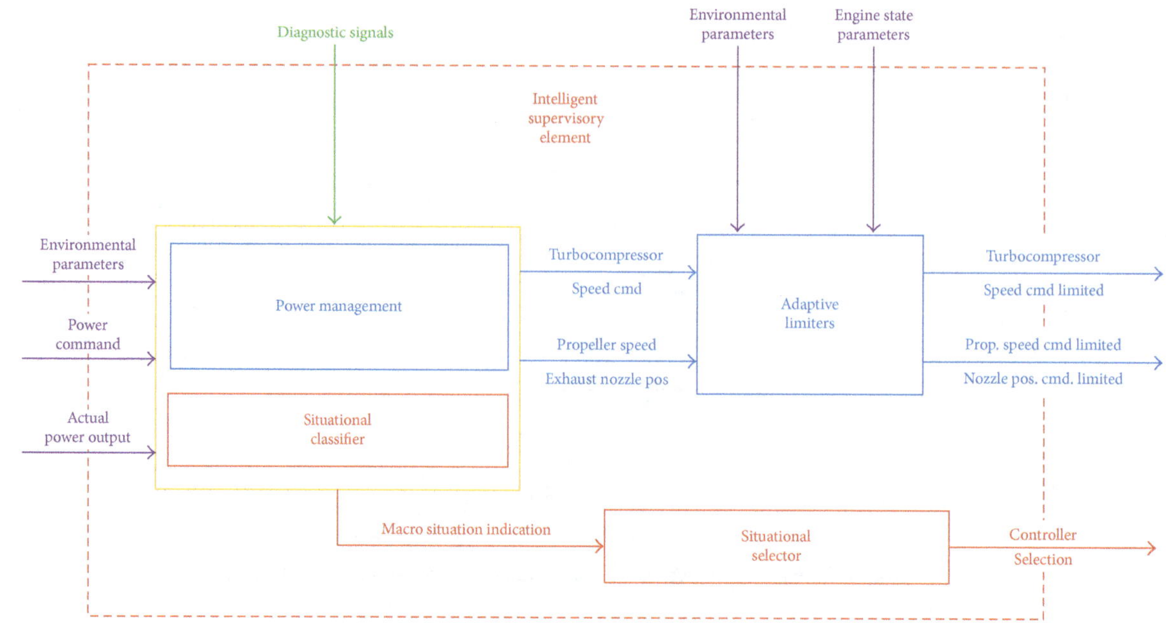

Figure 3: Intelligent supervisory element.

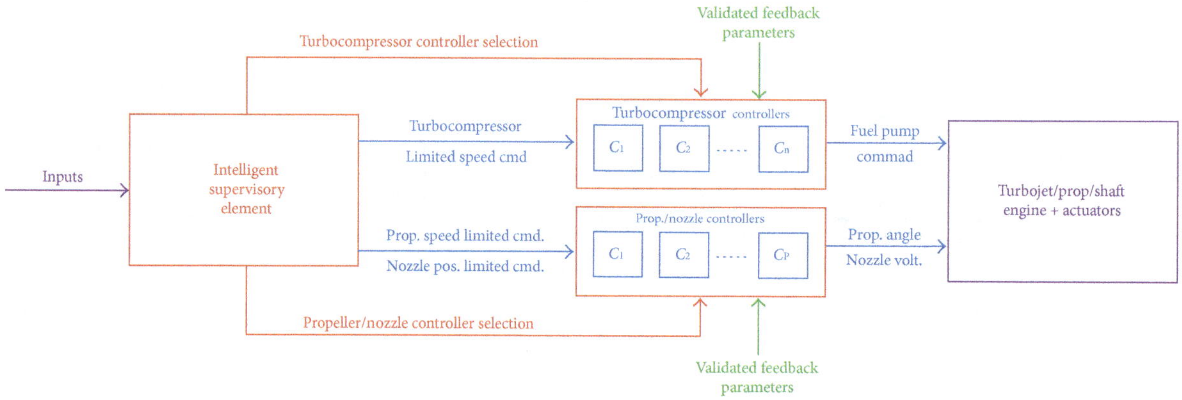

Figure 4: Connection of the intelligent supervisory element with lower level controllers.

environmental conditions. The expanded situational control system developed in Figures 2, 3, and 4, as framework architecture, is the foundation of an intelligent FADEC or i-FADEC to be implemented and tested on a small turbojet engine. In order to design similar situational control frameworks and to define its elements, the following design steps are proposed:

(i) Selection of operational parameters for individual data analyzers

(ii) Decomposition of all operational states of the system into situational frames

(iii) Selection of the methodologies used in the subsystems of the intelligent supervisory element

(iv) Selection of the methodologies for the diagnostic and backup system

These steps will be taken in order to use this framework for design of a situational control system for a small turbojet engine iSTC-21v.

3. A Small Turbojet Engine: An Experimental Object

Small turbojet engines present an ideal platform for development and testing of advanced control algorithms for gas turbine engines as mentioned in the introduction chapter. An experimental small turbojet engine iSTC-21v has been developed from the turbostarter TS-21 used in turboshaft configuration for start-up of normal-sized aircraft engines, used in legacy aircraft utilizing engines Lyulka AL-21F and Tumansky R-29, characteristics of it being described in [46]. It is a single spool, single stream engine with a radial compressor and a single-stage noncooled turbine of old design in a standard configuration of a small gas turbine engine as described for example in [2, 3]. It has been modified by the authors having implemented a digital control system with a direct control of fuel flow supply utilizing a BLDC oil/fuel pump as an actuator in a similar arrangement to the most modern small turbine engines like the TJ-100 engine together with digital data acquisition system [5, 46]. Moreover, the engine was expanded and redesigned with a digitally controlled variable exhaust nozzle, which is a unique design in this class of turbojet engines [46]. The engine on a test bench as used in laboratory conditions is shown in Figure 5. It has to be noted that its turbocompressor core components have unknown history as it was salvaged from a flight unworthy aircraft. Using auxiliary power units or turbostarters from flight unworthy aircraft is a cost-effective solution for testing and design of progressive algorithms for this class of engines.

The following basic engine parameters are measured at the basic sampling frequency of $f_s = 10$ Hz using a National Instruments Compact DAQ system and are used in control of the engine in individual data analyzers, the measurement system being described in [45, 46]:

(i) Outside air temperature T_0 (°C) and atmospheric pressure P_0 (Atm)

(ii) Total air temperature at the inlet of the radial compressor T_{1C} (°C)

(iii) Total air temperature at the outlet from the diffuser of the radial compressor T_{2C} (°C)

(iv) Total gas temperature at the inlet of the gas turbine T_{3C} (°C)

(v) Total gas temperature at the outlet of the gas turbine T_{4C} (°C)

(vi) Total pressure of air at the outlet of the compressor P_{2c} (Atm)

(vii) Total pressure of gases at the inlet of the gas turbine P_{3c} (Atm)

(viii) Fuel flow supply FF (l/min)

(ix) Thrust T_h (kg)

(x) Shaft speed of the turbine/compressor, n_1 (rpm)

(xi) Exhaust nozzle diameter A_5 (mm)

The temperatures and pressures points of the engine measured are the standard measurement points [1, 2] and are specifically for a small turbojet engine in Figure 6.

One run of the engine in laboratory conditions is shown in Figure 7. The graph shows operation of the engine with a stable command of fuel flow supply set at $FF_{cmd} = \{0.9\,l/min\}$ without the situational control system using a traditional closed loop PI controller to meter the fuel flow at the desired level utilizing the electromechanical servo valve LUN 6743 [46]. This is basically showing a state of the engine with only one available setpoint preselected before starting the engine without situational control or other complex control algorithm. The results are illustrative showing the operational parameters of the engine running at a shaft speed setpoint $n_{cmd} = \{40{,}000\text{ rpm}\}$.

Fluctuations and disturbances in shaft speed, temperatures, and pressures, as well as other dynamic parameters, can be seen because of the old construction and the technical state of the engine's core components. The aim of the proposed situational control system is to considerably improve the engine's operational qualitative characteristics by application of the situational control methodology.

4. Situational Control System for a Small Turbojet Engine

4.1. Situational Control System Architecture. In order to design a situational control system with an integrated backup diagnostic module, the general framework as shown and designed in Figure 2 has been applied. The whole operation of the engine is systematically decomposed into the following macrosituational frames as envisioned in [44, 45]:

(1) Prestart control

(2) Start-up control

(3) Operational control

(4) Shutdown

Macrosituational frames are the frames, which incorporate several situational frames. The scheme in Figure 8 shows a set of eleven controllers specifically designed to control iSTC-21v engine in the same number of situational frames expanding and reevaluating the situational frames designed in [44]. The resulting architecture of the control system depicted in Figure 8 is highly modular although specifically designed for a small turbojet engine. Modularity of the control system is one of its key design points, and controller elements and other algorithms can be easily added without disrupting the functionality of other subsystems. This design will be further expanded by implementation of the exhaust nozzle controller and implementation of thrust management algorithms.

4.2. Control Strategies and Situational Decomposition. In design of the control system for a small turbojet engine with fixed exhaust, only the turbocompressor controllers block (see Figure 2 and Figure 8) will be defined. The performed situational decomposition consists of situational frames designated as S_i ($i = 1,\ldots,n$; n is the number of situational frames) with the corresponding controllers designated as C_j ($j = 1,\ldots,n$; n is the number of situational frames). If each situational frame has a controller directly assigned to it, then $i = j$ and the controller can be designated as C_i. The situational frames are organized in a horizontal decomposition and cover the following situations in control of the small turbojet engine iSTC-21v.

The situations S_7, S_8, S_9, and S_{10} represent atypical operational conditions with specialized control strategies to handle them and are shown in grey in Figure 8 as well as the

FIGURE 5: iSTC-21v during operation in laboratory conditions.

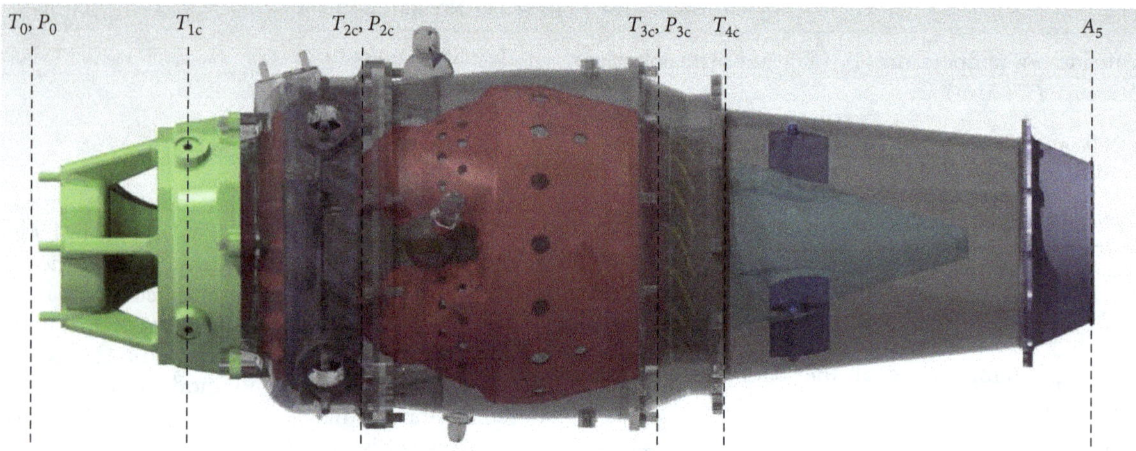

FIGURE 6: Measurement points on a small turbojet engine.

Table 1. All feedback and command signals are transferred through the diagnostic block, which validates them and in case of any sensor's failure is able to replace its value with a computed synthetic value also being able to trigger atypical situations for the situational selector block [44, 45].

The following standard operational situational frames and the corresponding control strategies have been implemented and tested so far in order to demonstrate the functionality of the envisioned design:

(i) Prestart control strategy C_1: the control system performs a check of all aggregates and powers up all corresponding subsystem needed to be powered for launch [44, 45]

(ii) Launch control C_2: preignition control—spinning the electric starter up to 12,000 rpm, open the fuel valve, and energize spark plugs [44, 45]

(iii) Postignition control C_3: control of startup, during postignition—a feedback adaptive fuzzy controller is used to meter the fuel in combustion chamber in order to minimize EGT peak during startup [43]

(iv) Restricted acceleration/deceleration C_4: a PID controller is used to handle the fastest possible acceleration of the engine without exceeding its operational envelope

(v) Stable operation of the engine C_5: a specialized PID controller is used for handling of the constant speed operations of the engine

(vi) Idle control C_6: precise fuel metering near idle speeds of the engine where flameout might happen

(vii) Engine shutdown C_{11}: disabling fuel flow supply and closing of electromagnetic fuel and oil valves [45]

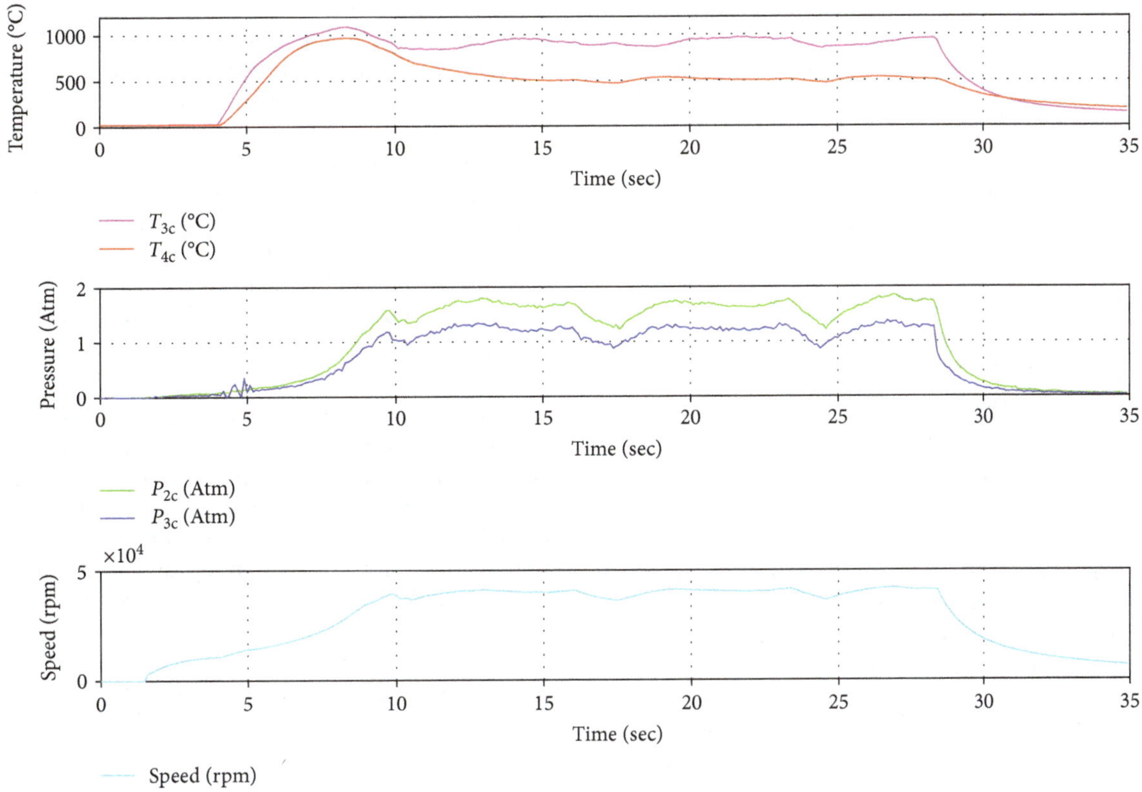

FIGURE 7: Dynamic engine data using the nonsituational PI control system.

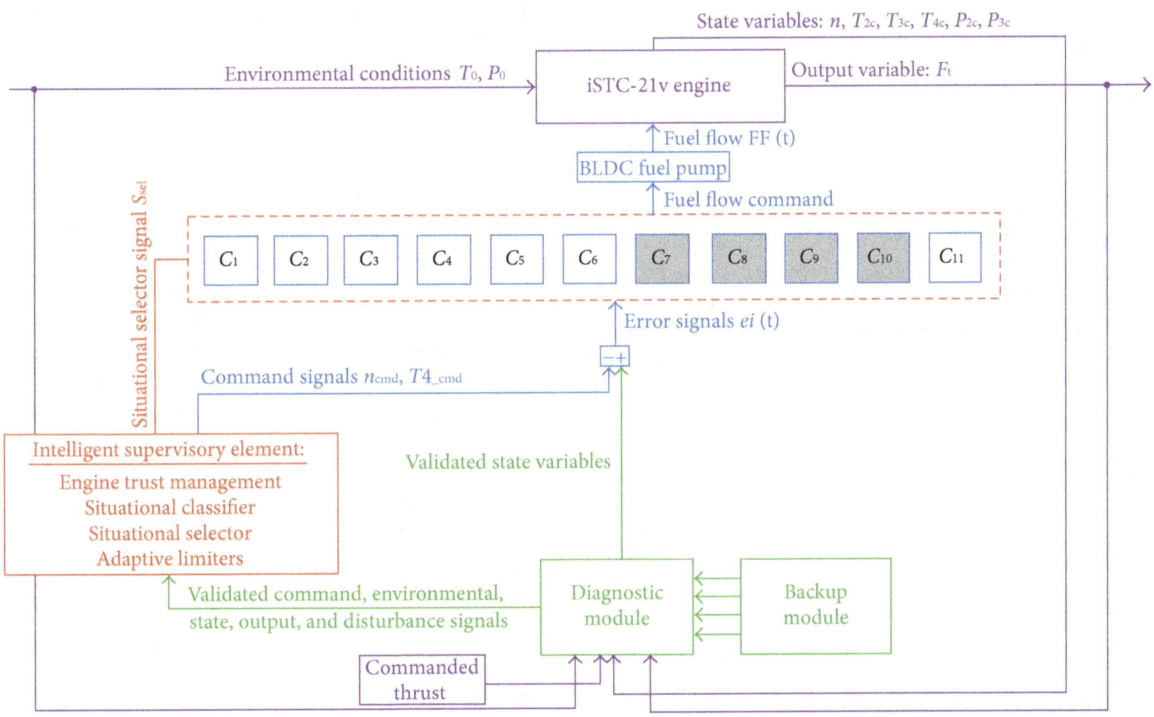

FIGURE 8: The situational control system of the iSTC-21v engine.

Table 1: Definition of situational frames and the corresponding controllers.

Situation	Strategy	Controller
S_1	Prestart control	C_1
S_2	Launch control	C_2
S_3	Ignition control	C_3
S_4	Acceleration/deceleration	C_4
S_5	Stable operation of the engine	C_5
S_6	Idle control	C_6
S_7	Compressor stall	C_7
S_8	Turbocompressor over speed	C_8
S_9	Turbine overheat	C_9
S_{10}	Unspecific degraded mode of operation	C_{10}
S_{11}	Engine shutdown	C_{11}

4.3. Intelligent Supervisory Element. This subsystem is the core element of the complex situational control system; it is acting as a supervisory control system as well as a decision-making element. It is designed to solve the following tasks:

(i) Compute commands for the engine's speed and exhaust nozzle diameter in case of optimal thrust control.

(ii) Evaluate the current state of the engine using all observed and measured parameters and perform situational classification.

(iii) Create switching signals for fluent transitions between controllers and assign controllers to corresponding situational frames—performed by the situational selector.

(iv) Limiting of computed commands and actions for certain control strategies is performed by the block of adaptive limiters.

In order for the concept of the situational control system to work, situational classification and interconnection of control strategies to situational frames are the key factors; therefore, these will be addressed further. The whole operational space of the engine can be described by the vector of parameters defined as $O_s = [E, S, O, I, C]$, which are the inputs of the situational classifier and represent the following parameters:

(i) E is the set of environmental parameters: T_0 and P_0

(ii) S is the set of state parameters: T_{1C}, T_{2C}, T_{3C}, T_{4C}; P_{2C}; P_{3C}; n_1

(iii) O is the set of output parameters: T_h, EPR

(iv) I is the set of input control parameters FF, A_5

(v) C is the set of command parameters: n_{1_cmd};T_{h_cmd}

The purpose of the situational classifier is to transform this multidimensional vector of parameters into a signal which will indicate a situational frame as its output. An ith situational frame S_i either occurs or not, using Boolean logic this means $S_i = \{0 ; 1\}$. Fuzzy logic can however also be used, where indication of the ith situational frame would be represented by a closed interval, where $S_i = <0 ; 1>$. The resulting output of the situational classifier can be formalized as a vector of indicated situational frames: $S_{cl} = [S_1, S_2, S_3, \ldots, S_i, \ldots, S_n]$, $i = 1, \ldots, n$, where n is the number of situational frames. The situational classifier is then a function which transforms the multidimensional vector of the engine's parameters into the output classifier signal at any given time:

$$S_{\text{classifier}}(t)\colon O_s(t) \rightarrow S_{\text{cl}}(t). \quad (1)$$

Any decision-making algorithm can be used in the role of the situational classifier, be it a rule-based system or a neural network. In the case of situational control of a small turbojet engine, a suitable and computational less intensive approach was taken in the form of combination of a rule-based expert system for indication of start-up and shutdown situations S_1, S_2, S_3, and S_{11} and a fuzzy inference system for indication of the remaining operational situations $S_4, \ldots, S_{10}$.

This combination is necessary in order to secure fluent transitions between the controllers during operational feedback control of the engine, where Boolean indication is enough for start-up and shutdown operational states, where direct feedback control signals are not used. Start-up and shutdown situational frames with the corresponding control strategies are described in [44, 45]. To further enhance the quality of switching of individual controllers, the output of the situational classifier is transformed through time-based first-order differential equations, by the situational selector, output of which is designated as $S_{\text{sel}}(i)$. The selector then transforms the indication-gating signals for the ith situational frame according to the following differential equation in time t:

$$T\frac{\mathrm{d}S(i,t)}{\mathrm{d}t} + S(i,t) = S_{\text{sel}}(i,t), \quad (2)$$

where T is an arbitrary time constant.

The time constant defines the speed of situational frame switch; for a fast system like a small turbojet, values from the interval $T = <0.1, 0.7>$ are suitable.

The resulting architecture of the situational classifier and selector systems suitable for small turbojet engines is shown in Figure 9. The situational frames correspond to the designed situational frames for the small turbojet engine iSTC-21v as described in the previous chapter.

The output of the situational selector is used to gate the individual situational controllers, thus fluently switching them on or off. An output from the ith controller can be expressed by its transfer function expressed after Laplace transformation:

$$\mathrm{FF}_i(s) = C_i(s)\left(S_{\text{sel}}(i)e_i(s)\right), \quad (3)$$

where in (3), $\mathrm{FF}_i(s)$ is the fuel flow supply calculated by the ith situational controller, $C_i(s)$ is the transfer function of the ith controller, and $e_i(s)$ is the control error expressed as a difference between the commanded and actual value of

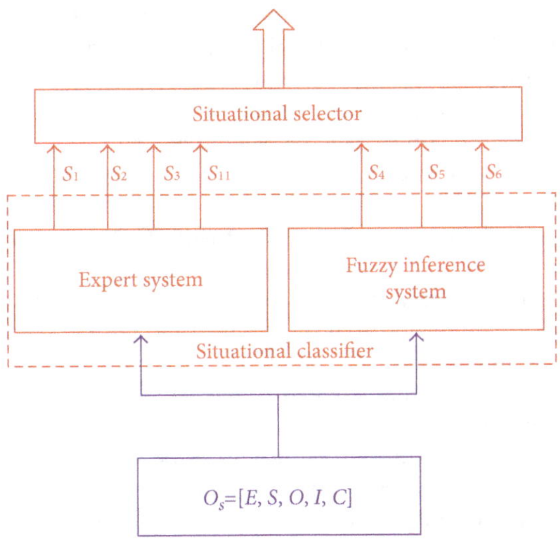

FIGURE 9: The situational classifier and the situational selector for the iSTC-21v engine.

the controlled parameter. The final commanded fuel FF_{cmd} flow for the digitally controlled fuel pump of the engine is then a sum of all fuel flows computed by the individual situational controllers expressed as

$$FF_{cmd}(s) = \sum_{i=1}^{n} S_{sel}(i) FF_i(s). \tag{4}$$

It needs to be noted here that both the input and the output of the controllers are multiplied by a signal from the situational selector in order to stop individual integrators at their inputs and to reset them at their outputs.

4.4. Design of Individual Controllers. For operational control after ignition and startup of the small turbojet engine, the controllers, C_4—acceleration control, C_5—constant shaft speed control, and C_6—idle control, have been designed and tested. In order to compute parameters of these regulators, linearized dynamic models of the iSTC-21v engine have been used as well as nonlinear dynamic models used for testing and tuning as described in [47]. The simplest controller of these is the idle regime controller designated as C_6. The control law of this controller is defined according to (5), which represents a feedback PI controller operating in a fuel flow feedback loop, maintaining fuel flow at the lowest preset possible value for the iSTC-21v engine:

$$FF_6(s) = (FF_{idle}(s) - FF_{act}(s))\left(K + T_i\frac{1}{s}\right), \tag{5}$$

where $FF_{act}(s)$ is the actual value of the current fuel flow, K and T_i are the coefficients of the PI controller. The idle fuel flow is set to $FF_{idle} = 0.75$ l/min for zero speed and zero altitude conditions. The resulting designed controller has been computed by the Naslin method [48] using a dynamic transfer function model of the fuel supply system. The Naslin method provided the best control quality during tests of the fuel flow control feedback loop. The resulting controller has the following control law:

$$FF_6(s) = (0.75 - FF_{act}(s))\left(0.1 + 0.3\frac{1}{s}\right). \tag{6}$$

In operational control, the acceleration and constant speed controllers are the most important ones, as they are responsible for tracking of command signals [1, 2, 20]. Both controllers can be also designed as a single robust feedback controller [30, 31] or fuel flow scheduler [19, 20]; however, the situational control approach allows to design them as two specialized feedback controllers aimed to improve the resulting control quality. The situational control system with intelligent switching can fluently switch between those controllers and even have them operate in cooperative mode to further enhance control quality in order to remove any transients, which could appear during a switch between the acceleration and constant speed controllers. The acceleration/deceleration controller C_4 is defined as a feedback PID controller with the control law defined as follows:

$$FF_4(s) = (n_{cmd}(s) - n_{act}(s))\left(K + T_i\frac{1}{s} + T_d s\right), \tag{7}$$

where $n_{cmd}(s)$ represents the commanded shaft speed of the engine by the operator or the higher level engine thrust management algorithm and $n_{act}(s)$ represents the actual shaft speed. K, T_i, and T_d are the coefficients of the PID controller. The controller is limited in computation of fuel flow constrained in the interval $FF_4(s) = <0.75, 1.6>$ l/min. Anti-windup algorithm is also implemented for the integrator of the controller. In order to tune the controller, Ziegler-Nichols algorithm has been used [49], as the engine is a nonlinear system with complex dynamics and hysteresis. Individual C_i controllers have been afterwards validated using a simulation nonlinear model based on transfer functions obtained by means of experimental identification [47], with follow-up experimental tuning of the computed PID coefficient during operation of the engine in the region of ±5% from the computed design point.

The resulting acceleration controller has the following transfer function:

$$FF_4(s) = (n_{cmd}(s) - n_{act}(s)) \left(0.000042 + 0.00006\frac{1}{s} + 0.000015s\right). \tag{8}$$

The constant speed controller C_5 has been also designed by means of the Ziegler-Nichols tuning rule resulting in a different tuning of the PID controller corresponding to the different control strategy of constant speed hold, where the resulting fuel flow supply command is a sum of a direct setpoint of the fuel flow supply and the action hit of a less aggressive PID controller with its derivative gain equal to zero:

$$FF_5(s) = FF_{setpoint}(s) + FF_{PID}(s), \tag{9}$$

$$FF_{setpoint}(s) = a_2 n^2(s) + a_1 n(s) + a_0, \tag{10}$$

$$FF_{PID}(s) = (n_{cmd}(s) - n_{act}(s))\left(K + T_i\frac{1}{s} + T_d s\right), \tag{11}$$

$$FF_{PI}(s) = <-0.15, 0.15> l/\min. \quad (12)$$

The setpoint fuel flow $FF_{setpoint}$ is a direct linear inverse model of the engine and the PI controller is tuned for limited response in the interval $FF_{PI} = <-0.15,015>$ l/min with antiwindup protection. The resulting controller's output with computed coefficients is defined as follows:

$$FF_5(s) = \left(3.6e^{-09}n^2(s) - 0.000282n(s) + 6.37\right) + \left(n_{cmd}(s) - n_{act}(s)\right)\left(0.00005 + 0.000015\frac{1}{s}\right). \quad (13)$$

5. Experimental Evaluation of the Designed Control System

The proposed framework has been tested in laboratory experiments in a pilot testing program, to verify that the situational control system with situational frame switching is working as designed. In these experiments, a small turbojet engine iSTC-21v has been employed with exhaust nozzle fixed [46]. These tests have been executed as a proof concept of the designed control system and compared to experimental results measured with the TJ-100 engine in our laboratory, which represents a current modern production small turbojet engine with fixed exhaust nozzle and a state-of-the-art FADEC control system [5]. The functionality of the situational control system is demonstrated in Figures 10 and 11 with the iSTC-21v operating from start-up to shutdown with all situational controllers employed. The figures show efficient control of the engine's shaft speed and exhaust gas temperature.

In order to further experimentally evaluate efficiency of the situational control system, the following experimental pilot testing set-up has been designed:

(i) The operational controllers are tested at the commanded speeds defining three operational points $n_{cmd} = \{40{,}000\text{ rpm}\,;\, 45{,}000\text{ rpm}\,;\, 50{,}000\text{ rpm}\}$.

(ii) Each operational point is held by a command for time $T = 15$ sec.

(iii) The resulting acceleration, rise and settling times between the situationally controlled iSTC-21v and FADEC controlled TJ-100, is compared.

(iv) The resulting constant speed deviations between the situationally controlled iSTC-21v and FADEC controlled TJ-100 are compared.

Experimental results as measured in laboratory conditions are shown in Figures 12–14. The figures illustrate that control systems of both engines have acceptable control quality, as there are no undamped oscillations or large overshoots. Figure 12 is showing acceleration of the engine from the stable operating $n_{cmd} = \{40{,}000\text{ rpm}\}$ to the operating point $n_{cmd} = \{45{,}000\text{ rpm}\}$; the time plot depicts dynamic characteristics of the situationally controlled iSTC-21v engine and the TJ-100 engine. The TJ-100 is slightly slower in acceleration and has a small overshoot and undershoot, while it is very good in maintaining the steady shaft speed. In comparison, the situational controller of the iSTC-21v engine is quite fluent in accelerations and decelerations with a bit quicker rise and settling times.

The second comparison shown in Figure 13 depicts accelerations from the operating point $n_{cmd} = \{40{,}000\text{ rpm}\}$ to the operating point $n_{cmd} = \{50{,}000\text{ rpm}\}$, which is the operating point near the maximum speed of the iSTC-21v engine. The situational control system employed on the iSTC-21v has very good acceleration characteristics here; its acceleration is considerably faster than the TJ-100, and both control systems are fluent; however, the settling characteristics of the iSTC-21v are qualitatively better here, as the controller does not have any overshoots.

In order to quantify the differences between control systems of both engines, mean absolute errors and standard deviations [50] of both engines have been compared during the steady state of their operation with results depicted in Figure 14. The steady state is defined as a state where the actual shaft speed is in the interval $n_{act} = <-200, 200\text{ rpm}>$.

As can be seen in Figure 14, the iSTC-21v engine performs slightly worse at the stable operating points $n_{cmd} = \{40{,}000\text{ rpm}\}$ and $n_{cmd} = \{45{,}000\text{ rpm}\}$ as it has larger deviation from the commanded setpoint speed. However, it performs better at the operating point $n_{cmd} = \{50{,}000\text{ rpm}\}$ where it has lower average deviation from the setpoint speed. Standard deviations at these setpoints are nearly equal; this means that the controlled shaft speeds oscillate in a similar way. It can be concluded that both control systems are comparable in maintaining the steady operational point.

In order to compare and evaluate acceleration times, rise and settling times of both engines in accelerations to the previously defined operating points $n_{cmd} = \{45{,}000\text{ rpm}, 50{,}000\text{ rpm}\}$ have been measured. The rise time is defined as the time it takes for the response to rise from 10% to 90% of the steady-state response. The settling time is defined as the time it takes for the shaft speed error to fall to into the interval $n_{act} = <-200, 200>$ rpm. The results of the average measured acceleration times during three performed accelerations as shown in Figures 12 and 13 are presented in Table 2.

The table shows that the situational control system of the iSTC-21v engine is dynamically better than the acceleration control of the TJ-100 engine. This can be also attributed to the fact that the control system of the TJ-100 engine has to be more robust as it is also tuned during flight conditions. In overall evaluation, it can be stated that both control systems are comparable in quality, but a fact is that iSTC-21v engine is using a turbocompressor core, which is of old construction and a worse technical state than the TJ-100 engine, which uses state-of-the-art technical and design solutions in its construction; moreover, it is an engine with only 2 hours of runtime in laboratory conditions; thus, its technical state is flawless. In this regard, the results obtained with the intelligent situational control system employed on iSTC-21v engine can be considered as very positive. The control system of the iSTC-21v has to exert much more effort just to keep a stable operational

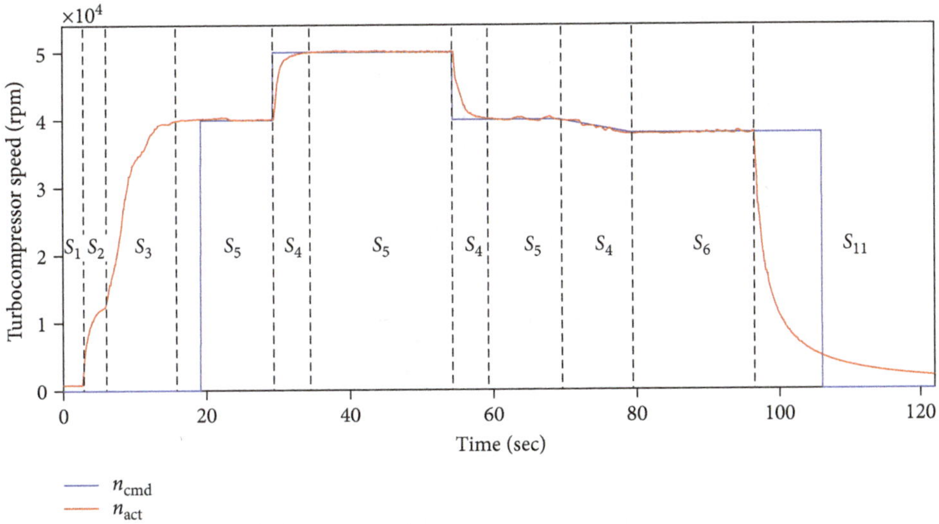

FIGURE 10: A single run of the iSTC-21v engine with the situational control system.

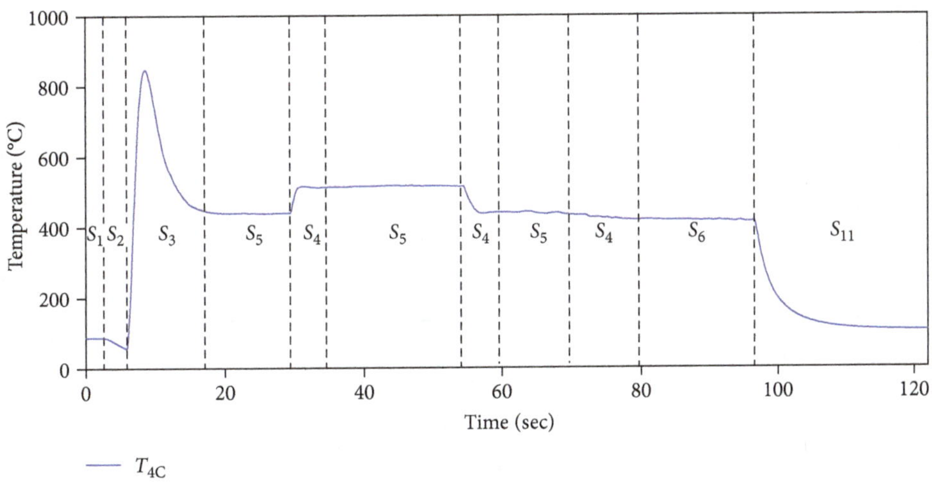

FIGURE 11: Exhaust gas temperature during operation of the situational control system.

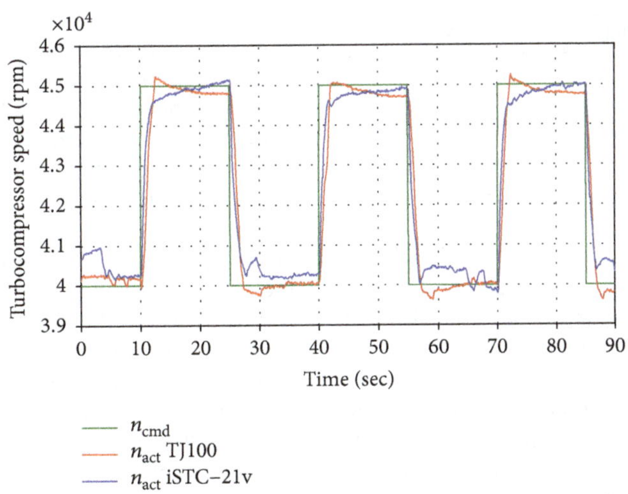

FIGURE 12: Acceleration from 40,000 rpm to 45,000 rpm.

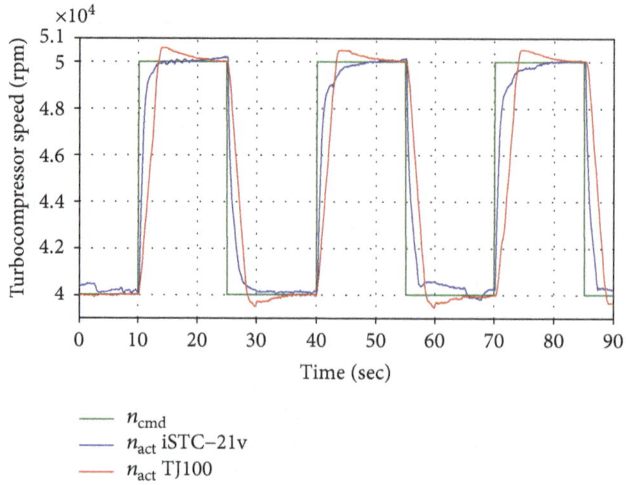

FIGURE 13: Acceleration from 40,000 rpm to 50,000 rpm.

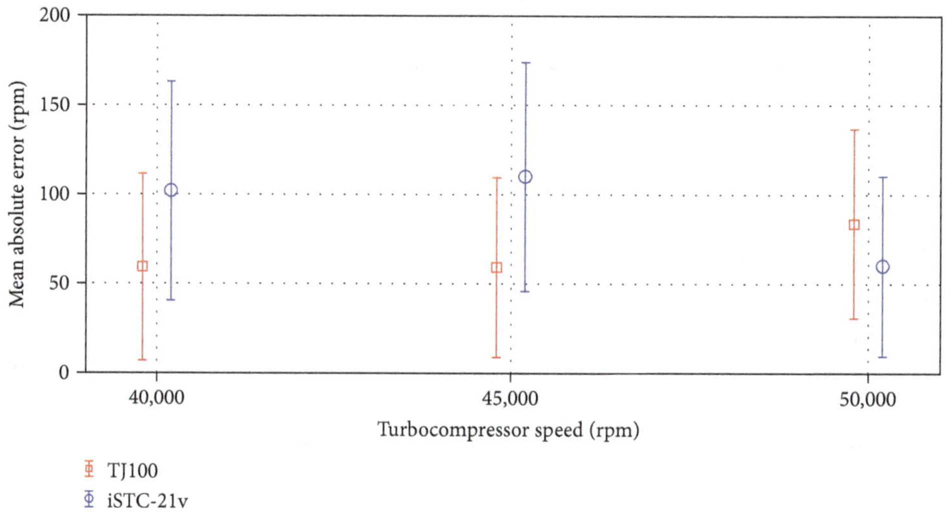

FIGURE 14: The comparison of mean absolute error and standard deviations of the error during the steady state.

TABLE 2: Comparison of rise and settling times.

Rise/settling times	n_{cmd} = 45,000 rpm	n_{cmd} = 50,000 rpm
Rise time: iSTC-21v	1.7 sec	2.1 sec
Rise time: TJ-100	1.8 sec	3.1 sec
Settling time: iSTC-21v	4.5 sec	6 sec
Settling time: TJ-100	5 sec	8.5 sec

point in order to compensate for its old constructional and material faults. In order to illustrate this point, the situational control system is compared with the classical nonsituational control system in Figure 15 with the engine accelerating to the shaft speed $n_{cmd} = \{40{,}000\text{ rpm}\}$.

As can be seen in the figure, the engine iSTC-21v with the situational control system has much better characteristics during its startup with a lower exhaust gas temperature T_{4c} by 200°C, speeds and pressures without oscillations, which makes its running a lot smoother.

6. Conclusions

The paper presents a comprehensive description of an approach, which can be taken in design of a complex intelligent control system suitable for small turbojet engines. The system can be considered as an evolution of FADEC systems, and it can be designated as an intelligent i-FADEC. Its main asset is its modularity, as each module can be further improved by utilizing better or more advanced algorithms, be it algorithms of the situational classifier or improved individual controller modules. The main aim of the paper was to present the framework architecture as a working solution, and operability and efficiency of which has been demonstrated in real-world laboratory conditions with a small turbojet engine, not just as a simulation example. This aim has been fulfilled; moreover, the obtained control quality equals or even surpasses the control quality of a modern turbojet engine TJ-100 lowering the acceleration times of the engine up to 30%.

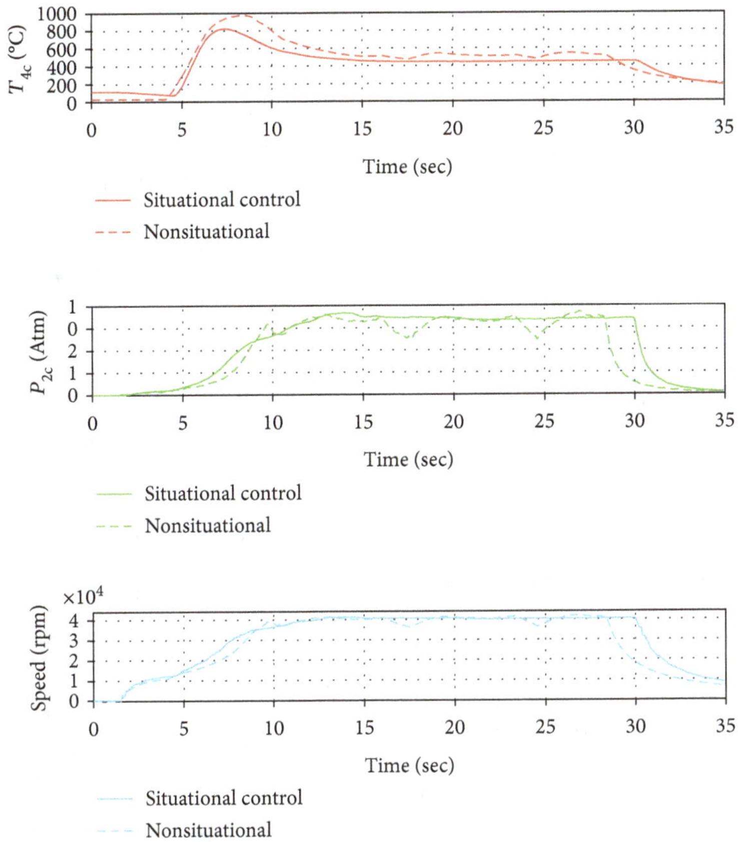

FIGURE 15: Comparison of the situational control system on the iSTC-21v with the nonsituational PI control system.

The designed control system needs to be taken as a proof of concept, and it is expected that a similar or even higher increase in control quality can be achieved using a state-of-the-art engine like TJ-100, this being one of the next research aims. The efficiency of the control system can be further enhanced by follow-up design of different control strategies for individual situational states, utilization of methodologies from the areas of robust or LQ control, focusing on control strategies during atypical modes of operation as well as cooperative control strategies with intelligent controller switching, where several cooperating controllers can be used to handle a certain situation.

In conclusion, the proposed situational control system has to cope with an engine core from the TS-21 engine, which is a very old design from the last century; it is hard to control being in a flawed technical condition, the engine not being previously designed as a multiregime engine. It can be concluded that the designed intelligent FADEC situational control system was able to bring its dynamic characteristics comparable to modern standards, which can be considered as a substantial engineering and design success. The core control system is a real-time software implementation, and our results show that by application of progressive control methodologies, the efficiency and reliability of an old turbojet engine can be considerably improved and the methodology of situational control can be applied and operated successfully in real-world conditions.

Nomenclature

A_5: Exhaust nozzle diameter, mm
BLDC: Brushless DC motor
C_i: Transfer function of the ith controller
EGT: Exhaust gas temperature
EPR: Engine pressure ratio
FADEC: Full authority digital engine control
FF: Fuel flow, l/min
i-FADEC: Intelligent full authority digital engine control
iSTC-21v: Intelligent small turbocompressor engine-21 with variable exhaust nozzle
n_1: Turbocompressor shaft speed, rpm
N_dot: Derivative of shaft speed stabilization algorithm
PID: Proportional integral derivative controller
P_0: Atmospheric pressure, Atm
P_2: Compressor outlet pressure, Atm
P_3: Turbine inlet pressure, Atm
P_4: Turbine outlet pressure, Atm
S_{cl}: A vector of indicated situational frames
S_i: ith situational frame
S_{sel}: Output of the situational selector
T_0: Outside air temperature, °C
T_{2C}: Air temperature at the outlet from the diffuser of the radial compressor, °C
T_{3C}: Gas temperatures in front of the gas turbine, °C
T_{4C}: Total gas temperature aft of the gas turbine, °C
T_h: Thrust, kg

TJ-100: Turbojet 100
TS-21: Turbostarter-21.

Conflicts of Interest

The authors declare that they have no conflicts of interest.

Acknowledgments

The work was supported by the project ESPOSA—Efficient Systems and Propulsion for Small Aircraft, funded by the European Commission in the Seventh Framework Programme under Grant Agreement no. ACP1-GA-2011-284859-ESPOSA and Slovak Research and Development Agency APVV under Grant Agreement no. DO7RP-0023-11.

References

[1] M. Boyce, *Gas Turbine Engineering Handbook*, Elsevier, Oxford, UK, Third edition, 2006.

[2] Rolls-Royce, *The Jet Engine*, Wiley-Blackwell, London, UK, 5th edition, 2015.

[3] E. Benini and S. Giacometti, "Design, manufacturing and operation of a small turbojet-engine for research purposes," *Applied Energy*, vol. 84, no. 11, pp. 1102–1116, 2007.

[4] B. C. Min, C. H. Cho, K. M. Choi, and D. H. Kim, "Development of a micro quad-rotor UAV for monitoring an indoor environment: advances in robotics," in *Advances in Robotics. FIRA 2009*, J. H. Kim, S. S. Ge, P. Vadakkepat, N. Jesse, A. Manum, K. Sadasivan Puthusserypady, U. Rückert, J. Sitte, U. Witkowski, R. Nakatsu, T. Braunl, J. Baltes, J. Anderson, C.-C. Wong, I. Verner, and D. Ahlgren, Eds., vol. 5744 of Lecture Notes in Computer Science, pp. 262–271, Springer, Berlin, Heidelberg, 2009.

[5] PBS Velká Bíteš Aircraft Engines websitehttp://www.pbsvb.com/customer-industries/aerospace/aircraft-engines.

[6] J. Pecinka and A. Jilek, "Preliminary design of a low-cost mobile test cell for small gas turbine engines," in *Proceedings of ASME Turbo Expo 2012: Turbine Technical Conference and Exposition*, pp. 471–478, Copenhagen, Denmark, 2012.

[7] A. J. B. Jackson, P. Laskaridis, and P. Pilidis, "A test bed for small aero gas turbines for education and for university: industry collaboration," in *Proceedings of ASME Turbo Expo 2004: Power for Land, Sea, and Air*, pp. 901–909, Vienna, Austria, 2004.

[8] C. R. Davison and A. M. Birk, "Set up and operational experience with a micro-turbine engine for research and education," in *Proceedings of ASME Turbo Expo 2004: Power for Land, Sea, and Air*, pp. 849–858, Vienna, Austria, 2004.

[9] M. Badami, P. Nuccio, and A. Signoretto, "Experimental and numerical analysis of a small-scale turbojet engine," *Energy Conversion and Management*, vol. 76, pp. 225–233, 2013.

[10] M. Badami, P. Nuccio, D. Pastrone, and A. Signoretto, "Performance of a small-scale turbojet engine fed with traditional and alternative fuels," *Energy Conversion and Management*, vol. 82, pp. 219–228, 2014.

[11] M. Makida, Y. Kurosawa, H. Yamada et al., "Emission characteristics through rich-lean combustor development process for small aircraft engine," *Journal of Propulsion and Power*, vol. 32, no. 6, pp. 1315–1324, 2016.

[12] J. Michalek and P. Straka, "A comparison of experimental and numerical studies performed on a low-pressure turbine blade cascade at high-speed conditions, low reynolds numbers and various turbulence intensities," *Journal of Thermal Science*, vol. 22, no. 5, pp. 413–423, 2013.

[13] S. Guo, F. Duan, H. Tang, S. C. Lim, and M. S. Yip, "Multi-objective optimization for centrifugal compressor of mini turbojet engine," *Aerospace Science and Technology*, vol. 39, pp. 414–425, 2014.

[14] A. O. Pugachev, A. V. Sheremetyev, V. V. Tykhomirov, and O. I. Shpilenko, "Structural dynamics optimization of rotor systems for a small-size turboprop engine," *Journal of Propulsion and Power*, vol. 31, no. 4, pp. 1083–1093, 2015.

[15] F. K. Lu and E. M. Braun, "Rotating detonation wave propulsion: experimental challenges, modeling, and engine concepts," *Journal of Propulsion and Power*, vol. 30, no. 5, pp. 1125–1142, 2014.

[16] S. Adibhatla and T. Lewis, "Model-based intelligent digital engine control (MoBIDEC)," in *33rd Joint Propulsion Conference and Exhibit*, pp. 1–10, Seattle, WA, U.S.A, 1997.

[17] L. C. Jaw and J. D. Mattingly, *Aircraft Engine Controls - Design, System Analysis, and Health Monitoring*, American Institute of Aeronautics and Astronautics, Reston, VA, USA, 2009.

[18] A. L. Diesinger, *Systems of Commercial Turbofan Engines*, Springer, Berlin, Heidelberg, 2008.

[19] J. Csank, R. May, J. Litt, and T.-H. Guo, *Control Design for a Generic Commercial Aircraft Engine*, NASA, Glenn Research Center, Cleveland, OH, USA, 2010.

[20] J. W. Connolly, J. Csank, A. Chicatelli, and K. Franco, "Propulsion controls modeling for a small turbofan engine," in *AIAA Propulsion and Energy Forum - 53rd AIAA/SAE/ASEE Joint Propulsion Conference*, pp. 1–15, Atlanta, GA, USA, 2017.

[21] J. S. Litt, D. L. Simon, S. Garg et al., "A survey of intelligent control and health management technologies for aircraft propulsion systems," *Journal of Aerospace Computing, Information, and Communication*, vol. 1, no. 12, pp. 543–563, 2004.

[22] J. S. Litt, J. Turso, N. Shah, T. Sowers, and A. Owen, "A demonstration of a retrofit architecture for intelligent control and diagnostics of a turbofan engine," in *Infotech@Aerospace Conferences*, pp. 1–18, Arlington, Virginia, 2005.

[23] I. D. Landau, R. Lozano, M. M'Saad, and A. Karimi, *Adaptive Control - Algorithms, Analysis and Applications*, Springer, London, UK, 2nd edition, 2011.

[24] R. C. Roman, M. B. Radac, R. E. Precup, and E. M. Petriu, "Data-driven model-free adaptive control tuned by virtual reference feedback tuning," *Acta Polytechnica Hungarica*, vol. 13, no. 1, pp. 83–96, 2016.

[25] S. Garg, "Controls and health management technologies for intelligent aerospace propulsion systems," in *NASA, Glenn Research Center*, Cleveland, OH, USA, 2004.

[26] Y. Diao and K. M. Passino, "Stable fault-tolerant adaptive fuzzy/neural control for a turbine engine," *IEEE Transactions on Control Systems Technology*, vol. 9, no. 3, pp. 494–509, 2001.

[27] J. Mu, D. Rees, and G. P. Liu, "Advanced controller design for aircraft gas turbine engines," *Control Engineering Practice*, vol. 13, no. 8, pp. 1001–1015, 2005.

[28] T. A. Várkonyi, J. Tar, and I. Rudas, "Improved stabilization for robust fixed point transformations-based controllers," *Journal of Advanced Computational Intelligence and Intelligent Informatics*, vol. 17, no. 3, pp. 418–424, 2013.

[29] J. K. Tar, J. F. Bitó, and I. J. Rudas, "Contradiction resolution in the adaptive control of underactuated mechanical systems evading the framework of optimal controllers," *Acta Polytechnica Hungarica*, vol. 13, no. 1, pp. 97–121, 2016.

[30] D. K. Frederick, S. Garg, and S. Adibhatla, "Turbofan engine control design using robust multivariable control technologies," *IEEE Transactions on Control Systems Technology*, vol. 8, no. 6, pp. 961–970, 2000.

[31] Z. Knoll and G. Tao, "Multivariable adaptive LQ control of jet engines," in *2015 American Control Conference (ACC)*, pp. 1193–1198, Chicago, IL, USA, 2015.

[32] I. Kisszolgyemi, K. Beneda, and Z. Faltin, "Linear quadratic integral (LQI) control for a small scale turbojet engine with variable exhaust nozzle," in *2017 International Conference on Military Technologies (ICMT)*, pp. 507–513, Brno, Czech Republic, 2017.

[33] B. J. Brunell, R. R. Bitmead, and A. J. Connolly, "Nonlinear model predictive control of an aircraft gas turbine engine," in *Proceedings of the 41st IEEE Conference on Decision and Control, 2002*, pp. 4649–4651, Las Vegas, NV, USA, 2002.

[34] M. Montazeri-Gh and A. Safari, "Tuning of fuzzy fuel controller for aero-engine thrust regulation and safety considerations using genetic algorithm," *Aerospace Science and Technology*, vol. 15, no. 3, pp. 183–192, 2011.

[35] W. Yan, C. J. Li, and K. F. Goebel, "A multiple classifier system for aircraft engine fault diagnosis," in *Proceedings of the 60th Meeting of the Society For Machinery Failure Prevention Technology (MFPT)*, pp. 271–279, Virginia Beach, Virginia, 2006.

[36] J. B. Armstrong and D. L. Simon, *Implementation of an Integrated OnBoard Aircraft Diagnostic System*, NASA, Glenn Research Center, Cleveland, OH, USA, 2012.

[37] A. Kyriazis and K. Mathioudakis, "Gas turbine fault diagnosis using fuzzy-based decision fusion," *Journal of Propulsion and Power*, vol. 25, no. 2, pp. 335–343, 2009.

[38] J. Seok, I. Kolmanovsky, and A. Girard, "Coordinated model predictive control of aircraft gas turbine engine and power system," *Journal of Guidance, Control, and Dynamics*, vol. 40, no. 10, pp. 2538–2555, 2017.

[39] J. Hu and H. Gu, "Survey on flight control technology for large-scale helicopter," *International Journal of Aerospace Engineering*, vol. 2017, Article ID 5309403, 14 pages, 2017.

[40] I. Moir, A. Seabridge, and M. Jukes, *Civil Avionics Systems*, AIAA Education Series, Reston, VA, USA, Second edition, 2013.

[41] X. Wang, J. Zhao, and X. Sun, "Overshoot-free acceleration of aero-engines: an energy-based switching control method," *Control Engineering Practice*, vol. 47, pp. 28–36, 2016.

[42] L. Madarász, R. Andoga, L. Fozo, and T. Lazar, "Situational control, modeling and diagnostics of large scale systems," in *Towards Intelligent Engineering and Information Technology*, I. J. Rudas, J. Fodor, and J. Kacprzyk, Eds., vol. 243 of Studies in Computational Intelligence, pp. 153–164, Springer, Berlin, Heidelberg, 2009.

[43] L. Madarász, R. Andoga, and L. Főző, "Intelligent technologies in modeling and control of turbojet engines," in *New Trends in Technologies: Control, Management, Computational Intelligence and Network Systems*, M. J. Er, Ed., pp. 17–38, Sciyo, Rijeka, Croatia, 2010.

[44] R. Andoga, L. Főző, L. Madarász, and T. Karo, "A digital diagnostic system for a small turbojet engine," *Acta Polytechnica Hungarica*, vol. 10, no. 4, pp. 45–58, 2013.

[45] R. Andoga, L. Madarász, L. Főző, T. Lazar, and V. Gašpar, "Innovative approaches in modeling, control and diagnostics of small turbojet engines," *Acta Polytechnica Hungarica*, vol. 10, no. 5, pp. 81–99, 2013.

[46] L. Főző, R. Andoga, L. Madarász, J. Kolesár, and J. Judičák, "Description of an intelligent small turbo-compressor engine with variable exhaust nozzle," in *2015 IEEE 13th International Symposium on Applied Machine Intelligence and Informatics (SAMI)*, pp. 22–24, Herl'any, Slovakia, 2015.

[47] L. Főző, R. Andoga, K. Beneda, and J. Kolesár, "Effect of operating point selection on non-linear experimental identification of iSTC-21v and TKT-1 small turbojet engines," *Periodica Polytechnica Transportation Engineering*, vol. 45, no. 3, pp. 141–147, 2017.

[48] P. Naslin, *Essentials of Optimal Control*, Boston Technical Publishers, Boston, MA, USA, 1969.

[49] J. Graf, *PID Control: Ziegler-Nichols Tuning*, CreateSpace Independent Publishing Platform, Cambridge, MA, USA, 2013.

[50] R. R. Wilcox, *Basic Statistics: Understanding Conventional Methods and Modern Insights*, Oxford University Press, Hannover, Germany, 1st edition, 2009.

13

New Integrated Guidance and Control of Homing Missiles with an Impact Angle against a Ground Target

Shengjiang Yang,[1,2] **Jianguo Guo**,[1] **and Jun Zhou**[1]

[1]*Institute of Precision Guidance and Control, Northwestern Polytechnical University, Shaanxi 710072, China*
[2]*Beijing Aerospace Technology Institute, Beijing 100074, China*

Correspondence should be addressed to Jianguo Guo; guojianguo@nwpu.edu.cn

Academic Editor: Giovanni Palmerini

A new integrated guidance and control (IGC) law is investigated for a homing missile with an impact angle against a ground target. Firstly, a control-oriented model with impact angle error of the IGC system in the pitch plane is formulated by linear coordinate transformation according to the motion kinematics and missile dynamics model. Secondly, an IGC law is proposed to satisfy the impact angle constraint and to improve the rapidity of the guidance and control system by combining the sliding mode control method and nonlinear extended disturbance observer technique. Thirdly, stability of the closed-loop guidance and control system is proven based on the Lyapunov stability theory, and the relationship between the accuracy of the impact angle and the estimate errors of nonlinear disturbances is derived from stability of the sliding mode. Finally, simulation results confirm that the proposed IGC law can improve the performance of the missile guidance and control system against a ground target.

1. Introduction

It is well known that the guidance and control system plays a key role in realizing the flight mission of missiles, and it becomes more and more important to improve the whole performance of the flight control system. The traditional guidance and control systems are separately designed based on the principle of separate frequency spectrum [1]. The effectiveness of the conventional design method has been confirmed in many engineering applications. However, it is argued that the design method may not fully exploit the synergistic relationships between these two interacting subsystems. As a result, the performance of the overall system may be constrained [2–4].

In order to improve the whole performance of the guidance and control system, an IGC method was proposed; that is, a whole model combining the guidance system and the autopilot system can be directly designed [5]. However, the increased dimension of the IGC model makes the controller design more difficult, and it is noted that the IGC model is a nonlinear model with mismatched uncertainties. Some control methods, such as the sliding mode control [6], adaptive control [7], and θ-D method [8], have been employed to design the IGC law in recent literatures. In general, IGC methods could be classified into the following two categories.

The first category method mainly focuses on the two-loop control structure in an IGC framework, and the outer loop and inner loop are separated in the guidance and control system [9–12] due to inherent time scale separation between the faster and slower dynamics. Although the outer loop takes the body rates as the virtual control input to directly yield the body rate command and the inner loop tracks the body rate commands with fin deflections as the control inputs, the settling time of response of different loops affects the performance of the whole system due to the delay problem. In fact, the first method is similar to the traditional design method of the guidance and control system.

The second category method for the IGC system combines the backstepping technique and the other control methods [13–19], such as the small-gain method [14], input-to-state stability theory [15], and dynamic surface control technique [16–19]. The model of the IGC system can be written as a strict feedback system under reasonable assumptions, and some virtual control inputs can be obtained by the backstepping technique [19]. However, the filters/differentiators have to be utilized to overcome the calculating expansion problem caused by analytic differentiation of virtual control inputs. The rapidity of the IGC system decreases because the delay time increases by introducing the filters/differentiators.

In order to overcome the delay problem, a new IGC law is presented for a homing missile with an impact angle against a ground target, motivated by the sliding surface technique consisting of the system states and the estimated states [20, 21] derived from the nonlinear disturbance observers. The contributions of this paper lie in the following aspects.

(a) A new control-oriented model with impact angle error of the IGC system in the pitch plane is built by combining the missile dynamics model and motion kinematics between the missile and target.

(b) A novel IGC law is proposed for a homing missile to satisfy the impact angle constraint and to improve rapidity of the IGC system by using the sliding mode control and the nonlinear disturbance observer technique.

(c) The stability of the closed-loop system is proven by using the Lyapunov stability theory, and the relationship between the accuracy of the impact angle and the estimate error of nonlinear disturbances is derived from the stability of the sliding mode.

(d) The proposed IGC law is evaluated by comparison with the classic guidance and control law and the other IGC law for a homing missile with an impact angle against a ground target.

The rest of this paper is organized as follows: in Section 2, the control-oriented model of the IGC system for the missile is established by linear coordinate transformation. A new IGC law is presented and the stability of the closed-loop system is proven in Section 3. Finally, the numerical simulation results are provided in Section 4, and conclusions are reported in Section 5.

2. Control-Oriented Dynamics Model

2.1. Nonlinear Motion Kinematics. The two-dimensional engagement dynamics model is described in Figure 1.

$$R\dot{q} = V_M \sin(q-\theta) + V_{T1}, \tag{1}$$

$$\dot{R} = -V_M \cos(q-\theta) + V_{T2}, \tag{2}$$

$$V_M\dot{\theta} = n_{yM} - g\cos\theta, \tag{3}$$

where q is the line-of-sight (LOS) angle and R, θ, V_M, and n_{yM} are the relative distance, the flight path angle, the velocity, and the acceleration of the missile, respectively.

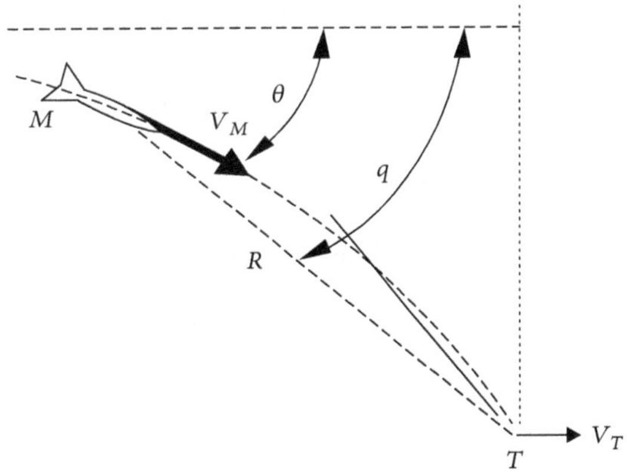

FIGURE 1: Two-dimensional engagement geometry.

V_{T1} and V_{T2} are unknown velocities with respect to the maneuver target on the ground, and g is acceleration of gravity.

By differentiating (1), and substituting (1), (2), and (3) into the derivative, it is easy to obtain

$$\ddot{q} = a_{11}\dot{q} + a_{12}n_{yM} + \Delta_q, \tag{4}$$

where $a_{11} = \dot{V}_M/V_M - 2\dot{R}/R$, $a_{12} = -\dot{R}/RV_M$, and $\Delta_q = -(\dot{V}_M/RV_M)V_{T1} + \dot{V}_{T1}/R + (V_{T2}/RV_M)n_{yM} - ((-\dot{R} + V_{T2})/RV_M)g\cos\theta$ are unknown uncertainties due to maneuver targets on the ground.

2.2. Missile Dynamics Model in the Pitch Plane. The longitude model of a missile in the pitch plane is described as [22]

$$\begin{aligned}\dot{\alpha} &= w_z - \frac{57.3QSc_y^\alpha + P}{mV_m}\alpha + \Delta_\alpha,\\ \dot{w}_z &= \frac{QSL^2m_z^{\bar{w}_z}}{J_zV_m}w_z + \frac{57.3QSLm_z^\alpha}{J_zV_m}\alpha + \frac{57.3QSLm_z^{\delta_z}}{J_z}\delta_z + \Delta_w,\end{aligned} \tag{5}$$

where α is the angle of attack, w_z is the angular pitch rate, δ_z is the deflection angle for pitch control, m is the missile mass, Q is the dynamic pressure, S is the aerodynamic reference area, L is the reference length, P is the thrust of the missile, J_z is the moment of the inertia about z-axis, c_y^α is the lift force derivative with respect to the attack angle, and m_z^α, $m_z^{\bar{w}_z}$, and $m_z^{\delta_z}$ represent the pitch moment derivatives with respect to the attack angle, the nondimensional angular pitch rate, and the deflection angle for pitch control, respectively. Δ_α and Δ_w are unknown-bounded uncertainties of the missile-related parameters [22], for example, the aerodynamic coefficients.

2.3. Control-Oriented Model of the IGC System. The control-oriented model of the IGC system is built to design the IGC law. Assuming that q_M is the desired impact LOS angle at the engagement time, the following equation and inequality hold:

$$\begin{aligned} R\dot{q} &= V_M \sin(q-\theta) + V_{T1} = 0, \\ |q_M - \theta_M| &< \frac{\pi}{2}, \end{aligned} \tag{6}$$

where θ_M is the desired impact flight path angle at the engagement time.

Noting that $V_M \gg V_{T1}$, it is easy to obtain

$$q_M \approx \theta_M. \tag{7}$$

Let

$$e_q = q - q_M \approx q - \theta_M. \tag{8}$$

Differentiation (8) yields

$$\dot{e}_q = \dot{q}. \tag{9}$$

The acceleration of the missile is described as

$$n_{yM} = \frac{57.3QSc_y^{\alpha} + P}{m}\alpha + \Delta n_{yM}, \tag{10}$$

where Δn_{yM} is the unknown-bounded uncertainty of the missile-related parameters [23].

It is obtained from (4), (8), and (9) that

$$\ddot{e}_q = \ddot{q} = a_{11}\dot{q} + a_{13}\alpha + \Delta_e, \tag{11}$$

where $a_{13} = ((57.3QSc_y^{\alpha} + P)/m)a_{12}$ and $\Delta_e = a_{12}\Delta n_{yM} + \Delta_q$ are unknown uncertainties due to Δn_{yM} and Δ_q.

According to the above analysis, the control-oriented model of the IGC system for a homing missile against a ground target in the pitch plane can be described as

$$\dot{x}_1 = x_2, \tag{12}$$

$$\dot{x}_2 = x_3 + d_2, \tag{13}$$

$$\dot{x}_3 = x_4 + d_3, \tag{14}$$

$$\dot{x}_4 = a_{14}x_3 + a_{15}x_4 + bu + d_4, \tag{15}$$

where $x_1 = e_q$, $x_2 = \dot{e}_q$, $x_3 = a_{13}\alpha$, $x_4 = a_{13}w_z$, $u = \delta_z$, $a_{14} = 57.3QSLm_z^{\alpha}/J_zV_m a_{13}$, $a_{15} = QSL^2 m_z^{\bar{\omega}_z}/J_zV_m a_{13}$, $b = (57.3QSLm_z^{\delta_z}/J_z)a_{13}$, $d_2 = a_{11}\dot{q} + \Delta_e$, $d_3 = -((57.3QSc_y^{\alpha} + P)/mV_m)a_{13}\alpha + a_{13}\Delta_\alpha$, and $d_4 = a_{13}\Delta_w$.

Remark 1. a_{12}, a_{13}, a_{14}, a_{15}, and b are known constants due to aerodynamic data. According to the definitions of the disturbance d_i $(i = 2, 3, 4)$ in the models (12), (13), (14), and (15), from a practical point of view, they represent the uncertainties coming from the missile-target motion kinematics and aerodynamic forces and moments of the missile.

Assumption 1. The disturbance d_i $(i = 2, 3, 4)$ and the first-order derivative of d_2 in the systems (12), (13), (14), and (15) are assumed to be bounded by known constants.

3. Control Design and Stability Analysis

Note that the control-oriented model of the IGC system for a homing missile is affected by the mismatched uncertainties; a new sliding mode controller is designed for the IGC system to exploit the synergistic relationship between guidance and autopilot subsystems. Meanwhile, the mismatched disturbances in the control-oriented model are estimated by using the nonlinear disturbance observer to compensate the sliding mode controller.

3.1. New IGC Law Design. Considering the control-oriented models (12), (13), (14), and (15) of the IGC system, a sliding surface with disturbance estimates is chosen as

$$s = c_1x_1 + c_2x_2 + c_3x_3 + x_4 + \hat{\dot{d}}_2 + c_3\hat{d}_2 + \hat{d}_3, \tag{16}$$

where $c_1 > 0$, $c_2 > 0$, and $c_3 > 0$ are chosen to satisfy the Hurwitz polynomial $g(\lambda) = c_1 + c_2\lambda + c_3\lambda^2 + \lambda^3$. $\hat{\dot{d}}_2$, $\hat{d}_2$, and $\hat{d}_3$ are the estimates of $\dot{d}_2$, d_2, and d_3, respectively.

In order to obtain the abovementioned estimates, three nonlinear disturbance observers based on [24] for the IGC models (12), (13), (14), and (15) are designed as follows:

$$\hat{d}_2 = p_{21} + l_{21}x_2, \tag{17}$$

$$\dot{p}_{21} = -l_{21}\left(x_3 + \hat{d}_2\right) + \hat{\dot{d}}_2, \tag{18}$$

$$\hat{\dot{d}}_2 = p_{22} + l_{22}x_2, \tag{19}$$

$$\dot{p}_{22} = -l_{22}\left(x_3 + \hat{d}_2\right), \tag{20}$$

$$\hat{d}_3 = p_{31} + l_{31}x_3, \tag{21}$$

$$\dot{p}_{31} = -l_{31}\left(x_4 + \hat{d}_3\right), \tag{22}$$

$$\hat{d}_4 = p_{41} + l_{h41}x_4, \tag{23}$$

$$\dot{p}_{41} = -l_{41}\left(a_{14}x_3 + a_{15}x_4 + bu + \hat{d}_4\right), \tag{24}$$

where p_{21}, p_{22}, p_{31}, and p_{41} are auxiliary variables and l_{21}, l_{22}, l_{31}, and l_{41} are the elements of the positive observer gain matrix to be designed.

The estimate errors are also defined as $\tilde{e}_{d2} = [\tilde{d}_2, \tilde{\dot{d}}_2]$, $\tilde{e}_{d3} = \tilde{d}_3$, and $\tilde{e}_{d4} = \tilde{d}_4$ where $\tilde{d}_2 = d_2 - \hat{d}_2$, $\tilde{\dot{d}}_2 = \dot{d}_2 - \hat{\dot{d}}_2$, $\tilde{d}_{3h} = d_{3h} - \hat{d}_{3h}$, and $\tilde{d}_3 = d_3 - \hat{d}_3$. It is also obtained from [24] that $\|\tilde{e}_{d2}\| \le \lambda_2$, $|\tilde{e}_{d3}| \le \lambda_3$, and $|\tilde{e}_{d4}| \le \lambda_4$, where $\lambda_i > 0$, for $i = 2, 3, 4$. It is easily obtained as

$$\begin{aligned} \left|\tilde{d}_2\right| &\le \lambda_2, \\ \left|\tilde{\dot{d}}_2\right| &\le \lambda_2, \\ \left|\tilde{d}_3\right| &\le \lambda_3, \\ \left|\tilde{d}_4\right| &\le \lambda_4. \end{aligned} \tag{25}$$

Meanwhile, it can be also obtained from the disturbance observers (17), (18), (19), (20), (21), (22), (23), and (24) that

$$\dot{\hat{\hat{d}}}_2 = \dot{p}_{22} + \dot{l}_{22}x_2 = l_{22}\left(d_2 - \hat{d}_2\right) = l_{22}\tilde{d}_2, \tag{26}$$

$$\dot{\hat{d}}_2 - \dot{\hat{\hat{d}}}_2 = \dot{p}_{21} + l_{21}\dot{x}_2 - \dot{\hat{\hat{d}}}_2 = l_{21}\left(d_2 - \hat{d}_2\right) = l_{21}\tilde{d}_2, \tag{27}$$

$$\dot{\hat{d}}_3 = \dot{p}_{31} + l_{31}\dot{x}_3 = l_{31}\left(d_3 - \hat{d}_3\right) = l_{31}\tilde{d}_3. \tag{28}$$

Computing the derivative of $\dot{s}$ along systems (12), (13), (14), and (15), it can be obtained as

$$s = c_1x_2 + c_2(x_3 + d_2) + c_3(x_4 + d_3) + a_{14}x_3 + a_{15}x_4 + bu + d_4 + \dot{\hat{\hat{d}}}_2 + c_3\dot{\hat{d}}_2 + \dot{\hat{d}}_3. \tag{29}$$

Accordingly, the new sliding controller for the control-oriented models (12), (13), (14), and (15) is obtained as

$$u = -\frac{1}{b}\left[c_1x_2 + c_2x_3 + c_3x_4 + a_{14}x_3 + a_{15}x_4 + c_2\hat{d}_2 + c_3\hat{\hat{d}}_2 + c_3\hat{d}_3 + \hat{d}_4 + k_hs + \varepsilon_h \operatorname{sgn}(s)\right], \tag{30}$$

where $k > 0$ and $\varepsilon > \eta = (c_2 + l_{22} + c_3l_{21})\lambda_2 + (c_3 + l_{31})\lambda_3 + \lambda_4$.

3.2. Stability Analysis. In order to illustrate the effectiveness of the proposed IGC, the stability of a closed system is analyzed as follows.

Choosing a Lyapunov function candidate as $V_1 = s^2/2$ and computing the derivative of $\dot{V}_1$ along the system (16), it is obtained that

$$\dot{V}_1 = s\dot{s} = c_2\left(d_2 - \hat{d}_2\right)s + c_3\left(d_3 - \hat{d}_3\right)s + \left(d_4 - \hat{d}_4\right)s + \dot{\hat{\hat{d}}}_2 s + c_3\left(\dot{\hat{d}}_2 - \dot{\hat{\hat{d}}}_2\right)s + \dot{\hat{d}}_3 s - ks^2 - \varepsilon|s|. \tag{31}$$

According to (26), (27), and (28), the following inequality is obtained as

$$\dot{V}_1 = (c_2 + l_{22} + c_3l_{21})\tilde{d}_2s + (c_3 + l_{31})\tilde{d}_3s + \tilde{d}_4s - ks^2 - \varepsilon|s_h| \le -ks^2 - (\varepsilon - \eta)|s|. \tag{32}$$

It is also obtained from (32) that system signals in system (16) can reach the sliding mode manifolds in finite time due to Lyapunov theory; that is, $s = 0$.

It is obtained that $x_4 = -c_1x_1 - c_2x_2 - c_3x_3 - \hat{\hat{d}}_2 - c_3\hat{d}_2 - \hat{d}_3$ when $s = 0$, and the system dynamic in the sliding mode will be governed as

$$\dot{x}_1 = x_2, \tag{33}$$

$$\dot{x}_2 = x_3 + d_2, \tag{34}$$

$$\dot{x}_3 = -c_1x_1 - c_2x_2 - c_3x_3 - \hat{\hat{d}}_2 - c_3\hat{d}_2 - \hat{d}_3 + d_3. \tag{35}$$

A corresponding Lyapunov function candidate is chosen as

$$V_2 = \mathbf{x}^T\mathbf{T}\mathbf{x}, \tag{36}$$

where $\mathbf{x} = [x_1\ x_2\ x_3 + d_2]^T$ and $\mathbf{T}$ is a positive matrix such that

$$\mathbf{A}^T\mathbf{T} + \mathbf{T}\mathbf{A} = -\mathbf{Q}, \tag{37}$$

for any given positive defined matrix $\mathbf{Q}$, where

$$\mathbf{T} = \begin{bmatrix} T_{11} & T_{12} & T_{13} \\ T_{12} & T_{22} & T_{23} \\ T_{13} & T_{23} & T_{33} \end{bmatrix}, \tag{38}$$

$$\mathbf{A} = \begin{bmatrix} 0 & 1 & 0 \\ 0 & 0 & 1 \\ -c_1 & -c_2 & -c_3 \end{bmatrix}, \tag{39}$$

are stable matrices. It is easily obtained that $\mathbf{T}$ is satisfied as the following conditions:

$$T_{11} = T_{13}c_2 + T_{23}c_1, \tag{40}$$

$$T_{12} = \frac{-q_2 + 2T_{23}c_2}{2} = T_{13}c_3 + T_{33}c_1, \tag{41}$$

$$T_{13} = \frac{q_1}{2c_1}, \tag{42}$$

$$T_{22} = -T_{13} + T_{23}c_3 + T_{33}c_2, \tag{43}$$

$$T_{23} = \frac{-q_3 + 2T_{33}c_3}{2}, \tag{44}$$

when $\mathbf{Q} = \operatorname{diag}(q_1, q_2, q_3)$, where $q_1 > 0$, $q_2 > 0$, and $q_3 > 0$.

Differentiating $\dot{V}_2$ along (33), (34), and (35) is

$$\begin{aligned}\dot{V}_2 = & -2T_{13}c_1x_1^2 + (2T_{12} - 2T_{23}c_2)x_2^2 \\ & + (2T_{23} - 2T_{33}c_3)(x_3 + d_2)^2 \\ & + (2T_{11} - 2T_{13}c_2 - 2T_{23}c_1)x_{1h}x_{2h} \\ & + (2T_{12} - 2T_{13}c_3 - 2T_{33}c_1)x_1(x_3 + d_2) \\ & + (2T_{13} + 2T_{22} - 2T_{23}c_3 - 2T_{33}c_2)x_{2h}(x_3 + d_{2h}) \\ & + 2T_{13}x_1\left(c_3\tilde{d}_2 + \tilde{d}_3 + \dot{\tilde{d}}_2\right) + 2T_{23}x_2\left(c_3\tilde{d}_2 + \tilde{d}_3 + \dot{\tilde{d}}_2\right) \\ & + 2T_{33}(x_3 + d_2)\left(c_3\tilde{d}_2 + \tilde{d}_3 + \dot{\tilde{d}}_2\right).\end{aligned} \tag{45}$$

Due to (37), it can be obtained as

$$\begin{aligned}\dot{V}_2 = & -q_1x_1^2 - q_2x_2^2 - q_3(e_3 + d_2)^2 + 2T_{13}x_1\left(c_3\tilde{d}_2 + \tilde{d}_3 + \dot{\tilde{d}}_2\right) \\ & + 2T_{23}x_2\left(c_3\tilde{d}_2 + \tilde{d}_3 + \dot{\tilde{d}}_2\right) + 2T_{33}(x_3 + d_2)\left(c_3\tilde{d}_2 + \tilde{d}_3 + \dot{\tilde{d}}_2\right) \\ \le & -|x_1|(q_1|x_1| - 2T_{13}\mu) - |x_2|(q_2|x_2| - 2T_{23}\mu) \\ & - |x_3 + d_2|(q_3|x_3 + d_2| - 2T_{33}\mu),\end{aligned} \tag{46}$$

where $\mu = |c_3\tilde{d}_2 + \tilde{d}_3 + \dot{\tilde{d}}_2|$. It is easily obtained from (46) that

$$\mu \le (c_3 + 1)\lambda_2 + \lambda_3. \tag{47}$$

From inequality (47), the error states of the velocity subsystem are bounded, that is,

$$|x_1| \le \frac{2T_{13}\mu}{q_1}, \tag{48}$$

$$|x_2| \le \frac{2T_{23}\mu}{q_2}, \tag{49}$$

$$|x_3 + d_2| \le \frac{2T_{33}\mu}{q_3}. \tag{50}$$

Remark 2. It is worth noting that the state x_1 can be reduced by increasing q_1 or decreasing λ_j $(j = 2, 3)$ according to (47), (48), (49), and (50). With increasing q_1, the T_{13} increases due to (40), (41), (42), (43), and (44). Only when the estimate accuracy of d_j $(j = 2, 3)$ is improved by the nonlinear disturbance observers (17), (18), (19), (20), (21), (22), (23), and (24) can λ_j be reduced and the accuracy of the state x_1 be also increased. In fact, the impact of the LOS angle error is satisfied with $|e_q| \le 2T_{13}\mu/q_1$ due to the definition of the state x_1; that is, it is obtained from the relationship between the accuracy of the LOS angle error and the estimate error of nonlinear disturbances. In other words, it is obtained from the relationship between the accuracy of impact angle and the estimate errors of nonlinear disturbances from (48), (49), and (50).

4. Simulation

In this section, the effectiveness of the proposed IGC law is verified by the nonlinear numerical simulations. For the nonlinear numerical simulations, the original nonlinear motion model of the missile given in [23] and the relative motion model between the missile and target given in [25] are adopted, and the aerodynamic forces, moments, and nominal parameters are obtained in [4].

The simulation step is 0.001 s; the initial values of the simulation are assumed to be $q_M = -90$ deg, $V_M = 200$ m/s, $\alpha(0) = 0$ deg, $w_z(0) = 0$ rad/s, and $R(0) = 3000$ m. Parameters of the proposed IGC law are $c_1 = 9$, $c_2 = 11$, $c_3 = 8$, $k = 0.975$, and $\varepsilon = 0.2$.

In addition, the uncertainties of missile-related aerodynamic parameters can include 20% variations with respect to their normal values; some requirements for the guidance and control system are listed as follows:

(1) The blind area for the homing guidance seeker is 50 m.

(2) The miss distance is no more than 0.5 m.

(3) The absolute error of terminal impact angle distance is no more than 5 deg.

(4) The angle of attack is less than 10 deg.

(5) The fin deflection limit and fin rate limit are restricted as $|\delta| \le 10$ deg and $|\dot{\delta}| \le 100$ deg/s.

(6) The states of the missile are bounded.

For comparison studies, two control laws are introduced to the nonlinear numerical simulations. The first is the classical guidance and control (CGC) law [1], which treats the guidance and control system as two separate processes, and the second is the partial integrated guidance and control (PIGC) law [11], which combines the backstepping technique and sliding mode control method.

In order to demonstrate the performance of the proposed IGC law in the presence of uncertainties in the parameters, three cases are considered for a homing missile against a fixed target, a moving target, and a maneuvering target on the ground.

Case 1. Variable response curves of the guidance and control system under the three control laws against a fixed target on the ground are shown in Figure 2. The estimate errors of disturbances coming from the nonlinear disturbance observer are shown in Figure 3. Missile/target trajectories under the three control laws are shown in Figure 4(a). Miss distance and impact angle under the three control laws are shown in Tables 1 and 2, respectively.

As presented in Figure 2, the LOS angle rate under the CGC law slowly converges to zero in the terminal phase, while the LOS angle rate under the IGC law rapidly converges to zero in Figure 2(a). Meanwhile, the angle of attack, pitch rate, and elevator deflection under the PIGC law vary more acutely than the other laws at two seconds from

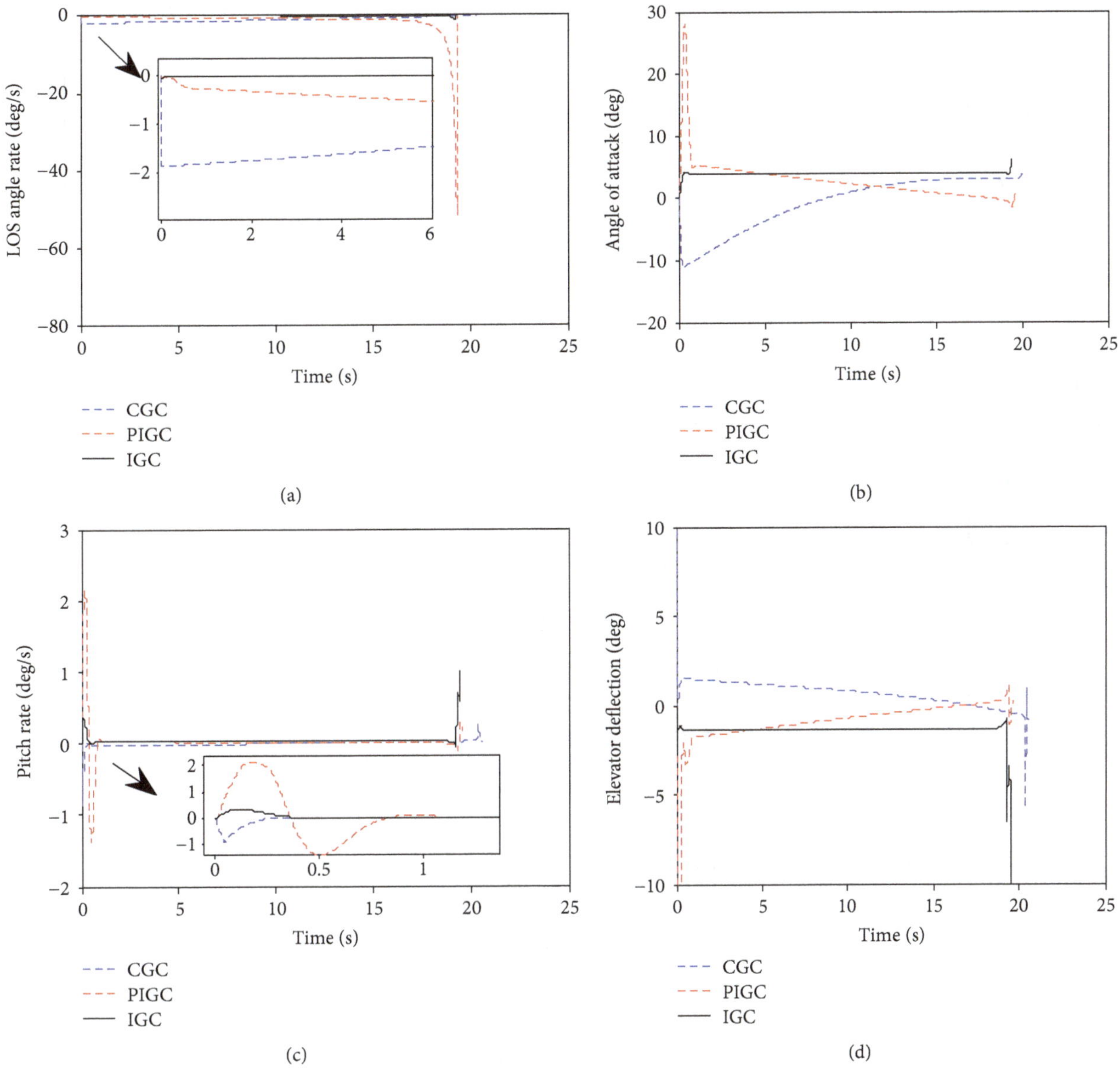

FIGURE 2: Variable response curves under the three control laws against a fixed target.

Figures 2(b)–2(d), and the angle of attack is more than 10 deg under the PIGC law. Even when the target enters the homing head blind zone, the elevator deflection still works under the IGC law as shown as Figure 2(d). The states have been stabilized more rapidly under the IGC law than the other control laws because the disturbance observer technique is employed to estimate the unknown uncertainties, and the estimate errors quickly converge to zero under the designed nonlinear disturbance observer as shown in Figure 3. Both miss distance and impact angle under the IGC law achieve the satisfactory performance.

Moreover, the time under the IGC law is about 0.1 s shorter than that of the PIGC law and about 0.6 s shorter than that of the CGC law against a fixed target on the ground.

Case 2. To make the work more challenging, the target moves at 20 m/s on the ground in this case. Missile/target trajectories under the three control laws are shown in Figure 4(b), and variable response curves of the missile guidance and control system under the three control laws against a moving target are shown in Figure 5. Miss distance and impact angle under the three control laws are shown in Tables 3 and 4, respectively.

From the simulation results among the three control laws, the states are stabilized most rapidly under the IGC; both miss distance and impact angle under the IGC law also achieve satisfactory performance. The time in the simulation experiment under the IGC law is about 0.1 s shorter than that of the PIGC law and about 0.5 s shorter than that of the CGC law against a moving target on the ground.

Case 3. In this case, the target on the ground moves with a maneuver of $2.4\cos(0.1t)$ m/s; the variable response curves of the missile guidance and control system under the three control laws are shown in Figure 6. Miss distance and impact angle under the three control laws are

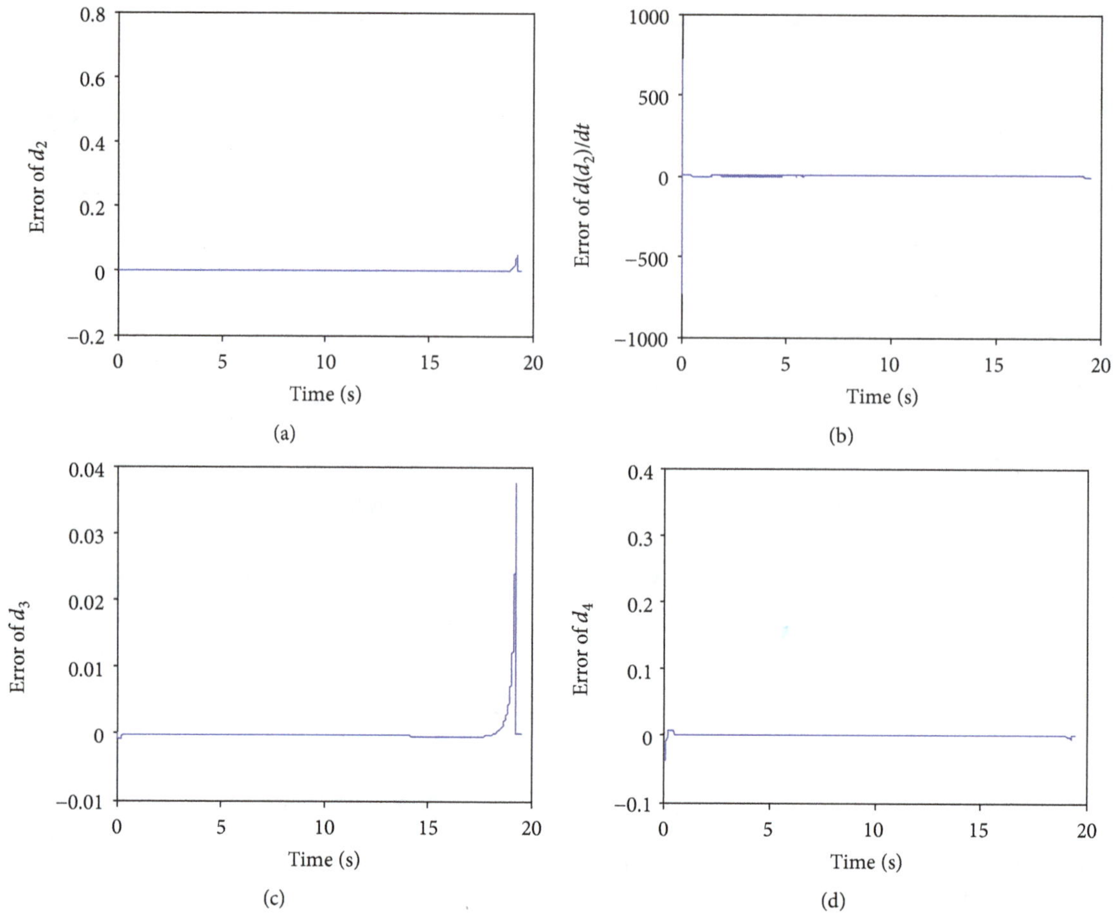

FIGURE 3: Estimate curves of uncertainties under the nonlinear disturbance observer.

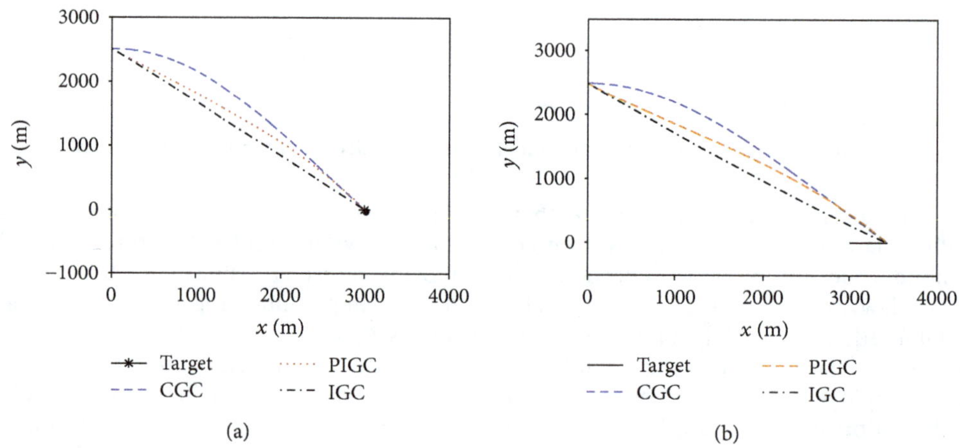

FIGURE 4: Missile/target trajectories under the three control laws against a fixed target and a moving target.

TABLE 1: Miss distance under the three control laws against a fixed target.

CGC law	PIGC law	IGC law
0.09943 m	5.611 m	0.2059 m

TABLE 2: Impact angle under the three control laws against a fixed target.

CGC law	PIGC law	IGC law
−62.8 deg	−89.7 deg	−88.7 deg

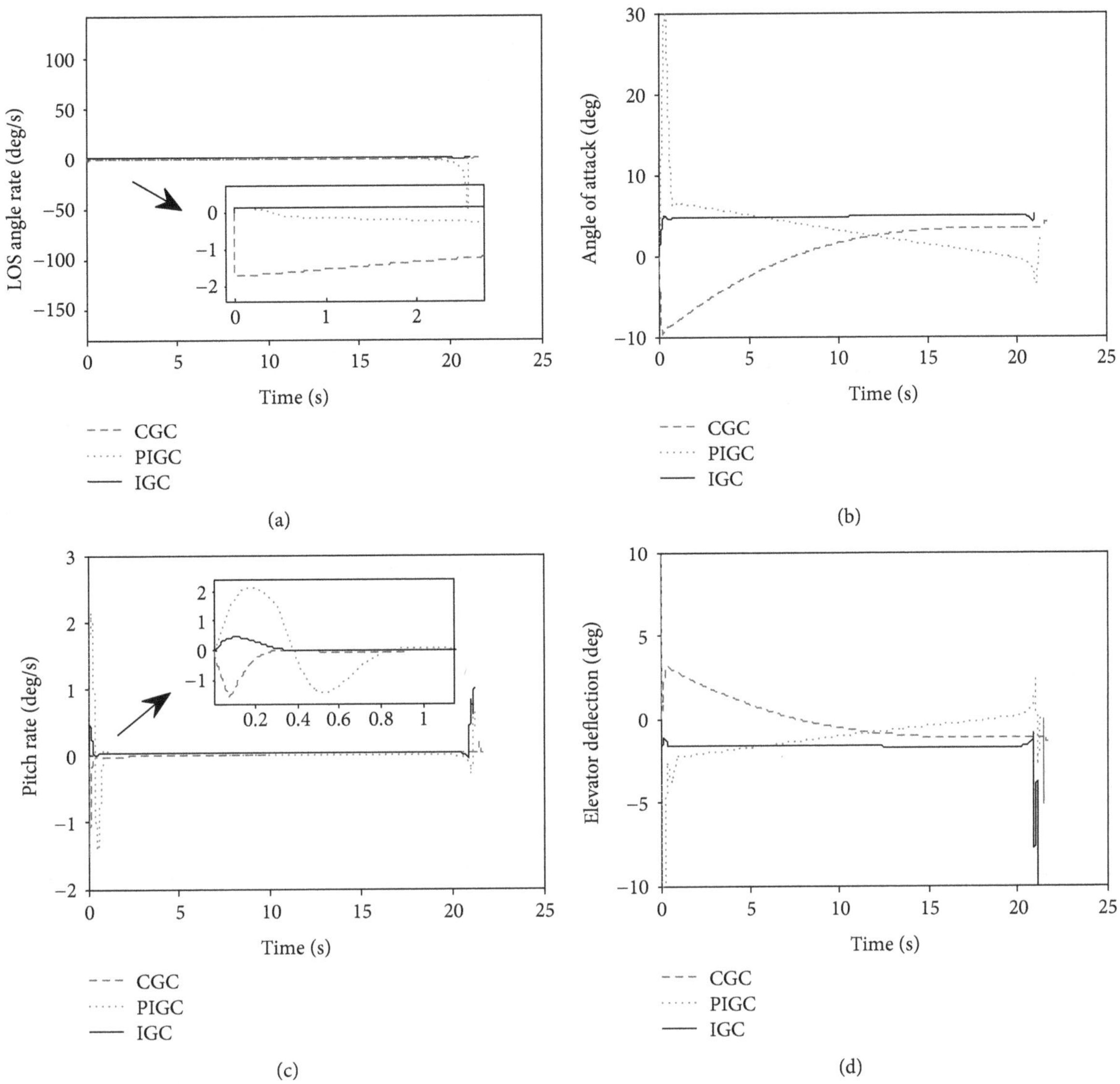

FIGURE 5: Variable response curves under the three control laws against a moving target.

TABLE 3: Miss distance under the three control laws against a moving target.

CGC law	PIGC law	IGC law
0.093 m	11.4 m	0.4426 m

TABLE 4: Impact angle under the three control laws against a moving target.

CGC law	PIGC law	IGC law
−68.1 deg	−89.9 deg	−86.2 deg

also shown in Tables 5 and 6, respectively. The same results can be also obtained from the simulation results. The time in the simulation experiment under the IGC law is about 0.1 s shorter than that of the PIGC law, and it is about 1.4 s shorter than that of the CGC law against a maneuvering target.

In short, it is easily obtained that the states are stabilized most rapidly; both the miss distance and the impact angle can be satisfied under the IGC law, while only partial index can be satisfied under the CGC law and the PIGC law. Meanwhile, the flight time of engagement with a fixed target, a moving target, and a maneuvering target is shortened by utilizing the proposed IGC law, compared with the CGC law and the PIGC law.

5. Conclusion

A novel IGC law for a homing missile with impact angle constraint is proposed against a ground target to improve the rapidity of the missile guidance and control system. A new control-oriented model with impact angle error of the IGC system in the pitch plane is built, and an IGC law is

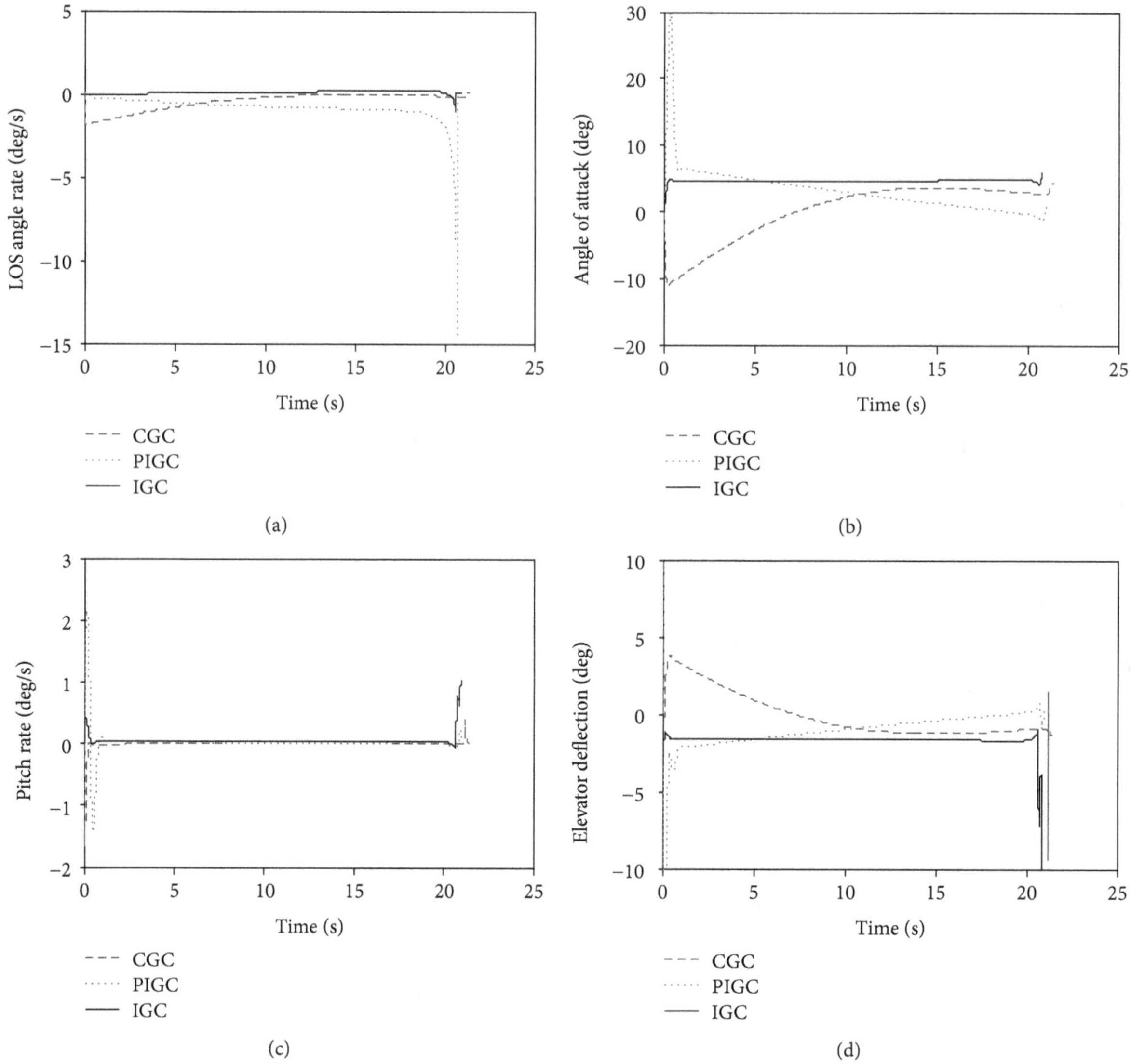

Figure 6: Variable response curves under the three control laws against a maneuvering target.

Table 5: Miss distance under the three control laws against a maneuvering target.

CGC law	PIGC law	IGC law
0.067 m	3.176 m	0.4221 m

Table 6: Impact angle under the three control laws against a maneuvering target.

CGC law	PIGC law	IGC law
−67.6 deg	−89.4 deg	−84.9 deg

designed by utilizing the sliding mode control and the nonlinear disturbance observer. The relationship between the accuracy of impact angle and the estimate error of mismatched uncertainties can be obtained from the stability of the system. Simulation results have shown that the proposed IGC approach achieves good performance and shortens the time of engagement.

Conflicts of Interest

The authors declare that they have no conflicts of interest.

Acknowledgments

This work is supported by the National Nature Science Foundation of China under Grant 61703339.

References

[1] M. Levy, T. Shima, and S. Gutman, "Linear quadratic integrated versus separated autopilot-guidance design," *Journal of Guidance, Control, and Dynamics*, vol. 36, no. 6, pp. 1722–1730, 2013.

[2] P. K. Menon and E. J. Ohlmeyer, "Integrated design of agile missile guidance and autopilot systems," *Control Engineering Practice*, vol. 9, no. 10, pp. 1095–1106, 2001.

[3] N. F. Palumbo, B. E. Reardon, and R. A. Blauwkamp, "Integrated guidance and control for homing missiles," *Johns Hopkins APL Technical Digest*, vol. 25, no. 2, pp. 121–139, 2004.

[4] M. Z. Hou, X. L. Liang, and G. R. Duan, "Adaptive block dynamic surface control for integrated missile guidance and autopilot," *Chinese Journal of Aeronautics*, vol. 26, no. 3, pp. 741–750, 2013.

[5] D. E. Williams, J. Richman, and B. Friedland, "Design of an integrated strapdown guidance and control system for a tactical missile," in *Guidance and Control Conference*, pp. 57–66, Gatlinburg, TN, USA, August 1983.

[6] T. Shima, M. Idan, and O. M. Golan, "Sliding mode control for integrated missile autopilot-guidance," in *AIAA Guidance, Navigation, and Control Conference and Exhibit*, Providence, Rhode Island, August 2004.

[7] D. Chwa, J. Y. Choi, and S. G. Anavatti, "Observer-based adaptive guidance law considering target uncertainties and control loop dynamics," *IEEE Transactions on Control Systems Technology*, vol. 14, no. 1, pp. 112–123, 2006.

[8] M. Xin, S. N. Balakrishnan, and E. J. Ohlmeyer, "Integrated guidance and control of missiles with θ-D method," *IEEE Transactions on Control Systems Technology*, vol. 14, no. 6, pp. 981–992, 2006.

[9] R. Padhi and C. Chawla, "Neuro-adaptive augmented dynamic inversion based PIGC design for reactive obstacle avoidance of UAVs," in *AIAA Guidance, Navigation, and Control Conference*, Portland, Oregon, August 2011.

[10] X. Wang and J. Wang, "Partial integrated guidance and control for missiles with three-dimensional impact angle constraints," *Journal of Guidance, Control, and Dynamics*, vol. 37, no. 2, pp. 644–657, 2014.

[11] X. Wang and J. Wang, "Partial integrated guidance and control with impact angle constraints," *Journal of Guidance, Control, and Dynamics*, vol. 38, no. 5, pp. 925–936, 2015.

[12] F. B. Ibarrondo and P. Sanz-Aránguez, "Integrated versus two-loop guidance-autopilot for a dual control missile with high-order aerodynamic model," *Proceedings of the Institution of Mechanical Engineers, Part G: Journal of Aerospace Engineering*, vol. 230, no. 1, pp. 60–76, 2016.

[13] X. Liu, W. Huang, and L. du, "An integrated guidance and control approach in three-dimensional space for hypersonic missile constrained by impact angles," *ISA Transactions*, vol. 66, pp. 164–175, 2017.

[14] H. Yan, S. P. Tan, and Y. Z. He, "A small-gain method for integrated guidance and control in terminal phase of reentry," *Acta Astronautica*, vol. 132, pp. 282–292, 2017.

[15] H. Yan, X. Wang, B. Yu, and H. Ji, "Adaptive integrated guidance and control based on backstepping and input-to-state stability," *Asian Journal of Control*, vol. 16, no. 2, pp. 602–608, 2014.

[16] S. Wang, W. H. Wang, and S. F. Xiong, "Impact angle constrained three-dimensional integrated guidance and control for STT missile in the presence of input saturation," *ISA Transactions*, vol. 64, pp. 151–160, 2016.

[17] H. Song and T. Zhang, "Fast robust integrated guidance and control design of interceptors," *IEEE Transactions on Control Systems Technology*, vol. 24, no. 1, pp. 349–356, 2016.

[18] Z. Cong and W. Yun-jie, "Non-singular terminal dynamic surface control based integrated guidance and control design and simulation," *ISA Transactions*, vol. 63, pp. 112–120, 2016.

[19] S. He, T. Song, and D. Lin, "Impact angle constrained integrated guidance and control for maneuvering target interception," *Journal of Guidance, Control, and Dynamics*, vol. 40, no. 10, pp. 2653–2661, 2017.

[20] S. M. He, J. Wang, and W. Wang, "A novel sliding mode guidance law without line-of-sight angular rate information accounting for autopilot lag," *International Journal of Systems Science*, vol. 48, no. 16, pp. 3363–3373, 2017.

[21] J. G. Guo, Y. Xiong, and J. Zhou, "A new sliding mode control design for integrated missile guidance and control system," *Aerospace Science and Technology*, vol. 78, pp. 54–61, 2018.

[22] H. Mingzhe and D. Guangren, "Integrated guidance and control of homing missiles against ground fixed targets," *Chinese Journal of Aeronautics*, vol. 21, no. 2, pp. 162–168, 2008.

[23] X. F. Qian, R. X. Lin, and Y. N. Zhao, *Missile Flight Mechanics*, Beijing Institute of Technology Press, Beijing, 2000.

[24] D. Ginoya, P. D. Shendge, and S. B. Phadke, "Sliding mode control for mismatched uncertain systems using an extended disturbance observer," *IEEE Transactions on Industrial Electronics*, vol. 61, no. 4, pp. 1983–1992, 2014.

[25] D. Zhou, C. D. Mu, and W. L. Xu, "Adaptive sliding-mode guidance of a homing missile," *Journal of Guidance, Control, and Dynamics*, vol. 22, no. 4, pp. 589–594, 1999.

14

Investigation of an Autofocusing Method for Visible Aerial Cameras based on Image Processing Techniques

Zhichao Chen[1,2] **and Tao Zhang**[1,2]

[1]*Changchun Institute of Optics, Fine Mechanics and Physics, Chinese Academy of Sciences, Changchun 130033, China*
[2]*University of Chinese Academy of Sciences, Beijing, China*

Correspondence should be addressed to Zhichao Chen; 17522332@qq.com

Academic Editor: Xuesong Li

In order to realize the autofocusing in aerial camera, an autofocusing system is established and its characteristics such as working principle and optical-mechanical structure and focus evaluation function are investigated. The reason for defocusing in aviation camera is analyzed and several autofocusing methods along with appropriate focus evaluation functions are introduced based on the image processing techniques. The proposed autofocusing system is designed and implemented using two CMOS detectors. The experiment results showed that the proposed method met the aviation camera focusing accuracy requirement, and a maximum focusing error of less than half of the focus depth is achieved. The system designed in this paper can find the optical imaging focal plane in real-time; as such, this novel design has great potential in practical engineering, especially aerospace applications.

1. Introduction

The image quality is directly related to the optical measurement and reconnaissance equipment measurement result. However, the performance of this equipment can be largely impacted by surrounding environment [1]. For example, because of the change of local temperature, atmospheric pressure, and focal distance from the target, the aerial camera would easily become out of focus [1]; therefore, a robust system which can address this issue is crucial in aerial camera applications. Autofocusing is a key for eliminating the defocusing effect of an aerial camera to ensure high quality of the image. The precision of autofocusing has a direct effect on the imaging quality of the aerial camera. With the improvement of aerial camera's resolution, autofocusing technique has been greatly desired to improve the overall image quality of aerial imaging [2].

Autofocusing of aerial camera is realized by the following three methods [3–6]. (1) Ground calibration method [7], which is also called program control method: according to the actual measured disturbances, such as temperature, atmosphere pressure, and distance to the ground, which would change the location of the focal plane, this method uses a preinstalled program which is calibrated in advance to adjust the actual position of the focal plane. The advantage of the ground calibration method is that it is simple and is ready to be implemented. However, the practical situation has a certain deviation from the ideal situation, and the ground calibration method only takes into account a limited number of possible disturbances, which could bring in a large error. (2) Optoelectronic autocollimation method [7]: for long focal length aviation cameras with photographic film as a recording medium, optoelectronic autocollimation method is preferred. Optoelectronic autocollimation method can satisfy the requirement of high-precision focusing of aerial imaging. This method can accurately reflect the actual optical system imaging position of infinity target. But the method requires the use of a large, high-quality plane reflector, whose diameter is usually larger than the clear aperture of the camera. Even though optoelectronic autocollimation method is an established technology, this application is too complex to be realized. Furthermore, it can only calibrate the position of infinity target, and it cannot calibrate the position of actual target, which further limits the application of such technique [8–10]. (3) Image processing methods [6], in which the position of the focal plane will be estimated from the actual images: the focusing is an "intelligent" method, which is to say the focusing criterion is flexible, and different focus evaluation function can be automatically chosen according to the requirement. Furthermore, it employs the actual images as the evaluation subjects, so the image processing method is more practical and reliable. In this paper, autofocusing using image processing method is investigated further.

When the aerial camera focuses accurately, the edges and details of the aerial image should be well defined in the spatial domain, and the high-frequency component of the aerial image should be adequate in the frequency domain [11]. To achieve this goal, the differential of image is needed to obtain image edges and detailed information, while for the latter image spectrum information can be obtained by the Fast Fourier Transform (FFT) to convert the image into frequency domain signals. Image differential algorithm can accurately obtain image edges and necessary information, but it cannot filter out the noise of the image. Through image FFT, high-frequency information of the image can be preserved while simultaneously high-frequency noise is cancelled out. However, the FFT algorithm is complicated and time consuming. Aerial systems have to make real-time observations of the targets and, as such, this computationally expensive method is not desirable. The fundamental principle of the automatic focusing system based on image processing method is to count corresponding image focus evaluation function (FEF) value to each focal plane position, and the corresponding position of the maximum FEF value is the best imaging focal plane position [4].

An autofocusing method based on image processing is applicable for processing an image sequence, such as taking images of the target with multiple focal plane positions. According to the focus evaluation function value of image sequence data, the optimal cost function value is found, and the corresponding focal plane position is the best imaging focal plane position. However, due to rapid and continuous movement of aircraft, it is difficult for an aerial camera to take multiple images of the same target. So it is necessary to design an autofocusing method which can obtain the image sequence data to the same ground target with a different focal plane position in real-time. In this paper, an autofocusing method with two CMOS detectors that are located on both sides of the symmetry plane of the imaging CCD detector on the optical path is proposed. The method is applicable to aerial camera with a line scan CCD detector or an area scan CCD detector.

The rest of the paper is organized as follows. The reason of aviation camera defocusing and the theory of autofocusing are analyzed in Section 2. Sections 3–7 introduce the method of autofocusing, including working principle of autofocusing system, soft flow diagram, optical-mechanical structure, the choice of focus evaluation function, and image preprocess. Section 9 gives the experiment result and analysis. Conclusions are given in Section 10.

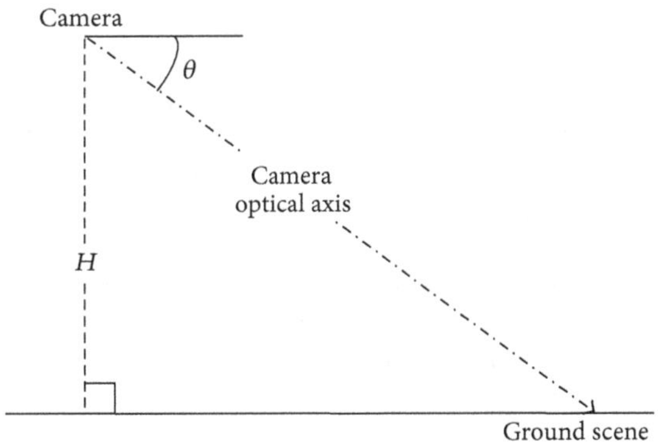

FIGURE 1: Defocus of focal length caused by photographic distance.

2. Analysis of Aviation Camera Defocusing

The disturbances that affect the focus of aerial camera such as temperature, atmosphere pressure, and distance to the ground are studied in this section [12].

(a) Defocus Caused by Temperature Change. For single lens, for example, defocus caused by temperature change can be approximately calculated as follows:

$$\Delta f_t \approx f \cdot \Delta T \left(\frac{\eta_a n_g - \eta_g}{n_g - 1} \right), \quad (1)$$

where Δf_t is defocus of the focal length caused by temperature change; f is focal length of the lens; ΔT is temperature variation; n_g is refractive index of glass; η_g is n_g variation under each Celsius; η_a is refractive index of air variation under each Celsius.

(b) Defocus Caused by Atmosphere Pressure Change. For single lens, if the temperature change with height at the same time is ignored, the defocus caused by atmospheric pressure change is given by the following equation:

$$\Delta f_p = n_g f \left(\frac{n_a - 1}{n_g - 1} \right) \cdot \frac{p - p_0}{p_0}, \quad (2)$$

where Δf_p is defocus of focal length caused by atmospheric pressure change; p is atmospheric pressure at certain altitude; p_0 is atmospheric pressure when camera is assembled on the ground; n_a is refractive index of air; n_g is refractive index of glass.

(c) Defocus Caused by the Change in Photographic Distance. Defocus caused by the photographic distance is given by the following equation, as shown in Figure 1:

$$\Delta f_h = \frac{f^2}{H} \sin\theta, \quad (3)$$

where Δf_h is defocus of focal length caused by photographic distance change; f is focal length of lens; H is relative height between aerial camera and ground scene; θ is the angle between camera optical axis and horizontal plane:

$$\Delta f = \Delta f_t + \Delta f_p + \Delta f_h. \quad (4)$$

As shown in (4), total defocus is the sum of the defocus caused due to individual component. The defocus due to three components can be obtained according to the camera's optical analysis and simulation results, but the aerial camera in actual use is quite complex. For example, the actual optical system of aerial camera is not a single lens but multiple lens or mirrors, and the lens' refractive index, curvature radius, thickness, and the gap between the lenses, stress in lens material, and the expansion of metal frame will change due to the change of environment. So it is difficult to understand the focus of aerial camera based on the result from the optical analysis.

The results from the optical analysis are applied to find the focus of lens approximately, which can help in narrowing the scope of focusing and save expired time.

3. Analysis of Aviation Camera Autofocusing

According to optical principle [13], to see a clear image of any object through a thin lens, the Gaussian optics formula must be satisfied:

$$\frac{1}{f} = \frac{1}{u} + \frac{1}{v}, \quad (5)$$

where u is the object distance from the lens, v is image distance from the lens, and f is the focal length of the lens.

The focus adjustment mechanism is to adjust the object distance and focal length such that the Gaussian optics formula is satisfied.

The thin lens model of optical imaging system is shown in Figure 2, where P is a dot on the subject and P' is the image of P. When focusing is accurate, P and P' will satisfy (5). When P content focuses accurately, the image of light spot size is minimal.

If the camera is not focused accurately, in other words, if the reaction surface of the detector is not at P', the image of P is a light spot at P'', it is a blurred image, and its shape is similar to the camera lens aperture. Since the aperture of the camera (especially aviation camera) is circular, the object cannot be focused accurately and the image will form as a fuzzy circle with a radius of RP' on the sensor.

According to the property of similar triangles, the scaling relation between the fuzzy image point P'' and the subject P is as follows:

$$q = \frac{2R}{D} = \frac{d}{v}. \quad (6)$$

So the radius of fuzzy circle R is given by the following expression:

$$R = q \cdot \frac{D}{2} = \frac{D}{2} \cdot \frac{d}{v}. \quad (7)$$

The sensor is located in the clear image plane "P," that is, distance "d" from the former position, as shown in Figure 3. According to the properties of similar triangle, the relation between the fuzzy image point P' and the image point P is as follows:

$$q = \frac{2R}{D} = \frac{d}{v}. \quad (8)$$

Hence, the fuzzy circle radius R is given as follows:

$$R = q \cdot \frac{D}{2} = \frac{D}{2} \cdot \frac{d}{v}. \quad (9)$$

According to the above analysis, when the image detector in the clear image is at the same distance "d" before and after P', the corresponding blur circle radius will be the same. This shows that the fuzzy degree of two images is the same, and hence, the corresponding focus evaluation function value remains unchanged.

From the result, if one can find two images of different locations and their focus evaluation function value is the same, then the position of the clear image will be in between two positions.

Therefore, this paper puts forward a method of autofocusing with two detectors. Two identical CMOS detectors are used to detect the image on the ground scene for autofocusing, and an imaging detector is located in the middle of the two CMOS detectors.

When the focus evaluation function values of two CMOS detectors on the same ground scenery are the same, that is, the difference between the two values is less than a certain threshold, the imaging detector will capture a clear image as it is located in the clear image position. Thus, the autofocusing will be achieved.

4. Working Principle

Autofocusing system consists of atmosphere pressure sensors, temperature sensors, power amplifier, stepper motor, position encoder, focusing mirror, prefocal CMOS detector, postfocal CMOS detector, and so forth, as shown in Figure 4. Atmosphere pressure sensors and temperature sensors are located on the optical lens of aerial camera and these sensors return the atmosphere pressure and temperature of the aerial camera. According to the relationship between temperature data and focal plane position provided by optical analysis (1), an estimated position of the focal plane is obtained. Atmosphere pressure has a similar impact on the lens and can be analyzed in the same approach. Therefore, the autofocusing controller provides a comprehensive estimated position of the focal plane with atmosphere pressure and temperature inputs. The autofocus controller drives a stepper motor by the power amplifier to move the focusing mirror to an appropriate position.

As forward motion compensation mirror works, the camera controller sends the focus checkup instruction to the autofocus controller. The autofocus controller starts to control and receive the resulting images of the prefocal CMOS detector and postfocal CMOS detector. The autofocus controller captures and averages the gray values of the images from the same detector. After preprocessing, autofocus controller computes the focus evaluation function (FEF) value of these images and drives the stepper motor to move the focusing mirror to the desired position according to the autofocusing algorithm that is shown in Figure 5, until the FEF value of images exported by two CMOS detectors is the same; that is, the difference between two values is less than the threshold value.

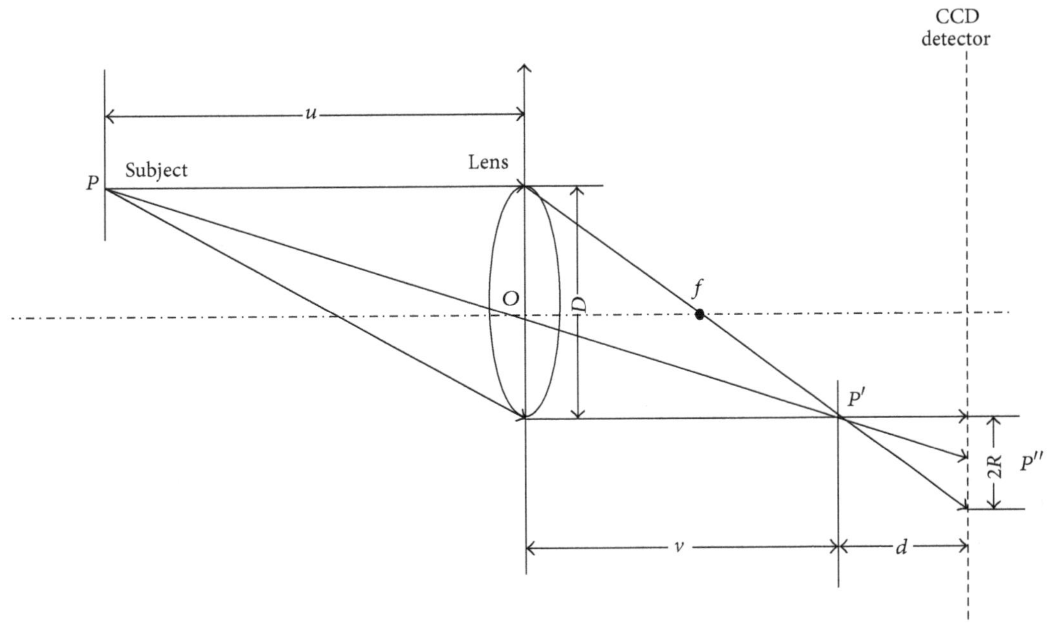

FIGURE 2: Optical imaging system model.

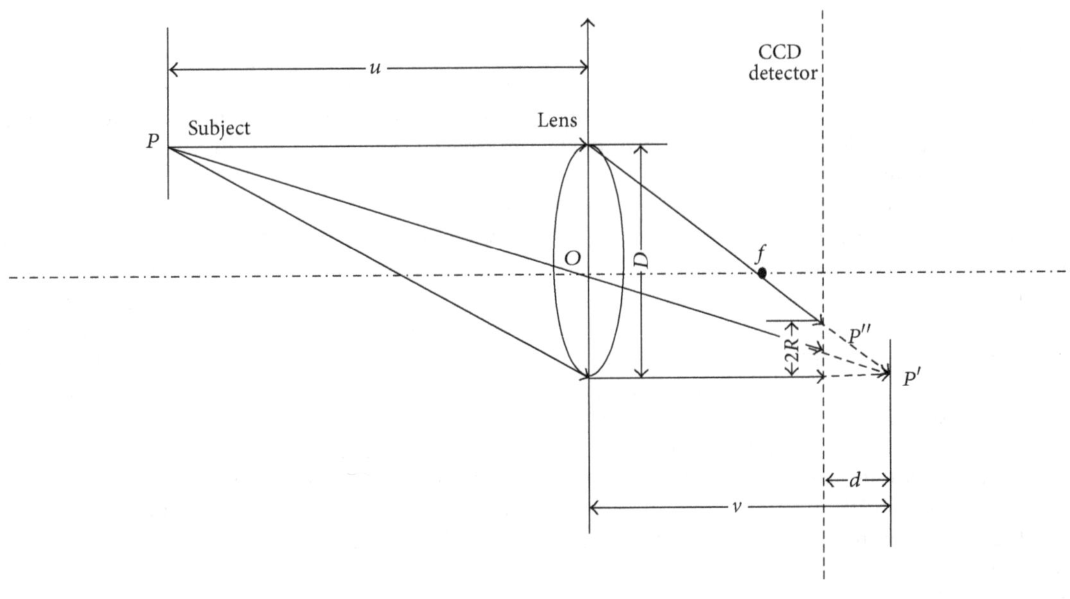

FIGURE 3: Optical imaging system model diagram with CCD detector before P''.

As is shown in Figure 6, because the imaging CCD detector is located in the middle of the prefocal CMOS detector and postfocal CMOS detector, according to the characteristic of FEF, the position of the imaging CCD detector is the position where the values of FEF are maximum. Therefore, the imaging CCD detector is located in the best position of the focal plane.

Thus, through the acquisition of the focusing images and control of focusing mirror, the real-time automatic focusing process is realized. The flowchart diagram of the autofocusing system is shown in Figure 5.

This autofocus method combines Phase-Difference AF and Contrast AF. Compared with typical Canon Phase-Difference AF method, this AF method possesses high AF precision and relatively slow AF speed, and it is suitable for low light situations.

5. Autofocusing Algorithm

As shown in Figure 5, autofocus algorithm carries the preliminary focusing according to the surrounding environment, to narrow the autofocusing range.

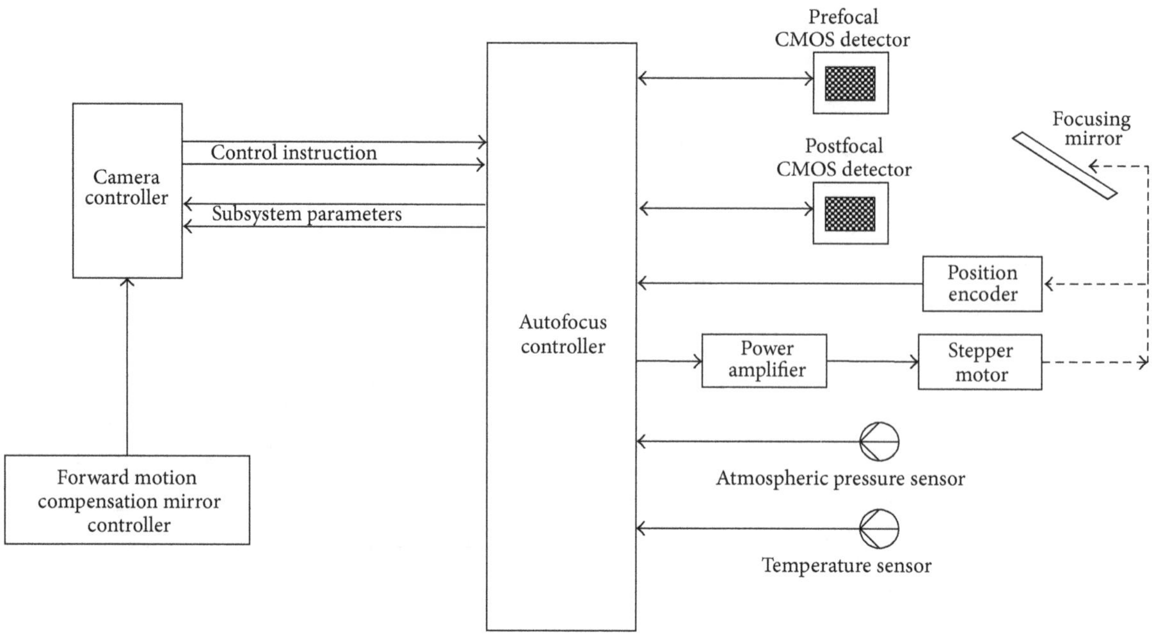

FIGURE 4: Framework for the autofocusing system.

Firstly, the camera collects the actual temperature, atmospheric pressure, and imaging distance between the aerial camera and target and computes the defocus quantity of the focal plane according to the focal plane position under different condition of temperature and pressure provided by optical analysis. Then the defocus quantity is added to the focal plane position which calibrated in the lab. Finally, a rough focal plane position is obtained. The camera drives the focusing mirror to the approximate focal plane position [10, 14].

General aviation camera has 2 mm range for focusing [15]. According to the camera's optical analysis and simulation results, the possible focus range will be narrowed down to (1/4)th of the original scope. It shortens the time needed for image inspection method and also for the efficient working of the focus evaluation function [15].

Then, when the scanning mirror works, namely, the camera works in the mode of forward direction image motion compensation (FIMC), ground scenery is relatively stationary to the camera. Now autofocusing controller starts focusing based on image, commanding the postfocal CMOS detector and prefocal CMOS detector image on the ground scenery at the same time. Next autofocusing controller preprocesses the images and then calculates the value of the focus evaluation function of the two images, numerical F_1 and F_2. Autofocusing controller compares F_1 and F_2, if $F_1 > F_2$, and $(F_1 - F_2) > 0.02 * (F_1 + F_2)$, it is illustrated that FEF value of image obtained by prefocal CMOS detector is bigger than the postfocal CMOS detector, and the best focal plane is near the prefocal CMOS detector. So autofocusing controller drives the focusing lens to move a step backward.

Previous steps are repeated and the sizes of F_1 and F_2 are compared until $(F_1 - F_2) \leq 0.02 * (F_1 + F_2)$. The difference between F_1 and F_2 is less than 1%; that is, F_1 and F_2 are mainly the same. As described in Analysis of Aviation Camera Autofocusing above, the best focal plane is in the middle of two CMOS detectors.

Because the imaging detector is placed in the middle of the two CMOS detectors on the optical path, the best focal plane of the imaging detector is in the actual position.

The above work is carried out in front of the aerial camera taking pictures. In the process of aviation camera taking pictures, when the camera is in a state of IMC, focusing controller also controls two CMOS detectors and performs the required numerical calculation. Since the photo has been checked for the focal work before, so at this point the defocus quantity of camera is not big. Generally two to five times of the function comparison can give a best approximate of the actual focal plane; this ensures that the imaging detector is located in the best focal plane, achieving the goal of real-time focusing.

Since this method does not depend on the imaging detector, the line array and area array can be used as imaging detectors. Hence, this method can be adopted universally.

6. Optical-Mechanical Structure of Autofocusing System

The optical-mechanical structure of autofocusing system is shown in Figure 6; postfocal CMOS detector is installed in the optical system after the imaging CCD detector. The distance between photosensitive surface of postfocal CMOS detector and imaging CCD detector is D. Before the postfocal CMOS detector, a transflective mirror is placed and prefocal CMOS detector is located in the left side of the transflective mirror. The distance between photosensitive surface

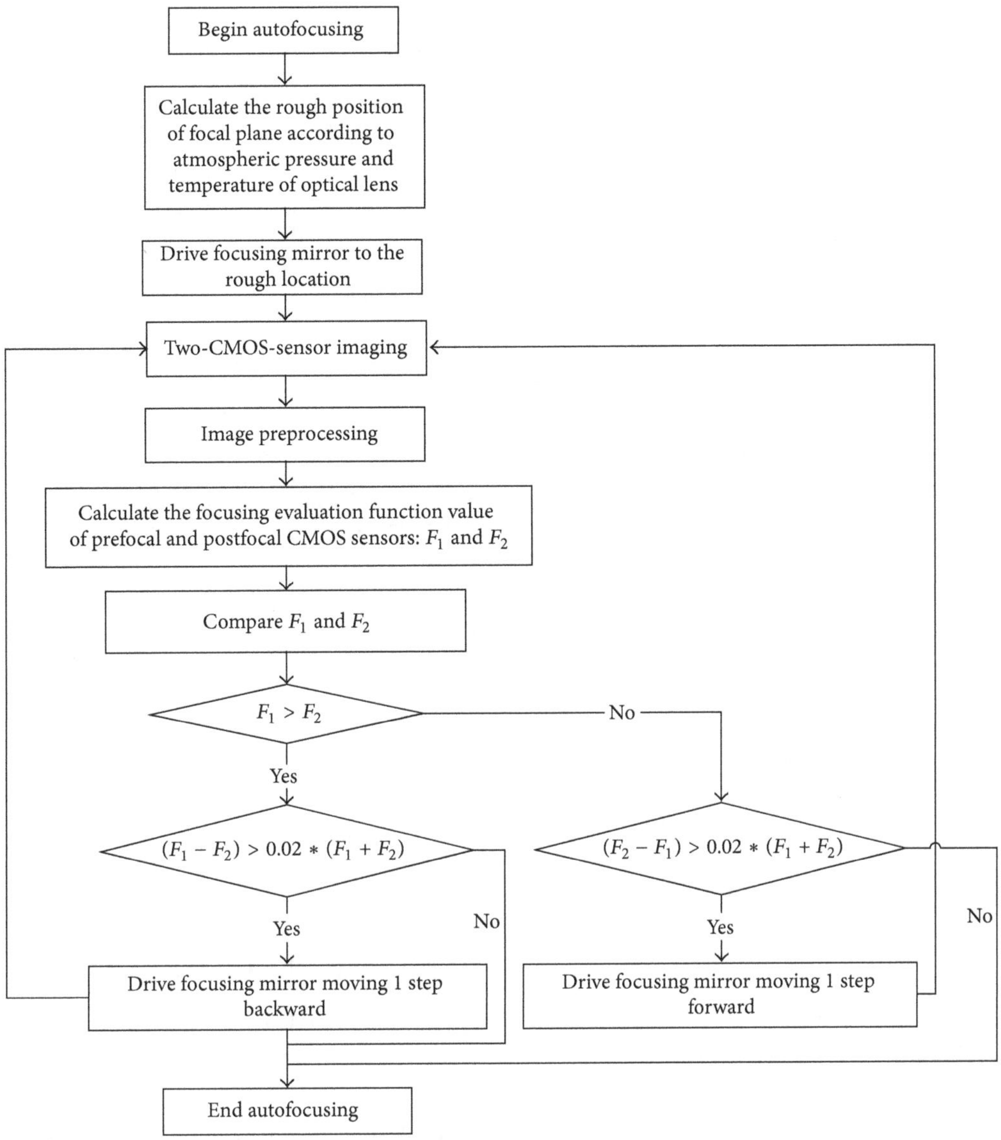

FIGURE 5: Soft flow diagram for autofocusing.

of prefocal CMOS detector and imaging CCD detector is also D in aplanatic focal plane. The difference is that two CMOS detectors locate at both sides of the imaging CCD detector.

Thus, ground scenery image through reflection of forward motion compensation mirror, refraction of optical lens, and reflection of focusing mirror arrives at transflective mirror. The transflective mirror reflects half of the light to the prefocal CMOS detector, and half of the light is transferred to the postfocal CMOS detector. So the same ground scenery finally images on the two CMOS detectors at the same time.

When the AF CMOS detector is off the optical axis of the system, along with the increase of angle of field, the lens curvature of field will reduce the position accuracy of optimal image plane. So it is necessary according to the astigmatic field curve provided by optical design to place the AF CMOS detectors on the angle of field where the field curvature is minimum.

7. Choice of Focus Evaluation Function

In this paper, the focus evaluation function (FEF) refers to the mathematical expression that is used to obtain the information about image resolution [6]. FEF can roughly boil down to gray level gradient function, the frequency domain function, informatics function, statistics function, and so on. The gray level gradient function is more practical for judging the focusing complexity and stability than other categories functions [3].

The following are four commonly used gray gradient functions.

(1) Gradient square sum function is

$$E_k = \sum_{i=1}^{M-1} \sum_{j=1}^{N-1} \left([f(i+1, j) - f(i, j)]^2 + [f(i, j+1) - f(i, j)]^2 \right). \tag{10}$$

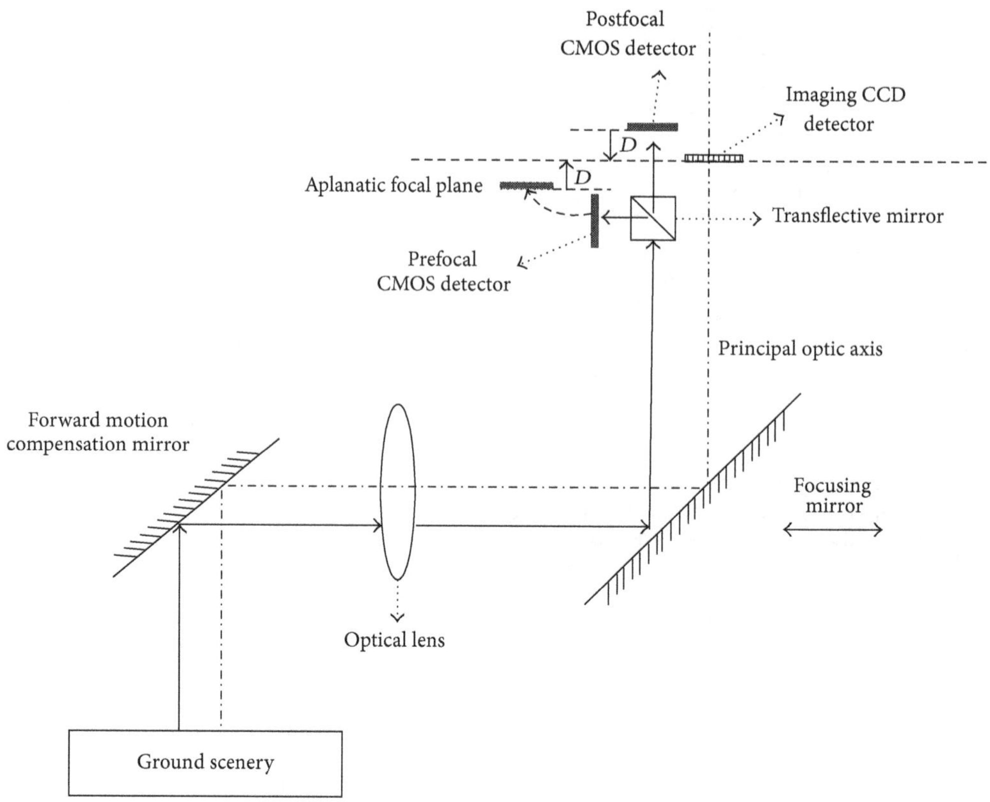

FIGURE 6: Optical-mechanical structure of autofocusing system.

(2) Laplace operator sum function is

$$L_K = \sum_{i=2}^{M-1} \sum_{j=2}^{N-1} |8f(i,j) - f(i,j-1) - f(i-1,j) - f(i+1,j) - f(i,j+1) - f(i-1,j-1) - f(i-1,j+1) - f(i+1,j-1) - f(i+1,j+1)|. \tag{11}$$

(3) Roberts gradient sum of squares function is

$$R_K = \sum_{i=1}^{M-1} \sum_{j=1}^{N-1} \sqrt{|f(i,j) - f(i+1,j+1)|^2 + |f(i,j+1) - f(i+1,j)|^2}. \tag{12}$$

(4) Secondary polynomial function is

$$S = \sum_{i=1}^{M} (f_{\max}(i) - f_{\min}(i)) \sum_{j=1}^{N} (f_{\max}(j) - f_{\min}(j)) + \frac{1}{M(N-1)} \sum_{i=1}^{M} \sum_{j=1}^{N-1} [f(i,j+1) - f(i,j)]^2, \tag{13}$$

where i, j are the column and row numbers of the pixel in spatial domain and $f(i,j)$ is the gray value of the pixel.

For autofocusing system, it is very important to select appropriate focus evaluation function. The ideal focus evaluation function should have the following characteristics: (1) it should be unbiased and have maximum value only when object plane and focal plane overlap; (2) it should have a single-peak: that is, focus evaluation function in the whole range has only one extreme value; (3) it should have high signal-to-noise ratio; in certain noise interference, autofocusing system can accurately detect defocused signal; (4) the computation burden should be as minimum as possible; this is a prerequisite for quick automatic focusing; (5) it should have high sensitivity; that is, autofocusing system should distinguish between focus and defocus; this is an important requirement for focusing precisely [4].

Experiments show that a secondary polynomial function with an approximately linear curvature is suitable for the larger focusing range. But due to its low sensitivity, it is

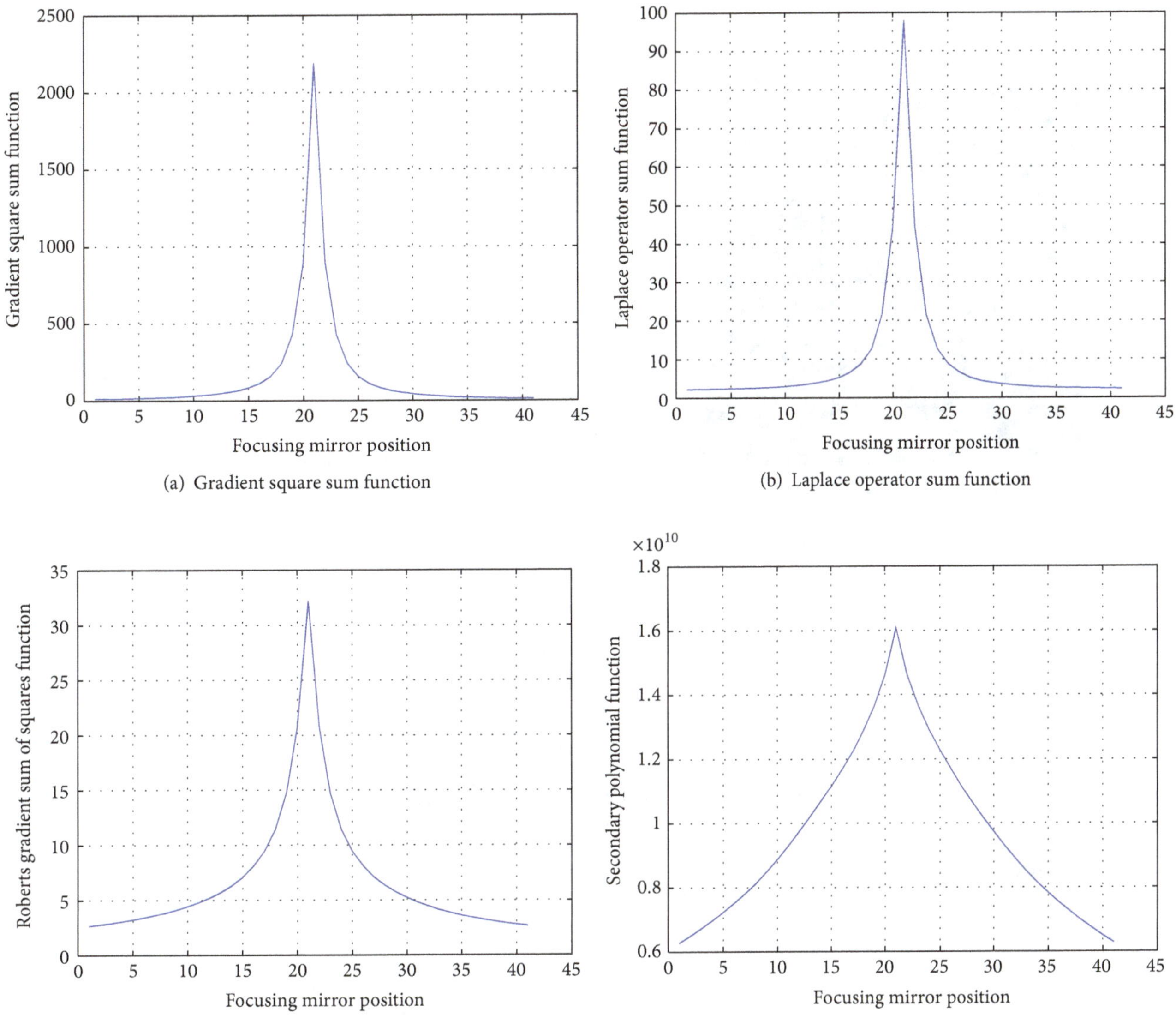

(a) Gradient square sum function

(b) Laplace operator sum function

(c) Roberts gradient sum of squares function

(d) Secondary polynomial function

FIGURE 7: Variation trend of focus evaluation function.

suitable for coarse focusing in a large range. Roberts gradient sum of squares function, gradient square sum function, and Laplace operator sum function are suitable for focusing accurately because of their high sensitivity and high stability in a small range [16].

Comparing several small range focusing functions, Roberts gradient sum of squares function has advantages in terms of minimal computational burden and stability. At the same time, before the application of proposed image-focusing algorithm, an approximate focusing is carried out considering the environmental conditions, so that the possible focal range is narrowed down to a quarter of original. Then, the focus evaluation function which has high sensitivity but fits for small focal range can be used. Hence, the Roberts gradient sum of squares function is chosen as the focus evaluation function. The variation trend of various focus evaluation function is presented in Figure 7.

8. Image Preprocessing

In automatic focusing of the imaging system, along with the constant change of the camera position and the shaking of view axis or fluctuations of the illumination, average brightness of the received image will change. Since most of the imaging systems possess automatic lighting function, even in the same position, the two-frame image's brightness is probably different. The difference will affect the output of the focus evaluation function because gray level values of the image are computational basis. The focus evaluation functional values of a clear image with low brightness are less than that of an unclear image with higher brightness. In such cases, focus evaluation functional curve can have multiple peaks. Therefore, image must be normalization in the gray scale.

Figure 8(a) shows 20 frames of images of the same view that are continuously collected. Figure 8(b) shows

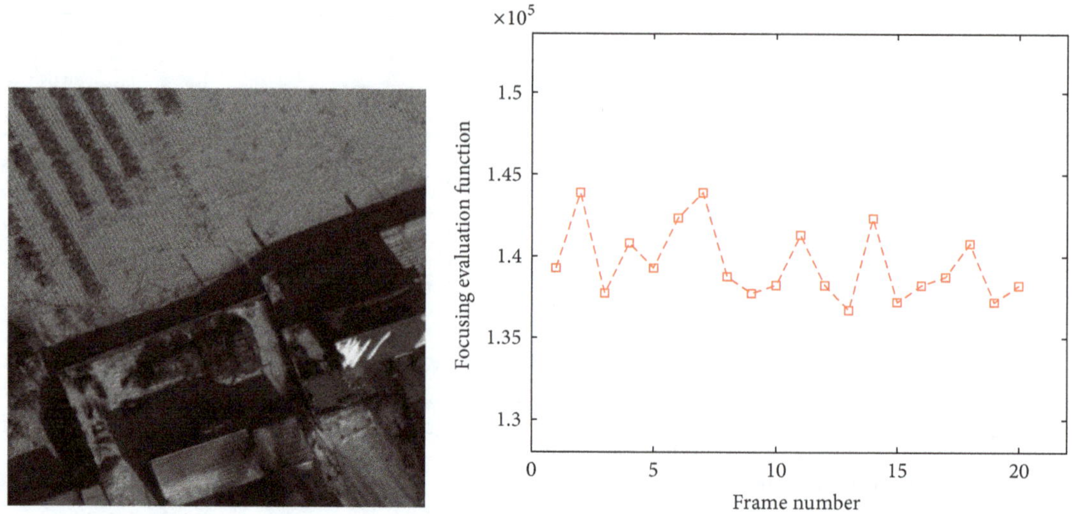

(a) Visible image collection (b) Focus evaluation functional values

FIGURE 8: Visible image collection and the corresponding values of focus evaluation function.

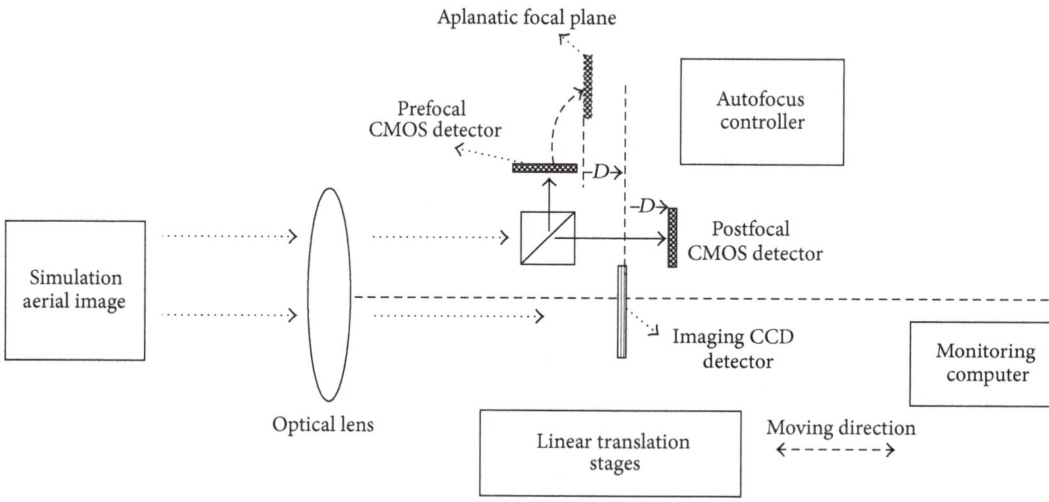

FIGURE 9: The schematic diagram of autofocusing experiment prototype.

the corresponding focus evaluation function values. From the experiment results, one can observe the fluctuations in the image; meanwhile, the fluctuations will switch the stability of focus evaluation function. So the image preprocessing refers to 10 frames of the same ground scenery and takes the average of the grayscale values of 10 frames point by point, to reduce the interference mentioned above.

9. Experiment Result and Analysis

In order to validate the performance of autofocusing system, few aerial images in the laboratory are considered. The hardware design of autofocusing system and error analysis are discussed in this section [16].

In order to verify the feasibility of this method, a focus experiment is conducted. The test principle is illustrated in Figure 9, and the material object is shown in Figures 10 and 11. Test apparatus includes simulation air images, optical lens, motorized linear translation stage, one imaging CCD detector and two CMOS detectors for focusing, transflective mirror, autofocusing controller, and monitor computer [6, 16].

The imaging CCD detector is an aerial scan CCD detector, Gazelle GZL-CL-41C6M with a resolution of 2048 × 2048, pixel size of 8 μm × 8 μm, and a frame rate of 30 fps. The optical lens is SMC PENTAX 67 with a focal length of 200 mm, relative aperture of 1 : 6, and an operating wavelength of 0.4~0.9 μm. The focusing detector is an area scan CMOS detector, OV9650, with a resolution of 1280 × 1024, pixel size of 3.18 μm × 3.18 μm, and a frame rate of 15 fps. Autofocusing controller used is S3C6410, ARM11 series processor. KSA300-11/12-X is used as a high-precision electronic linear translation stage with a closed-loop resolution 1 μm.

The target image and the lens are fixed on the desktop. Imaging CCD detector, two focusing CMOS detectors,

FIGURE 10: The autofocusing experiment system.

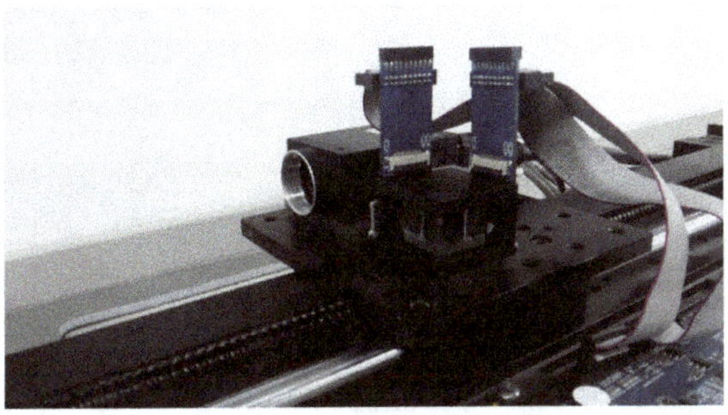

FIGURE 11: The detectors and transflective mirror.

and a transflective mirror are fixed to electronic linear translation stage. The transflective mirror is placed as shown in Figure 11, to ensure that the two CMOS detectors are in front of the focal imaging detectors and focal plane, respectively, after the same distance "D." In the experiment, the autofocusing controller manages the electronic linear translation stage and controls the two focusing CMOS detectors. When the electronic linear translation stage moves to a certain position, two focusing CMOS detectors take photos of the test aerial images. The autofocusing controller preprocesses these images and calculates the focus evaluation function value of images. According to the algorithm presented in Figure 5, the control electronic linear translation stage moves to the next position until focusing ends.

(1) The Experiment Verification of Parameter "D." At first, autofocusing controller controls the electronic linear translation stage to move to a certain position till the image CCD detector captures an image clearly. The location is marked as L_0.

To verify this parameter "D," near L_0 with a precision of ±0.5 mm, the electronic line translation stage is moved to a random location and the imaging detector is intentionally defocus. According to the proposed focusing algorithm, the autofocusing controller readjusts the electronic line translation stage and controls the two focusing CMOS detectors to capture the image until the end. The location is recorded as L.

To compute the parameter "D" accurately, a different numerical is selected as shown in Table 1. A city is selected as the test aerial image as shown in Figure 12(b). In Table 1, "L_0" refers to the location where the imaging CCD detector captures the image clearly, which is a location found from the algorithm. "Error" which is given by the difference between the absolute values of "L" and "L_0" is computed for each case and results are presented in Table 1. It can be seen from the experimental result that when the parameter "D" is equal to 0.10 mm, the autofocusing error is minimum.

(2) Autofocusing Experiment. After finding the parameter "D," three aerial images are used as the target images, as is shown in Figure 12; the above-mentioned experiment method is repeated. The obtained experimental results are shown in Table 2.

When the aerial camera is working, the error of defocus allowed is less than a half of focal depth of the optical system. It is defined as follows:

$$\delta = 2F^2\lambda. \tag{14}$$

δ is the error of defocus allowed; F is the reciprocal of relative aperture of optical lens; λ is central wavelength of optical lens.

According to (14), the test equipment's error of defocus allowed can be calculated. For an F value of 6, the center wavelength of optical system is 550 nm, and the obtained error of defocus allowed is 0.0396 mm.

From Table 2, it can be concluded that the proposed method meets the requirements of the aviation camera focusing accuracy; the maximum error of focusing is 0.029 mm, which is less than the error of defocus allowed, that is, 0.0396 mm. The average error of focusing is 0.017 mm, which is less than a half of the error of defocus allowed, 0.0198 mm. Although three example images are different in content and richness, our focusing method can accurately find the best focal plane.

10. Conclusions

In this paper, a new autofocusing approach of aerial camera was proposed. The proposed autofocusing method is based on the actual image, so it is direct and accurate. Because the ground target is imaged on two CMOS detectors simultaneously, the autofocusing method can reduce the aircraft attitude stabilization accuracy requirements. It can be applied to an aerial camera based on line scan TDICCD detector or area scan CCD detector. When the aerial camera captures the images of the ground scenery, the autofocusing system can work simultaneously. So it is a real-time autofocusing method.

To validate the performance of the proposed autofocusing method, few aerial images in laboratory are considered. Results show that the average error of focusing is 0.017 mm, which is less than a half of the error of defocus allowed, that is, 0.0198 mm. So the proposed method can meet the needs of aviation camera; the error of defocus is within the allowed half depth of defocus. In the future, the autofocusing method based on 2 differential CMOS detectors will be investigated for the better and precise aerial imaging.

Competing Interests

The authors declare that they have no competing interests.

(a) River and tree (b) City (c) Suburb

FIGURE 12: Test aerial images.

TABLE 1: Parameter "D" experiment result.

Measurement times	D = 0.03 mm			D = 0.05 mm			D = 0.10 mm			D = 0.15 mm		
	L (mm)	L_0 (mm)	Error (mm)	L (mm)	L_0 (mm)	Error (mm)	L (mm)	L_0 (mm)	Error (mm)	L (mm)	L_0 (mm)	Error (mm)
1	61.381	61.355	0.026	61.333	61.355	0.022	61.323	61.355	0.032	61.383	61.355	0.028
2	61.398	61.355	0.043	61.314	61.355	0.041	61.372	61.355	0.017	61.323	61.355	0.032
3	61.326	61.355	0.029	61.387	61.355	0.032	61.334	61.355	0.021	61.319	61.355	0.036
4	61.392	61.355	0.037	61.329	61.355	0.026	61.368	61.355	0.013	61.376	61.355	0.021
5	61.316	61.355	0.039	61.386	61.355	0.031	61.374	61.355	0.019	61.328	61.355	0.027
Average	—	—	0.035	—	—	0.030	—	—	0.020	—	—	0.029

TABLE 2: Autofocusing experiment result.

Measurement times	Figure 12(a) river and tree			Figure 12(b) city			Figure 12(c) suburb		
	L (mm)	L_0 (mm)	Error (mm)	L (mm)	L_0 (mm)	Error (mm)	L (mm)	L_0 (mm)	Error (mm)
1	61.372	61.355	0.017	61.333	61.355	0.022	61.372	61.355	0.017
2	61.379	61.355	0.024	61.343	61.355	0.012	61.334	61.355	0.021
3	61.326	61.355	0.029	61.368	61.355	0.013	61.341	61.355	0.014
4	61.371	61.355	0.016	61.361	61.355	0.006	61.339	61.355	0.016
5	61.346	61.355	0.009	61.366	61.355	0.011	61.36	61.355	0.005
6	61.381	61.355	0.026	61.338	61.355	0.017	61.377	61.355	0.022
7	61.367	61.355	0.012	61.366	61.355	0.011	61.342	61.355	0.013
8	61.341	61.355	0.014	61.367	61.355	0.012	61.348	61.355	0.007
9	61.363	61.355	0.008	61.34	61.355	0.015	61.371	61.355	0.016
10	61.342	61.355	0.013	61.368	61.355	0.013	61.376	61.355	0.021
Average	—	—	0.017	—	—	0.013	—	—	0.015

References

[1] B. G. Tao, Z. Hua, and T. Wei, "Algorithm's study on automatic focusing for image measurement technology," *Journal of Shanghai Jiaotong University*, vol. 39, no. 1, pp. 121–124, 2005.

[2] T. Hu, S.-Z. Chen, G.-D. Liu, and Z.-B. Pu, "Selection of auto-focus function in micro visual system," *Semiconductor Optoelectronics*, vol. 27, no. 2, pp. 216–220, 2006.

[3] H. Y. Liu and Q. H. Wan, "Elements of automatic focusing based on image processing method and development of system," *Micromechanical Electronics Information*, vol. 24, no. 5, pp. 39–40, 2008.

[4] Q. Li, H.-J. Feng, and Z.-H. Xu, "Image pre-processing techniques for auto focusing," *Opto-Electronic Engineering*, vol. 31, no. 9, pp. 66–68, 2004.

[5] S.-G. Ren, J.-W. Li, and L.-L. Xie, "Automatic focusing technique based on gray scale difference method," *Opto-Electronic Engineering*, vol. 30, no. 2, pp. 53–55, 2003.

[6] L. Q. Wang, L. Ben, X. D. Xin, and Z. Lu, "An automatic focusing system based on USB video camera," *Opto-Electronic Engineering*, vol. 28, no. 5, pp. 32–34, 2001.

[7] Z. Wang, L.-P. Zhang, Z.-H. Li, and X.-C. Xun, "Design of focusing mechanism of space tridimensional mapping camera," *Optics and Precision Engineering*, vol. 17, no. 5, pp. 1051–1056, 2009.

[8] C. C. Liebe, R. Pollock, B. Hannah et al., "System for establishing best focus for the Orbiting Carbon Observatory instrument," *Optical Engineering*, vol. 48, no. 7, Article ID 073605, 2009.

[9] M. E. Bravo-Zanoguera, C. A. Laris, L. K. Nguyen, M. Oliva, and J. H. Price, "Dynamic autofocus for continuous-scanning time-delay-and-integration image acquisition in automated microscopy," *Journal of Biomedical Optics*, vol. 12, no. 3, Article ID 034011, 2007.

[10] L. Zhang, H. Zhao, W. Dong, N. Cao, and P. Zhang, "Design of an optical system consisting of a special telecentric lens for side-scattering measurement on individual cells," *Optical Engineering*, vol. 49, no. 5, Article ID 053001, 2010.

[11] R. C. Gonzalez, *Digital Image Processing*, Pearson, London, UK, 2008.

[12] L. Liu, "Focusing range of space off-axial TMA optical camera," *Optics and Precision Engineering*, vol. 21, no. 3, pp. 631–636, 2013.

[13] M. Subbarao, T.-C. Wei, and G. Surya, "Focused image recovery from two defocused images recorded with different camera settings," *IEEE Transactions on Image Processing*, vol. 4, no. 12, pp. 1613–1628, 1995.

[14] V. Aslantas and D. T. Pham, "Depth from automatic defocusing," *Optics Express*, vol. 15, no. 3, pp. 1011–1023, 2007.

[15] C.-Z. Chen, G.-J. Zhao, X.-X. Zhang, E. Lu, and J.-Y. Ren, "A calculating method for temperature tolerance of space telescope," *Optics and Precision Engineering*, vol. 15, no. 5, pp. 668–673, 2007.

[16] J. Yan, L. J. Chen, and Y. B. Xin, *Development of DSP Embedded Application System and Typical Example*, China-Electric Power Press, Beijing, China, 2007.

15

Survey on Flight Control Technology for Large-Scale Helicopter

Jinshuo Hu and Hongbin Gu

College of Civil Aviation, Nanjing University of Aeronautics and Astronautics, Nanjing 211100, China

Correspondence should be addressed to Jinshuo Hu; jinshuohu@nuaa.edu.cn

Academic Editor: Christopher J. Damaren

A literature review of flight control technology is presented for large-scale helicopter. Challenges of large-scale helicopter flight control system (FCS) design are illustrated. Following this, various flight control methodologies are described with respect to their engineering implementation and theoretical developments, whose advantages and disadvantages are also analyzed. Then, the challenging research issues on flight control technology are identified, and future directions are highlighted.

1. Introduction

Large-scale helicopter has its unique characteristics of maneuverability and low speed performance compared to fixed-wing aircraft, which significantly extends its application in both military and civil. The complicated dynamic system of large-scale helicopter brings to the FCS a great deal of design considerations and updates.

Early helicopters were primarily controlled through mechanical linkages, and then, electronic stabilization was implemented via analog systems to provide electronic feedback. As technology developed, digital systems, with advantages in development flexibility, complex function support, and power reduction, started to replace analog systems. The most currently used FCS is digital Fly-By-Wire (FBW) system, designed to improve handling qualities, alleviate failure transients, and reduce maintenance activities [1]. This technique has been adopted by JUH-60A [2], AH-64D [3], CH-47B [4], RAH-66 [5], Bell-205 [6–8], NH-90 [9], and so forth. Besides, Fly-By-Light (FBL) system, which has demonstrated its superiority in saving weights and reducing electromagnetic interference susceptibility problems, was already applied to helicopter in-flight simulators, such as UH-60A [10], EC-135 [11, 12], and BO-105 [13].

With the improvement of helicopter multirole capability, FCS design has become more demanding, requiring large-scale helicopter capable of executing multitasks in adverse flying conditions. The advances of flight control methodologies play a central role in helicopter FCS design on improving helicopter dynamic performance and handling qualities. Moreover, the application of advanced flight control methodologies directly contributes to the rapid development of helicopter flight simulation technique which paves the way from advanced control theories to practical implementation. Therefore, a highly reliable flight control methodology considerably improves helicopters' performance and stability for implementing their various properties safely and effectively, within the flight envelop.

The rest of this paper is organized as follows. Section 2 illustrates the primary challenges of designing helicopter FCS. Section 3 gives an elaborate survey of flight control methodologies on classical control, modern control, and intelligent control. Core research questions are summarized and future developments of helicopter flight control technology are discussed in Section 4.

2. Challenges of Large-Scale Helicopter FCS Design

Large-scale helicopters are strong nonlinear, highly coupled, time-varying dynamic systems, which makes the FCS design a difficult task due to the following aspects.

(1) Complicated Dynamic Response. The nonlinear helicopter system is not affine in control inputs, which causes unwanted controller response if actuator and main-rotor dynamics are considered. Besides, the strong dynamic coupling effect makes the FCS design challenging, since each control input

not only affects the state variables of interest but also produces unintended secondary responses [14].

(2) Multiple Flight Modes. According to ADS-33E, military helicopters are capable of activating different flight modes, such as RC (Rate Command), ACAH (Attitude Command Attitude Hold), and TRC (Translational Rate Command) [15]. These advanced flight modes are designed to improve handling qualities in various flight conditions when executing military tasks. However, the complexity of controller design is increased, since quantities of switching logics are involved in flight mode transitions. Moreover, flight mode transitions may result in a few problems, such as unsmooth transition trajectory and actuator saturation.

(3) System Uncertainties. Model uncertainties primarily arise from the complicated aerodynamic nature of thrust generation. Unmodeled rotor dynamics (flapping and lagging dynamics) and high frequency dynamics associated with actuators make the controller sensitive to uncertainties. Besides, neglected dynamics, such as ground effect, near-ground wind gust and atmospheric turbulence, tend to reduce the controller efficiency as well.

(4) Rapid Varying Flight Conditions. Large-scale helicopters are capable of performing highly flexible maneuvers and producing multiple flight mode transitions within the full flight envelop, in which case flight parameters and dynamic states undergo intensive changes. The conventional linear controller is not able to stabilize the helicopter across various flight regimes.

3. Various Flight Control Methodologies for Large-Scale Helicopter

3.1. Classical Control. In general, the current helicopter FCS is designed either using PID-based feedback techniques or with model following architecture. Both of the control methods are elaborately reviewed in Sections 3.1.1 and 3.1.2. The application of gain-scheduling control policy is proposed in Section 3.1.3.

3.1.1. PID-Based Feedback Control. The design of primary flight control law is often accomplished using PID-based feedback controllers due to their easiness of realization and application. Without considering helicopter dynamics, the expected dynamic response is achieved by tuning PID parameters manually and by designing decoupling gains empirically. An overview of the PID FCS design for ARH-70 is provided in Figure 1. The helicopter was made to follow angular rate signals in pitch, roll, and yaw axes. The reference signals are provided by input filters and rate gyros are applied to detect angular rate feedback signals. Variables δ_e, δ_a, δ_r, and δ_c correspond to the longitudinal cyclic input, lateral cyclic input, tail rotor collective input, and main-rotor collective input, respectively. The mechanical mixer is designed to remove the inherent pitch/roll helicopter coupling. A limiting block is used to prevent actuator activities from reaching saturation limit.

From the late 1970s, large quantities of researches were conducted to evaluate PID-based feedback controllers with different levels of complexity and to investigate the influence of flight control methods on helicopter handling qualities [16–18]. Besides, the US Army initiated the Advanced Digital Optical Control System (ADOCS) program to investigate modern FCS concepts and conducted four simulations on Vertical Motion Simulator (VMS) from 1981 to 1985 [19, 20]. The partial authority FCS was tested on VMS since 1990s [21], and several simulation studies were conducted on VMS to evaluate the CH-47F Digital Automatic Flight Control System (DAFCS) in degraded visual environment from 2004 to 2005 [22].

In 2011, a novel automatic control system, using lagged-gains and scheduled SCAS (Stability and Control Augmentation System) gains, was tested on OH-58D helicopter to evaluate handling qualities in the good visual environment (GVE) and degraded visual environment (DVE) [23]. During the development of ARH-70, piloted tests of the optimized PID SCAS have been accomplished through ground-based and airborne simulations. Moreover, the development of ARH-70 FCS stands out as an excellent example of how linear modeling, gain optimization, and simulation can deliver the best possible flight control design with a minimal amount of flight testing required for design validation [24].

With the application of multiobjective parametric optimization approach, the performance of PID-based feedback control methods has been enhanced more significantly than before. However, the achievable performance is rather limited when applying classical PID control approach to multivariable, time-varying, highly coupled helicopter dynamics, since high frequency characteristic and disturbance rejection are not considered.

3.1.2. Explicit Model Following Control. Explicit model following control performs good command tracking and control decoupling properties, and this control method has been widely adopted in designing helicopter FCS.

From the 1980s to the beginning of 21st century, NASA-Ames and NASA-Army conducted several experiments to evaluate the explicit model following control technology on BO-105 [25] and UH-60A [2, 10]. This control architecture was also applied to the Core Automatic Flight Control System (CAFCS) on the RAH-66 Comanche [5]. In 2000, Modernized Control Laws (MCLAWS), designed to satisfy the ADS-33 DVE requirements using the existing limited authority actuators, began to replace the legacy explicit model following control architecture [26–28].

A schematic of the MCLAWS architecture is illustrated in Figure 2. This control architecture uses a model following approach to provide ACAH control mode in pitch and roll axes. ACAH means that the attitude of helicopter is proportional to pilot stick positions. The LVDT block represents Linear Variable Differential Transducers, which measure the pilot's stick positions and provide input signals to the control system. The command model transforms pilot inputs to desired attitudes, and the sensor data of attitudes are provided by vertical gyros. An inverse plant works to cancel

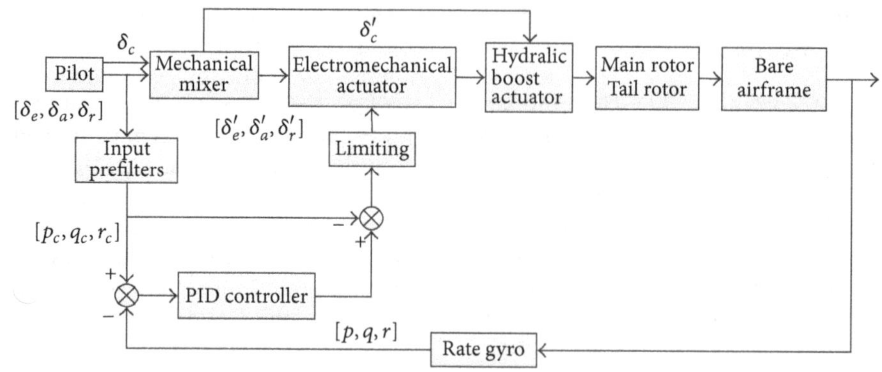

FIGURE 1: Implementation of RC mode into ARH-70's PID control architecture.

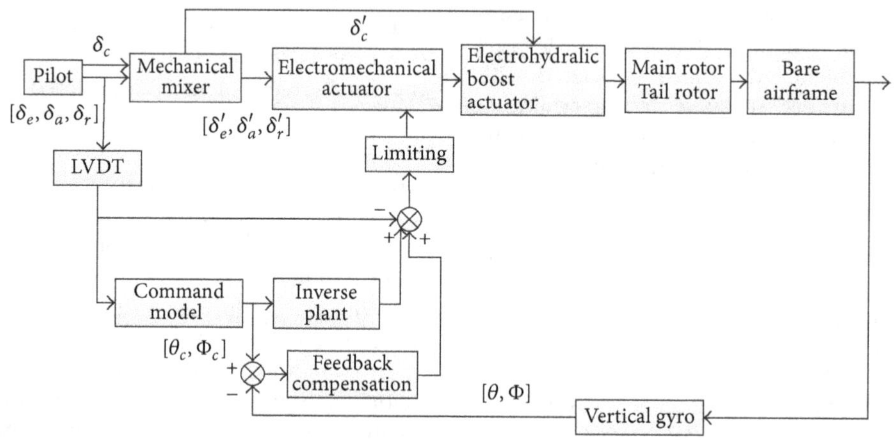

FIGURE 2: Implementation of ACAH mode into UH-60's MCLAWS model following control architecture.

the helicopter internal dynamics. Disturbance rejection and plant stabilization are achieved using feedback compensation.

The MCLAWS has a model following type architecture with the following unique characteristics different from the legacy explicit model following concept: (1) instead of just using a commanded attitude, both rate and attitude commands are generated and used; (2) unlike a full authority system, the MCLAWS has to contend with a full authority mechanical system and to achieve desired performance with partial authority AFCS; (3) the current implementation uses a low order inverse plant model. The MCLAWS architecture provides a high level of modularity and enhances its practicability in helicopter FCS design.

In 2003, the MCLAWS was tested on VMS and results indicated that improved handling qualities and reduced pilot workload are achieved in DVE [26]. Besides, piloted simulations also demonstrated that the MCLAWS performs better than the UH-60A legacy control laws when operating in wind and turbulence. Several upgrades were accomplished to enhance the performance of the standard MCLAWS in the following years. An improved MCLAWS was evaluated on an EH-60L research helicopter for near-earth operations as the pilot's visual cues degrade [27]. Two characteristics make this MCLAWS different from [26]: (1) this MCLAWS implements an ACAH response type using only the 10% SAS (Stability and Augmentation System) series servos with no trim actuators; (2) the application of a modern integrated toolset (MATLAB/Simulink, CIFER, CONDUIT, etc.) plays a central role in this MCLAWS design process and facilitates the MCLAWS development. Afterwards, the modern integrated package is widely used as design and analysis tool for AH-64D [3, 29]. In 2012, related works of the UH-60 MCLAWS continued to incorporate updates upon flight control design optimization methods and handling quality researches, and the FCS improvements have been addressed over two new iterations of the UH-60 MCLAWS [28, 30]. In 2016, an outer-loop position hold with velocity command mode and landing logic were integrated into the standard MCLAWS to further improve handling qualities in simulated DVE [31].

The testing results substantiate that explicit model following control guarantees excellent command tracking performance and control input decoupling, but the low order linear inverse plant model is invariant with flight conditions, which is incapable of canceling high frequency dynamics such as those associated with the main-rotor and actuators, illustrating the importance of accurate modeling of high frequency characteristics for high-bandwidth FCS [32]. Besides, unmodeled rotor dynamics and high frequency structural modes make the controller sensitive to uncertainties, which points out that controller robustness should be improved against model mismatches and neglected dynamics to obtain good model following performance.

3.1.3. Gain-Scheduling Control. Large-scale helicopter has a strongly nonlinear behavior and is running at different operating points, resulting in nonuniformly desired control response through the entire flight envelope. Therefore, the operating-point-oriented controller parameters are required to be optimal to keep the dynamic response within the neighborhood of each fixed operating point.

To cope with nonlinear process and large parameter varying, gain-scheduling control policy is proposed. This strategy grids the flight envelope into several operating points based on which the linear controller is designed, and interpolation algorithm is adopted so that smooth transitions are possible between the operating points. The realization of gain-scheduling strategy is essentially an engineering problem, since the applied control technique is mainly based on linear control methods (PID control or explicit model following control) and the primary task is to tune the gain-scheduled control parameters that are scheduled as a function of flight conditions (airspeed, altitude, dynamic pressure, etc.) in modern helicopter. Manual tuning of control system parameters to meet handling qualities and performance specifications has always been time-consuming and complicated. Instead, the control engineers resort to engineering tools and take advantage of the automated tuning feature, making the design applicable to nonlinear helicopters.

A state-of-the-art engineering tool, the Control Designer's Unified Interface (CONDUIT) software package, allows the designer to tune selected design parameters for the optimal performance against helicopter-oriented specifications [33]. Since the late 1990s, many FCS designs, such as those of the RASCAL JUH-60A Black Hawk and AH-64 Longbow Apache and OH-58D Kiowa Warrior helicopters, have been implemented using the CONDUIT. In 2016, the CONDUIT toolset is integrated into the recently developed SIMPLI-FLYD framework (a collection of tools) which is designed to perform modeling and analysis for the assessment of flight dynamics and control aspects of the handling qualities of rotorcraft conceptual designs [34]. Moreover, the SIMPLI-FLYD evaluation process also provides pathways to decrease the computational cost of running complete CONDUIT optimization.

In the multivariable control system design, a coupling numerator theory and quantitative feedback theory based precompensation technique is applied to minimize the helicopter cross-coupling response [35]. The resulting dynamic crossfeeds are used for gain-scheduled controller designs. To implement such a technique in existing helicopters, the designer selects the desired time domain/frequency domain coupling specifications (e.g., a time domain pitch/roll coupling specification is CouPRH1 [36]) and other specifications from the CONDUIT libraries to incorporate them in the FCS design problem, and then, the designer graphically selects gain-scheduled control parameters and crossfeeds that will be used by CONDUIT in the tuning process. These design parameters are optimized against a combination of specifications, such as stability, coupling, bandwidth, and actuator activity. Finally, the performance of the helicopter FCS is evaluated and ensures that Level 1 design specifications can be achieved.

The overall physical configuration for the implementation of gain-scheduling control method is deployed in a hierarchical architecture. Controller tuning is executed in the CONSOL-OPTCAD environment which is already integrated into the CONDUIT, and the CONDUIT resides in the computer operating system on engineering workstations. An automatic coding of the optimized control law is employed to facilitate piloted simulations following CONDUIT analyses. The pilot-in-the-loop evaluations are conducted on engineering simulators (e.g., VMS and RACAS [37]), including inceptors with control loading, to further evaluate the gain-scheduling flight control laws. Moreover, the gain-scheduling needs only small capacity of the dual Interl® Xeon® processors and requires relatively less calculating time.

Gain-scheduling is a beneficial strategy to adjust controllers to each operating point, if a nonlinear system exhibits different behaviors across different flight regimes. However, the robustness properties of the control system are not guaranteed only by designing each controller for its best performance, since no satisfactory mathematical theory approves that the sum of local optimal results will definitely contribute to the system's global optimum. To cope with this problem, the engineering tool CONDUIT uses multiobjective optimization, which allows the relative importance of each design specification to be determined, to ensure that the best feasible solution of the proposed control system is achieved within all robustness constraints. Besides, other techniques being developed are also applied to study multidisciplinary design optimization of FCS parameters [38] and automated procedure for gain-scheduled flight control law design [39]. Future development of the engineering toolset will directly provide designers rapid insight on control system response and accelerate the R&D process for helicopter industry.

3.2. Modern Control. This part reviews the modern control methods, with an emphasis on discussing those that have been tested through engineering simulations. These methods are H_∞ optimal control and Incremental Nonlinear Dynamic Inversion (INDI) control. The primary implementation of observer feedback control is presented in Section 3.2.3. Other representative modern control methods (see Table 1) are still in developing stage and only verified via desktop simulation.

3.2.1. H_∞ Optimal Control. H_∞ optimal control method is designed to minimize the infinity norm of weighted transfer functions on satisfying important frequency domain requirements often referred to as loop shaping. Loop shaping is used to shape sensitivity function, complementary sensitivity function, and control activity by selecting frequency dependent weights.

In the 1990s, McFarlane and Glover introduced a particular loop shaping method to guarantee close-loop stability and a level of robust stability at all frequencies against a

TABLE 1: Summary on modern control methodologies.

Control methods	Advantages	Disadvantages
LQR/LQG Control	Guaranteed stability margins for state feedback LQR; good transient response and disturbance rejection.	Full state feedback requires significant usage of sensors; Kalman filter degrades the frequency domain properties.
Eigenstructure Assignment Control	Close-loop performance index can be designed directly; the control method is relatively simple to implement without requiring high order dynamic compensators.	An analytical linear plant is required; stability and robustness is not guaranteed; dynamic cross-coupling between inputs and outputs cannot be canceled.
Adaptive Control	Control gains can be adjusted automatically using on-line adaption; robustness is achieved within bounded system uncertainties; the regulated output dynamics are allowed to be non-minimum phase.	Parameter drift is caused by process noise; high frequency flexible modes are difficult to control; large rates of adaption are practically non-realizable.
LPV Control	Nonlinear natures of the plant are captured accurately using LPV modeling; controller robustness is improved over conventional gain-scheduled controllers.	The complexity of LPV controller is significantly increased; system uncertainties caused by rapid varying flight parameters cannot be completely eliminated.
Sliding Mode Control	Different control laws can be adopted corresponding to the changes of system trajectories; the controller performance is insensitive to modeling errors and parameter uncertainties.	Chattering is induced by frequent controller switches; highly nonlinear sliding mode surface results in difficulties of analyzing system stability and reachability.
Backstepping Control	The exponential stability of dynamic errors guarantees asymptotic tracking of the desired command.	Calculating the derivative of pseudo control inputs results in huge computation.
MPC	External constraints are incorporated into the design process naturally; improved transient response is achieved using prediction models.	Close-loop property depends on the performance of reference model; on-line optimization is time-consuming.

coprime factor uncertainty [40]. The coprime factor robust stabilization problem is illustrated in Figure 3.

A linear, time-invariant plant is given in the following:

$$\Sigma_G : \begin{cases} \dot{x} = Ax + Bu + Ew, \\ z = Cx + D_1 u + D_2 w. \end{cases} \tag{1}$$

It is required to design the dynamic compensators W_1 and W_2 of the form

$$\Sigma_W : \begin{cases} \dot{x}_c = Hx_c + Fy, \\ u = Mx_c + Ny, \end{cases} \tag{2}$$

such that after applying the feedback Σ_W to the system Σ_G, the resulting H_∞ norm of the closed-loop system is less than a positive bound γ. In Figure 3, the nominal plant is

$$G = M^{-1}N \tag{3}$$

and a perturbed plant G_Δ is written as

$$G_\Delta = (M + \Delta_M)^{-1}(N + \Delta_N), \tag{4}$$

where M, N are a left coprime factorization of G and Δ_M, Δ_N are uncertainties. The nominal plant G and the dynamic

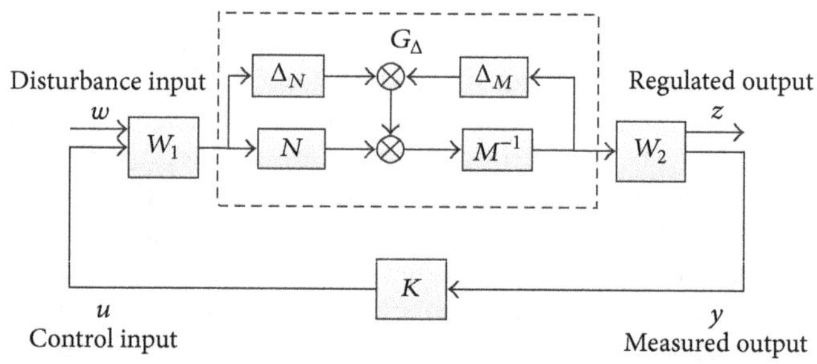

FIGURE 3: The Glover-McFarlane H_∞ loop shaping design.

compensators W_1 and W_2 are combined to form the shaped plant $G_s = W_2 G W_1$. The shaped open loop frequency response G_s is then used to seek a static feedback controller K such that

$$\left\| \begin{bmatrix} K \\ I \end{bmatrix} (I + G_s K)^{-1} M^{-1} \right\|_\infty \le \gamma. \tag{5}$$

This can be rewritten in the framework of an LMI based H_∞ optimization problem [8] and ensure that the closed-loop H_∞ optimal index $\gamma_{\min}$ is small enough. The final feedback

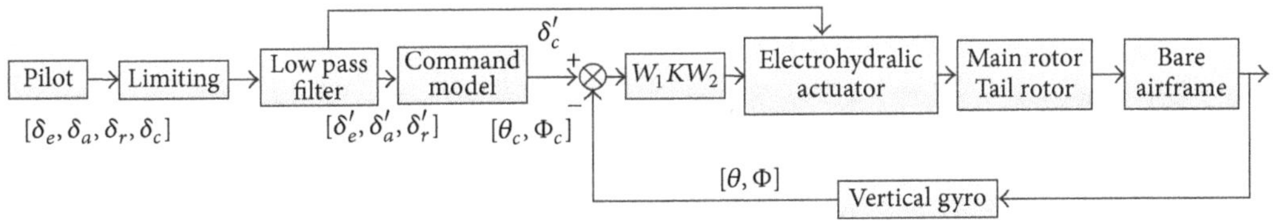

FIGURE 4: Implementation of ACAH mode into Bell 205's H_∞ control architecture.

controller K_∞ is then constructed using the static output feedback controller K with the compensators W_1 and W_2 such that $K_\infty = W_1KW_2$.

The Glover-McFarlane H_∞ loop shaping method guarantees that the nominal input-output characteristics are kept unchanged. Besides, an unstable system with right half plane poles can be canceled out using coprime factorization, and the existence of solution for compensators is guaranteed corresponding to particular robustness conditions. These advantages make this coprime factorization based H_∞ control better than the standard H_∞ optimal control strategy.

In 1993, Mammar and Duc proposed a loop shaping procedure using a normalized coprime factors robust stabilization. This design approach has been successively verified in solving the stabilization problem of a helicopter [41]. The first H_∞ controlled helicopter was reported in 1997 [6]. In order to implement an ACAH flight mode (see Figure 4) in hover and low speed, University of Leicester designed H_∞ controller against coprime factor uncertainties using a loop shaping procedure followed by a robust stabilization. Piloted simulations were conducted on the ground-based Large Motion System (LMS) simulator and in-flight evaluation was performed on the National Research Council of Canada's (NRCC) Bell 205 airborne simulator. The piloted simulations were a milestone in the development of H_∞ control policy, and positive evaluations on flying the H_∞ controlled system are provided. After that, the robustness properties of the nominal H_∞ control architecture were improved by including high-order rotor dynamics, used to eliminate the need for designing predictor-type filters, in the controller design process [7]. In 2005, Prempain and Postlethwaite proposed a static LMI based H_∞ loop shaping controller for the NRCC FBW Bell 205 in-flight simulator, and flight tests proved to be successful in achieving high-level performance [8].

The implementation of H_∞ control theory makes considerable efforts to close the gap between robust control theory and practical implementation and stimulates the theoretical development of multivariable H_∞ control methodology. However, H_∞ control method has deficiencies in system's dynamic response: (1) the controller generates slow response, especially in the yaw channel; (2) interaxis coupling is present, such as pitch-yaw coupling and heave-yaw coupling; (3) the H_∞ control method causes large undershoot in acceleration response if the plant is nonminimum phase.

On the contrary, H_2 control strategy enjoys its advantage over dynamic performance, but it has drawbacks in robust stability and disturbance rejection. In this case, a mixed H_2/H_∞ control strategy combines superiorities of both approaches. Extensive theoretical studies have been conducted on H_2/H_∞ strategy since 1990s [42–46]. Only a limited number of researches applied this control theory to air vehicles [47–53], and few achievements have been engineering tested.

In 2013, Rigsby et al. introduced a mixed H_2/H_∞ robust feedback controller, utilizing measurements of the external load dynamic states, for helicopters handling externally slung loads [54]. Piloted simulations were conducted on the Sikorsky simulation facility, and test results indicated that this control methodology is effective in reducing load swing and pilot workload [55]. This novel FCS, augmented with active external load controller, addresses special requirements for engineering simulation of flight control technology and provides insights on mission-oriented FCS design.

In general, most controller designs, including H_∞ and H_2/H_∞, are based on linear models utilizing the widely adopted concept of stability derivatives. Conservativeness of linear controller design lies in the assumption of small deviations from nominal conditions. This assumption is not always satisfied, which results in significant gain-scheduling, linear interpolation, and model switches. In this case, the airspeed dependent dynamics are difficult to capture under all circumstances, resulting in degraded dynamic response. Therefore, investigations on nonlinear controllers have been making contributions to understanding limitations of linear controllers and to analyzing the nonlinear nature of helicopter dynamics.

3.2.2. INDI Control. The nonlinear controller designs are mostly valued for their theoretical contribution to the helicopter flight control problem, but their applicability is an open challenge mainly due to the increased order and nonlinear structure of the controller [14]. It is the cooperative program between TU Delft and Boeing Mesa on developing advanced control laws for AH-64D helicopter that significantly contributes to the practical implementation of a nonlinear control scheme, the INDI methodology [56]. Engineering simulations of INDI were tested on SIMONA Research Simulator (SRS), indicating that this advanced nonlinear control strategy is potential for further practical implementation to helicopters.

The INDI is an incremental version of NDI. Basically, the NDI allows generating a control input via nonlinear feedback control and state transformation such that, when it is applied to the nonlinear system, the relation between a virtual control input and the output of the system becomes linear [57]. However, the high sensitivity to model uncertainties is the

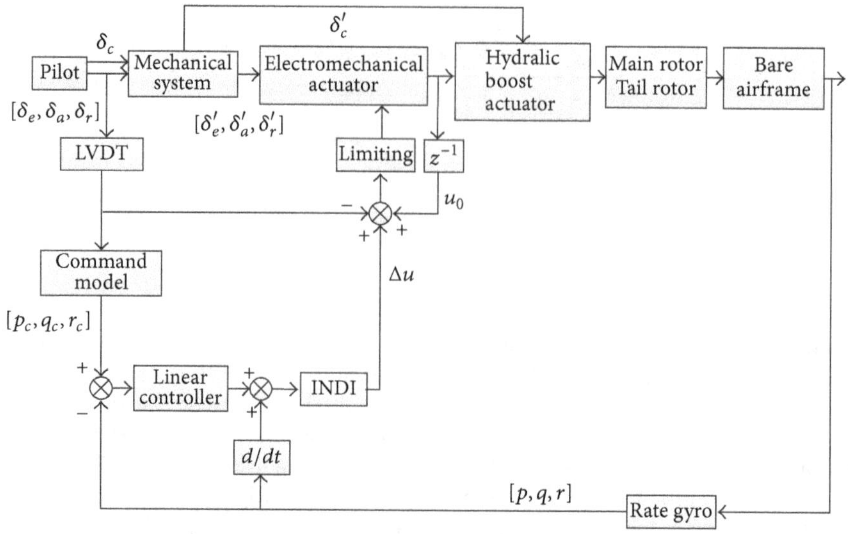

FIGURE 5: Implementation of RC mode into AH-64's INDI control architecture.

main shortcoming of the NDI [58], which motivates the development of INDI. In the past few years, many achievements have been applied to solve flight control problems for fixed-wing UAV [59], spacecraft [60], large-scale helicopter [61], tiltrotor UAV [62], quadrotor MAV [63], and so forth.

According to Figure 5, the INDI control architecture implements an RC mode using Apache's limited authority setting. Instead of using the nominal virtual input to compute the complete control input, it is possible to determine only the required variation with respect to the previous inputs. In order to form the required command signals, a unit delay operator z^{-1} is used to access the current values of the control inputs. The implementation of INDI is based on the onboard measurements of state derivatives without acquiring the state-dependent part of the plant.

However, accurate and almost instantaneous measurements or estimations of acceleration are not readily available. It is very to employ sensors to feedback angular/translational acceleration, but this approach introduces measurement noise and time delays, which results in additional designs of compensators and filters. Besides, it is also feasible to integrate observers into the INDI controller, but full order observer may degrade stability robustness properties of controller designs. These technical concerns bring to the application of INDI some difficulties.

3.2.3. Observer Feedback Control. In many practical helicopter FCS designs, the problem with state feedback control is that all of the state variables are rarely accessible for measurement. On the other hand, a large number of transduces are used to sense the state variables, making the application of the state feedback control strategy costly and impractical. To implement the state feedback control scheme, observers are required to reconstruct the states using available helicopter sensors. An observer is a dynamic system that takes the sensor signals as inputs and produces state estimates as outputs. The form of the observer is closely related to the particular complement of sensors available and often comprises the most complicated part of the controller architecture [64].

The Kalman filter is widely used to solve state estimation problems and it has numerous applications in technology. However, when used to estimate the state in output feedback control design problems, the optimal state estimator may not exhibit the best overall control properties [65]. Therefore, a tuning process called Loop Transfer Recovery (LTR) is designed to asymptotically recover the state feedback frequency domain properties. The LQG/LTR framework is verified to be a useful tool for linear control law design of a combat helicopter [66].

For the problem of nonlinear estimator designs, the Kalman filter is extended to nonlinear systems to obtain the Extended Kalman Filter (EKF). In 1986, an EKF was developed to track the states of Bell Jet Ranger helicopter during flight test. In particular, this EKF can also supply filtered state variables to a full state feedback framework or stability augmentation system [67]. In 1990, the University of Alabama Flight Dynamics Lab (UAFDL) proposed a variation of a Linearized EKF to accurately recreate the performance of UH-60 helicopter which was flown with stability augmentation system and flight path stabilization system operational [68]. Piloted simulations were conducted on the UAFDL helicopter simulator and experimental results indicated that the proposed compensator exhibits precise state tracking abilities and robustness to measurement sample rate and initial error conditions. Abraham and Costello applied the EKF framework to provide real-time estimates of weight and mass center for OH-6A helicopter, and the resulting in-flight estimates are used to improve state feedback flight control system performance [69].

It is denoted that, for helicopter systems, the EKF has proven to be a useful method of obtaining accurate estimates of the state variables. However, the EKF is obtained using a linear approximation of a nonlinear system, no guarantees of optimality are ensured in the sense of a mean squared error, and the convergence problem relies on how close we can drive the linearized point to the actual operating condition of the nonlinear system. Therefore, some reduced observer designs are applied to replace the complex structure of Kalman filter-type observers and to increase the need for an enhanced sensor complement.

In 2005, a viable practical feedback controller, using inertial rate feedback sensors coupled with a Luenberger reduced order observer, was designed to improve the helicopter handling qualities in pitch and roll axes [70]. The observer-based flight control law was excised on the BO-105 helicopter and it achieved good robustness to structured model perturbations. Utilizing the actual flight data of the Rooivalk attack helicopter, Hager et al. applied a reduced Brunovsky canonical observer to improve prediction capabilities in pitch axis [71]. In 2015, Greiser introduced a disturbance observer-based controller to suppress the roll oscillation for a highly modified research helicopter ACT/FHS [72]. The reduced order observer with disturbance compensation was tested on a ground-based simulator which simulates a highly sophisticated nonlinear model of ACT/FHS, and the desired experimental results will definitely motivate flight tests with the ACT/FHS for further validation.

To summarize, most of the observer designs use a Kalman filter-type scheme and experimental results show that the controller performance is highly dependent on the observation signal. However, a complete Kalman filter is complex, since models of the random process and measurement noise are required to perturb the helicopter system. Future developments of real-time helicopter state estimators would aim for the simplified observer structure which is particularly well suited to state feedback control problems where the Kalman filter-type filter is costly and additional sensors are demanded.

3.2.4. Other Modern Control Methodologies. LQR/LQG and Eigenstructure Assignment control approaches are linear control strategies. The former one is applied to adjust penalty matrices against a particular performance index by solving Algebraic Riccati Equations (ARE), and the latter one is used to assign the dominant eigenvalues and associated eigenvectors to achieve the desired transfer functions between commands and outputs. The effectiveness and reliability of both methods are demonstrated on controlling linearized helicopter dynamics [73–77]. However, both control approaches provide inadequate robustness to uncertainties and nonlinear effects begin to dominate the system response once actuator is saturated.

Adaptive control methods are able to control both linear and nonlinear systems with nonlinear uncertainties in their dynamics. This online control policy is effective to solve flight control problems where the aircraft aerodynamics are analytically nontractable and to capture rapid varying dynamic responses which are caused by system failure, flexibility, and battle damage. In the past two decades, newly developed adaptive controllers have been evaluated on nonlinear large-scale rotorcraft models, including UA-60A [78], BELL-412EP [79], BO-105 [80], PUMA-XW241 [81], UH-1H [82], and XV-15 [83–85]. Adaptive control provides the possibility of reducing the cost of gain-scheduling and to better understand helicopter aerodynamics across full flight envelop using online identification.

Another widely accepted approach to deal with large parameter varying is linear parameter varying (LPV) control. LPV controller can be synthesized as a single controller (control gains are continuous functions of scheduled variables) rather than a group of controllers. The LPV controller operates over the whole range of flight conditions without having to create a scheduling scheme. There are several researches applied LPV control method to fixed-wing aircraft [86–91], and few investigations are conducted on rotorcrafts [92, 93] because of the following reasons: (1) the nonlinear helicopter model is not affine in control inputs because of the main-rotor dynamics; (2) the rapid variation of flight parameters requires fast transient response, which provides rigorous constraints for controller synthesis; (3) the selection of effective scheduling variables is challenging, since inner dynamics, such as inflow dynamics and blade flapping dynamics, have to be considered.

In addition, other nonlinear control methodologies, including sliding mode control, backstepping control, and model predictive control (MPC), are also designed for controlling full-size helicopters, but only a few researches are conducted on sliding mode control and backstepping control [94–97].

MPC is an optimization based framework for learning a stabilizing control sequence which minimizes a specified cost function [98]. MPC has been successfully used in many fields, and its application to piloted rotorcrafts is increasing in the past five years.

Bogdanov et al. designed a neural-MPC trajectory tracking controller for a high-fidelity helicopter model [99]. He et al. developed a constraint reduction algorithm for rotorcraft MPC [100]. Ngo and Sultan studied ship landing control for a full-size helicopter model using MPC [101]. In 2016, Cooper et al. developed a partial authority Receding Horizon Optimal (RHO) controller to improve rotorcraft handling qualities in DVE [102]. The control law was evaluated on a 22 states EH-60L rotorcraft model. Simulation results verified the efficacy of RHO control strategy and validated its candidacy as a potential rotorcraft control architecture.

3.3. Intelligent Control. Developing a high-fidelity nonlinear helicopter model is usually time-consuming. The intelligent control methodology requires no assumptions about the internal structure of the plant and the reliance on exact model can be alleviated. Two dominant intelligent control methodologies are based on either fuzzy logic or artificial neural networks (ANN). Further illustrations are given in the following.

3.3.1. Fuzzy Control. Fuzzy control, also called rule-based control, adjusts its outputs on errors by applying fuzzy logic, linguistic variables and fuzzy inference to imitate the thinking process of human brains. The advantages of fuzzy control are as follows: (1) it can be built up from rules of logic that have physical meaning to the designer; (2) this control policy relaxes the requirements of obtaining an accurate and analytical model.

The work on designing fuzzy controller for helicopters dates back to the 1990s. Jiang et al. combined a fuzzy controller with a nonlinear model-based controller to track attitude commands for the AH-64 helicopter [103]. Numerical simulations indicated that the inversion error is canceled out and close-loop system stability is guaranteed. Mulgund and Zacharias developed a hybrid neural network/fuzzy logic control approach for detection and prevention of helicopter flight limit exceedances, and it was tested on the numerical model of the UH-1H helicopter [104]. Amaral and Crisóstomo proposed a neurofuzzy inference controller used to control a Sikorsky S-61 full-size helicopter in hover and forward flight [105]. Simulation results showed that fuzzy logic theory is applicable to control nonlinear systems with very promising results. A fuzzy switching controller, using fuzzy logic theory and H_{∞} control, was designed for the UH-60A helicopter to satisfy ADS-33D level 1 requirements in the full flight envelop [106, 107]. Yu designed a LQ optimal controller with T-S fuzzy gain-scheduler and applied this control strategy to the nonlinear BO-105 helicopter [108]. The turning rate regulation and forward velocity tracking response demonstrate the effectiveness of fuzzy gain-scheduling. Lower et al. developed a fuzzy regulator based on expert knowledge for hover control of a single-rotor helicopter, PZL Kania [109]. In 2015, Sunberg et al. studied the autorotation control using a new fuzzy logic-based controller [110]. This control law was validated on a high-fidelity model of full-size attack helicopter, and simulation results proved that safely decent landing was achieved successfully under the vast majority of flight conditions.

It is concluded that the application of a single fuzzy controller is not adequate to stabilize a helicopter because of its complex nonlinear dynamics and dramatic changes of flight conditions, which addresses the point that fuzzy logic controller should be augmented with model-based control, robust control, neural network control, or other control methods to enhance its nominal performance. Most of the FCS designs using fuzzy logic have been tested only through digital simulations, indicating that there still exists a big gap between fuzzy control theory and its practical implementation. Moreover, the fuzzy treatment of knowledge reduces the controller efficiency and results in degraded dynamic response.

3.3.2. ANN Control. ANN is a distributed parallel computational system comprised of interconnected neurons, capable of self-learning and self-adapting, for information processing. The computational requirements for executing trained ANN are generally small and can be readily implemented for real-time control. ANN can also be trained to perform a task without having to exactly model the underlying dynamics [111]. This is tempting for controlling complicated nonlinear systems whose characteristics are poorly understood.

The ANN control strategy has been applied to unmanned small-scale helicopters since 1990s [112–117], and several researches were conducted on AH-64 helicopter to evaluate the ANN control policy [84, 118–121]. In 1994, Calise et al. proposed a neural network controller based on feedback linearization approaches to compensate for inversion error and approximate dynamic changes induced by imprecise modeling [118]. Kim and Calise developed a training algorithm based on Lyapunov stability, which is used to train the NN online to approximate the inversion errors [84]. Leitner et al. designed an online adaptive NN control architecture with the adaptation gain parameter to improve command tracking performance in the presence of significant modeling errors for complicated maneuvers [119]. They also developed a robust trajectory tracking controller, enhanced with a simple two layer adaptive NN, to mitigate uncertainties and rapid varying nonlinear dynamics [120]. A model-free online learning control scheme, neural dynamic programming, was introduced in [121], and it was applied to stabilize a high-fidelity Apache helicopter in the presence of turbulence and gusts.

In 2001, the NASA-Ames conducted a research on designing a close-loop optimal NN controller (ONNC) to optimize rotorcraft aeromechanical behavior. A modern constrained optimization method is used to determine and update the constants in the NN plant model. Moreover, the ONNC was coded and loaded to digital processors [122]. Sahani and Horn designed a limit prediction algorithm based on neural networks, which provides envelope limiting avoidance through the AFCS by constraining the collective input. Simulations were conducted on a high-fidelity UH-60A model, indicating that this technique provides potential for implementing helicopter flight envelope protection system [123].

Previous researches indicate that ANN can act as excellent function interpolators but cannot be applied to regions where they have not been trained [111]. In other words, the controller performance largely depends on the training algorithm, which is still an open issue to scholars since the quantity of training data, effectiveness of training strategy, and computational efficiency have to be considered. Moreover, the ANN significantly increases the complexity of controller structure and proved to be challenging for practical implementation.

4. Conclusions

To sum up, various state-of-the-art control technologies assume an indispensable role in research and development for large-scale helicopters since they directly benefit the current and ongoing FCS designs. The design and application of the helicopter FCS was driven by various aspects, such as control theories, the complexity of controller architecture, simulation devices, and helicopter mission elements. However, only a few controller designs are applicable when considering engineering implementation for actual helicopters. In this survey, some important research issues are discovered. The further research is being conducted to incorporate the following topics.

(1) Parameter Design upon Command Model. Command models are required whether linear or nonlinear control policy is adopted. Engineers mostly design the command model manually and they mainly refer to the ADS-33E.

Further researches can be conducted on investigating the design criteria and tuning strategies to help engineers reduce their workload.

(2) Problems of Actuator Saturation during Automatic Flight Mode Transitions. Actuator saturation could occur during flight mode transitions, especially for highly maneuverable attack helicopters which produce flight transitions among various flight modes. However, only a few surveys are conducted on actuator saturation problems without considering the influence of multimode transitions. An efficient switching control logic that aims to eliminate actuator saturation for high-bandwidth helicopter FCS is worth investigating.

(3) Helicopter Dynamic Modeling. A high-fidelity nonlinear helicopter model gives a more accurate description on dynamic responses and input-output relations than a linear model, but it makes the controller design more challenging because of the rapid changing parameters and system nonlinearities. The LPV model is able to approximate the original system quite closely, but only the affine format of LPV model is appropriate for controller synthesis, which is not always the case for large-scale helicopter. Besides, the LPV model also increases the controller complexity within constraints. Thus, further investigations of simplified nonlinear modeling techniques would provide insights to controller design and evaluation.

(4) Gaps between Optimized Gains and Unexpected Simulation Results. Simulation studies present a problem that optimized control gains results in less than optimum performance in piloted simulations. It seems that, only through trial and error that could close this gap, different design criteria lead to various testing results. Detailed analysis and investigations should be conducted on this discrepancy to shorten the distance between simulation and application.

The future implementation of advanced control methodologies, such as model predictive control and intelligent control, is expected to provide potential efficacies as practical helicopter flight control strategies. Besides, newly developed ground-based simulation devices, with complex mechanical structures, are designed to perform more complicated maneuvers and evaluate more sophisticated control algorithms. Furthermore, with the development of avionics, FBL control systems, and sensors, novel in-flight simulations will permit more realistic representation of helicopter aerodynamic response and shed light on FCS design as well.

Competing Interests

The authors declare that there is no conflict of interests regarding the publication of this paper.

References

[1] K. H. Landis, J. M. Davis, C. Dabundo, and J. F. Keller, "Advanced flight control technology achievements at Boeing Helicopters," *International Journal of Control*, vol. 59, no. 1, pp. 263–290, 1994.

[2] C. R. Frost, W. S. Hindson, E. Moralez III, G. E. Tucker, and J. B. Dryfoos, "Design and testing of flight control laws on the rascal research helicopter," in *Proceedings of the Modeling and Simulation Technologies Conference and Exhibit (AIAA '02)*, Monterey, Calif, USA, August 2002.

[3] J. Harding, S. Moody, G. Jeram et al., "Development of modern control laws for the AH-64D in hover/low speed flight," in *Proceedings of the American Helicopter Society 62nd Annual Forum*, Phoenix, Ariz, USA, May 2006.

[4] W. S. Hindson, "Past applications and future potential of variable stability research helicopters," in *Proceedings of the Specialists Meeting on Helicopter Handling Qualities Sponsored by the NASA Ames Research Center and the American Helicopter Society and Held at NASA Ames Research Center*, vol. 2219, p. 209, Moffett Field, Calif, USA, April 1982.

[5] B. D. Kothmann and J. Armbrust, "RAH-66 Comanche core AFCS control law development: DEMVAL to EMD," in *Proceedings of the American Helicopter Society 58th Annual Forum*, vol. 1, pp. 523–546, Montreal, Canada, June 2002.

[6] I. Postlethwaite, A. Smerlas, D. J. Walker et al., "H∞ control of the NRC Bell 205 fly-by-wire helicopter," *Journal of the American Helicopter Society*, vol. 44, no. 4, pp. 276–284, 1999.

[7] A. J. Smerlas, D. J. Walker, I. Postlethwaite, M. E. Strange, J. Howitt, and A. W. Gubbels, "Evaluating H_∞ controllers on the NRC Bell 205 fly-by-wire helicopter," *Control Engineering Practice*, vol. 9, no. 1, pp. 1–10, 2001.

[8] E. Prempain and I. Postlethwaite, "Static H_∞ loop shaping control of a fly-by-wire helicopter," *Automatica*, vol. 41, no. 9, pp. 1517–1528, 2005.

[9] *NATO Helicopter NH90 TTH*, Revell, Bünde, Germany, 2007.

[10] E. Aiken, R. Jacobsen, M. Eshow, W. Hindson, and D. Doane, "Preliminary design features of the RASCAL—a NASA/Army rotorcraft in-flight simulator," in *Proceedings of the AIAA Flight Simulation Technologies Conference*, Hilton Head Island, SC, USA, August 1993.

[11] S. Klaes, "ATTAS and ACT/FHS system simulation for preflight software and hardware testing," in *Proceedings of the AIAA Modeling and Simulation Technologies Conference and Exhibit*, Monterey, Calif, USA, August 2002.

[12] C. Ockier and U. Butter, "ACT/FHS-an airborne rotorcraft simulator for technology development and research," in *Proceedings of the AIAA Modeling and Simulation Technologies Conference*, Denver, Colo, USA, August 2000.

[13] W. Grunhagen, G. Bouwer, H. J. Pausder et al., "Helicopter in-flight simulator ATTHeS—a multipurpose testbed and its utilization," *International Journal of Control*, vol. 59, no. 1, pp. 239–261, 1994.

[14] I. A. Raptis and K. P. Valavanis, *Linear and Nonlinear Control of Small-Scale Unmanned Helicopters*, Springer, London, UK, 2011.

[15] B. J. Baskett, "Aeronautical design standard performance specification handling qualities requirements for military rotorcraft," Tech. Rep. ADS-33E-PRF, Army Aviation and Missile Command, Redstone Arsenal, Ala, USA, 2000.

[16] R. T. N. Chen, P. D. Talbot, R. M. Gerdes et al., "A piloted simulator study on augmentation systems to improve helicopter flying qualities in terrain flight," Tech. Rep. NASA-TM-78571, 1979.

[17] J. V. Lebacqz, R. D. Forrest, and R. M. Gerdes, "A piloted simulator investigation of stability and control, display and crew-loading requirements for helicopter instrument approach. Part 1: technical discussion and results," Tech. Rep. NASA-TM-84258-PT-1, 1982.

[18] K. H. Landis and E. W. Aiken, "Simulator investigations of side–stick controller/stability and control augmentation systems for night nap–of–earth flight," *Journal of the American Helicopter Society*, vol. 29, no. 1, pp. 56–65, 1984.

[19] K. H. Landis and S. I. Glusman, *Development of ADOCS Controllers and Control Laws. Volume 3: Simulation Results and Recommendations*, NASA-CR-177339-VOL-3, NASA, 1985.

[20] B. Aponso, D. Tran, J. Schroeder et al., "Rotorcraft research at the NASA Vertical Motion Simulator," in *Proceedings of the American Helicopter Society 64th Annual Forum*, vol. 63, p. 2390, Montreal, Canada, April 2008.

[21] M. Whalley and J. Howitt, "Optimization of partial authority automatic flight control systems for hover/low-speed maneuvering in degraded visual environments," *Journal of the American Helicopter Society*, vol. 47, no. 2, pp. 79–89, 2002.

[22] J. G. Irwin, P. G. Einthoven, D. G. Miller et al., "ADS-33E predicted and assigned low-speed handling qualities of the CH-47F with digital AFCS," in *Proceedings of the American Helicopter Society 63rd Annual Forum*, vol. 63, p. 809, Virginia Beach, Va, USA, May 2007.

[23] T. Berger, M. B. Tischler, C. L. Blanken et al., "Improved handling qualities for the OH-58D Kiowa Warrior in the degraded visual environment," in *Proceedings of the 67th American Helicopter Society International Annual Forum*, pp. 1337–1357, Virginia Beach, Va, USA, May 2011.

[24] K. T. Christensen, K. G. Campbell, C. D. Griffith, C. M. Ivler, M. B. Tischler, and J. W. Harding, "Flight control development for the ARH-70 armed reconnaissance helicopter program," in *Proceedings of the American Helicopter Society International 63rd Annual Forum*, pp. 839–863, Virginia Beach, Va, USA, May 2007.

[25] K. B. Hilbert, "The design of a model-following control system for helicopters 84-1 941," in *Proceedings of the 17th Fluid Dynamics, Plasma Dynamics, and Lasers Conference*, Snowmass, Colo, USA, June 1984.

[26] V. Sahasrabudhe, E. Melkers, A. Faynberg et al., "Piloted evaluation of modernized limited authority control laws in the NASA-Ames Vertical Motion Simulator (VMS)," in *Proceedings of the American Helicopter Society 59th Annual Forum*, Phoenix, Ariz, USA, May 2003.

[27] M. B. Tischler, C. L. Blanken, K. K. Cheung, S. S. M. Swei, V. Sahasrabudhe, and A. Faynberg, "Modernized control laws for UH-60 BLACK HAWK optimization and flight-test results," *Journal of Guidance, Control, and Dynamics*, vol. 28, no. 5, pp. 964–978, 2005.

[28] B. T. Fujizawa, M. B. Tischler, C. L. Blanken et al., "UH-60 partial authority modernized control laws for improved handling qualities in the degraded visual environment," in *Proceedings of the American Helicopter Society 70th Annual Forum*, Montreal, Canada, May 2014.

[29] J. W. Harding, M. H. Mansur, M. B. Tischler, S. J. Moody, and R. C. McCann, "Optimization and piloted simulation results of the AH-64D modern control laws," in *Proceedings of the American Helicopter Society International 63rd Annual Forum*, pp. 883–896, Virginia Beach, Va, USA, May 2007.

[30] C. L. Blanken, M. B. Tischler, J. A. Lusardi et al., "Aeronautical design standard-33 (ADS-33)... past, present, and future," in *Proceedings of the Rotorcraft Handling Qualities Specialists' Meeting*, Huntsville, Ala, USA, February 2014.

[31] B. T. Fujizawa, M. B. Tisceler, and M. A. J. J. S. Minor, "Outer-loop development and DVE flight test assessment of a partial authority," in *Proceedings of the American Helicopter Society 72th Annual Forum*, West Palm Beach, Fla, USA, May 2016.

[32] R. W. Prouty and H. C. Curtiss Jr., "Helicopter control systems: a history," *Journal of Guidance, Control, and Dynamics*, vol. 26, no. 1, pp. 12–18, 2003.

[33] M. B. Tischler, J. D. Colbourne, M. R. Morel, D. J. Biezad, W. S. Levine, and V. Moldoveanu, "Conduit—a new multidisciphnary integration environment for flight control development," in *Proceedings of the Guidance, Navigation, and Control Conference*, pp. 1759–1781, New Orleans, La, USA, August 1997.

[34] B. Lawrence, T. Berger, M. B. Tischler et al., "Integrating flight dynamics & control analysis and simulation in rotorcraft conceptual design," in *Proceedings of the American Helicopter Society 72nd Annual Forum*, West Palm Beach, Fla, USA, May 2016.

[35] D. R. Catapang, M. B. Tischler, and D. J. Biezad, "Robust crossfeed design for hovering rotorcraft," *International Journal of Robust and Nonlinear Control*, vol. 4, no. 1, pp. 161–180, 1994.

[36] J. D. Colbourne, D. J. Biezad, C. R. Frost et al., "Control law design and optimization for rotorcraft handling qualities criteria using CONDUIT," in *Proceedings of the American Helicopter Society 55th Annual Forum*, Montreal, Canada, May 1999.

[37] G. J. Jeram, G. Baldwin, R. C. McCann, and J. K. McKay, "Handling qualities prototyping for improvements to the ah-64d across heterogeneous simulation environments," in *Proceedings of the 62nd American Helicopter Society International Annual Forum*, May 2006.

[38] D. Fusato and R. Celi, "Design sensitivity analysis for helicopter flight dynamic and aeromechanic stability," *Journal of Guidance, Control, and Dynamics*, vol. 26, no. 6, pp. 918–927, 2003.

[39] M. Oosterom, G. Schram, R. Babuška, and H. B. Verbruggen, "Automated procedure for gain scheduled flight control law design," in *Proceedings of the 18th Applied Aerodynamics Conference*, Denver, Colo, USA, August 2000.

[40] D. McFarlane and K. Glover, "A loop-shaping design procedure using H_∞ synthesis," *IEEE Transactions on Automatic Control*, vol. 37, no. 6, pp. 759–769, 1992.

[41] S. Mammar and G. Duc, "Loop shaping H_∞ design: application to the robust stabilization of a helicopter," *Control Engineering Practice*, vol. 2, no. 2, pp. 349–356, 1993.

[42] P. P. Khargonekar and M. A. Rotea, "Mixed H2/H∞ control: a convex optimization approach," *IEEE Transactions on Automatic Control*, vol. 36, no. 7, pp. 824–837, 1991.

[43] K. Zhou, K. Glover, B. Bodenheimer, and J. Doyle, "Mixed H2 and H∞ performance objectives. I. Robust performance analysis," *IEEE Transactions on Automatic Control*, vol. 39, no. 8, pp. 1564–1574, 1994.

[44] K. Zhou, K. Glover, B. Bodenheimer et al., "Mixed H_2 and H_∞ performance objectives. II. Optimal control," *IEEE Transactions on Automatic Control*, vol. 39, no. 8, pp. 1575–1587, 1994.

[45] E. Schoemig, M. Sznaier, and U.-L. Ly, "Mixed H2/H-∞ control of multimodel plants," *Journal of Guidance, Control, and Dynamics*, vol. 18, no. 3, pp. 525–531, 1995.

[46] K. B. Datta and A. RaiChaudhuri, "H_2/H_∞ Control of discrete singularly perturbed systems: the state feedback case," *Automatica. A Journal of IFAC, the International Federation of Automatic Control*, vol. 38, no. 10, pp. 1791–1797, 2002.

[47] Y. Fan, E. M. Cliff, F. H. Lutze, and M. R. Anderson, "Mixed H2/H∞ optimal control for an elastic aircraft," *Journal of Guidance, Control, and Dynamics*, vol. 19, no. 3, pp. 650–655, 1996.

[48] S.-P. Shue, M. E. Sawan, and K. Rokhsaz, "Mixed H2/H∞ method suitable for gain scheduled aircraft control," *Journal of Guidance, Control, and Dynamics*, vol. 20, no. 4, pp. 699–706, 1997.

[49] S.-P. Shue and R. K. Agarwal, "Design of automatic landing systems using mixed H2/H∞ control," *Journal of Guidance, Control, and Dynamics*, vol. 22, no. 1, pp. 103–114, 1999.

[50] M. Trentini and J. K. Pieper, "Mixed norm control of a helicopter," *Journal of Guidance, Control, and Dynamics*, vol. 24, no. 3, pp. 555–565, 2001.

[51] M. Lungu and R. Lungu, "Application of H_2/H_∞ technique to aircraft landing control," *Asian Journal of Control*, vol. 17, no. 6, pp. 2153–2164, 2015.

[52] W. Jiang, H. Wang, J. Lu, and Z. Xie, "HOSVD-based LPV modeling and mixed robust H_2/H_∞ control design for air-breathing hypersonic vehicle," *Journal of Systems Engineering and Electronics*, vol. 27, no. 1, Article ID 7424934, pp. 183–191, 2016.

[53] H.-N. Wu, S. Feng, Z.-Y. Liu, and L. Guo, "Disturbance observer based robust mixed H2/H∞ fuzzy tracking control for hypersonic vehicles," *Fuzzy Sets and Systems*, vol. 306, pp. 118–136, 2017.

[54] J. Rigsby, C. H. Hung, and J. F. Horn, "Design and evaluation of a robust feedback controller for helicopters handling externally slung loads," in *Proceedings of the American Helicopter Society 69th Annual Forum*, Phoenix, Ariz, USA, May 2013.

[55] J. Rigsby and J. F. Horn, "Piloted simulation evaluation of a robust feedback controller for helicopters handling externally slung loads," in *Proceedings of the American Helicopter Society 70th Annual Forum*, Montreal, Canada, May 2014.

[56] M. D. Pavel, P. Shanthakumaran, O. Stroosma et al., "Development of advanced flight control laws for the AH-64 apache helicopter-sketches from the work of TU Delft-Boeing project in SIMONA simulator," in *Proceedings of the American Helicopter Society 72th Annual Forum*, Palm Beach, Fla, USA, May 2016.

[57] P. V. M. Simplıcio, *Helicopter Nonlinear Flight Control Using Incremental Nonlinear Dynamic Inversion*, University of Technology, Delft, The Netherlands, 2011.

[58] H. B. Chen and S. G. Zhang, "Robust dynamic inversion flight control law design," in *Proceedings of the 2nd International Symposium on Systems and Control in Aerospace and Astronautics (ISSCAA '08)*, Beijing, China, December 2008.

[59] S. Sieberling, Q. P. Chu, and J. A. Mulder, "Robust flight control using incremental nonlinear dynamic inversion and angular acceleration prediction," *Journal of Guidance, Control, and Dynamics*, vol. 33, no. 6, pp. 1732–1742, 2010.

[60] P. B. Acquatella, W. Falkena, E.-J. van Kampen, and Q. P. Chu, "Robust nonlinear spacecraft attitude control using incremental nonlinear dynamic inversion," in *Proceedings of the AIAA Guidance, Navigation, and Control Conference*, Minneapolis, Minn, USA, August 2012.

[61] P. Simplício, M. D. Pavel, E. van Kampen, and Q. P. Chu, "An acceleration measurements-based approach for helicopter nonlinear flight control using incremental nonlinear dynamic inversion," *Control Engineering Practice*, vol. 21, no. 8, pp. 1065–1077, 2013.

[62] G. D. Francesco and M. Mattei, "Modeling and incremental nonlinear dynamic inversion control of a novel unmanned tiltrotor," *Journal of Aircraft*, vol. 53, no. 1, pp. 73–86, 2016.

[63] E. J. Smeur, Q. Chu, and G. C. H. E. de Croon, "Adaptive incremental nonlinear dynamic inversion for attitude control of micro aerial vehicles," in *Proceedings of the AIAA Guidance, Navigation, and Control Conference*, p. 1390, San Diego, Calif, USA, January 2016.

[64] R. M. McKillip Jr., "Kinematic observers for rotor vibration control," in *Proceedings of the American Helicopter Society 42nd Annual Forum*, Washington, DC, USA, June 1986.

[65] E. Lavretsky and K. A. Wise, *Robust and adaptive control—with aerospace applications*, Advanced Textbooks in Control and Signal Processing, Springer, London, UK, 2013.

[66] J. J. Gribble, "Linear quadratic Gaussian/loop transfer recovery design for a helicopter in low-speed flight," *Journal of Guidance, Control, and Dynamics*, vol. 16, no. 4, pp. 754–761, 1993.

[67] E. Hendricks and J. Holst, "An extended kalman filter applied to helicopter state and parameter estimation," in *Proceedings of the 13th Atmospheric Flight Mechanics Conference*, Williamsburg, Va, USA.

[68] M. Whorton, "UH-60 flight data replay and refly system state estimator analysis," in *Proceedings of the 28th Aerospace Science Meeting*, Reno, Nev, USA, January 1990.

[69] M. Abraham and M. Costello, "In-flight estimation of helicopter gross weight and mass center location," *Journal of Aircraft*, vol. 46, no. 3, pp. 1042–1049, 2009.

[70] S. Srinathkumar, "Design of helicopter stability augmentation system using eigenstructure assignment and minimal order functional observers," in *Proceedings of the American Helicopter Society 61st Annual Forum*, Grapevine, Tex, USA, May 2005.

[71] L. vS. Hager, K. R. Uren, and G. van Schoor, "Series-parallel approach to on-line observer based neural control of a helicopter system," *IFAC Proceedings Volumes*, vol. 47, no. 3, pp. 2545–2550, 2014.

[72] S. Greiser, "Disturbance observer-based control to suppress air resonance for the EC135 ACT/FHS research helicopter," in *Proceedings of the AIAA Guidance, Navigation, and Control Conference*, Kissimmee, Fla, USA, January 2015.

[73] L. E. Ghaoui, A. Arrier, and A. E. Ryson, "Linear quadratic minimax controllers," *Journal of Guidance, Control, and Dynamics*, vol. 15, no. 4, pp. 953–961, 1992.

[74] J. K. Pieper, S. Baillie, and K. R. Goheen, "Linear-quadratic optimal model-following control of a helicopter in hover," in *Proceedings of the American Control Conference*, vol. 3, pp. 3470–3474, Baltimore, Md, USA, June 1994.

[75] W. L. Garrard, E. Low, and S. Prouty, "Design of attitude and rate command systems for helicopters using eigenstructure assignment," *Journal of Guidance, Control, and Dynamics*, vol. 12, no. 6, pp. 783–791, 1989.

[76] E. Low and W. L. Garrard, "Design of flight control systems to meet rotorcraft handling qualities specifications," *Journal of Guidance, Control, and Dynamics*, vol. 16, no. 1, pp. 69–78, 1993.

[77] Y. Aslan, D. Beauvois, and J. A. Rossiter, "Eigenstructure approximations and their use for a commutative controller strategy: application to a helicopter," *Control Engineering Practice*, vol. 1, no. 2, pp. 357–363, 1993.

[78] N. A. Sahani and J. F. Horn, "Adaptive model inversion control of a helicopter with structural load limiting," *Journal of Guidance, Control, and Dynamics*, vol. 29, no. 2, pp. 411–420, 2006.

[79] T. G. Amaral, M. M. Crisóstomo, and V. F. Pires, "Helicopter motion control using adaptive neuro-fuzzy inference controller," in *Proceedings of the 28th Annual Conference of the IEEE Industrial Electronics Society*, pp. 2090–2095, IEEE, Sevilla, Spain, November 2002.

[80] T. Ertuğrul, M. A. Adli, and M. U. Salamci, "Model reference adaptive control design for helicopters using gain scheduled reference models," in *Proceedings of the 17th IEEE International Carpathian Control Conference (ICCC '16)*, pp. 182–187, Tatranska Lomnica, Slovakia, May 2016.

[81] Z. Jia, C. Shenggong, and S. Gongzhang, "Novel control scheme for helicopter flight: fuzzy immune adaptive model inversion control," *IFAC Proceedings Volumes*, vol. 41, no. 2, pp. 15046–15051, 2008, Proceedings of the 17th IFAC World Congress, Seoul, South Korea, July 2008.

[82] M. Bichlmeier, F. Holzapfel, E. Xargay et al., "L_1 adaptive augmentation of a helicopter baseline controller," in *Proceedings of the AIAA Guidance, Navigation and Control Conference*, Boston, Mass, USA, August 2013.

[83] R. T. Rysdyk and A. J. Calise, "Adaptive model inversion flight control for tilt-rotor aircraft," *Journal of Guidance, Control, and Dynamics*, vol. 22, no. 3, pp. 402–407, 1999.

[84] B. S. Kim and A. J. Calise, "Nonlinear flight control using neural networks," *Journal of Guidance, Control, and Dynamics*, vol. 20, no. 1, pp. 26–33, 1997.

[85] J. Horn, A. J. Calise, and J. V. R. Prasad, "Flight envelope limit detection and avoidance for rotorcraft," *Journal of the American Helicopter Society*, vol. 47, no. 4, pp. 253–262, 2002.

[86] J.-Y. Shin, G. J. Balas, and M. A. Kaya, "Blending methodology of linear parameter varying control synthesis of F-16 aircraft system," *Journal of Guidance, Control, and Dynamics*, vol. 25, no. 6, pp. 1040–1048, 2002.

[87] B. Lu and F. Wu, "Probabilistic robust linear parameter-varying control of an F-16 aircraft," *Journal of Guidance, Control, and Dynamics*, vol. 29, no. 6, pp. 1454–1460, 2006.

[88] B. Lu, F. Wu, and S. Kim, "Switching LPV control of an F-16 aircraft via controller state reset," *IEEE Transactions on Control Systems Technology*, vol. 14, no. 2, pp. 267–277, 2006.

[89] C. Westermayer, A. Schirrer, M. Hemedi et al., "Linear parameter-varying control of a large blended wing body flexible aircraft," *IFAC Proceedings Volumes*, vol. 43, no. 15, pp. 19–24, 2010.

[90] L. Wu, X. Yang, and F. Li, "Nonfragile output tracking control of hypersonic air-breathing vehicles with an LPV model," *IEEE/ASME Transactions on Mechatronics*, vol. 18, no. 4, pp. 1280–1288, 2013.

[91] W. Jiang, C. Dong, and Q. Wang, "Smooth switching linear parameter-varying control for hypersonic vehicles via a parameter set automatic partition method," *IET Control Theory & Applications*, vol. 9, no. 16, pp. 2377–2386, 2015.

[92] A. Fujimori and K. Miura, "Autonomous flight control system for longitudinal motion of a helicopter," in *Proceedings of the SICE(Society of Instrument and Control Engineers)Annual Conference (SICE '07)*, pp. 2258–2263, IEEE, Takamatsu, Japan, September 2007.

[93] C. Malpica, C. R. Theodore, B. Lawrence et al., "Handling qualities of large rotorcraft in hover and low speed," Tech. Rep. NASA-TP-2015-216656, 2015.

[94] S. K. Spurgeon, C. Edwards, and N. P. Foster, "Robust model reference control using a sliding mode controller/observer scheme with application to a helicopter problem," in *Proceedings of the IEEE International Workshop on Variable Structure Systems*, pp. 36–41, Tokyo, Japan, December 1996.

[95] D. J. McGeoch, E. W. McGookin, and S. S. Houston, "Sliding mode implementation of an attitude command flight control system for a helicopter in hover," *Acta Polytechnica*, vol. 45, no. 4, pp. 88–95, 2005.

[96] F. Chen, R. Jiang, C. Wen, and R. Su, "Self-repairing control of a helicopter with input time delay via adaptive global sliding mode control and quantum logic," *Information Sciences*, vol. 316, pp. 123–131, 2015.

[97] I. A. Raptis, K. P. Valavanis, and W. A. Moreno, "A novel nonlinear backstepping controller design for helicopters using the rotation matrix," *IEEE Transactions on Control Systems Technology*, vol. 19, no. 2, pp. 465–473, 2011.

[98] E. S. Meadows and J. B. Rawlings, *Model Predictive Control*, Prentice-Hall, Upper Saddle River, NJ, USA, 1997.

[99] A. A. Bogdanov, E. A. Wan, M. Carlsson, Y. Zhang, R. Kieburtz, and A. Baptista, "Model predictive neural control of a high-fidelity helicopter model," in *Proceedings of the AIAA Guidance, Navigation, and Control Conference and Exhibit*, Québec, Canada, August 2001.

[100] M. Y. He, M. Kiemb, A. L. Tits, A. Greenfield, and V. Sahasrabudhe, "Constraint-reduced interior-point optimization for model predictive rotorcraft control," in *Proceedings of the American Control Conference (ACC '10)*, July 2010.

[101] T. D. Ngo and C. Sultan, "Nonlinear helicopter and ship models for predictive control of ship landing operations," in *Proceedings of the AIAA Guidance, Navigation, and Control Conference*, pp. 2014–1298, National Harbor, Md, USA, January 2014.

[102] J. Cooper, P. Thompson, C. Schulze et al., "Robust partial authority model predictive control in degraded visual environments," in *Proceedings of the 72nd Annual Forum of the American Helicopter Society*, West Palm Beach, Fla, USA, May 2016.

[103] T. Y. Jiang, J. V. R. Prasad, and A. J. Calise, "Adaptive fuzzy logic flight controller for rotorcraft," in *Proceedings of the Guidance, Navigation, and Control Conference and Exhibit*, San Diego, Calif, USA, July 1996.

[104] S. S. Mulgund and G. L. Zacharias, "A hybrid neural network-fuzzy logic limit protection system for rotorcraft," in *Proceedings of the AIAA Guidance, Navigation, and Control Conference*, San Diego, Calif, USA, July 1996.

[105] T. G. B. Amaral and M. M. Crisóstomo, "Automatic helicopter motion control using fuzzy logic," in *Proceedings of the 10th IEEE International Conference on Fuzzy Systems*, pp. 860–863, Victoria, Australia, December 2001.

[106] J.-Y. Dai and C. Tan, "Helicopter flight control and simulation based on fuzzy smooth switching," *Journal of System Simulation*, vol. 23, no. 6, pp. 1273–1277, 2011.

[107] J. Dai, C. Tan, J. Ying, and G. Wu, "Fuzzy multi-model switching H-infinity control for helicopters in a full envelope," *Circuits, Systems, and Signal Processing*, vol. 32, no. 5, pp. 2185–2197, 2013.

[108] G.-R. Yu, "Nonlinear optimal control of helicopter using fuzzy gain scheduling," in *Proceedings of the IEEE International Conference on Systems, Man and Cybernetics*, pp. 3313–3318, Waikoloa, Hawaii, USA, October 2005.

[109] M. Lower, B. Szlachetko, and D. Krol, "Fuzzy flight control system for helicopter intelligence in hover," in *Proceedings of the 5th International Conference on Intelligent Systems Design and Applications (ISDA '05)*, pp. 370–374, Warsaw, Poland, September 2005.

[110] Z. N. Sunberg, N. R. Miller, and J. D. Rogers, "A Real-time expert control system for helicopter autorotation," *Journal of the American Helicopter Society*, vol. 60, no. 2, pp. 1–15, 2015.

[111] M. A. Garratt, *Biologically Inspired Vision and Control for an Autonomous Flying Vehicle*, The Research School of Biological Sciences, The Australian National University, Canberra, Australia, 2007.

[112] J. V. R. Prasad, A. J. Calise, Y. Pei, and J. E. Corban, "Adaptive nonlinear controller synthesis and flight test evaluation: on an unmanned helicopter," in *Proceedings of the IEEE International Conference on Control Applications*, pp. 137–142, Kohala Coast, Hawaii, USA, August 1999.

[113] K. P. Tee, S. S. Ge, and F. E. H. Tay, "Adaptive neural network control for helicopters in vertical flight," *IEEE Transactions on Control Systems Technology*, vol. 16, no. 4, pp. 753–762, 2008.

[114] C.-T. Lee and C.-C. Tsai, "Nonlinear adaptive aggressive control using recurrent neural networks for a small scale helicopter," *Mechatronics*, vol. 20, no. 4, pp. 474–484, 2010.

[115] M. Chen, S. S. Ge, and B. Ren, "Robust attitude control of helicopters with actuator dynamics using neural networks," *IET Control Theory & Applications*, vol. 4, no. 12, pp. 2837–2854, 2010.

[116] K. Dalamagkidis, K. P. Valavanis, and L. A. Piegl, "Nonlinear model predictive control with neural network optimization for autonomous autorotation of small unmanned helicopters," *IEEE Transactions on Control Systems Technology*, vol. 19, no. 4, pp. 818–831, 2011.

[117] D. Nodland, H. Zargarzadeh, and S. Jagannathan, "Neural network-based optimal adaptive output feedback control of a helicopter uav," *IEEE Transactions on Neural Networks and Learning Systems*, vol. 24, no. 7, pp. 1061–1073, 2013.

[118] A. J. Calise, B. S. Kim, J. Leitner, and J. V. R. Prasad, "Helicopter adaptive flight control using neural networks," in *Proceedings of the 33rd IEEE Conference on Decision and Control*, pp. 3336–3341, Lake Buena Vista, Fla, USA, December 1994.

[119] J. Leitner, A. Calise, and J. V. R. Prasad, "Analysis of adaptive neural networks for helicopter flight control," *Journal of Guidance, Control, and Dynamics*, vol. 20, no. 5, pp. 972–979, 1997.

[120] J. Leitner, A. Calise, and J. V. R. Prasad, "Full authority helicopter adaptive neuro-controller," in *Proceedings of the IEEE Aerospace Conference*, pp. 117–126, IEEE, Aspen, Colo, USA, March 1998.

[121] R. Enns and J. Si, "Apache helicopter stabilization using neural dynamic programming," *Journal of Guidance, Control, and Dynamics*, vol. 25, no. 1, pp. 19–25, 2002.

[122] J. A. Leyland, *A Closed-Loop Optimal Neural-Network Controller to Optimize Rotorcraft Aeromechanical Behaviour. Volume 1; Theory and Methodology*, NASA/TM-2001-209622, NASA, 2001.

[123] N. A. Sahani and J. F. Horn, "Neural network based algorithms for comprehensive collective axis limit avoidance on rotorcraft," *Journal of Aerospace Computing, Information and Communication*, vol. 1, no. 11, pp. 432–451, 2004.

16

Analysis of Global Sensitivity of Landing Variables on Landing Loads and Extreme Values of the Loads in Carrier-based Aircrafts

Jin Zhou, Jianjiang Zeng, Jichang Chen, and Mingbo Tong

College of Aerospace Engineering, Nanjing University of Aeronautics & Astronautics, Nanjing, China

Correspondence should be addressed to Mingbo Tong; tongw@nuaa.edu.cn

Academic Editor: Kenneth M. Sobel

When a carrier-based aircraft is in arrested landing on deck, the impact loads on landing gears and airframe are closely related to landing states. The distribution and extreme values of the landing loads obtained during life-cycle analysis provide an important basis for buffering parameter design and fatigue design. In this paper, the effect of the multivariate distribution was studied based on military standards and guides. By establishment of a virtual prototype, the extended Fourier amplitude sensitivity test (EFAST) method is applied on sensitivity analysis of landing variables. The results show that sinking speed and rolling angle are the main influencing factors on the landing gear's course load and vertical load; sinking speed, rolling angle, and yawing angle are the main influencing factors on the landing gear's lateral load; and sinking speed is the main influencing factor on the barycenter overload. The extreme values of loads show that the typical condition design in the structural strength analysis is safe. The maximum difference value of the vertical load of the main landing gear is 12.0%. This research may provide some reference for structure design of landing gears and compilation of load spectrum for carrier-based aircrafts.

1. Introduction

Carrier-based aircraft is the most important method for aircraft carrier strike group to control sea supremacy, and it is also an indispensable power for modern navies. Due to limited area in deck landing zone and the demand for bolting and go-around, carrier-based aircraft usually lands on deck via impact method under high sinking speed and high engaging speed along a fixed glide path angle [1]. The impact load, braking load of arresting cable [2, 3], and other loads at the moment when the aircraft touches deck put forward higher requirements for design and analysis of landing gears and airframe structure, especially for the structures closely related to landing [4].

At the primary design stage of an aircraft landing gear, the landing loads and the most severe landing conditions of the landing gear under different landing conditions should be determined according to design outline, which will be adopted in the parameter design of landing gear buffers and structural strength analysis. At the detailed design stage, optimization design will be carried out to balance performance and structure of the landing gear according to the relation between landing conditions and landing loads. Finally, the maneuvering envelope will be determined and the design of service load spectrum will be compiled for the assessment of fatigue life [5, 6].

At present, the research literatures and reports on definition of arrested deck landing conditions are mainly military standards and guides [7, 8]. MIL-A-8863C (AS) provides the landing variables, their distribution form, and empirical formula of mean value and standard deviation that should be considered in deck landing. Micklos [9] provided a measurement report on the landing variables of all kinds of carrier-based aircrafts landed on the Enterprise Aircraft Carrier in both day and night. The measurement of landing variables and loads is a costly and tedious task, thus a large amount of researches have been carried out on the simulation and analysis of landing dynamics to analyze different landing conditions and loads in recent years. Zhang [10] introduced deck motion into a dynamics model through the wheel-deck coordinate system and simulated three landing situations to verify the model. Mikhaluk [11] and Lihua et al. [12] used finite element model to analyze forces on cables under different initial velocities and masses. Sati et al. [13] built a detailed aircraft arresting system using bond graph approach and studied the landing performance at different engagement speeds and masses.

The studies mentioned above give a simple trend of the relationship between landing variables and loads under a few landing conditions. However, the conditions empirically selected are unrepresentative. During the compilation of design service load spectrum, representative values of typical landing conditions can be obtained by discretizing the interval of landing parameters through sensitivity analysis. Incorrect structural load analysis may lead to excessive design, which would induce large weight and high cost at the primary stage of structure design. A further determination of the relationship between landing variables as well as the extreme value of the loads is required. Sebastian [14] analyzed the condition of free flight engagement (FFE), and the results show that the most severe loading condition of the landing gear takes place under the condition of free flight engagement at small sinking speed. Chester [15] researched the response of the main and nose gears via simulation of landing impact considering pitching and heaving degrees of freedom of the aircraft motion. The simulation indicates that the maximum vertical loads of main gears are almost linearly dependent on the sinking speed, while the response of nose gear is very sensitive to the initial values of pitch angle and pitching inertia. Yunwen et al. [16] explored the effect of different landing variables on sinking velocity based on the landing data measured from an E-2C. The results show that aircraft path angle and deck pitch angle are highly correlated with sinking velocity. Although they have studied part of the relevance between landing variables and loads, there is no direct conclusion on method accurate enough for determination of the representative values that can reflect typical landing conditions. Therefore, sensitivity and extreme value analysis need to be studied and it provides an important basis for the parameter design of landing gear buffer, structural optimization design [17], and fatigue analysis. However, theoretical basis and analysis method still need to be further investigated.

The extended Fourier amplitude sensitive test (EFAST) analysis method [18] is adopted in this paper to study the coupling and sensitivity between landing variables and landing loads. The method has been well applied in the field of hydrology [19], physical model [20], and so on. Based on military standards and guides, the effects of the multivariate distribution on the probability density function of single landing variable were analyzed and the distribution of single landing variable was fitted. By using the simplified landing virtual prototype and analysis of a large amount of landing conditions with the multicondition automatic simulation technology, the first-order and global sensitivity coefficients of landing variables on landing loads were calculated quantitatively, and the extreme value conditions and the frequency curves of the landing load were obtained.

2. Studies on Landing Variables in Military Standards

Many factors would affect the carrier-based aircrafts during the arrested deck landing. The landing variables and their relationships are mainly stipulated by related military standards and specifications.

2.1. Landing Variables and Multivariate Distribution. The military standard MIL-A-8863C (AS) stipulates the landing variables and their distributions. The condition of the landing variables needs to satisfy the demands of the multivariate distributions. The joint probability of the eight landing variables P_T is calculated by

$$\begin{aligned} P_T = & P\left(V_{TD}{}^{>}/_{<}V_{TDi}\right)P\left(V_E{}^{>}/_{<}V_{Ei}\right)P\left(V_V{}^{>}/_{<}V_{Vi}\right) \\ & \cdot P\left(\theta_P{}^{>}/_{<}\theta_{Pi}\right)P\left(\theta_R{}^{>}/_{<}\theta_{Ri}\right)P\left(\dot{\theta}_R{}^{>}/_{<}\dot{\theta}_{Ri}\right) \\ & \cdot P\left(\theta_Y{}^{>}/_{<}\theta_{Yi}\right)P\left(d^{>}/_{<}d_i\right), \end{aligned} \tag{1}$$

where V_{TD} is the approaching speed, V_E is the engaging speed, V_V is the sinking speed, θ_P is the aircraft pitching angle, θ_R is the aircraft rolling angle, $\dot{\theta}_R$ is the aircraft rolling rate, θ_Y is the aircraft yawing angle, and d is the off-center arresting distance. The subscript i denotes the initial value of each landing variable. The symbol $^{>}/_{<}$ is chosen to be either > or < according to the initial value of the variable. When the initial value of any landing variable is greater than the average one, the symbol > is chosen and it represents that the cumulative probability of that landing variable is greater than the initial value. Conversely, when the symbol < is chosen, it represents that the cumulative probability of that landing variable is less than the initial value.

The evaluation of the landing variables is determined by the extreme value of the cumulative probability P_0, namely,

$$P_0 = P(x > x_{\max i}) = P(x < x_{\min i}), \tag{2}$$

where $x_{\max i}$ and $x_{\min i}$ are the maximum and minimum value of the variables. In this paper, only the touch-and-go and arrested landing conditions are considered and the landing variables follow a normal distribution. The cumulative probability $P_0 = 1 \times 10^{-3}$, and the multivariate joint probability $P_T = 7.8125 \times 10^{-6}$.

2.2. The Effects of the Multivariate Distribution on the Landing Variables. The relationships among landing variables are according to the multivariate distribution. The product of the joint probability of the landing variables does not reflect the effects on the distribution of the single landing variable directly. Therefore, the distribution of landing variables is investigated according to the multivariate distribution.

Take the approaching speed V_{TD} as an example. There exists a relationship:

$$P\left(V_{TD}{}^{>}/_{<}V_{TDi}\right) \leq P(\mu(V_{TD})) = 0.5, \tag{3}$$

where $\mu(V_{TD})$ is the average of the carrier engaging speeds. Take V_{TD} and V_E as an example. The extreme value of the joint probability of any two landing variables is given by inserting (3) into (1).

$$\begin{aligned} P_T' &= P\left(V_{TD} {}^{>}/_{<} V_{TDi}\right) P\left(V_E {}^{>}/_{<} V_{Ei}\right) \\ &\geq \frac{P_T}{0.5 \cdot 0.5 \cdot 0.5 \cdot 0.5 \cdot 0.5 \cdot 0.5} \\ &= \frac{P_T}{0.015625} = 0.0005. \end{aligned} \tag{4}$$

The distribution of the landing variable is mapped to a standard normal distribution in order to unify the range of the variables. The mutual effects among the value ranges of the landing variables are determined by the multivariate distribution [21]. In Figure 1, there are six equiprobable design envelopes of any two landing variables. The probability of each equiprobable envelope is shown in the figure.

The equiprobable envelope 6 represents the condition that the joint probability of two landing variables reaches the minimum, the value of which is $P_T' = 0.0005$. The equiprobable envelope 1 degenerates to a point which represents the condition that the joint probability of two landing variables reaches the maximum.

The evaluation of the eight landing variables of four groups on the equiprobable envelopes must ensure that the sum of their sequence number is nine, satisfying the demand of the joint probability. For example, when one group of variables is the equiprobable envelope 5, then other three groups must be the equiprobable envelope 1, 1 and 2.

It is difficult to make analytical calculations on the effects of the multivariate distribution on the single landing variable. Monte Carlo method is adopted to sample simulation for the complex probability constraints. Before sampling, the distribution of each variable should be mapped to the standard normal distribution. 10^5 simulate samples are extracted according to the distribution of each landing variable. The simulating sample screening is according to the multivariate distribution determined by (1). Figures 2 and 3 show the Monte Carlo sampling condition and the condition of applying multivariate distribution.

From the screening results, it can be seen that only 29,194 out of 10^5 samples remain after adopting multivariate distribution and the 71% of the samples cannot satisfy the requirements of the joint probability. By comparing Figure 2 with Figure 3, it can be known that the multivariate distribution deletes lots of samples that landing variables approaching the extreme value and keeps the samples that landing variables closing to the average. In Figure 3, the equiprobable envelope 6 is the envelope curve of the minimum value of the multivariate distribution shown in Figure 1.

The normal test is carried out by Kolmogorov-Smirnov method. $D = 0.005$, $P = 0.372 > 0.05$, and the landing variables after screening obey the normal distribution. The comparisons between distributions of landing variables before and after screening as well as the fitting conditions are shown in Figure 4. It is seen that after adopting the multivariate distribution, the probability density function of the single landing variable changes a lot.

From the analysis above, it is seen that the requirement of the multivariate distribution restricts the optional values

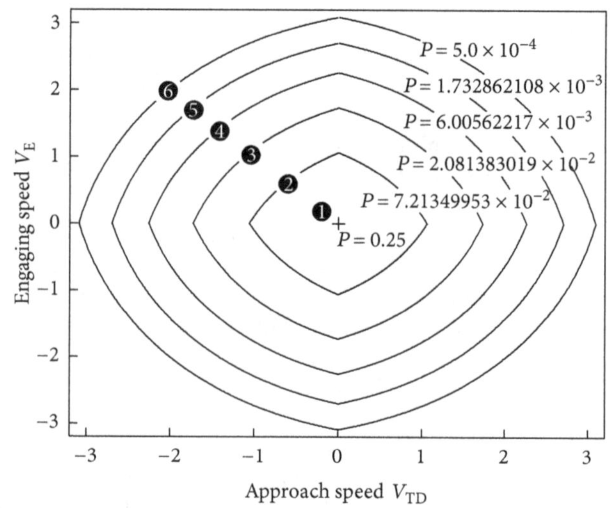

FIGURE 1: Equiprobable design envelope graph.

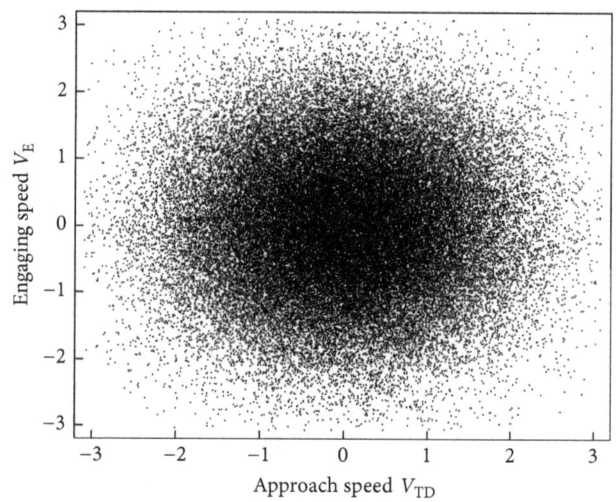

FIGURE 2: Multivariate Monte Carlo scatter diagram.

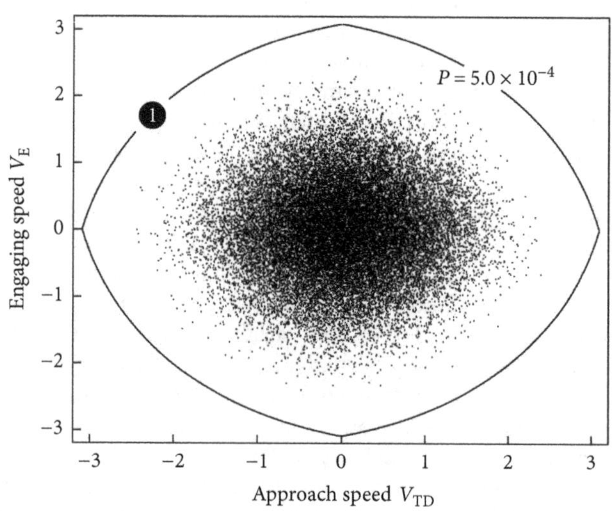

FIGURE 3: Multivariate distribution screening scatter diagram.

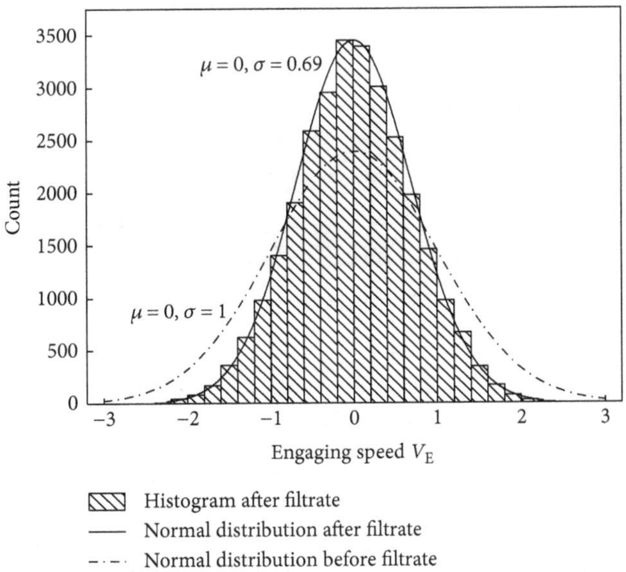

FIGURE 4: Comparisons of sample distributions before and after screening.

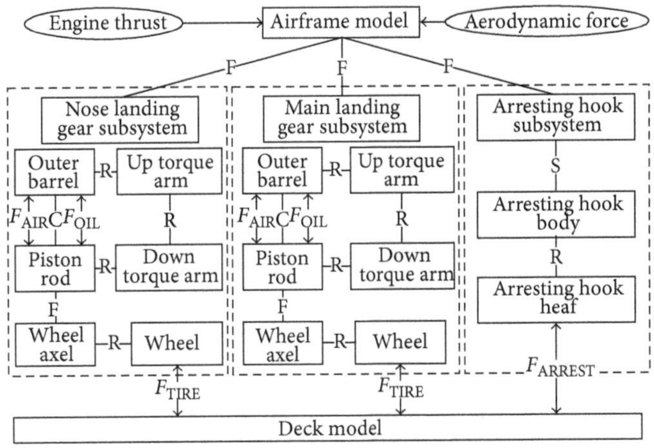

FIGURE 5: Landing virtual prototype constraints and loads. In the figure: R: swirl joint; C: cylindrical joint; S: spherical joint; F: fixed joint; F_{AIR}: air spring force; F_{OIL}: oil damping force; F_{TIRE}: tire force; F_{ARREST}: arresting force; omit the gravity.

among the variables and has a large effect on the standard deviation of the single landing variable. The results of the sensitivity analysis are closely related to the probability density function of the input variables. The adoption of the distribution of the landing variables before their multivariate distribution would overestimate the probability that landing variables reach the maximum or minimum values, which may cause the errors of the global sensitivity analysis of the landing load. In this section, based on the requirements of the multivariate distribution, the equiprobable design envelopes of the landing variables are obtained. The distribution of the single landing variable for global sensitivity analysis is studied by Monte Carlo sampling method.

3. The Landing Virtual Prototype of the Carrier-Based Aircraft and the Multicondition Automatic Simulation

The attitudes of the aircrafts differ a lot at the moment of deck landing. Under the effects of the deck reacting force, the arresting force of the arresting hook, the aerodynamic force, and other forces, the load-time history of the landing gear is rather complicated during the landing process. The establishment of the accurate landing simulation model of the aircraft is the basis of the sensitivity analysis of the landing load. The accuracy of the sensitivity analysis is related with the sample quantity. In order to improve the reliability of the sensitivity analysis, it is necessary to consider the computation efficiency during simulating different landing conditions. The constraints and loads of the aircrafts during landing are analyzed, and the landing virtual prototype of the carrier-based aircraft based on LMS Virtual.Lab is established. For the computation efficiency problem of the massive landing conditions, the secondary development technology is applied to realize the multicondition automatic simulation and data processing of the landing conditions.

3.1. Hypothesis of the Aircraft Virtual Prototype and Model Establishment. The aircraft virtual prototype consists of nose landing gear subsystem, main landing gear subsystem, arresting hook system, airframe model, and deck model. The external forces on the aircraft during landing are body load which includes aircraft gravity and engine thrust, aerodynamic force, acting force between tires, and deck and arresting force.

The virtual prototype has the following reasonable assumptions:

(1) Ignore the motion of the aircraft carrier and off-center landing.

(2) Assume that all components of the subsystems are rigid body, and the mass concentrates on the centroid.

(3) Ignore the effects of yawing landing on the arresting force curve.

According to the real motion of each component of the subsystems during the aircraft landing, the relating motion pairs such as swirl joint, cylinder joint, and spherical joint in subsystems are established. The constraints and load conditions of the subsystems are shown in Figure 5.

The initial condition of the virtual prototype is the moment that the aircraft lands on the deck, and the landing gears just touch the deck. During landing, the engine thrust maintains the maximum (twin engine). Arresting force-distance curve refers to military standards [8]. MK7-3 arresting machine is selected. The curve is obtained by weight interpolation and velocity compensation, shown in Figure 6.

The virtual prototype is established according to E-2 airborne early warning aircraft, shown in Figure 7.

3.2. Landing Multicondition Automatic Simulation. In order to improve the computation efficiency of the massive landing conditions, the automation technology is applied to realize the multicondition automatic simulation and data processing of the landing conditions. The auto-set, renewal, and analysis of the virtual prototype parameters are carried out by LMS

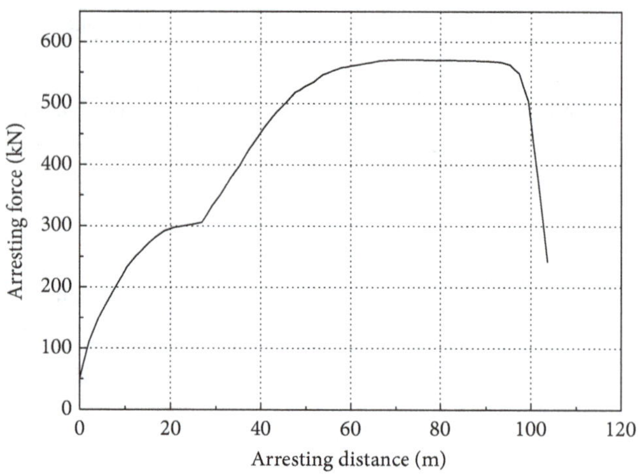

FIGURE 6: The arresting force-distance curve.

secondary development to realize the multicondition automatic simulation of the virtual prototype.

The virtual prototype of LMS Virtual.Lab motion can be divided into the product model and analysis model by function. The secondary development of the virtual prototype computes the initial landing attitude and location of the aircraft. The flow chart of the multicondition automatic simulation of the virtual prototype is shown in Figure 8.

FIGURE 7: Carrier-based aircraft landing virtual prototype.

The initial landing attitude of the carrier-based aircraft is determined by the multivariable pitching angle θ_P, rolling angle θ_R, and yawing angle θ_Y. The attitude transition matrix from airframe coordinate system to ground coordinate system is calculated by Tait-Bryan angles [22]. The transition matrix is used to update the aircraft attitude of the product model as follows:

$$L_{bg} = \begin{pmatrix} \cos\theta\cos\psi & \cos\theta\sin\psi & -\sin\theta \\ \sin\phi\sin\theta\cos\psi - \cos\phi\sin\psi & \sin\phi\sin\theta\sin\psi + \cos\phi\cos\psi & \sin\phi\cos\theta \\ \cos\phi\sin\theta\cos\psi + \sin\phi\sin\psi & \cos\phi\sin\theta\sin\psi - \sin\phi\cos\psi & \cos\phi\cos\theta \end{pmatrix}, \quad (5)$$

where the pitching angle $\theta = \theta_P$, rolling angle $\varphi = \theta_R$, and yawing angle $\psi = \theta_Y$.

When the aircraft attitude is determined, the aircraft location can be determined just with the constraints between the aircraft and the deck. When the rolling angle θ_R is positive, the aircraft tilts to the right and the right main landing gear just touches the deck, otherwise the opposite. The initial condition of the carrier-based aircraft includes the carrier engaging speed V_E, the sinking speed V_V, and the aircraft rolling rate $\dot{\theta}_R$. These variables can be set by analyzing the model. In the eight landing variables stipulated by military standards and guides, V_{TD} is not the variable at the moment of landing, so it cannot be considered in the virtual prototype. Besides, off-center condition is not taken into consideration. Therefore, both the ground critical velocity V_{TD} and off-center arresting distance d are simplified by using their average ones.

The sinking speed of the carrier-based aircraft is large during landing, and the landing gear bears rather large impact load. Referring to the analysis of the land-based landing load of the aircraft [23], the three directional loads of the landing gear and barycenter overload are selected as landing load to carry out the study. The course load and vertical load of the landing gear act on the center of the tire axle, while the lateral load acts on the contact point between the tires and the ground. The bounce landing is not permitted, so the maximum values of the three directional load and barycenter overload are selected as landing loads for analysis. After the BDF solver finishing solving the landing process, the extreme values of the three directional load and barycenter overload under this condition are output by calling the analysis model.

4. EFAST Global Sensitivity Analysis Method

During the aircraft landing, there are many landing variables and landing loads. The local sensitivity analysis method is limited by the variable selection and model, so the analytical results are usually not comprehensive or accurate enough. The variable range of the global sensitivity analysis can be extended to the entire variable domain, which can provide rather integrated quantitative analysis results. EFAST method is used to study the global sensitivity of the landing variables to the landing loads.

EFAST is the quantitative global sensitivity analysis method which is developed by Saltelli et al. [18]. Based on classical Fourier amplitude sensitivity test (FAST), Saltelli et al. combined Sobol variance decomposition thought to develop the method. This method maps the multidimensional variable to independent variable by appropriate searching curves. Then the search within the multidimensional variable

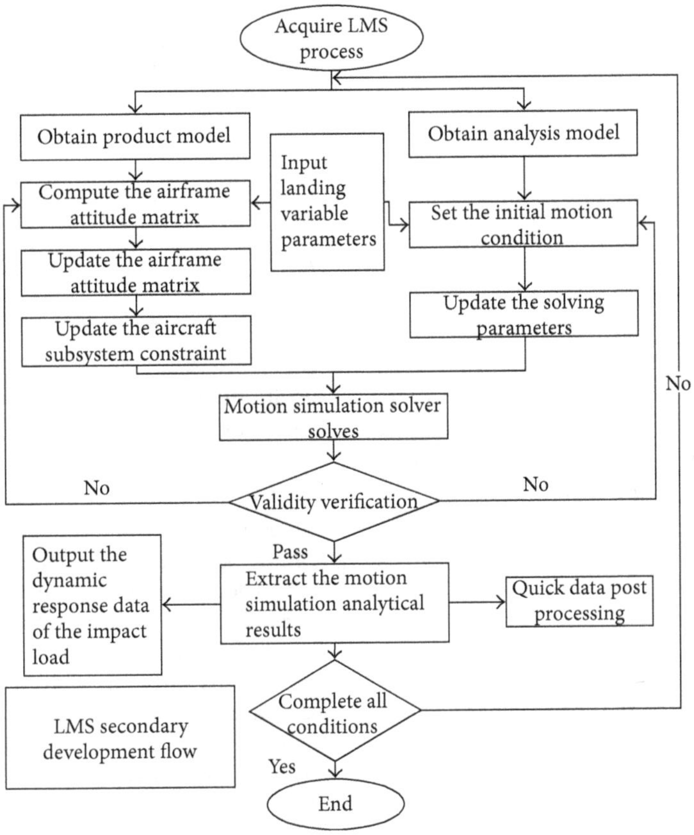

FIGURE 8: Flow chart of the virtual prototype multicondition automatic simulation.

space of the model is changed into the periodic function controlled by independent variable. Fourier transform is used to calculate the amplitude output by the model. The greater the change is, the more sensitive the variables are.

Assume that the landing analytical model is described as $y = f(X)$, where $X = X(x_1, x_2, \ldots, x_n)$, $n = 6$ is the combination of the landing variables. Before performing the global sensitivity analysis, all variables have to be mapped to the interval $[0, 1]$ to constitute the n-dimensional unit super cube space.

$$K^n = (X|0 \leq x_i \leq 1; i = 1, 2, \ldots, n). \quad (6)$$

Define an independent variable s and introduce the same mapping $s \rightarrow X$, so

$$x_i(s) = G_i(\sin \omega_i s), \quad \forall i = 1, 2, \ldots, n, \quad (7)$$

where $s(-\infty < s < +\infty)$ is the scalar variable, $\omega_i \in Z = \{-\infty, \ldots, -1, 0, 1, \ldots, +\infty\}$, and its value changes to suitable value according to x_i. With the change of s, all x_i change in the value space K^n. The oscillation frequency is ω_i. G_i is the searching curve and recommended by Sati et al. [13] as

$$x_i = \frac{1}{2} + \frac{1}{\pi}\arcsin(\sin \omega_i s + \varphi), \quad \forall i = 1, 2, \ldots, n, \quad (8)$$

where φ is the random value in $[0, 2\pi]$.

Expand $f(X)$ by Fourier series.

$$\begin{aligned} y = f(X) &= f(x_1, x_2, \ldots, x_n) \\ &= f(x_1(s), \ldots, x_n(s)) \\ &= \sum_{j=-\infty}^{j=+\infty} \{A_j \cos js + B_j \sin js\}, \end{aligned} \quad (9)$$

where

$$\begin{aligned} A_j &= \frac{1}{\pi}\int_{-\pi}^{\pi} f(s) \cos jsds, \\ B_j &= \frac{1}{\pi}\int_{-\pi}^{\pi} f(s) \sin jsds. \end{aligned} \quad (10)$$

$j \in z = \{-\infty, \ldots, -1, 0, 1, \ldots, +\infty\}$, and the amplitude of the Fourier series is $\Lambda_j = A_j^2 + B_j^2$.

From the properties of A_j and B_j, it is known that $A_{-j} = A_j$, $B_{-j} = -B_j$, and $\Lambda_{-j} = \Lambda_j$. By calculating all frequencies, the total variance caused by each variable x_i can be obtained, namely,

$$V(E(y|x_i)) = \sum_{p \in Z^0} \Lambda_{p\omega_i} = 2\sum_{p=1}^{+\infty} \Lambda_{p\omega_i}, \quad (11)$$

where $Z^0 = Z - \{0\}$.

The total variance of the model is

$$V(E(y)) = \sum_{j\in Z^0} \Lambda_j = 2\sum_{j=1}^{+\infty} \Lambda_j. \quad (12)$$

According to Sobol variance decomposition thought, the model can be decomposed into functions of single variable and combined variable, namely,

$$V(y) = \sum_{i=1}^{n} V_i + \sum_{i\neq j} V_{i,j} + \sum_{i\neq j\neq k} V_{i,j,k} + \cdots + V_{1,2,3,\ldots,n}, \quad (13)$$

where $V_i = V(E(y|x_i))$ which represents the variance of x_i, $V_{i,j} = V(E(y|x_i, x_j)) - V_i - V_j$ which represents the variance of x_i contributed by x_j, and $V_{i,j,k}\sim V_{1,2,\ldots,n}$ are by analogy. The first-order sensitivity coefficient S_{x_i} of the variable x_i is defined as below.

$$S_{x_i} = \frac{V_i}{V(y)}. \quad (14)$$

The total effects of the variable x_i include the single effect and interacting effects on the output of the model. The global sensitivity coefficient of the variable x_i is defined as

$$S_{x_i}^T = S_{x_i} + \sum_{i\neq j} S_{x_i x_j} + \sum_{i\neq j\neq k} S_{x_i x_j x_k} + S_{x_1,x_2,\ldots,x_n}. \quad (15)$$

The difference between global sensitivity coefficient $S_{x_i}^T$ and first-order sensitivity coefficient S_{x_i} reflects the degree of interaction between x_i and other variables.

5. Landing Variable Sensitivity and Extreme Value Analysis

When using EFAST method to analyze the global sensitivity, the input of probability distribution of the variable and the output of the model corresponding to the inputting samples are required [18]. The landing variables given by standards [7] should be modified according to practical situation. The type of the landing variable distribution maintains constant, then the average and standard deviations of the landing variables need to be modified to satisfy the design requirements. Table 1 shows the distribution of the six landing variables under the given design requirements. Due to the effects of the deck layout, the value range of the rolling angle and yawing angle are not bilateral symmetric.

The input parameters of EFAST analysis method use the distribution of the landing variable and its value range obtained after adopting multivariate distribution presented in Section 2.2. The EFAST sampling is carried out by using (8), then 6438 samples are obtained. The samples in the generated interval [0, 1] are mapped back to the distribution of the landing variable, used to input the variables of the virtual prototype. For all landing conditions, the virtual prototype and its secondary development technology are used to analyze the landing loads automatically. According to the analytical results of the landing load, the first-order and global sensitivity coefficients of landing variables to landing loads are calculated.

Table 1: Aircraft landing variables distribution.

Landing variables	Mean	Deviation
V_E	166.164	9.26
V_V	3.5052	0.9184
θ_P	4.5	0.85
θ_R	2.2	2.5
$\dot{\theta}_R$	0	2.8
θ_Y	0.8	0.7

5.1. The Sensitivity Analysis of Main Landing Gear Load. When the carrier-based aircraft lands on the deck, main landing gear touches the deck first in most cases. The vertical impact load is the main load that acts on the main landing gear. The sensitivity analysis results of the vertical load of the main landing gear are shown in Figure 9.

The analysis results show that sinking speed and rolling angle are the main affecting variables of the vertical load of the main landing gear. The total contribution of the two variables is more than 82%, while the remaining landing variables only have little coupling effects on the vertical load. The sinking speed has larger first-order sensitivity coefficient in the analysis of the left main landing gear while the rolling angle has larger first-order sensitivity coefficient in the analysis of the right main landing gear. When the right main landing gear touches the deck first, the rolling angle is large and the landing load is relatively severe. When the left main landing gear touches the deck first, the rolling angle is small and the sinking speed is the main effect of the vertical load.

Figure 9 also shows the distribution of the cumulative frequency of the main landing gear's vertical load. The largest vertical load (853 kN) condition of the left main gear is $V_V = 4.76$ m/s and $\theta_R = -1.99°$. The largest vertical load (1278 kN) condition of the right main gear is $V_V = 4.77$ m/s and $\theta_R = 7.09°$. The extreme value of the vertical load of the right landing gear is 1.5 times of the one of the left landing gear.

The sensitivity analysis results of the main landing gear's lateral load are shown in Figure 10. It is seen that the sinking speed, rolling angle, and yawing angle are the main affecting factors of the main landing gear's lateral load. The total contribution of the three variables is over 91% while remaining variables barely have direct effects on the lateral load.

From the cumulative frequency curve of the lateral load shown in Figure 10, it is seen that the extreme value of the lateral load of the left main landing gear is nearly the same as that of the right main landing gear. However, the average load of the right main landing gear is larger than that of the left main landing gear. Under the condition that the largest lateral load of the left main landing gear is 120 kN, the sinking speed $V_V = 4.62$ m/s, rolling angle $\theta_R = 7.36°$, and yawing angle $\theta_Y = 0.84°$. Under the condition that the largest lateral load of the right main landing gear is 122 kN, the sinking

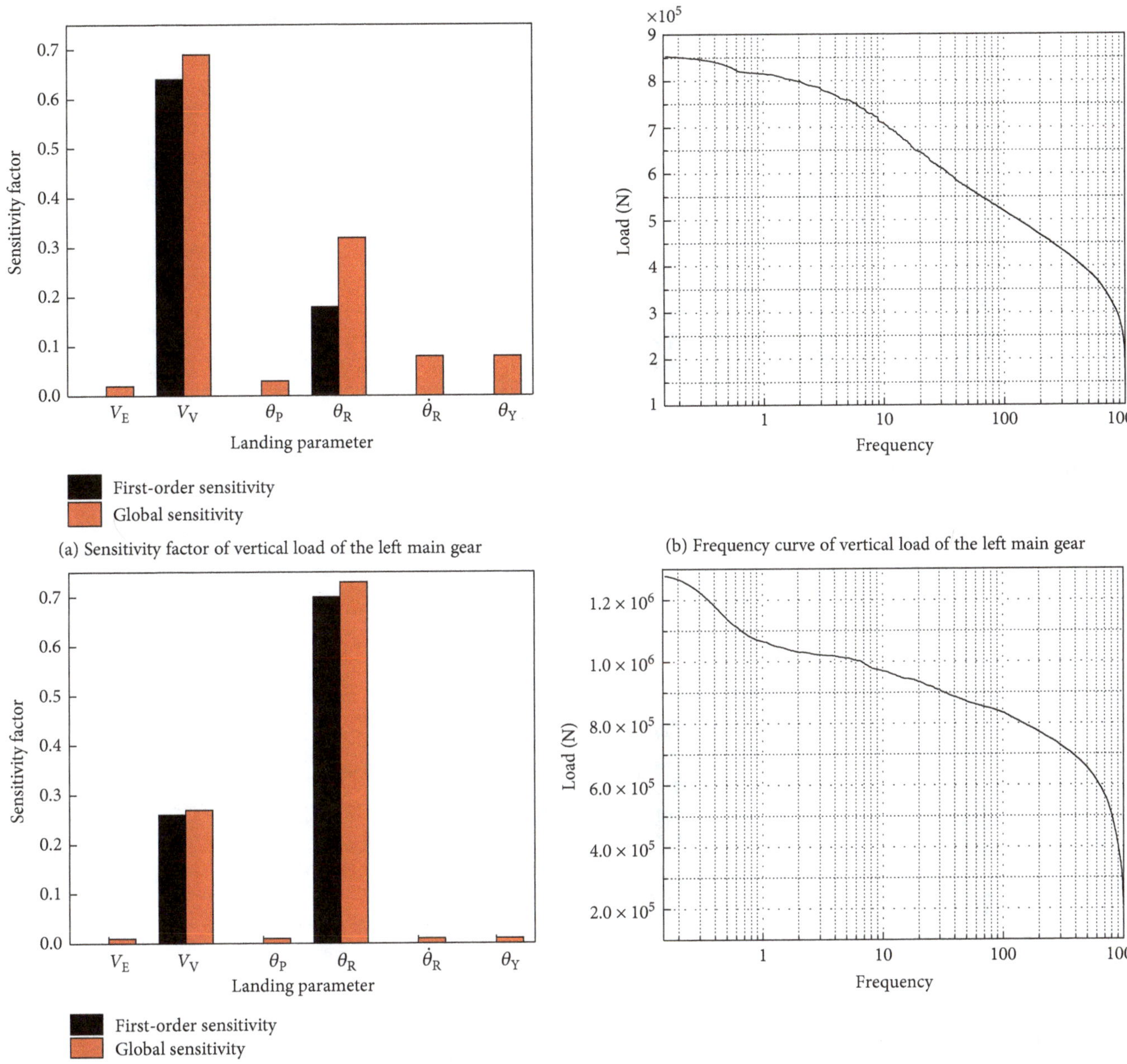

(a) Sensitivity factor of vertical load of the left main gear

(b) Frequency curve of vertical load of the left main gear

(c) Sensitivity factor of vertical load of the right main gear

(d) Frequency curve of vertical load of the right main gear

FIGURE 9: Sensitivity and cumulative frequency curve of vertical load of the main landing gear.

speed $V_V = 4.61\,\text{m/s}$, rolling angle $\theta_R = 7.36°$, and yawing angle $\theta_Y = 0.84°$.

The sensitivity analysis results of the main landing gear's course load are shown in Figure 11. The analysis results show that the sinking speed and rolling angle are the main affecting variables of the main landing gear's course load. The two variables' total contribution is 76% while the remaining landing variables have little coupling effects on the course load. Known from the cumulative frequency curve, the load distribution and extreme load of the left main landing gear is almost the same as that of the right main landing gear. Under the condition that the largest course load of the left main landing gear is 609 kN, the sinking speed $V_V = 4.59\,\text{m/s}$ and rolling angle $\theta_R = -2.04°$. Under the condition that the largest course load of the right main landing gear is 721 kN, the sinking speed $V_V = 4.76\,\text{m/s}$ and rolling angle $\theta_R = 7.09°$. The course loads of the left and right main landing gears have little difference. The extreme value of the course load also occurs under the condition of large sinking speed and large rolling angle.

Figures 9 and 11 show that the sensitivity analysis results of the course load are similar to that of the vertical load. The sensitivity coefficients of sinking speed and rolling angle are nearly the same. Besides, the course loads of the right main landing gear are all greater than that of the left main landing gear. The course load has rather high correlation with the vertical load.

5.2. Nose Landing Gear Load and Barycenter Overload Analysis. The sensitivity analysis results of the nose landing

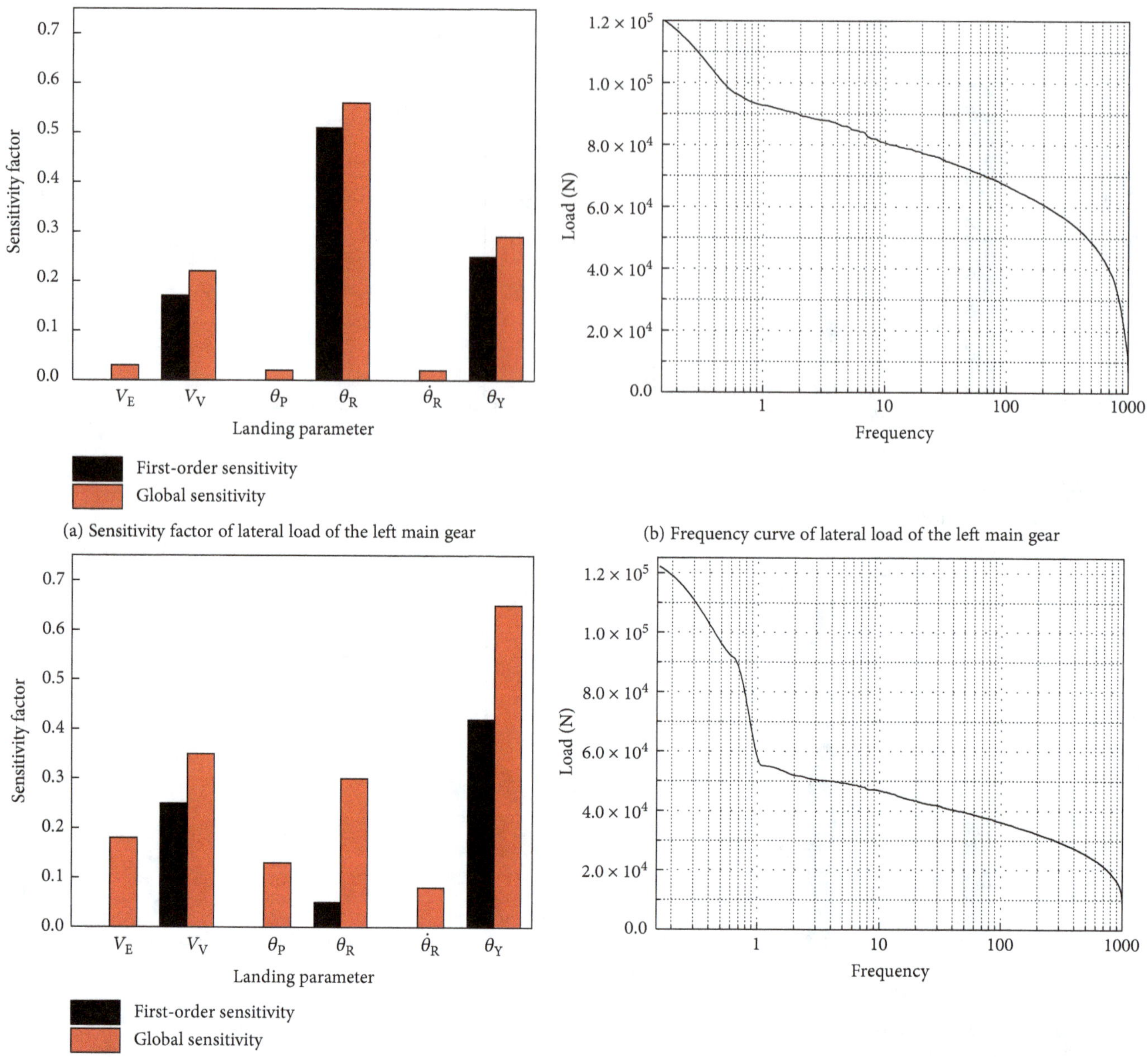

(a) Sensitivity factor of lateral load of the left main gear

(b) Frequency curve of lateral load of the left main gear

(c) Sensitivity factor of lateral load of the right main gear

(d) Frequency curve of lateral load of the right main gear

FIGURE 10: Sensitivity and cumulative frequency curve of lateral load of the main landing gear.

gear's three directional loads are shown in Figure 12. It can be known that the sinking speed and rolling angle are the main affecting variables of the nose landing gear's course load and their first-order sensitivity coefficients are 0.59 and 0.14, respectively. The pitching angle is the secondary affecting variable, and its first-order sensitivity coefficient is 0.004. Other variables have a few coupling effects on the course load.

For the lateral load of the nose landing gear, the rolling angle is the main affecting variable and its first-order sensitivity coefficient is 0.6. The sinking speed, pitching angle, and yawing angle have a few effects on the lateral load. For the vertical load of the nose landing gear, the pitching angle, rolling angle, and sinking speed are the main affecting variables and their first-order sensitivity coefficients are 0.53, 0.14, and 0.07, respectively. Other variables do not have appreciated effects on the vertical load.

From the cumulative frequency curve of the nose landing gear's three directional loads, the sinking speed $V_V = 5.41$ m/s, the pitching angle $\theta_P = 0.62^\circ$, and the rolling angle $\theta_R = 3.93^\circ$ under the condition that the largest course load of the nose landing gear is 133 kN. Under the condition that the largest lateral load of the nose landing gear is 93 kN, the sinking speed $V_V = 4.77$ m/s, the pitching angle $\theta_P = 3.09^\circ$, and the rolling angle $\theta_R = 7.09^\circ$, which is the large rolling angle condition. Under the condition that the largest vertical load of the nose landing gear is 196 kN, the sinking speed $V_V = 3.44$ m/s, the pitching angle $\theta_P = 6.23^\circ$, and the rolling

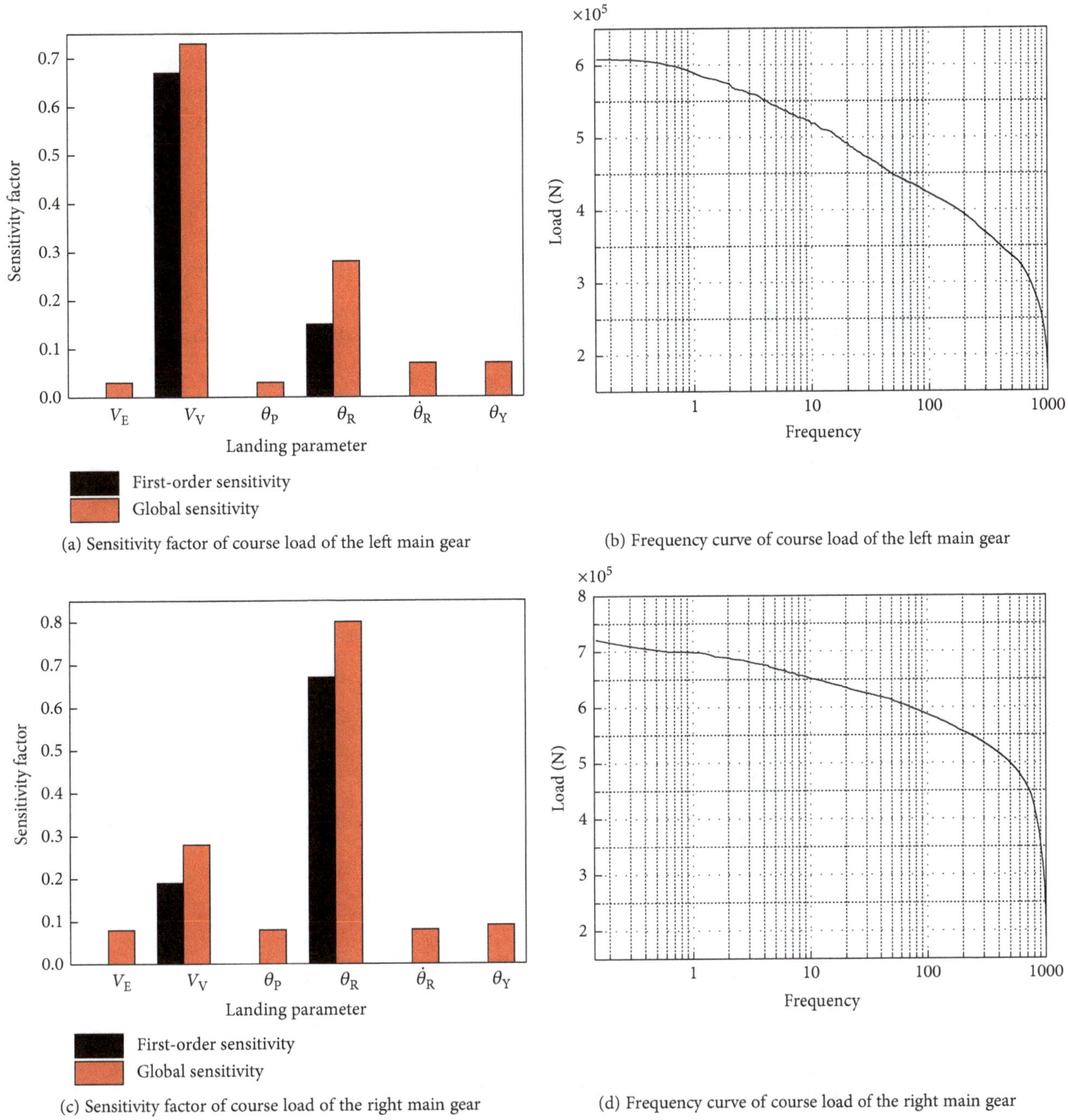

(a) Sensitivity factor of course load of the left main gear

(b) Frequency curve of course load of the left main gear

(c) Sensitivity factor of course load of the right main gear

(d) Frequency curve of course load of the right main gear

FIGURE 11: Sensitivity and cumulative frequency curve of course load of the main landing gear.

angle $\theta_R = 6.07^\circ$, which is the condition of medium sinking speed and large pitching angle.

The sensitivity analysis results of the barycenter overload of the airframe are shown in Figure 13. It is seen that sinking speed and rolling angle are the main affecting variables and their first-order sensitivity coefficients are 0.74 and 0.18, respectively. Under the condition that the largest barycenter overload is 5.11, the sinking speed $V_V = 4.77$ m/s and the rolling angle $\theta_R = 7.09^\circ$, which is the condition of large rolling angle and relatively large sinking speed. The barycenter overload of the airframe mainly concentrates between 2.0 and 4.0.

5.3. Sensitivity Analysis of the Landing Load of the Carrier-Based Aircraft. According to the sensitivity analysis results of the main landing gear, nose landing gear, and barycenter overload, the influence degree of the landing variables on landing loads can be classified as shown in Table 2. The sinking speed and pitching angle are the main affecting factors of the landing loads. The pitching angle and yawing angle have significant effects on some loads. The ground critical velocity and rolling rate have little effect on the landing loads, and they are influenced a lot by the change of the arresting cable load and the motion of the deck due to the restrictions of the simplified condition of the virtual prototype.

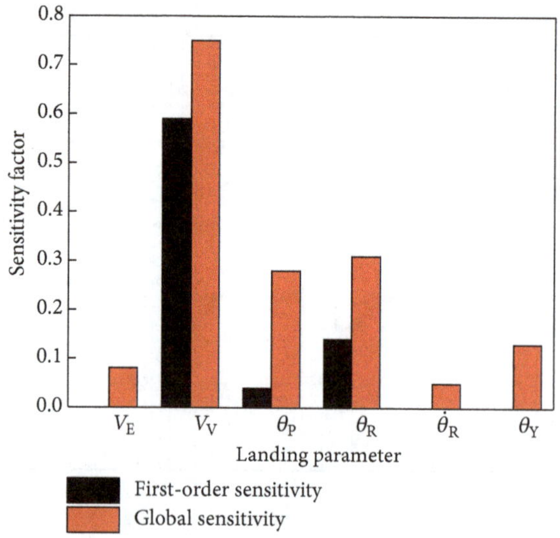

(a) Sensitivity factor of the course load of nose gear

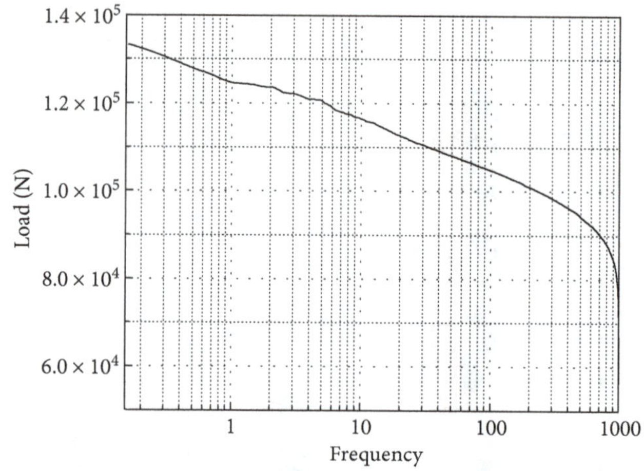

(b) Frequency curve of the course load of nose gear

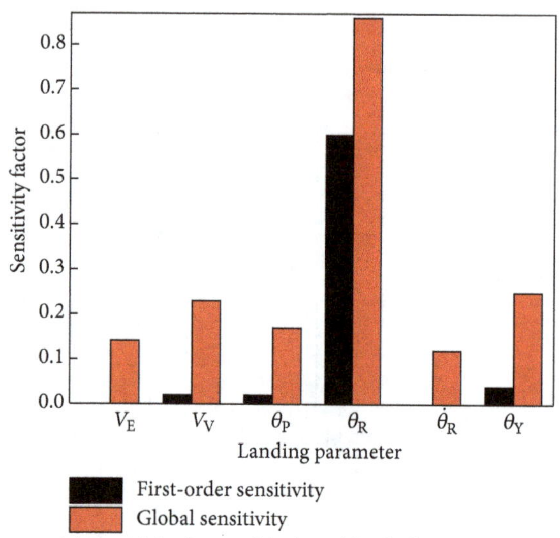

(c) Sensitivity factor of the lateral load of nose gear

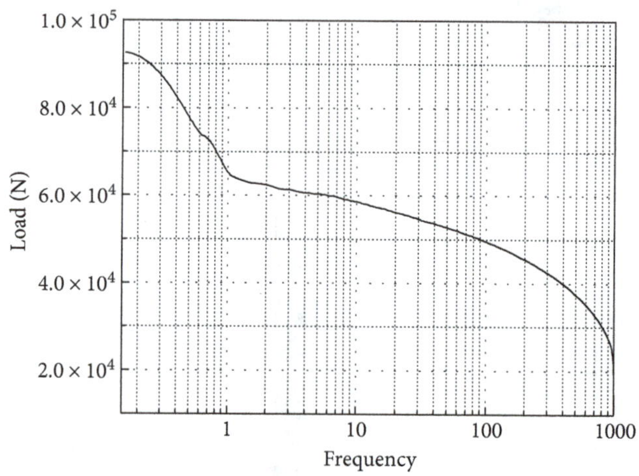

(d) Frequency curve of the lateral load of nose gear

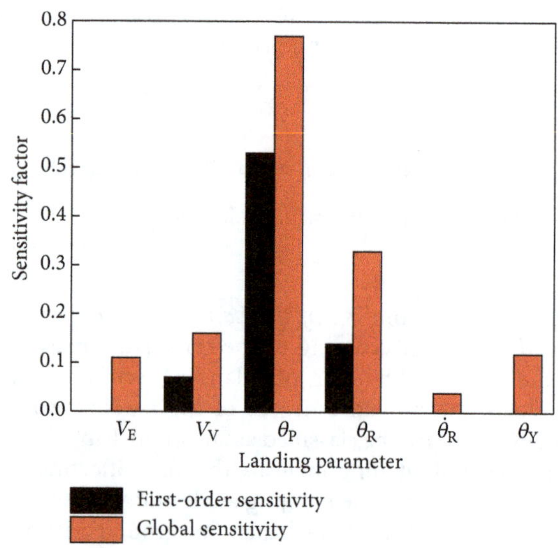

(e) Sensitivity factor of the vertical load of nose gear

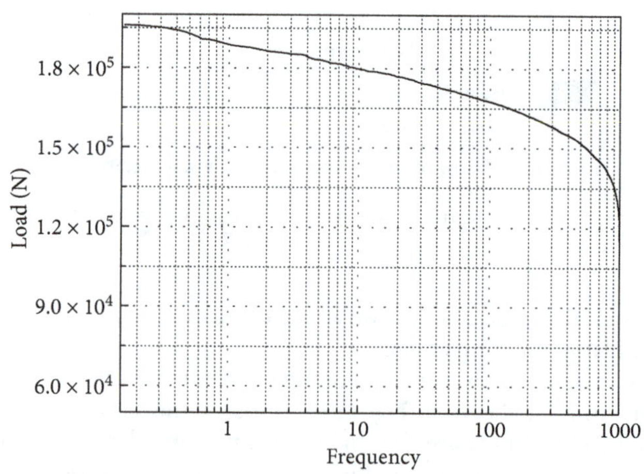

(f) Frequency curve of the vertical load of nose gear

FIGURE 12: Sensitivity and cumulative frequency curve of three directional loads of the nose landing gear.

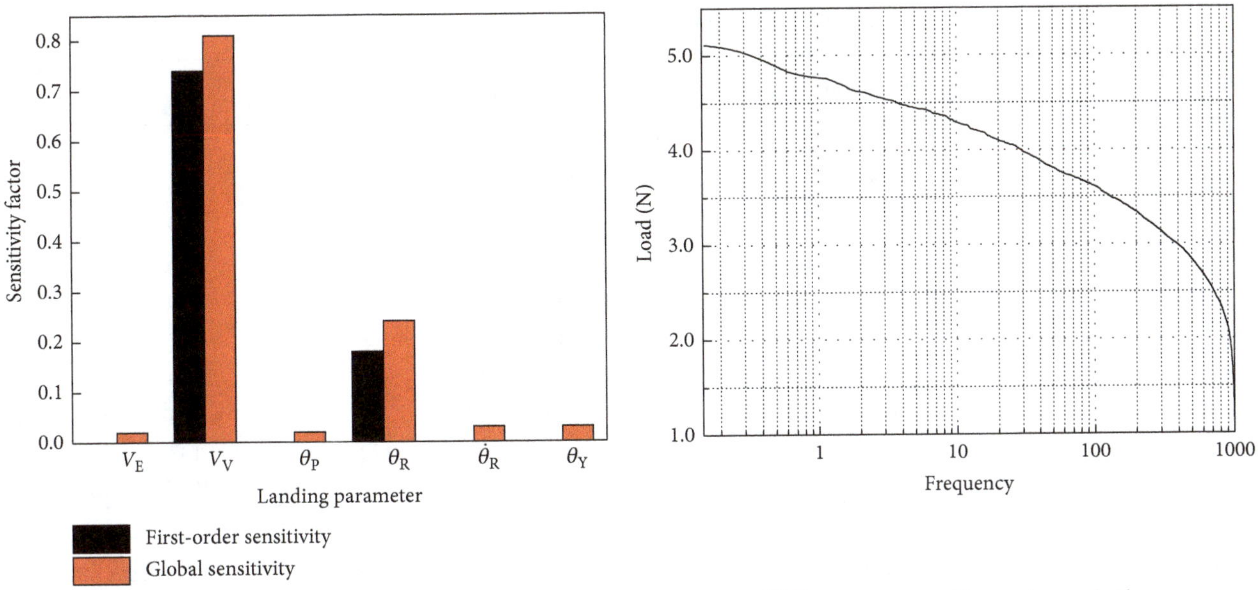

(a) Sensitivity factor of barycenter overload

(b) Frequency curve of barycenter overload

FIGURE 13: Sensitivity and cumulative frequency curve of the barycenter overload.

TABLE 2: Sensibility of landing variables to landing loads.

Type	High	Moderate	Low
Main gear	$V_V\theta_R$	θ_Y	$V_E\theta_P\dot{\theta}_R$
Nose gear	$V_V\theta_R\theta_P$	θ_Y	$V_E\dot{\theta}_R$
Gravity overload	$V_V\theta_R$	—	$V_E\theta_P\dot{\theta}_R\theta_Y$

When selecting the representative landing conditions of the carrier-based aircraft, the sinking speed, rolling angle, and pitching angle which have remarkable sensitivity should be discretized into several intervals. The intervals are selected to represent the values of variables. Different representative landing conditions are formed by the combination of the variables to reflect the landing load condition during the entire landing process.

5.4. Extreme Value Analysis of the Landing Loads of the Carrier-Based Aircraft. In the structure design stage, it is difficult to obtain the most severe load condition of the airframe during its service. When determining the most severe parameter combination, some conservative conditions are usually selected. For example, a certain or several extreme values of landing variables and the average of remaining landing variables are selected as extreme value conditions of the load for the researches. The conditions represent different landing conditions such as tail sinking, yawing landing, and landing with single landing gear. Some conditions will not happen during the service period, which leads to structural overdesign.

There are many landing samples selected in the sensitivity analysis of the landing load of the carrier-based aircraft. Approximately, the extreme load conditions of these samples can be used as extreme load conditions of all landing conditions to carry out the comparative analysis. From the computed results of the 6438 samples in EFAST sensitivity analysis method, the extreme value conditions of the three directional loads of the main and nose landing gear and barycenter overload are usually the condition that the sinking speed, rolling angle, or pitching angle are large and the landing variables do not pass the maximum of the value range. For further study in the rationality and security of the conservative condition design, typical condition design is selected for the comparative calculation.

Table 3 shows eight kinds of typical design conditions selected to compare with samples of EFAST sensitivity analysis in extreme loads of each direction. From the comparison, it is seen that the maximum gravity overload is slightly larger than that of the EFAST samples and the difference value is 23.5% approximately.

The extreme vertical and course loads of the landing gear of the typical design condition are nearly the same as the extreme values of EFAST samples. The maximum course load of the left and right main landing gear of the typical design condition is almost the same as the maximum load of the EFAST samples, and the percentage differences are about 6.2% and 16.0%, respectively. The maximum vertical load of the left and right main landing gear of the typical design condition is almost the same as the maximum load of the EFAST samples, and the percentage differences are about 2.5% and 12.0%, respectively. The maximum course and vertical loads of the nose landing gear of the typical design condition are quite different from the maximum loads of the EFAST samples, and the percentage differences are about 26.3% and 28.6%, respectively.

For the lateral load which is the secondary load, the maximum load of the typical design condition is quite different from that of the EFAST samples and the percentage difference is about 35.2%. From the overall analysis, the gravity overload and extreme load in each direction of the eight

Table 3: Extremum loads of typical landing conditions.

Condition of maximum variable (remaining variables are mean)	Gravity overload	Port main gear (kN)			STBD main gear (kN)			Nose gear (kN)		
		Course	Lateral	Vertical	Course	Lateral	Vertical	Course	Lateral	Vertical
V_V	5.5	564	81	787	692	40	1178	141	43	192
θ_R	3.9	340	63	398	586	52	1000	115	59	173
θ_P	2.8	328	53	371	459	26	645	113	33	195
$V_V\theta_R$	6.3	571	115	829	752	75	1431	166	74	219
$V_V\theta_P$	5.6	556	90	784	556	46	1166	140	46	215
$V_V\theta_R\theta_P$	6.2	564	118	827	649	75	1398	148	68	246
$V_V\theta_R\theta_Y$	6.2	564	158	832	836	69	1368	168	69	231
$V_V\theta_R\theta_P\theta_Y$	6.1	558	160	830	649	79	1333	154	84	252
EFAST samples	5.1	609	120	853	721	122	1278	133	93	196
Difference of maximum variable	23.5%	−6.2%	33.3%	−2.5%	16.0%	−35.2%	12.0%	26.3%	−9.7%	28.6%

kinds of typical design conditions are nearly the same as those of the EFAST samples, and the key loads of the typical design conditions are slightly larger. The extreme loads obtained by the conservative condition design are safe to be used in structure strength analysis and would not cause overdesign.

6. Conclusions

Based on the results reported, several conclusions can be drawn.

Firstly, the multivariate distribution has a large influence on the distribution of the landing variables. After multivariate distribution processing, the distribution of the landing variables is still close to the normal distribution, and the mean values of the variables maintain constant while the standard deviation decreases.

Secondly, the results of global sensitivity analysis show that all the six landing variables have effects on landing loads. The sinking speed and rolling angle almost influence all kinds of landing loads significantly and are the main affecting factors. The pitching angle and yawing angle may only influence some landing variables a lot, and they are the secondary affecting factors. The results of extreme value analysis show that it is safe to adopt typical condition design in the structural strength analysis.

Conflicts of Interest

The authors declare that there is no conflict of interest regarding the publication of this paper.

References

[1] S. A. Shappell and D. F. Neri, *The Effect of Combat on Aircrew Subjective Readiness and LSO Grades during Operation Desert Shield/Storm*, Naval Aerospace Medical Research Lab, Pensacola, FL, USA, 1992.

[2] B. N. Acharya, G. P. Gupta, S. Prakash, and M. P. Kaushik, "UV-resistant, water repellent and rodent repellent nylon tapes for aircraft arrester system," *Pigment & Resin Technology*, vol. 34, no. 5, pp. 270–274, 2005.

[3] D. T. Rusk, J. Pierce, W. Hoppe, B. Lancaster, R. Actis, and B. Szabo, *Analysis and Testing of Fleet Corroded F/A-18C/D Arrestment Shanks*, Naval Air Warfare Center Aircraft, Patuxent River, MD, USA, 2008.

[4] B. D. Flansburg, "Structural loads analysis of a carrier onboard delivery aircraft," in *57th AIAA/ASCE/AHS/ASC Structures, Structural Dynamics, and Materials Conference*, San Diego, CA, USA, 2016.

[5] N. S. Currey, *Aircraft Landing Gear Design: Principles and Practices*, American Institute of Aeronautics and Astronautics, Washington DC, USA, 1988.

[6] *Landing gear design loads*, Advisory Group for Aerospace Research and Development Neuilly-Sur-Seine, France, 1991, AD-A239 914.

[7] Naval Air Systems Command, *MIL-A-8863C (AS) Air-Plane Strength and Rigidity, Ground Loads for Navy Acquired Airplanes*, Patuxent River, MD, USA, Naval Air Systems Command, 1993.

[8] Naval Air Systems Command, *JSSG-2006 Aircraft Structures*, Patuxent River, MD, USA, Naval air systems command, 1998.

[9] R. P. Micklos, *Carrier Landing Parameters from Survey 45, Fleet and Training Command Aircraft Landing Aboard USS ENTERPRISE CVN-65 (Appendices B Through R)*, Naval Air Development Center, Warminster PA Air Vehicle and Crew Systems Technology Department, Warminster, PA, USA, 1991.

[10] Z. Wen, Z. Zhi, Z. Qidan, and X. Shiyue, "Dynamics model of carrier-based aircraft landing gears landed on dynamic deck," *Chinese Journal of Aeronautics*, vol. 22, no. 4, pp. 371–379, 2009.

[11] D. Mikhaluk, I. Voinov, and A. Borovkov, *Finite Element Modeling of the Arresting Gear and Simulation of the Aircraft Deck Landing Dynamics*, 2011.

[12] L. Lihua, W. Chen, and X. Panpan, "Dynamic analysis of aircraft arresting gear based on finite element method," in *2011 International Conference on System science, Engineering design and Manufacturing informatization*, pp. 118–121, Guiyang, China, 2011.

[13] S. C. Sati, A. S. Kanaskar, S. R. Kajale, and A. Mukherjee, "Modeling, simulation and analysis of aircraft arresting system using bond graph approach," *Simulation Modelling Practice and Theory*, vol. 19, no. 3, pp. 936–958, 2011.

[14] G. Sebastian, "Simulation of a free flight aircraft engagement during a carrier landing," in *35th Structures, Structural Dynamics, and Materials Conference*, Hilton Head, SC, USA, 1994AIAA-1994-1619- CP.

[15] D. H. Chester, "Aircraft landing impact parametric study with emphasis on nose gear landing conditions," *Journal of Aircraft*, vol. 39, no. 3, pp. 394–403, 2002.

[16] F. Yunwen, L. Sihong, X. Xiaofeng, C. Shuai, and P. Wenting, "Sinking velocity impact analysis of carrier-based aircraft based on test data," *Acta Aeronautica et Astronautica Sinica*, vol. 36, no. 11, pp. 3578–3585, 2015.

[17] M. Zhang, H. Nie, and Z. H. He, "Optimization parameters of nose landing gear considering both take-off and landing performance of catapult take-off carrier-based aircraft," *Transactions of Nanjing University of Aeronautics & Astronautics*, vol. 01, p. 12, 2016.

[18] A. Saltelli, S. Tarantola, and K. P.-S. Chan, "A quantitative model-independent method for global sensitivity analysis of model output," *Technometrics*, vol. 41, no. 1, pp. 39–56, 1999.

[19] M. Crosetto and S. Tarantola, "Uncertainty and sensitivity analysis: tools for GIS-based model implementation," *International Journal of Geographical Information Science*, vol. 15, no. 5, pp. 415–437, 2001.

[20] C. Xu and G. Z. Gertner, "A general first-order global sensitivity analysis method," *Reliability Engineering & System Safety*, vol. 93, no. 7, pp. 1060–1071, 2008.

[21] N. H. Sandlin, R. S. Ebers, and R. Black, "Ship motion effects on landing impact loads<149>V/STOL landing on aircraft carrier," in *20th Structures, Structural Dynamics, and Materials Conference*, St. Louis, MO, USA, 1979AIAA 79-0742.

[22] I. F. Mondragon, P. Campoy, C. Martinez, and M. A. Olivares-Mendez, "3D pose estimation based on planar object tracking for UAVs control," in *2010 IEEE International Conference on Robotics and Automation*, pp. 35–41, Anchorage, AK, USA, 2010.

[23] L. Kege and Y. Chuliang, "Load measurement and compilation of landing gear of airplane," *Acta Aeronautica et Astro-nautica Sinica*, vol. 32, pp. 841–848, 2011.

17

Active Vibration Control of Plate Partly Treated with ACLD Using Hybrid Control

Dongdong Zhang and Ling Zheng

State Key Laboratory of Mechanical Transmission, Chongqing University, Chongqing 400030, China

Correspondence should be addressed to Ling Zheng; zling@cqu.edu.cn

Academic Editor: N. Ananthkrishnan

A finite element model of plate partly treated with ACLD treatments is developed based on the constitutive equations of elastic, piezoelectric, viscoelastic materials and Hamilton's principle. The Golla-Hughes-Mctavish (GHM) method is employed to describe the frequency-dependent characteristics of viscoelastic material (VEM). A model reduction is completed by using iterative dynamic condensation and balance model reduction method to design an effective control system. The emphasis is concerned on hybrid (combined feedback/feedforward) control system to attenuate the vibration of plates with ACLD treatments. The optimal linear quadratic Gaussian (LQG) controller is considered as a feedback channel and the adaptive filtered-reference LMS (FxLMS) controller is used as a feedforward channel. They can be utilized individually or in a hybrid way to suppress the vibration of plate/ACLD system. The results show that the hybrid controller which combines feedback/feedforward together can reduce the displacement amplitude of plate/ACLD system subjected to a complicated disturbance substantially without requiring more control effort. Furthermore, the hybrid controller has more rapid and stable convergence rate than the adaptive feedforward FxLMS controller. Meanwhile, perfect robustness to phase error of the cancellation path in feedforward controller and the weight matrices in feedback LQG controller is demonstrated in proposed hybrid controller. Therefore, its application in structural engineering can be highly appreciated.

1. Introduction

Vibration and noise control is of great interest in many industrial structures, like airplane passenger cabin, vehicle body structure, and submarine hull. The thin-wall structure is a key part in the body/cabin/hull structure. The vibrating thin-wall structures, like plates and shells, which are mainly disturbed by the harmonic disturbance induced from the engine and other rotating machines, as well as stochastic disturbance, make radiation noise into the passenger cabin, influencing the performance of the structure.

Active vibration control (AVC) [1], which is well-known for its efficiency in low frequency band where traditional passive strategies cannot work well, has been applied to various structures over the past two decades. Active vibration control of simple structure, such as beams [2] and plates [3], was researched broadly in the open literatures. Thakkar and Ganguli [4] investigated the twist vibration control of helicopter rotor blade by using the piezoelectric actuators. Rao et al. [5] designed a H_∞ controller and demonstrated experimentally the effectiveness of multimode vibration control in the composite fin-tip of aircraft. Kwak and Yang [6] studied the suppression of vibration of ring-stiffened cylindrical shell in contact with external fluid using piezoelectric sensor and actuator.

Active constrained layer damping (ACLD) is a hybrid active-passive vibration control technique [7] which combines the advantages of conventional passive constrained layer damping (PCLD) and active vibration control (AVC). Therefore, in the same damping treatment, the broader band vibration suppression can be achieved through ACLD treatments. In recent decades, extensive efforts had been made to reduce the vibration of the thin-wall structure using ACLD treatments. Baz and Ro, as the pioneering researchers, demonstrated the feasibility of using the ACLD in controlling the vibration of rotating beams with the ultimate goal of extending its application of attenuating the structural vibration of rotor blades [8]. The vibration control of shells [9, 10] was also investigated to extend the application of ACLD in the cabin of aircraft and submarine. In these studies, the PVDF (polyvinylidene fluoride) was served as the sensors and actuators. They use the simple proportional and derivative feedback of the transverse deflection to suppress the vibration of the base structure theoretically and experimentally. To improve the force transmissibility between piezoelectric actuator and the base structure, Liao and Wang [11] proposed an enhanced active constraining layer (EACL), and Gao and Liao [12] extended this configuration. Kumar et al. [13] added the stand-off layer (SOL) between the viscoelastic layer and the base structure to increase the viscoelastic strain and enhance the effect of the active force. In recent years, piezoelectric fiber-reinforced composite (PFRC) materials [14] and piezoelectric composites (PZC) [15, 16], which have a wide range of effective materials properties, were introduced in the ACLD treatments to serve as active constrained layer. Shah and Ray [17] focused on investigating the effect of the piezoelectric fiber orientation angle in the constrained layer on the performance of ACLD patches. In these studies, the emphasis was placed on the vibration characteristics analysis/enhancement of the performance of the ACLD treatments.

In the open literatures, much attention was paid on active control methods in the structural active/active-passive vibration control. The optimal feedback control [18], velocity feedback control [19], fuzzy logic control [3], and the H_∞ control [5] were usually employed in active vibration control (AVC). Viswamurthy and Ganguli [20] compared the global harmonic control with local blade optimal control applied to vibration control of helicopter. For active-passive ACLD/structure, Baz [21] presented a variational mathematical model for beams and a globally stable boundary controller was investigated to get the high damping characteristics over broad frequency range. Meanwhile, Baz [22] developed a H_2 robust controller based on the transfer function model of the ACLD beam. The control strategy was stable in the presence of parameter uncertainty, as well as ensured optimal disturbance rejection capabilities. Furthermore, the H infinity robust control strategy was utilized by Crassidis et al. [23] to maximize disturbance rejection capabilities over a desired frequency band. Liu et al. [24] developed a H infinity controller based on the reduced finite element model. Rodriguez [25] made a comparison of sliding mode control (SMC) with state-feedback control (LQR) applied to a partially treated ACLD beam. In these studies, the control strategies can be classified into feedback control theories. These control strategies have been shown to be more suitable in applications where the structure is disturbed by the stochastic or impulsive disturbances.

Feedforward control, which is more effective in the case where the deterministic or correlated information about the disturbance is known, is another popular control strategy. It was widely used in active vibration and noise control field [26, 27]. However, only a few feedforward theories were utilized in structural vibration control with ACLD treatments. Cao [28] employed the feedforward LMS algorithm to control the vibration of a plate treated with ACLD treatments. The finite element model was developed and ADF method was used to identify the performance of VEM. Furthermore, hybrid control of feedback and feedforward control theories were also successfully used to noise and vibration suppression [29, 30]. However, it was rarely investigated for structural vibration control with ACLD treatments. Vasques validated the effectiveness of the hybrid control strategies for ACLD beam [31].

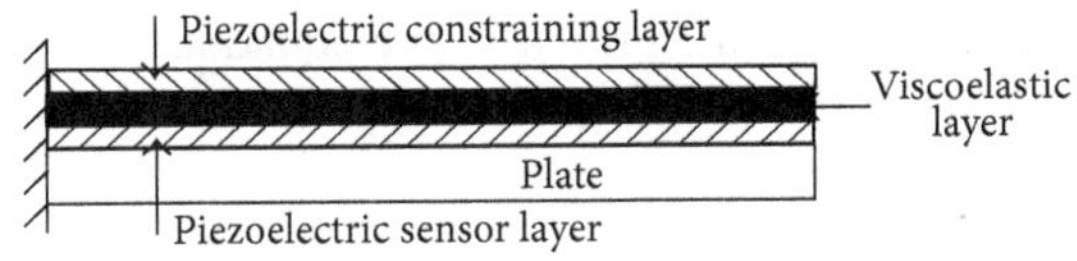

FIGURE 1: Schematic drawing of the ACLD/plate structure.

In this paper, the focus is placed on combining individual feedback control based on linear quadratic Gaussian (LQG) and feedforward control based on adaptive filtered-reference LMS (FxLMS), to develop a hybrid controller for the vibration control of plates treated with ACLD treatments. The performance of vibration suppression subjected to different control strategies is compared and the effect of parameters on vibration suppression of the plate with ACLD treatment is further investigated.

This paper is organized in the following five sections. A brief background is introduced in Section 1. The dynamic models including the finite element model of the plate treated with ACLD patches and GHM model of VEM are developed in Section 2. In Section 3, the feedback controller, the feedforward controller and the hybrid controller are designed. A numerical example is discussed in Section 4. Finally, the main results and future work are summarized in Section 5.

2. The ACLD/Plate System Model Construction

2.1. Finite Element Model. The ACLD/plate system is illustrated in Figure 1. The viscoelastic damping layer is sandwiched between two piezoelectric layers which serve as piezosensor and piezoactuator, respectively. The three layers are bonded to the base plate acting as a smart constraining layer damping. It is assumed that the piezosensor and the base plate are perfectly bonded together such that they can be reduced to a single equivalent layer. Some other assumptions are made in [32].

The ACLD/plate elements considered are four-node two-dimensional elements. Each node has seven degrees of freedom to describe the longitudinal displacements u_c and v_c of the constrained layer, the longitudinal displacements u_p and v_p of the base plate, the transverse deflection w, and the slopes x and y of the deflection line. The longitudinal displacement functions are expressed as 4-term polynomials, and transverse deflection w is expressed as a 12-term polynomial. Therefore, the element deflection vector is given by

$$\Delta^{(e)} = \{\Delta_1 \ \ \Delta_2 \ \ \Delta_3 \ \ \Delta_4\}^T, \tag{1}$$

where $\Delta_i = \{u_{ci} \ \ v_{ci} \ \ u_{pi} \ \ v_{pi} \ \ w_i \ \ \theta_{xi} \ \ \theta_{yi}\}^T$ for $i = 1, 2, 3, 4$.

Applying Hamilton's principle, the equations of motion for an ACLD element can be written as

$$\mathbf{M}^{(e)}\ddot{\Delta}^{(e)} + \mathbf{K}^{(e)}\Delta^{(e)} = \mathbf{F}_d^{(e)} + \mathbf{F}_c^{(e)}, \quad (2)$$

where $\mathbf{M}^{(e)} = \mathbf{M}_p^{(e)} + \mathbf{M}_c^{(e)} + \mathbf{M}_v^{(e)}$, $\mathbf{K}^{(e)} = \mathbf{K}_p^{(e)} + \mathbf{K}_c^{(e)} + \mathbf{K}_v^{(e)} + G_v\mathbf{K}_{v\beta}^{(e)}$, $\mathbf{M}_p^{(e)}$, $\mathbf{M}_c^{(e)}$, and $\mathbf{M}_v^{(e)}$ stand for the element mass matrices for base plate, the constrained layer, and the viscoelastic layer, $\mathbf{K}_p^{(e)}, \mathbf{K}_c^{(e)}$, and $\mathbf{K}_v^{(e)}$ stand for the element stiffness matrices for base plate, the constrained layer, and the viscoelastic layer, $\mathbf{K}_{v\beta}^{(e)}$ is the shear stiffness matrix for viscoelastic layer, G_v is the shear modulus of the viscoelastic materials, $\mathbf{F}_d^{(e)}$ is the external force vector, and $\mathbf{F}_c^{(e)}$ is the control force produced by the piezoelectric actuators.

Assembling the condensed system for all elements yields the dynamic equation of the plate with ACLD treatments:

$$\mathbf{M}\ddot{\Delta} + \left\{\mathbf{K} + G_v\mathbf{K}_{v\beta}\right\}\Delta = \mathbf{F}_d + \mathbf{F}_c, \quad (3)$$

where $\mathbf{K}$ is the global stiffness matrix without considering the shear stiffness of the VEM.

2.2. Viscoelastic Damping Model. The frequency- or time-dependent behavior of the viscoelastic material is characterized by using Golla-Hughes-Mctavish (GHM) method [24]. In this approach, the shear modulus of VEM is expressed as a series of "minioscillator" terms in the Laplace domain

$$sG(s) = G^{\infty}\left[1 + \sum_k \alpha_k \frac{s^2 + 2\varsigma_k\omega_k s}{s^2 + 2\varsigma_k\omega_k s + \omega^2}\right], \quad (4)$$

where G^{∞} corresponds to the equilibrium value of the modulus, constants α_k, ς_k, and ω_k govern the shape of the modulus function over the complex s-domain. A column matrix of dissipation coordinates is introduced:

$$Z'(s) = \frac{\omega^2}{s^2 + 2\varsigma\omega s + \omega^2}\Delta(s). \quad (5)$$

Considering a three-term GHM expression, (3) can be rewritten as follows:

$$\mathbf{M}'\ddot{\mathbf{X}} + \mathbf{C}'\dot{\mathbf{X}} + \mathbf{K}'\mathbf{X} = \mathbf{F}_\mathbf{d}' + \mathbf{F}_\mathbf{c}', \quad (6)$$

where each vector and matrix can be expressed as

$$\mathbf{M}' = \begin{bmatrix} \mathbf{M} & 0 & 0 & 0 \\ 0 & \frac{\alpha_1}{\omega_1^2}\mathbf{\Lambda} & 0 & 0 \\ 0 & 0 & \frac{\alpha_2}{\omega_2^2}\mathbf{\Lambda} & 0 \\ 0 & 0 & 0 & \frac{\alpha_3}{\omega_3^2}\mathbf{\Lambda} \end{bmatrix},$$

$$\mathbf{C}' = \begin{bmatrix} 0 & 0 & 0 & 0 \\ 0 & \frac{2\alpha_1\zeta_1}{\omega_1^2}\mathbf{\Lambda} & 0 & 0 \\ 0 & 0 & \frac{2\alpha_2\zeta_2}{\omega_2^2}\mathbf{\Lambda} & 0 \\ 0 & 0 & 0 & \frac{2\alpha_3\zeta_3}{\omega_3^2}\mathbf{\Lambda} \end{bmatrix},$$

$$\mathbf{K}' = \begin{bmatrix} G^{\infty}\mathbf{K}_{v\beta}\left(1 + \sum_{k=1}^{3}\alpha_k\right) + \mathbf{K}_v & -\alpha_1\mathbf{R} & -\alpha_2\mathbf{R} & -\alpha_3\mathbf{R} \\ -\alpha_1\mathbf{R}^T & \alpha_1\mathbf{\Lambda} & 0 & 0 \\ -\alpha_2\mathbf{R}^T & 0 & \alpha_2\mathbf{\Lambda} & 0 \\ -\alpha_3\mathbf{R}^T & 0 & 0 & \alpha_3\mathbf{\Lambda} \end{bmatrix},$$

$$\mathbf{X} = \begin{Bmatrix} \Delta \\ \mathbf{Z} \end{Bmatrix}, \quad \mathbf{F}_\mathbf{d}' = \begin{Bmatrix} \mathbf{F}_\mathbf{d} \\ \mathbf{0} \end{Bmatrix}, \quad \mathbf{F}_\mathbf{c}' = \begin{Bmatrix} \mathbf{F}_\mathbf{c} \\ \mathbf{0} \end{Bmatrix}, \quad (7)$$

where $\mathbf{K}_{v\beta} = \mathbf{R}_v\mathbf{\Lambda}_v\mathbf{R}_v^T$, $\mathbf{\Lambda} = G^{\infty}\mathbf{\Lambda}_v$, $\mathbf{R} = \mathbf{R}_v\mathbf{\Lambda}$, $\mathbf{Z} = \mathbf{R}_v^T\mathbf{Z}'$, $\mathbf{\Lambda}_v$ is a diagonal matrix of the positive eigenvalues of matrix $\mathbf{K}_{v\beta}$, and $\mathbf{R}_v$ is the eigenvectors matrix corresponding to the eigenvalues. For simplicity the global equation of motion can be rewritten as

$$\mathbf{M}\ddot{\mathbf{X}} + \mathbf{C}\dot{\mathbf{X}} + \mathbf{K}\mathbf{X} = \mathbf{F}_\mathbf{d} + \mathbf{F}_\mathbf{c}. \quad (8)$$

2.3. Model Reduction and State Space Design. As well known, the scale of the finite element model is increased largely due to the introduction of the dissipation coordinates from GHM model. This requires much computational effort to calculate the dynamic response of the plates with ACLD treatments. Otherwise, not all the state variables in the model are controllable and observable. Therefore, model reduction is necessary to design an effective controller. Here, an iterative dynamic condensation [33] is used to reduce the scale of the finite element model.

Equation (8) can be rewritten as

$$\begin{bmatrix} \mathbf{M}_{mm} & \mathbf{M}_{ms} \\ \mathbf{M}_{sm} & \mathbf{M}_{ss} \end{bmatrix}\begin{Bmatrix} \ddot{\mathbf{X}}_m \\ \ddot{\mathbf{X}}_s \end{Bmatrix} + \begin{bmatrix} \mathbf{C}_{mm} & \mathbf{C}_{ms} \\ \mathbf{C}_{sm} & \mathbf{C}_{ss} \end{bmatrix}\begin{Bmatrix} \dot{\mathbf{X}}_m \\ \dot{\mathbf{X}}_s \end{Bmatrix} + \begin{bmatrix} \mathbf{K}_{mm} & \mathbf{K}_{ms} \\ \mathbf{K}_{sm} & \mathbf{K}_{ss} \end{bmatrix}\begin{Bmatrix} \mathbf{X}_m \\ \mathbf{X}_s \end{Bmatrix} = \begin{Bmatrix} \mathbf{F}_m \\ \mathbf{F}_s \end{Bmatrix}. \quad (9)$$

Here, a matrix $\widehat{\mathbf{R}}$ is defined to relate the master d.o.f.s with the slave d.o.f.s and, after i iterations, the reduced dynamic equation is

$$\mathbf{M}_R^{(i)}\ddot{\mathbf{X}} + \mathbf{C}_R^{(i)}\dot{\mathbf{X}} + \mathbf{K}_R^{(i)}\mathbf{X} = \mathbf{F}_R^{(i)}, \quad (10)$$

where

$$\begin{aligned}
\mathbf{M}_R^{(i)} &= \mathbf{M}_{mm} + \mathbf{R}'^{(i)^T}\mathbf{M}_{sm} \\
&\quad + \mathbf{M}_{ms}\mathbf{R}'^{(i)} + \mathbf{R}'^{(i)^T}\mathbf{M}_{ss}\mathbf{R}'^{(i)}\mathbf{C}_R^{(i)} \\
&= \mathbf{C}_{mm} + \mathbf{R}'^{(i)^T}\mathbf{C}_{sm} + \mathbf{C}_{ms}\mathbf{R}'^{(i)} \\
&\quad + \mathbf{R}'^{(i)^T}\mathbf{C}_{ss}\mathbf{R}'^{(i)}, \\
\mathbf{K}_R^{(i)} &= \mathbf{K}_{mm} + \mathbf{R}'^{(i)^T}\mathbf{K}_{sm} \\
&\quad + \mathbf{K}_{ms}\mathbf{R}'^{(i)} + \mathbf{R}'^{(i)^T}\mathbf{K}_{ss}\mathbf{R}'^{(i)}\mathbf{F}_R^{(i)} \\
&= \mathbf{F}_m + \mathbf{R}'^{(i)^T}\mathbf{F}_s, \\
\mathbf{R}'^{(i+1)} &= -\mathbf{K}_{ss}^{-1}\left(\mathbf{M}_{ss}\mathbf{R}'^{(i)}\mathbf{M}_R^{(i)^{-1}}\mathbf{K}_R^{(i)} - \mathbf{K}_{sm}\right), \\
\mathbf{R}'^{(0)} &= -\mathbf{K}_{ss}^{-1}\mathbf{K}_{sm}.
\end{aligned} \quad (11)$$

Based on above reduced model, the state space form of the ACLD/plate is derived as follows:

$$\begin{aligned}
\dot{\mathbf{x}}(t) &= \mathbf{A}\mathbf{x}(t) + \mathbf{B}_d\mathbf{f}_d + \mathbf{B}_c\mathbf{f}_c, \\
\mathbf{y}(t) &= \mathbf{C}\mathbf{x}(t) + \mathbf{v},
\end{aligned} \quad (12)$$

where the state space vector $\mathbf{x(t)}$ is chosen as $\{\mathbf{X}_m \ \dot{\mathbf{X}}_m\}^T$, $\mathbf{A}$ is system matrix, and $\mathbf{B}_d$ and $\mathbf{B}_c$ are disturbance input matrices and the control force distribution matrices. $\mathbf{C}$ is the output matrix and $\mathbf{y(t)}$ is the output vector. $\mathbf{v}$ represents the measurement noise.

A further model reduction is performed with the balance method in the state space to guarantee the control system controllable and observable. The details are shown in [34].

3. Control System Architecture Design

3.1. Feedback Control. The ultimate aim of the feedback control is to reduce the motion of the system to the greatest possible extent; in that case, the control system is said to act as a regulator. Here, the linear quadratic regulator (LQR) is designed to control the vibration of plates with ACLD treatments. Cost function or performance index is given by

$$J = \int_0^{\infty}\left(\mathbf{x}(t)^T\mathbf{Q}_x\mathbf{x}(t) + \mathbf{f}_c^{\ T}(t)\,\mathbf{R}\mathbf{f}_c(t)\right)dt \quad (13)$$

or

$$J = \int_0^{\infty}\left(\mathbf{y}(t)^T\mathbf{Q}_y\mathbf{y}(t) + \mathbf{f}_c^{\ T}(t)\,\mathbf{R}\mathbf{f}_c(t)\right)dt, \quad (14)$$

where $\mathbf{Q}_x$, $\mathbf{Q}_y$, and $\mathbf{R}$ are the state variable, the output, and the control input weighting matrices, respectively.

The control input can be calculated using the state feedback so that

$$\mathbf{f}_c = -\mathbf{K}_c\mathbf{x}(t). \quad (15)$$

The optimal feedback gain matrix $\mathbf{K}_c$ is given by

$$\mathbf{K}_c = \mathbf{R}^{-1}\mathbf{B}_c^T\mathbf{P}, \quad (16)$$

where $\mathbf{P}$ is the solution of the famous Riccati equation.

Therefore, considering the control law in (15) and substituting it into the state space equation of the reduced model, the closed-loop state equation is given by

$$\mathbf{x}(t) = (\mathrm{A} - \mathbf{B}_c\mathbf{K}_c)\,\mathbf{x}(t) + \mathbf{B}_d\mathbf{f}_d. \quad (17)$$

In above equation, it is assumed that all the state variables can be measured. However that is not always available in a real system; an optimal design methodology that only utilizes noise, partial state information is required. The linear quadratic Gaussian (LQG) methodology provides a means of designing such controllers. The LQG controller is a combination of a LQR and a Kalman filter, a fact known as the separation principle.

The Kalman filter is an optimal state estimator which can minimize the effects of process and measurement noise. In (18), the process noise $\mathbf{f}_d$ and measurement noise $\mathbf{v}_d$ are both assumed to be white and have a Gaussian probability density function. Meanwhile, they are assumed to be uncorrelated with the inputs. The correlation properties of the plant and measurement noise vectors are given by the correlation matrices

$$E\left(\mathbf{f}_d\mathbf{f}_d^T\right) = Q_d, \qquad E\left(\mathbf{v}_d\mathbf{v}_d^T\right) = R_d, \quad (18)$$

where E denotes the expectation operator. These matrices are used as weighting matrices of the steady-state Kalman filter problem [20].

A state estimate $\mathbf{x}'(t)$ that minimizes the steady-state error covariance is constructed by the Kalman filter

$$P = \lim_{t\to\infty} E\left(\left[\mathbf{x}(t) - \mathbf{x}'(t)\right]\left[\mathbf{x}(t) - \mathbf{x}'(t)\right]^T\right). \quad (19)$$

In this way, a new state space model is generated

$$\begin{aligned}
\dot{\mathbf{x}}' &= \mathbf{A}\mathbf{x}' + \mathbf{B}_c\mathbf{f}_c^{fd} + \mathbf{L}\left(\mathbf{y}_v - \mathbf{C}\mathbf{x}'\right) \\
\begin{bmatrix}\mathbf{y}' \\ \mathbf{x}'\end{bmatrix} &= \begin{bmatrix}\mathbf{C} \\ \mathbf{I}\end{bmatrix}\mathbf{x}',
\end{aligned} \quad (20)$$

where $\mathbf{L}$ is the filter gain; it is determined by solving an algebraic Riccati equation; $\mathbf{y}_v$ is the sum of the output and the measurement noise.

3.2. Feedforward Control. The filtered-reference LMS (FxLMS) algorithm is widely studied and has been used in many active vibration control areas due to its easy implementation and remarkable performance. In a practical

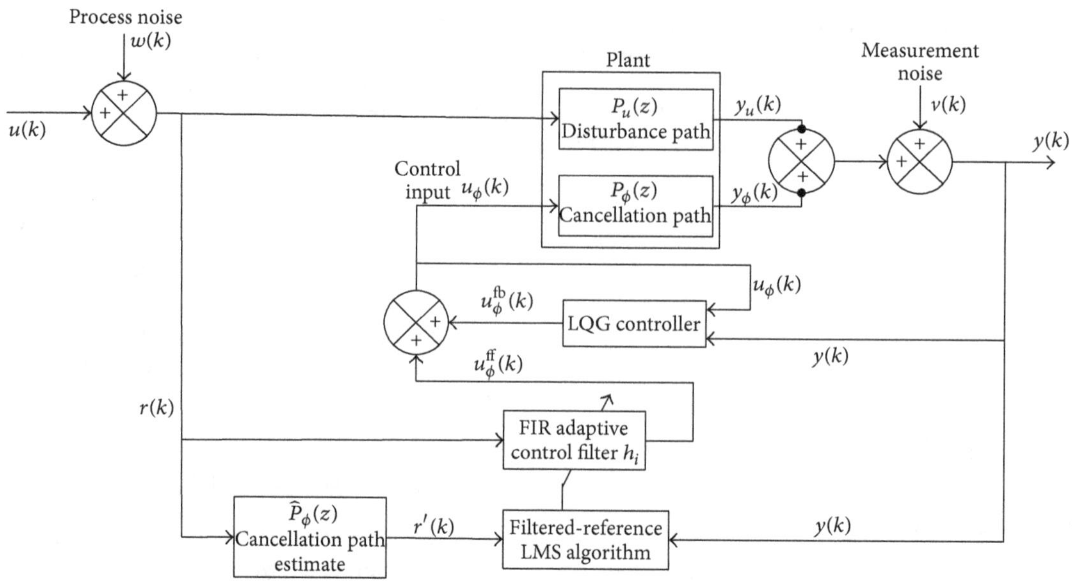

FIGURE 2: Schematic diagram of the generalized plant for hybrid (combined feedback/feedforward) controller.

vibration control system, a cancellation path exists, which has an important effect on the control results. In FxLMS algorithm, the model of the cancellation path is established. The primary disturbance $r(k)$ passes through the estimated model of the cancellation path, and a filtered-reference signal $r'(k)$, which is correlated with the disturbance, is produced. This signal is then adaptively filtered to generate the necessary control action f_c^{ff} to cancel the effect of the primary disturbance.

An adaptive finite impulse response (FIR) filter, whose ith coefficient at the kth sample time $h_i(k)$ is updated using the LMS algorithm, is used in filtering process. The coefficients vector $h_i(k)$ and the filter output $f_c^{ff}(k)$ are obtained from

$$h_i(k) = h_i(k-1) + 2\mu e(k-1) r'(k-1),$$
$$f_c^{ff}(k) = \sum_{i=0}^{N-1} h_i(k) r'(k-i), \tag{21}$$

where μ is the step size, which has a significant influence on the convergence rate of the LMS algorithm, and N is the number of filter coefficients (or the filter length).

The control signal has to pass through a part of the physical system before the error sensor measures the output. This physical path, $P_\phi(z)$, is called the cancelation path; this transfer characteristics also affect the performance of the FxLMS algorithm. Therefore, the output of the plant due to the feedforward control input only, $y_\phi(k)$, is given by

$$y_\phi(k) = \sum_{j=0}^{M-1} g_j \sum_{i=0}^{N-1} h_i(k) r'(k-i-j), \tag{22}$$

where g_i is the discrete impulse response of the cancelation path $P_\phi(z)$ which is assumed to be of order M. Considering the measurement noise $v(k)$, the output of the system, $y(k)$, can be written as

$$y(k) = y_u(k) + \sum_{j=0}^{M-1} g_j \sum_{i=0}^{N-1} h_i(k) r'(k-i-j) + v(k), \tag{23}$$

where $y_u(k)$ is the response due to the effects of the primary disturbance alone.

3.3. Hybrid Control. In general, the feedback control is more appropriate for suppression of stochastic disturbance. Meanwhile, it is susceptible to modeling errors. By contrast, the adaptive feedforward control is not susceptible to modeling errors and inherently stable, and it is more suitable to reduce the periodic or limited-band disturbance. Therefore, it is necessary to develop a hybrid (combined feedback/feedforward) control to attenuate vibration of plates with ACLD treatments against stochastic and harmonic disturbance simultaneously. The schematic diagram of the discrete-time generalized plant for a hybrid SISO control system is shown in Figure 2. It is seen that LQG control is considered as feedback channel and an adaptive filtered-reference LMS control is used as feedforward channel.

4. Numerical Simulation and Discussion

To the first bending mode of the plate, a cantilever ACLD/plate as shown in Figure 3 is used as numerical example to demonstrate the effectiveness of the above controllers. The emphasis is on the performance discussion of the controllers.

4.1. The Numerical Model of the ACLD/Plate System. The base plate, with the size of $0.2 \times 0.1\ \text{m}^2$, is partly treated with ACLD treatments. The main physical parameters of the base plate

TABLE 1: Geometrical and physical parameters of ACLD/plate system.

Aluminum		P-5H		ZN-1			
h_p	0.0008 m	h_c	0.0005 m	h_v	0.001 m	ς_1	148.0
E_p	69 GPa	E_c	74.5 GPa	μ_v	0.3	ς_2	12.16
μ_p	0.3	μ_c	0.32	ρ_v	789.5 kg/m^3	ς_3	810.4
ρ_p	2800 kg/m^3	ρ_c	7450 kg/m^3	G^∞	554200 Pa	ω_1	896200
		d_{31}, d_{32}	186×10^{-12} C/N	α_1	3.960	ω_2	927800
				α_2	65.69	ω_3	761300
				α_3	1.447		

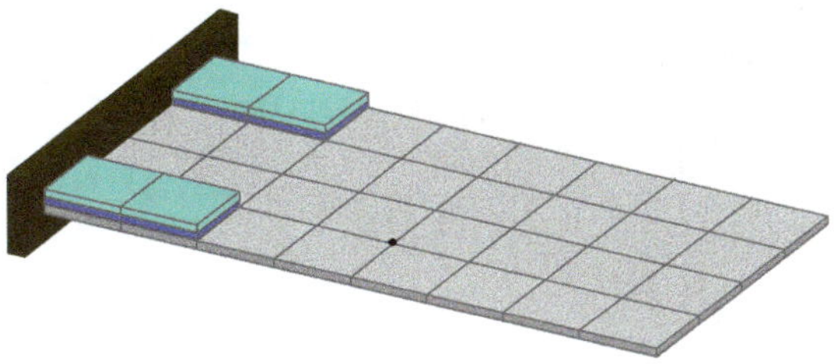

FIGURE 3: The schematic diagram of finite element grid for ACLD/plate.

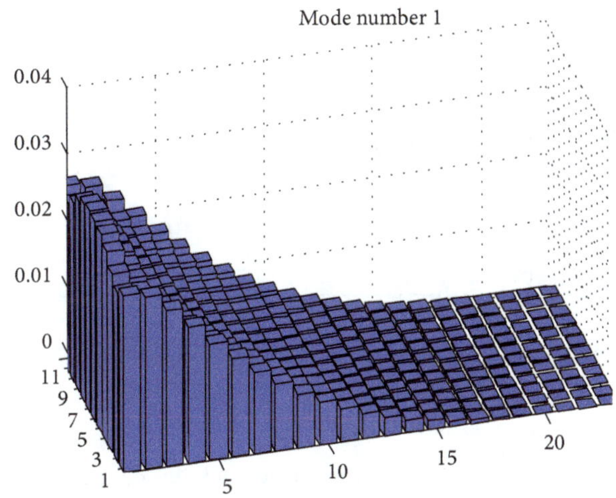

FIGURE 4: The modal strain energy of the first mode.

(aluminum), viscoelastic layer (ZN-1), and piezoelectric layer (P-5H) are listed in Table 1.

The cantilever plate/ACLD is divided into 8×4 elements. Four ACLD composite elements are applied in the root. The locations of the ACLD patches are decided according to the modal energy of the first mode [21], which is shown in Figure 4. A model reduction in which the deflection of the structure and displacements of the piezoelectric layer are selected to be master d.o.f.s and all others are slave d.o.f.s is performed according to the model reduction procedures in Section 3. The bold point in Figure 3 is used as the excited point and also the response point. The comparison of responses in frequency domain and time domain between the original model and the reduced model is presented in Figure 5. It is clear that the dynamic response of the reduced model is nearly coordinate with the dynamic response of the original model. A SISO control system is considered to design individual and hybrid controller. The input is the control voltage applied to the ACLD treatments, and the output is selected to be the displacement of the response point of the plate, which is assumed to be measurable.

4.2. The Control Results of the Feedback Controller. The LQG feedback controller is tested to mitigate the effects of stochastic disturbances. A white noise mechanical disturbance is modeled as process noise (input force disturbance). The process noise is such a signal with $E(ww^T) = 0.01N^2$ and the measurement noise is a signal with $E(ww^T) = 10^{-10}V^2$. The output weighting matrix is set to be $Q = 1 \times 10^6$ and the control weighting matrix $R = 1 \times 10^{-5}$. In order to avoid the depolarization, the maximum control voltage applied to piezoelectric actuators is set to be 150 V. The control results and the control voltage are presented in Figure 6.

As can be seen, when compared to the PCLD/plate system, the ACLD treatments can reduce the displacement of the response point of the plate significantly. The control voltage is limited within 30 V. This means that the proposed feedback control is effective against the stochastic disturbance.

4.3. The Control Results of the Feedforward Controller. The feedforward controller based on adaptive FxLMS algorithm is designed to attenuate the effects of the harmonic point-force disturbance on the ACLD/plate system in Figure 2. The disturbance frequency is close to the first natural frequency of the ACLD/plate system. For convenience, the estimation model of the plant cancellation path is substituted by the control path (control voltage to measured displacement) in the state space model. After that, the FxLMS algorithm is implemented. The filter length L is set to be 32 and the step size $u = 0.01$ in the FxLMS controller. The results are presented in Figure 7.

It can be seen that the displacement of the plate with ACLD treatment is attenuated from 9.4 mm to 0.08 mm at the end time of 20 s. The control voltage is close to 50 V. The vibration attenuation is improved substantially by ACLD treatments when the adaptive FxLMS algorithm is applied. It is clear that the FxLMS algorithm is very useful to attenuate the vibration of ACLD/plate system excited by harmonic disturbance.

TABLE 2: The convergence situation for hybrid controller and FxLMS controller under different step sizes.

	Step size μ							
	0.01	0.05	0.1	0.5	1.0	3.0	3.5	10.0
Hybrid controller	Y	Y	Y	Y	Y	Y	Y	Y
FxLMS controller	Y	Y	Y	Y	Y	Y	N	N

Y: convergence, N: no convergence.

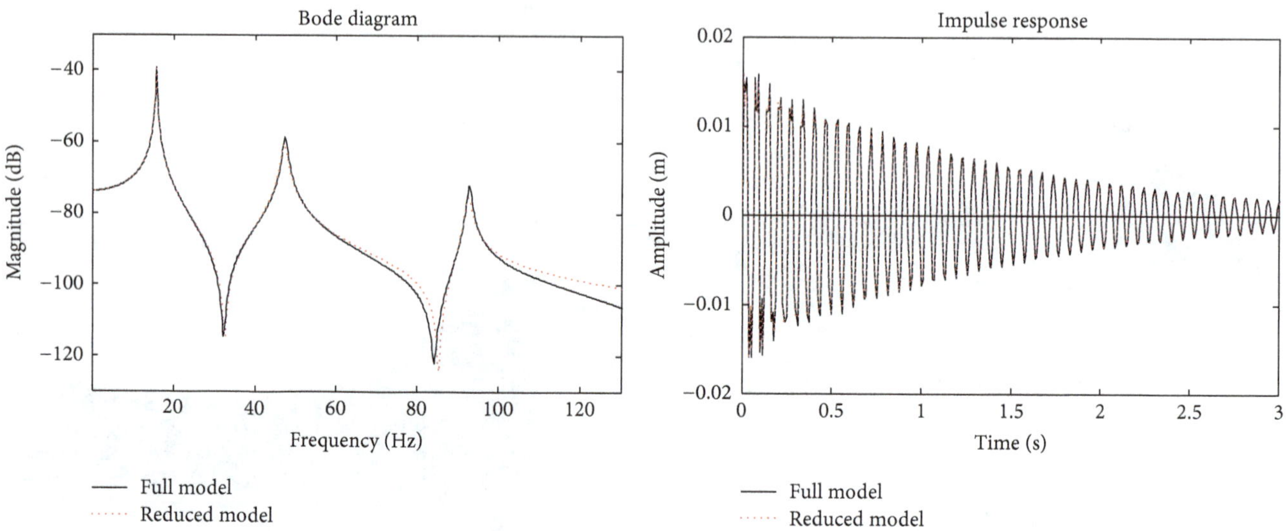

FIGURE 5: The frequency and time domain responses of the original and reduced model of the ACLD/plate.

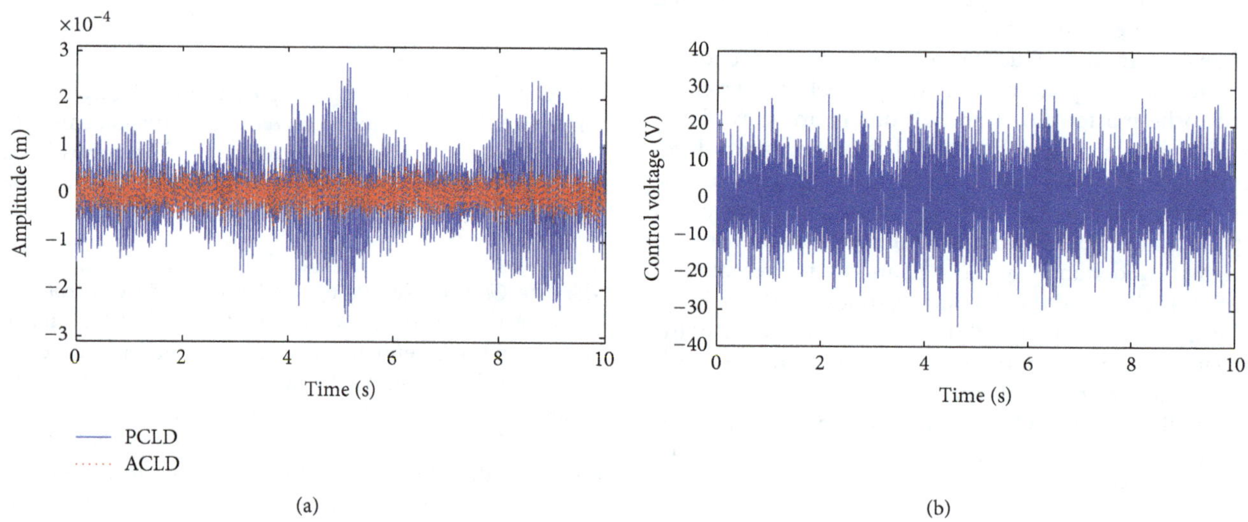

FIGURE 6: The displacement and control voltage for a white noise disturbance using the LQG feedback control: (a) the displacement response, (b) the control voltage.

4.4. The Control Results of the Hybrid Controller. Furthermore, it is particularly true that the system is often subject to both the harmonic and stochastic disturbances in engineering practice. Therefore, a hybrid controller, which combines the LQG feedback control and the FxLMS feedforward control together, is developed. Such a hybrid controller is further designed to attenuate the vibration of the plate/ACLD system subjected to the complicated disturbance. In the hybrid controller, the FxLMS control is employed as the feedforward channel, which aims to damp out the harmonic disturbance. Meanwhile, the LQG feedback control is used as a compensator. The mixed disturbance is the sum of

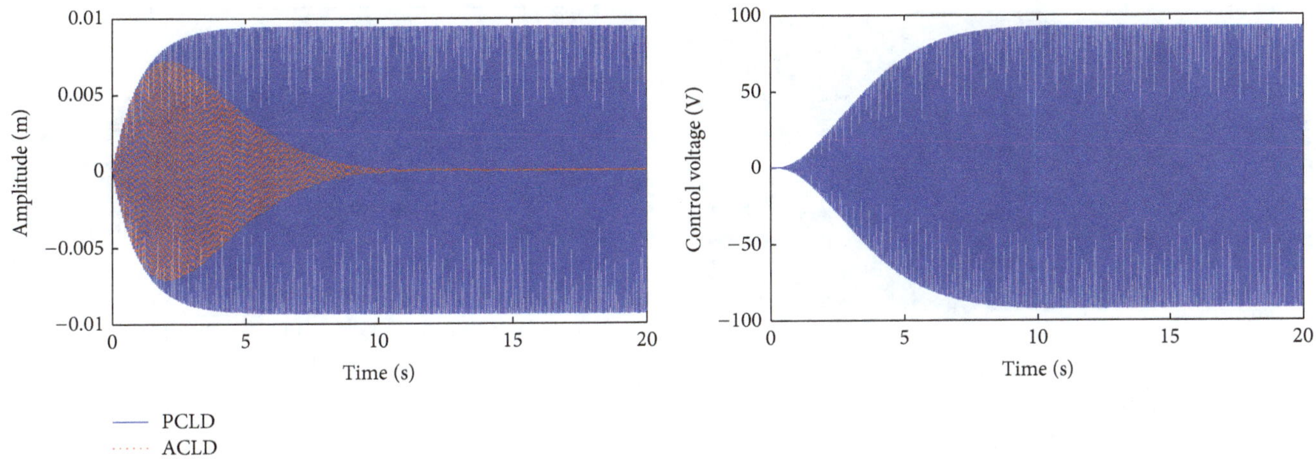

(a) (b)

FIGURE 7: The displacement, control voltage for a single frequency harmonic disturbance using the adaptive filtered-reference LMS feedforward control: (a) the displacement response, (b) the control voltage.

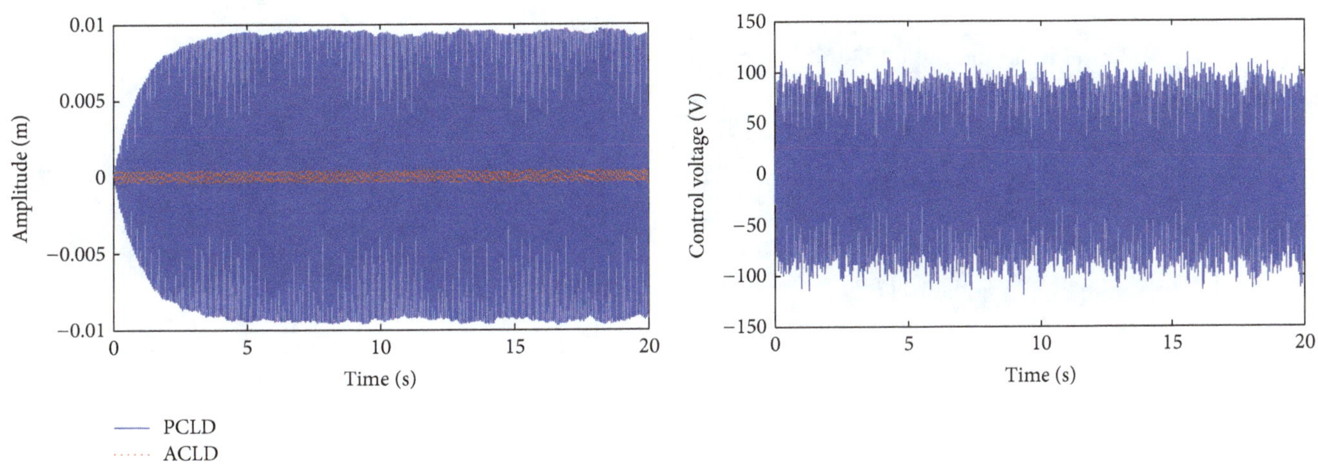

(a) (b)

FIGURE 8: The displacement response for a mixed disturbance using the LQG feedback controller: (a) the displacement response, (b) the control voltage.

the two previous disturbances. The results of individual LQG feedback control, feedforward control, and their combination are presented in Figures 8, 9, and 10, respectively.

When the hybrid control system is engaged, the feedback controller with the same weights and the feedforward controller with the same filter length are used, and the step size is set to be 0.5 to accelerate the convergence rate. It can be seen from Figures 8(a) and 10(a) that at the end of 20 s, the displacement amplitude produced by the hybrid controller is much smaller than that of the LQG feedback controller. It is also apparent that the fluctuation disappears during the convergence process and the convergence rate is improved from 10 s to 5 s when the hybrid controller is applied instead of the feedforward controller. Figure 11 shows that the hybrid control produces a smaller MSE than the feedforward control. The maximum control voltage in the feedforward controller achieves 210 V. However, the control effort in the hybrid controller is only about 110 V, a little larger than the stable control voltage of the two single controller.

4.5. The Performance Discussions to the Controllers. Table 2 lists the convergence situations for hybrid controller and FxLMS controller under different step sizes. It can be seen that the hybrid controller can converge over a larger range of step size. This implies that the hybrid controller has a better robustness than the FxLMS controller.

The effect of different step sizes on the performance of ACLD/plate with the FxLMS and hybrid controller is discussed. It is seen in Figure 12 that when the step size is increased, MSE decreases rapidly; meanwhile, the fluctuation increases. This means that convergence process is not smooth anymore, which results in the fluctuation of the control voltage shown in Figure 9(b). On the other hand, at the end

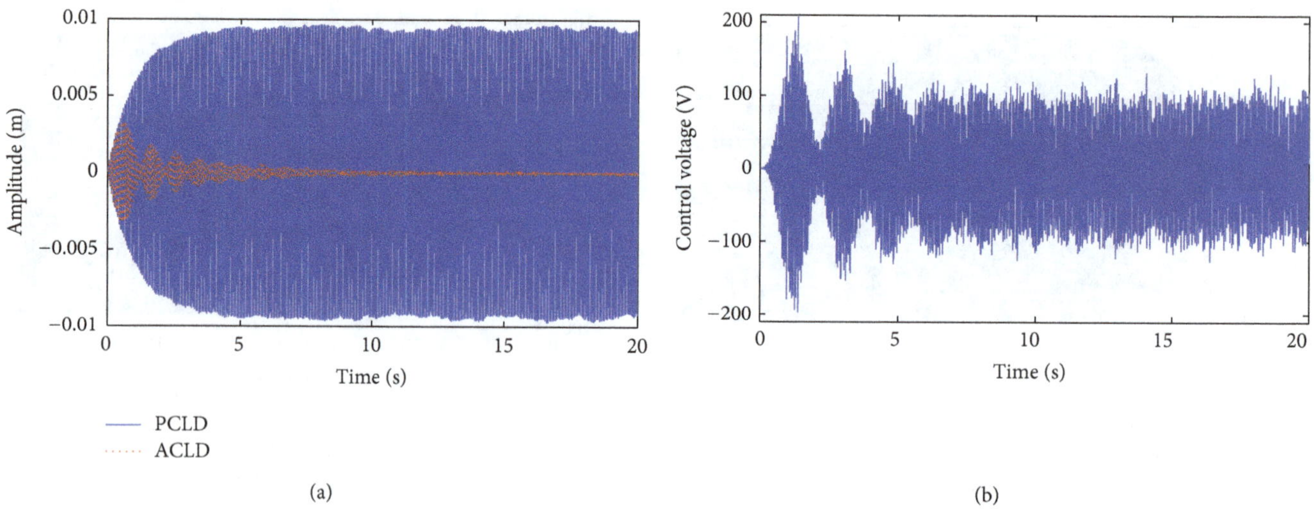

FIGURE 9: The displacement response for a mixed disturbance using the FxLMS feedforward controller: (a) the displacement response, (b) the control voltage.

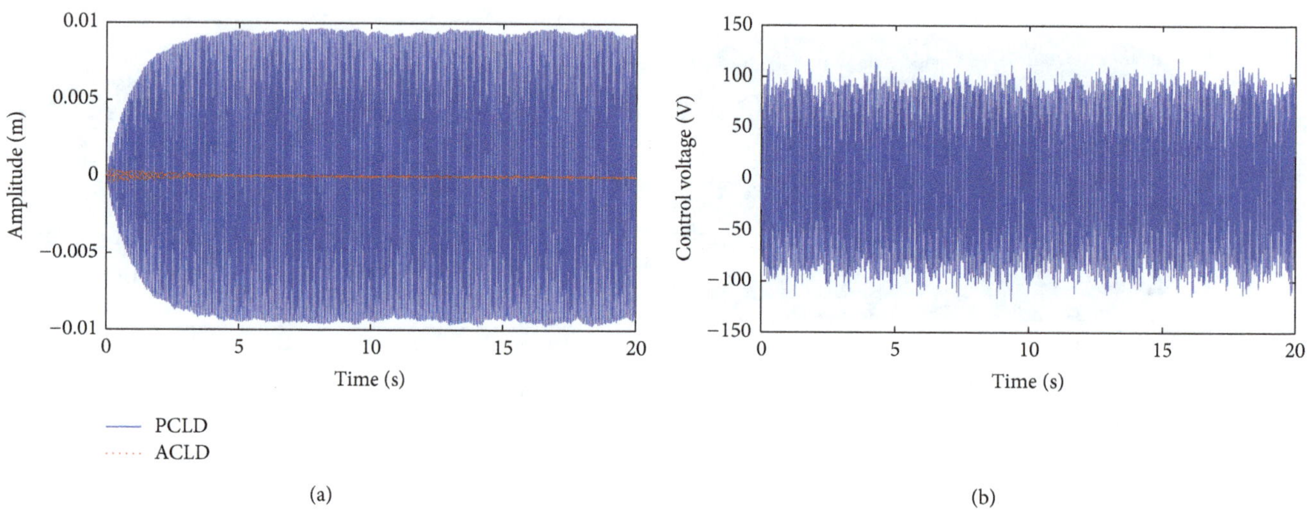

FIGURE 10: The displacement response for a mixed disturbance using the hybrid controller: (a) the displacement response, (b) the control voltage.

of the iteration process, the MSE approaches to the same value. Figure 13 shows that MSE varies with iteration number in hybrid control under different step sizes. It can be seen clearly that the fluctuation is reduced substantially in the convergence process and the convergence rate is improved greatly when the step size increases. This results in a stable control voltage in Figure 10(b).

In general, the cancellation path model of the feedforward controller is obtained by the system identification algorithm. Therefore, the errors including amplitude error and phase error in the cancellation path has a significant effect on the controller performance. Figures 14 and 15 show that the effect of phase errors of the cancellation path on the performance of ACLD/plate is considerable. It is seen that when the phase error achieves 30°, the control process in feedforward control will not converge anymore. However, when the phase error achieves 60°, the control process is still convergent in the hybrid control. It can be concluded that the hybrid control is not sensitive to the phase error, compared with the feedforward control.

Figure 16 shows the MSE using LQG feedback control under different Q matrix. Figure 17 demonstrates the MSE by using hybrid control under different Q matrix. It can be found that the performance of the feedback control has a distinct fluctuation when different Q matrix is applied. However, the performance of the hybrid control changes indistinctively. That implies that the hybrid control is more stable than individual LQG control.

Based on the above discussions, it can be inferred that the hybrid control will be a potential for vibration control of aerospace and other structures. The hybrid controller demonstrates a better performance than that of individual feedback/feedforward controller with different parameters. Meanwhile, even though one of individual controllers is out

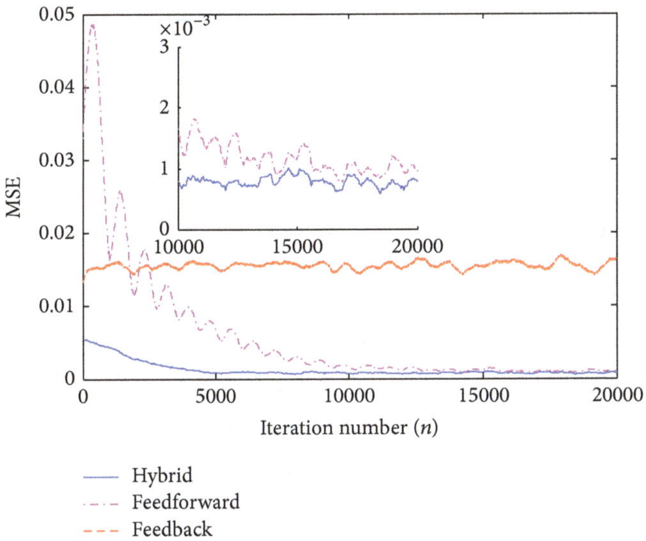

FIGURE 11: The MSE using the feedforward control, feedback control, and hybrid control.

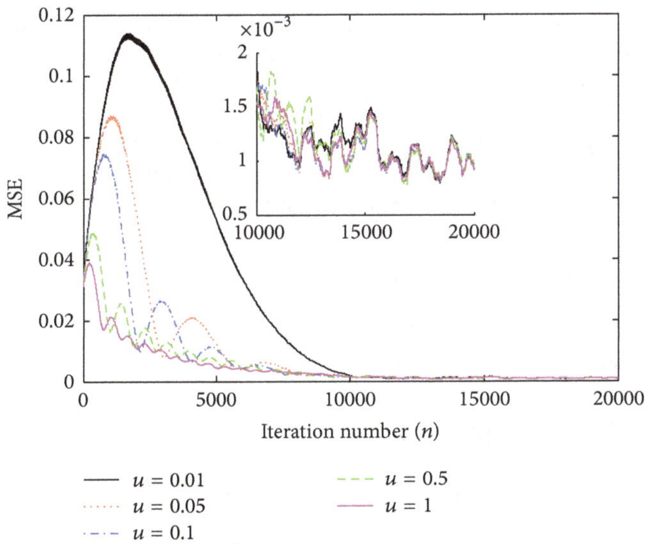

FIGURE 12: The MSE using the feedforward control under different step sizes.

of work, the other will still be effective on suppressing the vibration of structure. However, it is reasonable that the modeling accuracy of the cancellation path will make an influence on performance for vibration control in practical applications.

5. Conclusions

In this paper, a finite element model of ACLD/plate system is developed. The behavior of the viscoelastic material is modeled by the Golla-Hughes-Mctavish (GHM) method. The model reduction is carried out by means of dynamic condensation and balance reduction methods, respectively, to design an effective control system. The emphasis has been placed on vibration control of ACLD/plate using the hybrid controller which is composed of the LQG feedback controller and the adaptive FxLMS feedforward controller. The performance of these controllers with different parameters is further discussed in detail.

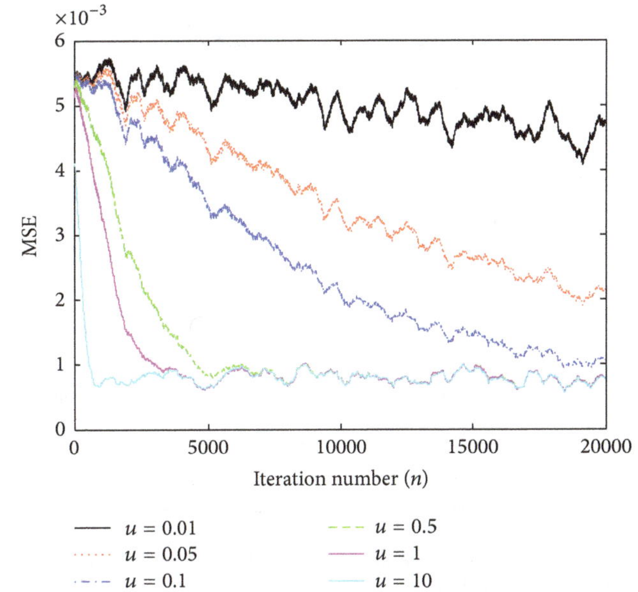

FIGURE 13: The MSE using the hybrid control under different step sizes.

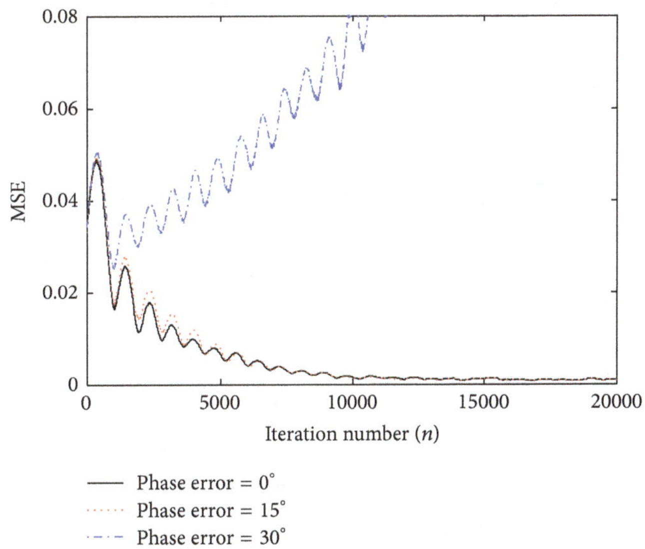

FIGURE 14: The MSE using the feedforward control under different phase errors of the cancellation path.

The results show that the reduction process is effective in low frequency range. The LQG feedback controller is very useful to attenuate the vibration of the ACLD/plate system against stochastic disturbance, while the adaptive FxLMS feedforward controller is more appropriate to attenuate the vibration of the ACLD/plate system induced by harmonic disturbance. Furthermore, the LQG feedback controller and

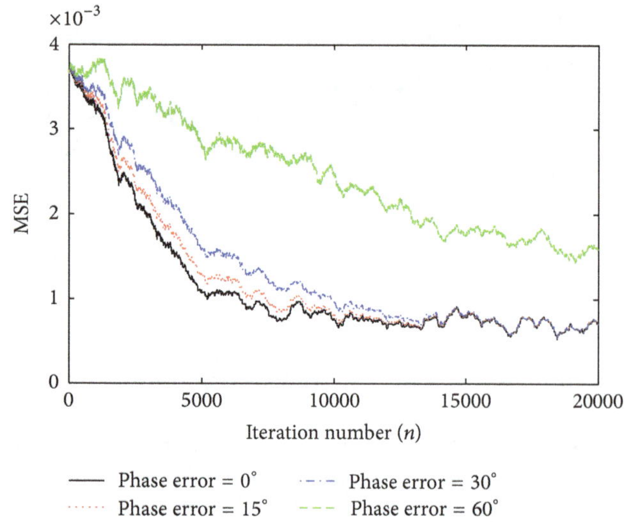

FIGURE 15: The MSE using the hybrid control under different phase errors of the cancellation path.

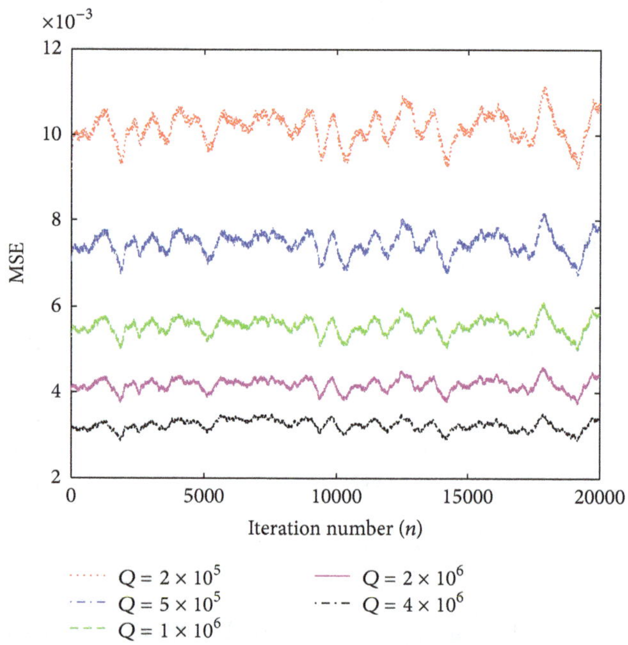

FIGURE 16: The MSE using the LQG feedback control under different Q matrix.

the adaptive FxLMS feedforward controller are combined as the hybrid controller, which can reduce substantially the displacement amplitude of plate/ACLD system subjected to a complicated disturbance without requiring more control effort and is more stable and smooth. On the other hand, the hybrid controller demonstrates a rapider and more stable convergence rate than the adaptive feedforward controller. Meanwhile, the hybrid controller demonstrates much better robustness than individual LQG feedback control or adaptive feedforward control.

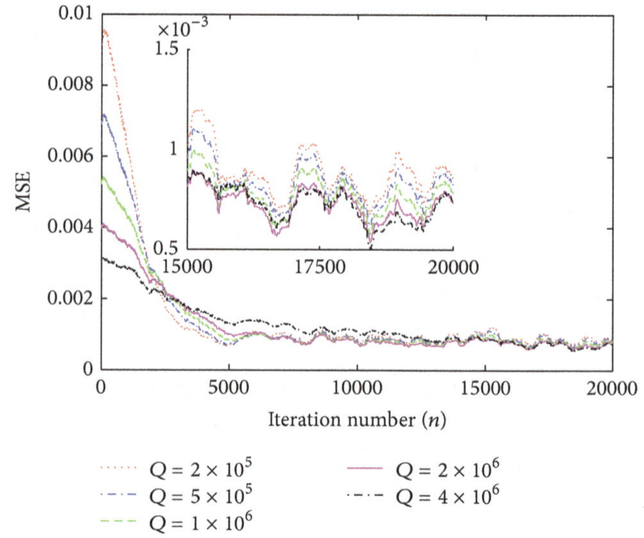

FIGURE 17: The MSE using the hybrid control under different Q matrix.

It is worth mentioning that although the hybrid controller presented is shown theoretically with better performance for a single mode vibration control, the research for SISO/MIMO vibration control of several modes using hybrid control strategies is more applicable in practical flexible structures of aerospace and other engineering fields. An attempt to address these issues is recommended to future studies.

Conflict of Interests

The authors declare that there is no conflict of interests regarding the publication of this paper.

Acknowledgments

This work was supported by Nature Science Foundation of China (no. 50775225), the State Key Laboratory of Mechanical Transmission, Chongqing University (no. 0301002109165). These financial supports are gratefully acknowledged.

References

[1] R. C. Fuller, J. S. Elliott, and A. P. Nelson, *Active Control of Vibration*, Academic Press, London, UK, 1997.

[2] C. M. A. Vasques and J. D. Rodrigues, "Active vibration control of smart piezoelectric beams: comparison of classical and optimal feedback control strategies," *Computers & Structures*, vol. 84, no. 22-23, pp. 1402–1414, 2006.

[3] A. H. N. Shirazi, H. R. Owji, and M. Rafeeyan, "Active vibration control of an FGM rectangular plate using fuzzy logic controllers," *Procedia Engineering*, vol. 14, pp. 3019–3026, 2011.

[4] D. Thakkar and R. Ganguli, "Induced shear actuation of helicopter rotor blade for active twist control," *Thin-Walled Structures*, vol. 45, no. 1, pp. 111–121, 2007.

[5] A. K. Rao, K. Natesan, M. S. Bhat, and R. Ganguli, "Experimental demonstration of H_∞ control based active vibration suppression in composite fin-tip of aircraft using optimally placed

piezoelectric patch actuators," *Journal of Intelligent Material Systems and Structures*, vol. 19, no. 6, pp. 651-669, 2008.

[6] M. K. Kwak and D. H. Yang, "Active vibration control of a ring-stiffened cylindrical shell in contact with unbounded external fluid and subjected to harmonic disturbance by piezoelectric sensor and actuator," *Journal of Sound and Vibration*, vol. 332, no. 20, pp. 4775–4797, 2013.

[7] A. Benjeddou, "Advances in hybrid active-passive vibration and noise control via piezoelectric and viscoelastic constrained layer treatments," *Journal of Vibration and Control*, vol. 7, no. 4, pp. 565–602, 2001.

[8] A. Baz and J. Ro, "Vibration control of rotating beams with active constrained layer damping," *Smart Materials and Structures*, vol. 10, no. 1, pp. 112–120, 2001.

[9] M. C. Ray, J. Oh, and A. Baz, "Active constrained layer damping of thin cylindrical shells," *Journal of Sound and Vibration*, vol. 240, no. 5, pp. 921–935, 2001.

[10] T. H. Chen, *Vibration control of cylindrical shells with active constrained layer damping [Ph.D. thesis]*, Catholic University, Washington, DC, USA, 1996.

[11] W. H. Liao and K. W. Wang, "On the active-passive hybrid control actions of structures with active constrained layer treatments," *Journal of Vibration and Acoustics, Transactions of the ASME*, vol. 119, no. 4, pp. 563–571, 1997.

[12] J. X. Gao and W. H. Liao, "Vibration analysis of simply supported beams with enhanced self-sensing active constrained layer damping treatments," *Journal of Sound and Vibration*, vol. 280, no. 1-2, pp. 329–357, 2005.

[13] S. Kumar, R. Kumar, and R. Sehgal, "Enhanced ACLD treatment using stand-off-layer: FEM based design and experimental vibration analysis," *Applied Acoustics*, vol. 72, no. 11, pp. 856–872, 2011.

[14] J. Shivakumar, M. H. Ashok, and M. C. Ray, "Active control of geometrically nonlinear transient vibrations of laminated composite cylindrical panels using piezoelectric fiber reinforced composite," *Acta Mechanica*, vol. 224, no. 1, pp. 1–15, 2013.

[15] P. H. Shah and M. C. Ray, "Active control of laminated composite truncated conical shells using vertically and obliquely reinforced 1-3 piezoelectric composites," *European Journal of Mechanics A/Solids*, vol. 32, pp. 1–12, 2012.

[16] M. C. Ray and J. N. Reddy, "Active damping of laminated cylindrical shells conveying fluid using 1–3 piezoelectric composites," *Composite Structures*, vol. 98, pp. 261–271, 2013.

[17] P. H. Shah and M. C. Ray, "Active structural-acoustic control of laminated composite truncated conical shells using smart damping treatment," *Journal of Vibration and Acoustics, Transactions of the ASME*, vol. 135, no. 2, Article ID 021001, 2013.

[18] M. V. Kumar, P. Sampath, S. Suresh, S. N. Omkar, and R. Ganguli, "Design of a stability augmentation system for a helicopter using LQR control and ADS-33 handling qualities specifications," *Aircraft Engineering and Aerospace Technology*, vol. 80, no. 2, pp. 111–123, 2008.

[19] J. Rohlfing, P. Gardonio, and D. J. Thompson, "Comparison of decentralized velocity feedback control for thin homogeneous and stiff sandwich panels using electrodynamic proof-mass actuators," *Journal of Sound and Vibration*, vol. 330, no. 5, pp. 843–867, 2011.

[20] S. R. Viswamurthy and R. Ganguli, "Performance sensitivity of helicopter global and local optimal harmonic vibration controller," *Computers and Mathematics with Applications*, vol. 56, no. 10, pp. 2468–2480, 2008.

[21] A. Baz, "Boundary control of beams using active constrained layer damping," *Journal of Vibration and Acoustics*, vol. 119, no. 2, pp. 166–172, 1997.

[22] A. Baz, "Robust control of active constrained layer damping," *Journal of Sound and Vibration*, vol. 211, no. 3, pp. 467–480, 1998.

[23] J. L. Crassidis, A. Baz, and N. Wereley, "H_∞ control of active constrained layer damping," *Journal of Vibration and Control*, vol. 6, no. 1, pp. 113–136, 2000.

[24] T. Liu, H. Hua, and Z. Zhang, "Robust control of plate vibration via active constrained layer damping," *Thin-Walled Structures*, vol. 42, no. 3, pp. 427–448, 2004.

[25] J. J. Rodriguez, *Comparison of sliding mode and state feedback control applied to a partilly treated actively damped beam [M.S. thesis]*, University of Texas, Austin, Tex, USA, 2006.

[26] P. Belanger, A. Berry, Y. Pasco, O. Robin, Y. St-Amant, and S. Rajan, "Multi-harmonic active structural acoustic control of a helicopter main transmission noise using the principal component analysis," *Applied Acoustics*, vol. 70, no. 1, pp. 153–164, 2009.

[27] L. Lei, Z. Gu, and M. Lu, "MIMO hybrid control of structural responses for helicopter," *Chinese Journal of Aeronautics*, vol. 16, no. 3, pp. 151–156, 2003.

[28] Y. Q. Cao, *Study on vibration and noise control for car body structure based on smart constrained layer damping [Ph.D. thesis]*, Chongqing University, Chongqing, China, 2011.

[29] W. R. Saunders, H. H. Robertshaw, and R. A. Burdisso, "A hybrid structural control approach for narrow-band and impulsive disturbance rejection," *Noise Control Engineering Journal*, vol. 44, no. 1, pp. 11–21, 1996.

[30] D. P. Man and A. Preumont, "Hybrid feedback-feedforward control for vibration suppression," *Journal of Structure Control*, vol. 3, no. 1-2, pp. 33–44, 1996.

[31] H. Kim and H. Adeli, "Hybrid feedback-least mean square algorithm for structural control," *Journal of Structural Engineering*, vol. 130, no. 1, pp. 120–127, 2004.

[32] Z. Ling, L. Zheng, and Y. Wang, "Vibration and damping characteristics of cylindrical shells with active constrained layer damping treatments," *Smart Materials and Structures*, vol. 20, no. 2, pp. 1–9, 2011.

[33] Z. Q. Qu, "A multi-step method for matrix condensation of finite element models," *Journal of Sound and Vibration*, vol. 214, no. 5, pp. 965–971, 1998.

[34] M. J. Lam, D. J. Inman, and W. R. Saunders, "Hybrid damping models using the Golla-Hughes-McTavish method with internally balanced model reduction and output feedback," *Smart Materials and Structures*, vol. 9, no. 3, pp. 362–373, 2000.

18

Active Control Optimization for Minimizing the Dynamic Response of Functionally Graded Plate

M. E. Fares, M. Kh. Elmarghany, and Doaa Atta

Department of Mathematics, Faculty of Science, Mansoura University, Mansoura 35516, Egypt

Correspondence should be addressed to M. E. Fares; me_fares@yahoo.com and Doaa Atta; doaaatta44@gmail.com

Academic Editor: Ronald M. Barrett

A control optimization problem for functionally graded (FG) plates is presented using a first-order shear deformation plate theory including through-the-thickness normal strain effect. The aim of the optimization is to minimize the vibrational response of FG plate with constraints on the control energy used in the damping process. An active control optimization is presented to determine the optimal level of a closed loop control function. Plate thickness and a homogeneity parameter of FG plates are used as design variables. Numerical results for the optimal control force and the total energy for a simply supported FG plate are given. The influence of through-the-thickness normal strain effect on the accuracy of the obtained results is illustrated. The effectiveness of the present control and design procedure is examined.

1. Introduction

The rapid development in various industrial fields requires new materials that can serve useful functions under certain conditions. In recent years, functionally graded materials (FGMs) have gained considerable attention in many engineering applications, and they are considered as a potential structural material for future high-speed spacecraft and power generation industries. Several studies have been performed to analyze the behavior of functionally graded plates [1].

In the aerospace industry and many other engineering applications, the suppression of excessive vibrations occurring in large structures represents one of the most pressing and difficult problems facing structural designers. An effective means of suppressing excessive vibrations is by active structural control. Thus, there is need for new light materials possessing a high degree of flexibility and with very low natural damping. These factors motivated the development for more accurate tools of analysis and rigorous design methods [2–7]. A series of publications has appeared concerning the vibration control and design optimization of composite structures. Grandhi [8] studied the structural and control optimization of space structures. Sloss et al. [9] presented an integrated approach to solve multiobjective design and control optimization for many composite structures. Adali et al. [10, 11], Sloss et al. [12], Sadek et al. [13], and Fares et al. [14–16] have studied the optimal design and control optimization of composite laminated plates and shells subjected to thermomechanical loadings.

In another work of Fares et al. [17], nonlinear optimization schemes are presented to solve multiobjective design and control problems for composite laminated plates. Other studies on this topic may be found in [18, 19].

Many studies indicate that the inclusion of the normal strain and the transverse shear effects are very important for predicting accurate vibrational responses of composite plates and shells, particularly, for those made of FGMs; see, for example, [20]. In the work [20], it is showed that the neglecting of these effects may cause errors in predicting the deflections and frequencies in FG plates reaching 30%. However, most studies related to the topic of design and control optimization of composite plates and shells were carried out based on classical theories which neglect the shear deformation and normal strain effects.

The current work deals with the active control of the dynamic response of a FG orthotropic rectangular plate. The present formulation is based on a consistent first-order plate theory including the shear deformation and through-the-thickness normal strain effects [20]. The design and control objective are to minimize the dynamic response with minimum expenditure of control energy. The total elastic energy is taken as a measure of the dynamic response. The expenditure of the control energy is limited by attaching to the control objective a functional including the control force. The necessary and sufficient condition for optimal stabilization in the Liapunov-Bellman sense [21] is used to determine the optimal control force and controlled deflections. Numerical examples are given to assess the efficiency of the present design control approach in the damping process for FG plate.

2. Formulation and Basic Equations

Consider an elastic plate of uniform thickness h, length a, and width b. The plate is composed of an orthotropic elastic material varying in the thickness direction. Let a Cartesian coordinates system (x_1, x_2, x_3) be taken such that the mid-plane coincides with the x_1x_2 plane and the plate occupies the following region:

$$0 \le x_1 \le a, \qquad 0 \le x_2 \le b, \qquad -\frac{h}{2} \le x_3 \le \frac{h}{2}. \tag{1}$$

Let the upper surface of the plate ($x_3 = h/2$) be subjected to a transverse distributed load $q(x_1, x_2, x_3)$ which may be taken as a control force.

The present formulation is based on a first-order theory including a normal strain and shear deformation effects derived using a mixed variational approach. This theory preserves the transverse shear stresses which vanish on the top and the bottom of the plate surfaces; therefore, the condition of introducing shear correction factors is not required. The displacement field is taken in the form [20]:

$$\begin{gathered} u_1 = u_{01} + \psi_1 x_3, \qquad u_2 = u_{02} + \psi_2 x_3, \\ w = w_0 + \alpha w_1 x_3 + \beta w_2 x_3^2, \end{gathered} \tag{2}$$

where (u_1, u_2, w) are the displacements along x_1, x_2, and x_3 directions, respectively, (u_{01}, u_{02}, w_0) are the displacements of a point on the mid-plane, and (ψ_1, ψ_2) are the slopes in the x_1x_3 and x_2x_3 planes due to bending only (slope rotations). The constants α and β may take the values 0 or 1 to study the effect of the normal strain on the control process.

The infinitesimal strains ε_{ij} associated with displacements (2) are given by

$$\begin{gathered} \varepsilon_{ii} = u_{0i,i} + \psi_{i,i} x_3, \\ \varepsilon_{i3} = \psi_i + w_{0,i} + \alpha w_{1,i} x_3 + \beta w_{2,i} x_3^2, \\ (i = 1, 2), \\ \varepsilon_{33} = \alpha w_1 + 2\beta w_2 x_3, \\ \varepsilon_{12} = \left(u_{01,2} + u_{02,1}\right) + \left(\psi_{1,2} + \psi_{2,1}\right) x_3, \end{gathered} \tag{3}$$

where the suffix (,i) denotes partial differentiation with respect to the coordinate x_i.

Moreover, the stresses are taken in the following form:

$$\begin{gathered} \sigma_{ij} = \frac{N_{ij}}{h} + \frac{12 M_{ij}}{h^3} x_3, \\ \sigma_{i3} = \frac{3 Q_{i3}}{2h}\left[1 - \left(\frac{x_3}{h/2}\right)^2\right], \\ (i, j = 1, 2), \\ \sigma_{33} = \left(\frac{3 N_{33}}{2h} + \frac{30 M_{33}}{h^3} x_3\right)\left[1 - \left(\frac{x_3}{h/2}\right)^2\right] \\ + \frac{q_1}{4}\left[\left(\frac{x_3}{h/2}\right)^2 - \left(\frac{x_3}{h/2}\right) - 2\right]\left(1 + \frac{x_3}{h/2}\right) \\ + \frac{q_2}{4}\left[\left(\frac{x_3}{h/2}\right)^2 + \left(\frac{x_3}{h/2}\right) - 2\right]\left(1 - \frac{x_3}{h/2}\right). \end{gathered} \tag{4}$$

The governing equations of this system are determined using the dynamic modified version of the mixed variational principle of Ressiner [20]:

$$\begin{gathered} \delta u_{0i} : N_{ii,i} + N_{ij,j} = I_1 \ddot{u}_{0i} + I_2 \ddot{\psi}_i, \\ \delta \psi_i : M_{ii,i} + M_{ij,j} - Q_{i3} = I_2 \ddot{u}_{0i} + I_3 \ddot{\psi}_i, \\ (i, j = 1, 2, \quad i \neq j), \\ \delta w_0 : Q_{13,1} + Q_{23,2} + q = I_1 \ddot{w}_0 + \alpha I_2 \ddot{w}_1 + \beta I_3 \ddot{w}_2, \\ \delta w_1 : hq - N_{33} = I_2 \ddot{w}_0 + \alpha I_3 \ddot{w}_1 + \beta I_4 \ddot{w}_2, \\ \delta w_2 : \frac{h^2\left(Q_{13,1} + Q_{23,2}\right)}{20} - 2M_{33} + \frac{9h^2 q}{20} \\ = I_3 \ddot{w}_0 + \alpha I_4 \ddot{w}_1 + \beta I_5 \ddot{w}_2. \end{gathered} \tag{5}$$

Moreover, the constitutive equations are

$$\begin{Bmatrix} N_{11} \\ N_{22} \\ N_{33} \\ N_{12} \\ M_{11} \\ M_{22} \\ M_{33} \\ M_{12} \end{Bmatrix} = \begin{bmatrix} A_{11} & A_{12} & A_{13} & 0 & B_{11} & B_{12} & B_{13} & 0 \\ A_{12} & A_{22} & A_{23} & 0 & B_{12} & B_{22} & B_{23} & 0 \\ A_{13} & A_{23} & A_{33} & 0 & \overline{B}_{13} & \overline{B}_{23} & B_{33} & 0 \\ 0 & 0 & 0 & A_{66} & 0 & 0 & 0 & B_{66} \\ B_{11} & B_{12} & \overline{B}_{13} & 0 & D_{11} & D_{12} & D_{13} & 0 \\ B_{12} & B_{22} & \overline{B}_{23} & 0 & D_{12} & D_{22} & D_{23} & 0 \\ B_{13} & B_{23} & B_{33} & 0 & D_{13} & D_{23} & D_{33} & 0 \\ 0 & 0 & 0 & B_{66} & 0 & 0 & 0 & D_{66} \end{bmatrix}^{-1} \begin{bmatrix} u_{01,1} - F_{13} \\ u_{02,2} - F_{23} \\ \alpha w_1 - F_{33} \\ u_{01,2} + u_{02,1} \\ \psi_{1,1} - K_{13} \\ \psi_{2,2} - K_{23} \\ 2\beta w_2 - K_{33} \\ \psi_{1,2} + \psi_{2,1} \end{bmatrix} \quad (6a)$$

$$Q_{i3} = \frac{1}{A_{kk}} \left[\psi_i + w_{0,i} + \left(\frac{\beta h^2 w_{2,i}}{20} \right) + H_i \right], \quad \left(i = 1, 2; k = i + \left(\frac{4}{i} \right) \right). \quad (6b)$$

The stress and moment resultants N_{ij}, M_{ij}, and Q_{ij} in (5) and (6a) and (6b) can be expressed as

$$\left(N_{ij}, M_{ij} \right) = \int_{-h/2}^{h/2} (1, x_3)\, \sigma_{ij}\, dx_3, \quad (i, j = 1, 2, 3)$$
$$\left(Q_{k3} \right) = \int_{-h/2}^{h/2} \sigma_{k3}\, dx_3, \quad (k = 1, 2), \quad (7)$$

and the stiffness coefficients A_{ij}, B_{ij}, and D_{ij} are given in Appendix A.

In the present control problem, all edges of the plate are taken simply supported (SSSS), and then, the following conditions must be satisfied at the plate edges:

$$u_{01} = w_0 = w_1 = w_2 = \psi_2 = N_{11} = M_{11} = 0, \qquad x_1 = 0, a$$
$$u_{02} = w_0 = w_1 = w_2 = \psi_1 = N_{22} = M_{22} = 0, \qquad x_2 = 0, b. \quad (8)$$

3. The Optimal Design and Control Problem

The objective of the present control problem is to minimize the dynamic response of the plate in a specified time $0 \le t \le \tau \le \infty$ with the minimum possible expenditure of force $q(x_1, x_2, t)$, when the plate is subjected to the following initial disturbances:

$$w(x_1, x_2, 0) = \overline{A}(x_1, x_2),$$
$$\dot{w}(x_1, x_2, 0) = \overline{B}(x_1, x_2). \quad (9)$$

The dynamic response of the plate is measured by a cost functional including the energy of the plate which is a function of displacements, its derivatives, and the velocity. The control force $q(x_1, x_2, t)$ is introduced into the formulation by taking a performance index which comprises a weighted sum of the energy (cost functional) and penalty functional involving the control energy. Then, the mathematical formulation of the control problem can be reduced to determine the optimization variable q that minimizes the following control objective:

$$J = \mu_1 J_1 + \mu_2 J_2 + \mu_3 J_3,$$
$$J_1 = \frac{1}{2} \int_0^{\infty} \int_0^b \int_0^a \int_{-h/2}^{h/2} \varepsilon_{ij} \sigma_{ij}\, dx_3 dx_1 dx_2 dt, \quad (i, j = 1, 2, 3),$$
$$J_2 = \frac{1}{2} \int_0^{\infty} \int_0^b \int_0^a \int_{-h/2}^{h/2} \rho \left(\dot{u}_1^2 + \dot{u}_2^2 + \dot{w}^2 \right) dx_3 dx_1 dx_2 dt,$$
$$J_3 = \int_0^{\infty} \int_0^b \int_0^a q^2 (x_1, x_2, t)\, dx_1 dx_2 dt, \quad (10)$$

where μ_1, μ_2, and μ_3 are positive constant weighting factors and J_1 and J_2 represent the strain and the kinetic energies of the plate, respectively. The functional J_3 is a penalty term involving the control function $q \in L^2$, where L^2 denotes the set of all bounded square integrable functions on the domain of solution $\{0 \le x_1 \le a, 0 \le x_2 \le b, 0 \le t \le \tau \le \infty\}$. Note that, the functional J is a quadratic and, hence, differentiable and positive definite on the domain of solution.

It is obvious that the solution of the present optimal control problem requires finding a solution of the initial boundary-value problem (5), (6a), (6b), (8), and (9) that minimizes the functional (10). Moreover, the design optimization procedure aims to minimize the actively controlled dynamic responses using geometry and material parameters as design variables.

4. Solution Procedure

The displacement functions $(u_{01}, u_{02}, w_0, \psi_1, \psi_2, w_1, w_2)$ and the mechanical load q may be taken for simply supported plate in the form:

$$(u_{0i}, \psi_i) = \sum_{m,n} (U_{imn}(t), \Psi_{imn}(t))\, X_{i,i} X_{i+1},$$

$$(i = 1 \longrightarrow i+1 = 2,\ i = 2 \longrightarrow i+1 = 1)$$

$$(w_j, q) = \sum_{m,n} \left(W_{jmn}(t), q_{mn}(t)\right) X_1 X_2, \quad (j = 0, 1, 2)$$

$$X_1(x_1) = \sin \mu_m x_1, \quad \mu_m = \frac{m\pi}{a},$$

$$X_2(x_2) = \sin \lambda_n x_2, \quad \lambda_n = \frac{n\pi}{b}, \tag{11}$$

where $U_{imn}(t)$, $\Psi_{imn}(t)$, $W_{jmn}(t)$, and $q_{mn}(t)$ are unknown functions of the time t.

Using (6a) and (6b) we can get the governing equations (5) in terms of the displacements $(u_{01}, u_{02}, w_0, \psi_1, \psi_2, w_1, w_2)$. For these equations, the in-plane inertia terms are neglected. Substituting expressions (11) into the resulting equations, we get the following time equations:

$$\begin{aligned} &U_{1i} U_{1mn} + U_{2i} U_{2mn} + \Psi_{1i} \Psi_{1mn} + \Psi_{2i} \Psi_{21mn} \\ &\quad + W_{0i} W_{0mn} + W_{1i} W_{1mn} + W_{2i} W_{2mn} \\ &\quad = \ddot{W}_{0mn} \overline{W}_i (\delta_{i3} + \delta_{i6} + \delta_{i7}) + Q_i q_{mn}, \quad (i = 1, 2, \dots, 7), \end{aligned} \tag{12}$$

where δ_{ij} is the kronecker delta and the coefficients U_{1i}, U_{2i}, Ψ_{1i}, Ψ_{2i}, W_{0i}, W_{1i}, W_{2i}, Q_i are given in Appendix B. Solving the system of (12), one gets an equation of the time-dependent functions q_{mn} and W_{0mn} only,

$$\ddot{W}_{0mn} + \omega^2 W_{0mn} = l q_0, \quad \omega^2 = \frac{\Delta_1}{\Delta_3}, \ l = \frac{\Delta_2}{\Delta_3}, \tag{13}$$

where Δ_1, Δ_2, and Δ_3 are given in Appendix B.

Again, using (2), (3), (4), and (11) with the objective functional (10), we get

$$\begin{aligned} J &= \sum_{mn} \int_0^\infty \overline{J}_{mn} dt, \\ \overline{J}_{mn} &= K_1 W_{0mn}^2 + K_2 W_{0mn} q_{mn} + K_3 q_{mn}^2 \\ &\quad + K_4 \dot{W}_{0mn}^2 + K_5 \dot{W}_{0mn} \dot{q}_{mn} + K_6 \dot{q}_{mn}^2, \end{aligned} \tag{14}$$

where the coefficients K_i $(i = 1, 2, \dots, 6)$ are given in Appendix B.

Now the optimal control problem is to find the control function $q_{mn}^{\text{opt}}(t)$ that minimizes the functional (14); that is,

$$J\left(q_{mn}^{\text{opt}}\right) \le J(q_{mn}), \quad \forall q_{mn}(t) \in L^2([0, \infty)); \tag{15}$$

that is,

$$\min_{q_{mn}} J = \min \sum J_{mn} = \sum \min_{q_{mn} \in L^2} J. \tag{16}$$

The minimization of the dynamic response using the control force q_0 can be carried out using the Liapunov-Bellman method [21] which is considered an effective approach for the solution. The necessary and sufficient condition for minimization of the functional (14) in the Liapunov-Bellman sense is

$$\min_q \left[\frac{\partial \overline{V}}{\partial W_{0mn}} \dot{W}_{0mn} + \frac{\partial \overline{V}}{\partial \dot{W}_{0mn}} \ddot{W}_{0mn} + \overline{J} \right] = 0; \tag{17}$$

provided that the Liapunov function $\overline{V}$ is a positive definite, then

$$\begin{gathered} \overline{V} = A W_{0mn}^2 + 2B W_{0mn} \dot{W}_{0mn} + C \dot{W}_{0mn}^2, \\ A > 0, \quad C > 0, \quad AC > B^2. \end{gathered} \tag{18}$$

Substituting expressions (13) and (14) into (17) one gets

$$\begin{aligned} \min_q \Bigg[&\frac{\partial \overline{V}}{\partial W_{0mn}} \dot{W}_{0mn} + \frac{\partial \overline{V}}{\partial \dot{W}_{0mn}} \left(l q_{mn} - \omega^2 W_{0mn} \right) \\ &+ K_1 W_{0mn}^2 + K_2 W_{0mn} q_{mn} + K_3 q_{mn}^2 \\ &+ K_4 \dot{W}_{0mn}^2 + K_5 \dot{W}_{0mn} \dot{q}_{mn} + K_6 \dot{q}_{mn}^2 \Bigg] = 0. \end{aligned} \tag{19}$$

This condition gives that

$$l \frac{\partial \overline{V}}{\partial \dot{W}_{0mn}} + K_2 W_{0mn} + 2 K_3 q_{mn}^{\text{opt}} = 0. \tag{20}$$

Substituting the Liapunov function (18) into the previous expression, we can get the minimum control force q_{mn}^{opt} in the form:

$$q_{mn}^{\text{opt}} = \frac{-1}{2K_3} (2Bl + K_2) W_{0mn} - \frac{1}{K_3} C l \dot{W}_{0mn}. \tag{21}$$

Again, using (18) with the expression (19), we obtain

$$\begin{aligned} \min_q \Big[&\left(2A W_{0mn} \dot{W}_{0mn} + 2B \dot{W}_{0mn}^2 \right) \\ &+ \left(2B W_{0mn} + 2C \dot{W}_{0mn} \right) \left(l q_{mn} - \omega^2 W_{0mn} \right) \\ &+ K_1 W_{0mn}^2 + K_2 W_{0mn} q_{mn} + K_3 q_{mn}^2 \\ &+ K_4 \dot{W}_{0mn}^2 + K_5 \dot{W}_{0mn} \dot{q}_{mn} + K_6 \dot{q}_{mn}^2 \Big] = 0. \end{aligned} \tag{22}$$

Using (21) with the last expression and equating the coefficients of $\dot{W}_{0mn}^2$, W_{0mn}^2, and $W_{0mn}\dot{W}_{0mn}$ by zeroes, the following system of algebraic equations is obtained:

$$C^2\left(a_1B^2+a_2B+a_3\right)+a_4B^2+a_5B+a_6=0,$$
$$C^2\left(a_7C^2+a_8B+a_9\right)+a_{10}B^2+a_{11}B+a_{12}=0,$$
$$a_{13}A+C\left(a_{14}C^2+a_{15}BC^2+a_{16}B^2+a_{17}B+a_{18}\right)=0, \tag{23}$$

where the coefficients a_i ($i = 1,2,\ldots,18$) are given in Appendix C.

Using (14) and (21), we obtain the following differential equation:

$$\ddot{W}_{0mn}+\alpha_{mn}\dot{W}_{0mn}+\Omega_{mn}^2W_{0mn}=0,$$
$$\alpha_{mn}=-lc_2,\quad \Omega_{mn}^2=\left(\omega^2-lc_1\right), \tag{24}$$
$$c_1=-\frac{(2Bl+K_2)}{(2K_3)},\qquad c_2=-\frac{Cl}{K_3}.$$

The solution of (24) for which $2\Omega_{mn}^2>\alpha_{mn}$ is given by

$$W_{0mn}^{\text{opt}}=e^{-\alpha_{mn}t/2}\left[\beta_{mn}\cos\left(\upsilon_{mn}t\right)+\gamma_{mn}\sin\left(\upsilon_{mn}t\right)\right],$$
$$\upsilon_{mn}=\Omega_{mn}^2-\frac{\alpha_{mn}}{4}, \tag{25}$$

where β_{mn} and γ_{mn} are unknown coefficients which may be obtained from conditions (9) as

$$\gamma_{mn}=\frac{\left(\alpha_{mn}\beta_{mn}+2\overline{A}_{mn}\right)}{\left(2\upsilon_{mn}\right)},$$
$$\left(\beta_{mn},\overline{A}_{mn}\right)=\frac{4}{ab\omega^2}\int_0^b\int_0^a\left(\overline{A},\overline{B}\right)X_1X_2\,dx_1dx_2. \tag{26}$$

Using (14), (21), and (25), we can get the controlled displacement, the total energy, and the optimal control force. Then we complete the minimization process by determining the optimal design of the plate using geometric and material parameters as design variables.

5. Homogenization of Material Properties and the Design Objective

Consider that a FG plate is made of two randomly distributed isotropic constituents (ceramic and metal) and the macroscopic response of the composite is isotropic. The matrix phase, denoted by the subscript 2, is reinforced by spherical particles as a particulate phase, denoted by the subscript 1. The homogeneity of this composition is taken to vary only in the thickness direction. The effective elastic moduli of the graded composite may be evaluated from the volume fraction of the constituents V_1 and V_2. The effective mass density ρ at a point is given by

$$\rho=\rho_1V_1+\rho_2V_2,$$
$$V_1=V_1^-+\left(V_1^+-V_1^-\right)\left(\frac{1}{2}+\frac{x_3}{h}\right)^p, \tag{27}$$

where V_1^+ and V_1^- are, respectively, the volume fractions of the particulate ceramic phase on the top and the bottom surfaces of the plate. The parameter p indicates the volume fraction profile through the thickness. In this case, the elastic coefficients C_{ij} become

$$C_{11}=C_{22}=C_{33}=\frac{1}{E},$$
$$C_{12}=C_{13}=C_{23}=-\frac{\nu}{E},$$
$$C_{44}=C_{55}=C_{66}=\frac{1}{G}, \tag{28}$$
$$K=\frac{E}{3(1-2\nu)},\qquad G=\frac{E}{2(1+\nu)},$$

where E is Young's modulus, ν is Poisson's ratio, G is shear modulus, and K is the effective bulk modulus.

The Mori and Tanaka [22] scheme is used to estimate the effective moduli for the regions of the graded microstructure which have a well-defined continuous matrix and a discontinuous particulate phase. The locally effective bulk modulus K and the effective shear modulus G of the composite are given by

$$\frac{K-k_1}{k_2-k_1}=\frac{V_2}{1+(1-V_2)\left((k_2-k_1)/(k_1+(4/3)G_1)\right)},$$
$$\frac{G-G_1}{G_2-G_1}=\frac{V_2}{1+(1-V_2)\left((G_2-G_1)/(G_1+f_1)\right)}, \tag{29}$$
$$f_1=\frac{G_1(9k_1+8G_1)}{(6k_1+12G_1)}.$$

The design procedure aims to determine the values of the design variable r ($=a/h$) and the index p that minimize the energy J; that is,

$$\hat{J}\left(q_{\text{opt}},r_{\text{opt}},p_{\text{opt}}\right)=\min_{r,p}J\left(q_{\text{opt}},r,p\right), \tag{30}$$

where q_{opt} is the optimal control force obtained from the previous control procedure.

6. Numerical Examples and Discussion

Numerical results are presented for the total energy J and the control force q for simply supported plates. Let the constituent materials of the FG plate be aluminum (Al) as

TABLE 1: Comparison of q_{opt}, $\tilde{q}_{opt}$, J, and $\tilde{J}$ of simply supported square Al/ZrO_2 FG plates with $m = n = 1$.

p	r	$\overline{q}_{opt}$	$\tilde{q}_{opt}$	J	$\tilde{J}$	$q^*\%$	$J^*\%$
1	5	0.0312	0.0295	0.1391	0.1401	5.5	0.7
	10	3.1358	3.0909	8.1959	8.2149	1.4	0.2
	15	43.401	43.133	103.64	103.79	0.61	0.1
	20	247.45	246.62	662.64	663.59	0.33	0.1
4	5	0.0344	0.0286	0.1062	0.1072	16.8	0.9
	10	3.0272	2.8914	6.1625	6.1720	4.5	0.2
	15	39.992	39.259	78.319	78.362	1.8	0.05
	20	221.95	219.87	499.76	500.30	0.9	0.1
10	5	0.0367	0.0284	0.0948	0.09580	22.6	1.0
	10	3.0335	2.8507	5.5484	5.5533	6.0	0.1
	15	39.201	38.246	70.938	70.456	2.4	0.01
	20	214.70	212.04	452.59	453.06	1.2	0.1

TABLE 2: Comparison of q_{opt}, $\tilde{q}_{opt}$, J, and $\tilde{J}$ of simply supported Al/ZrO_2 FG plates with $m = n = 1$, $a/h = 5$.

p	a/b	$\overline{q}_{opt}$	$\tilde{q}_{opt}$	J	$\tilde{J}$	$q^*\%$	$J^*\%$
1	1	0.0312	0.0295	0.1391	0.1401	5.5	0.7
	2	0.062	0.059	0.014	0.014	4.5	1.33
	3	0.021	0.020	0.003	0.003	4.8	1.73
4	1	0.344	0.286	0.106	0.107	16.8	0.94
	2	0.069	0.060	0.011	0.011	11.9	2.64
	3	0.023	0.021	0.003	0.003	10.1	2.28
10	1	0.367	0.284	0.095	0.096	22.7	1.06
	2	0.071	0.059	0.009	0.009	15.9	2.85
	3	0.024	0.021	0.002	0.002	11.7	2.84

TABLE 3: Comparison of q_{opt}, $\tilde{q}_{opt}$, J, and $\tilde{J}$ of simply supported square Al/ZrO_2 FG plates with $m = 2$, $n = 1$.

p	a/h	$\overline{q}_{opt}$	$\tilde{q}_{opt}$	J	$\tilde{J}$	$q^*\%$	$J^*\%$
1	5	0.0762	0.0599	0.0279	0.0282	21.5	1.3
	10	5.9923	5.6012	1.3429	1.3489	6.5	0.5
	15	88.508	85.897	14.884	14.912	3.0	0.2
4	5	0.1399	0.0603	0.0220	0.0223	56.9	1.7
	10	6.5968	5.3641	1.0158	1.0205	18.7	0.5
	15	81.835	80.583	11.203	11.201	8.5	0.02
10	5	0.2549	0.0597	0.0196	0.0199	76.6	1.7
	10	7.0887	5.3134	0.9099	0.9134	25.0	0.4
	15	89.668	79.529	10.088	10.070	11.3	0.2

a matrix phase and zirconia as particulate phase with material and geometrical properties given by

$$\begin{gathered} ZrO_2{:}E_1 = 200\,\text{GPa}, \qquad \nu_1 = 0.3, \\ \rho_1 = 5700\,\text{kg/m}^3, \qquad \text{Al}{:}E_2 = 70\,\text{GPa}, \\ \nu_2 = 0.3, \qquad \rho_2 = 2702\,\text{kg/m}^3, \\ a = b = 0.5\,\text{m}, \qquad h = 0.02\,\text{m}, \\ \mu_1 = \mu_2 = \mu_4 = 1, \qquad \mu_3 = 0.1, \\ \overline{A}(x_1, x_2) = 0.2X_1(x_1)X_2(x_2), \qquad \overline{B}(x_1, x_2) = 0, \\ \overline{q}_{opt} = \frac{a^3 q_{opt}}{(E_2 h^3)}, \qquad T = \left(\frac{ht}{a^2}\right)\sqrt{\frac{E_2}{\rho_2}}. \end{gathered} \tag{31}$$

Tables 1 and 2 contain numerical results for the controlled total energy J and the control force q_{opt} of simply supported Al/ZrO_2 FG plate with their counterparts $\tilde{J}$ and $\tilde{q}_{opt}$ which are calculated neglecting the effect of the normal strains ($\alpha = \beta = 0$). To study the influence of the normal strain effect on the control process, the following relative errors for the control force q and the total energy J are presented in Tables 1, 2, and 3 for various values of the ratios r, a/b and the index p. In these tables,

$$q^* = \left[\frac{(q_{opt} - \tilde{q}_{opt})}{q_{opt}}\right], \qquad J^* = \left[\frac{(J - \tilde{J})}{J}\right]. \tag{32}$$

Note that the relative errors in the control force decrease with the increase of ratio r, and these errors increase with increasing the index p of the volume fraction profile, so that

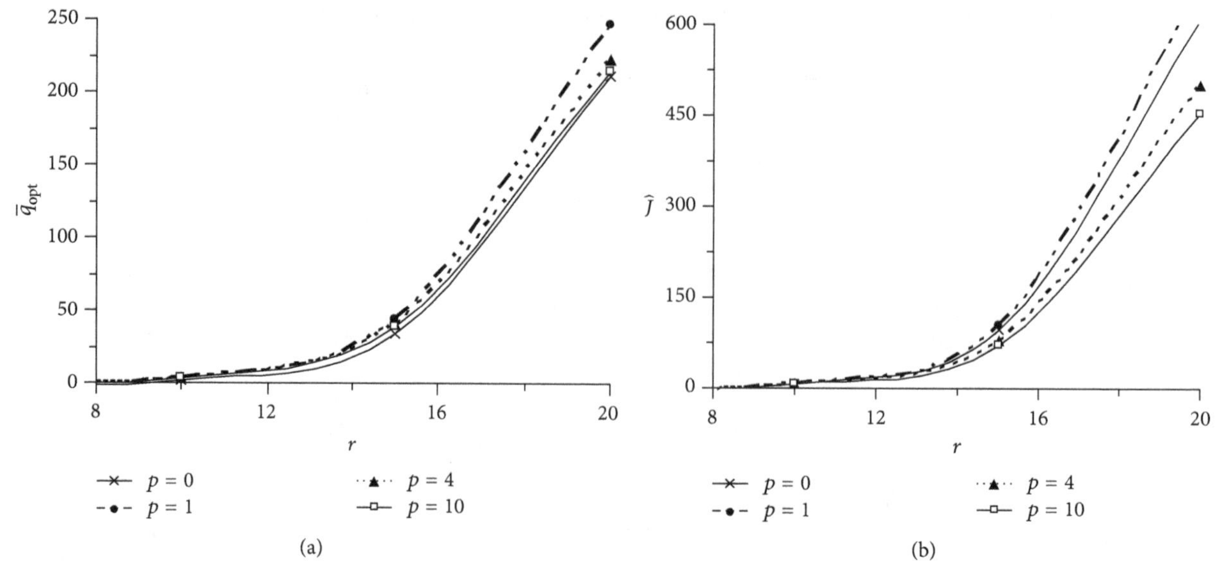

FIGURE 1: Effect of the thickness ratio r and the index p on curve for (a) the control force $\overline{q}_{\text{opt}}$, (b) for the energy $\hat{J}$, with $h = 0.02$ m, $a/b = 1$.

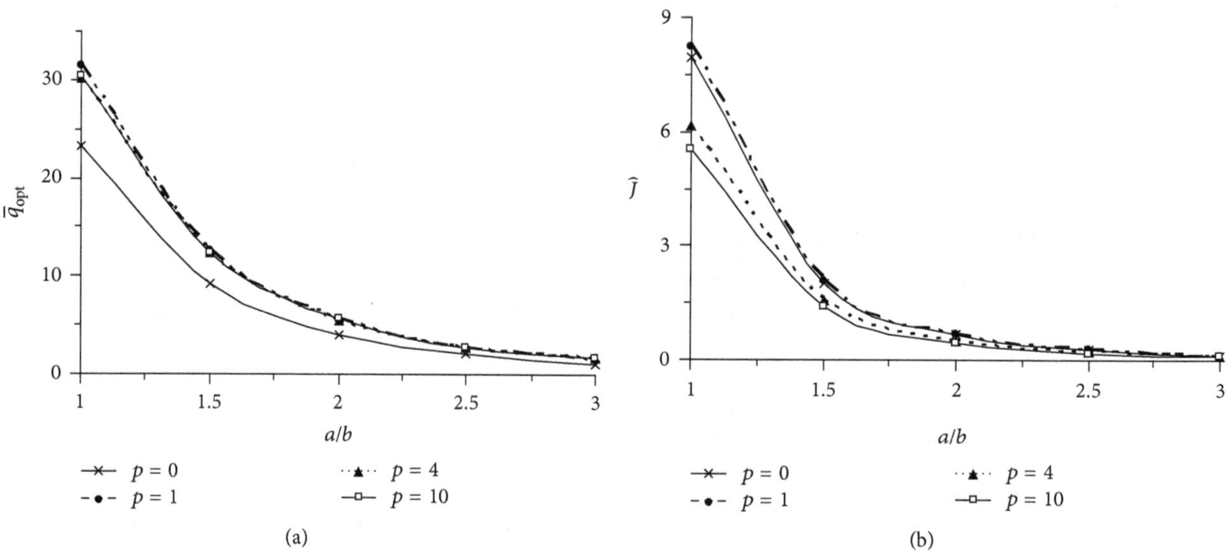

FIGURE 2: Effect of the aspect ratio a/b and the index p on curves for (a) the control force $\overline{q}_{\text{opt}}$, (b) for the energy $\hat{J}$, with $h = 0.02$ m, $r = 10$.

the errors reach more than 22% for plates with $a/b = 1$ and $p = 10$. This means that the effect of the normal strain is weak for thin FG plates with $r < 10$. Moreover, the inclusion of the effect of normal strain in the formulation with the cases $p > 1$ is required for obtaining more accurate results. This may be explained by the fact that, with increasing the index p, the homogeneity of the plate differs rapidly through the plate thickness. This requires the inclusion of the normal strain effect in the formulation to give more accurate description for the deformation process through the thickness. Table 3 contains similar numerical results as in Tables 1 and 2 for second mode $m = 2$, $n = 1$. These results show that the relative errors increase with the increase of the mode numbers m and n confirming the importance of the inclusion of the normal strain effect in the formulation of the problem to obtain more accurate results. In general, the effect of the normal strain on the accuracy of the numerical results of the total energy J is less significant than that on the control force q.

Figures 1 and 2 contain curves for the controlled force q and the total energy $\hat{J}$ $(= 100J)$ versus the ratios r and a/b for different values of index p for SSSS plates. These curves show that the volume fraction profile through the plate thickness has a significant effect on the control force and controlled energy, particularly for FG plates with $r \geq 15$ and square plates. Moreover, the curves show the high sensitivity of the total energy and control force of the FG plates versus the

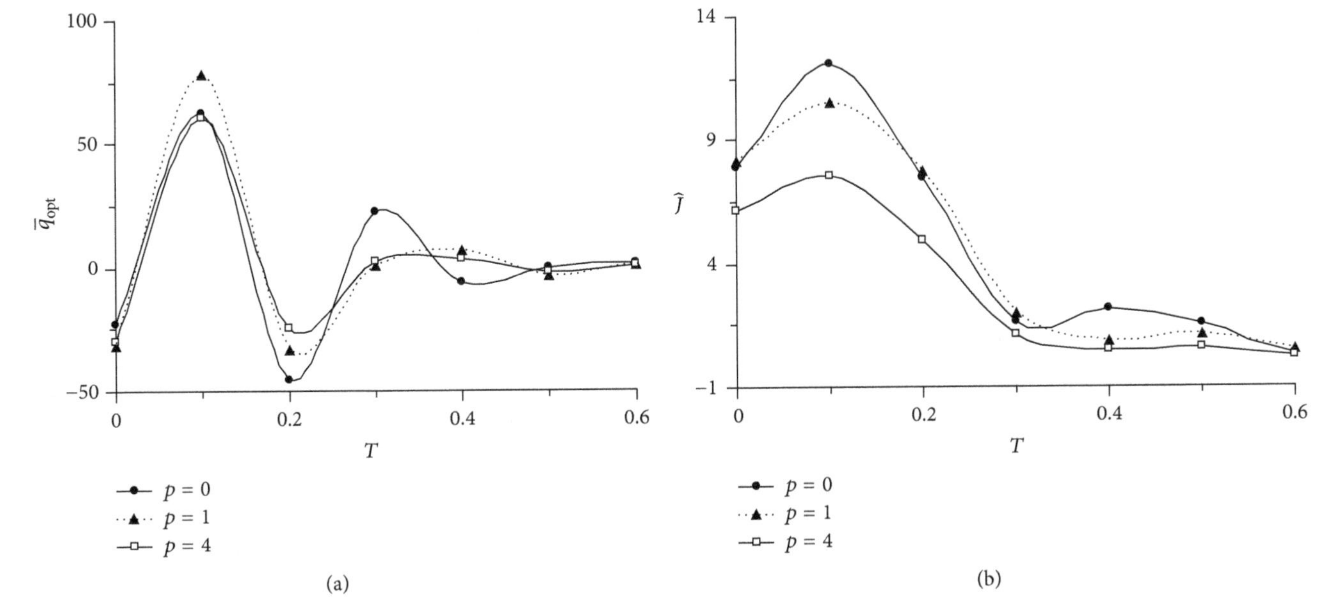

FIGURE 3: Effect of the index p on curves for (a) the optimal control force $\overline{q}_{opt}$, (b) for the energy $\hat{J}$, against the time T, $h = 0.02$ m and $a/b = 1$.

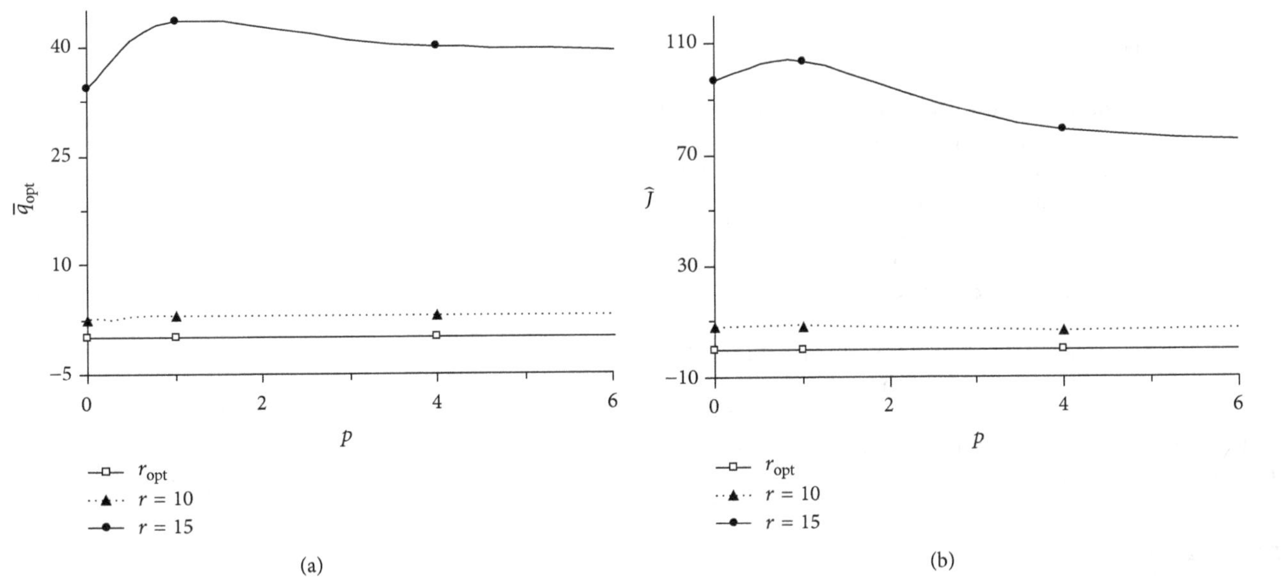

FIGURE 4: Effect of the thickness ratio r and the index p on curves of the control force $\overline{q}_{opt}$ and the energy $\hat{J}$ for SSSS plates with $a/b = 1$.

change of ratios r, a/b and the index p. This indicates that these design variables may play an important role in reducing the vibrational response of the FG plates. Figure 3 shows q-curves and $\hat{J}$-curves against the time parameter T. These curves show that the present design control procedure reduces the expenditure of the control energy and the time of the controlling process. Figure 4 contains curves for the control force q and the total energy $\hat{J}$ against the index number p for different values of the thickness ratio r. These curves confirm the previous discussion on the efficiency of the present procedure in damping process, where the use of the thickness ratio r as design variable reduces significantly the level of the total energy and decreases the consumed control energy. Figure 5 contains q-curves and $\hat{J}$-curves against the volume fraction V_1^+. Note that the design procedure over the plate thickness is more efficient than that over the volume fractional profile, while the design procedure over both r and p is the most efficient. In general the present control and design procedure strongly improves the performance of the FG plates.

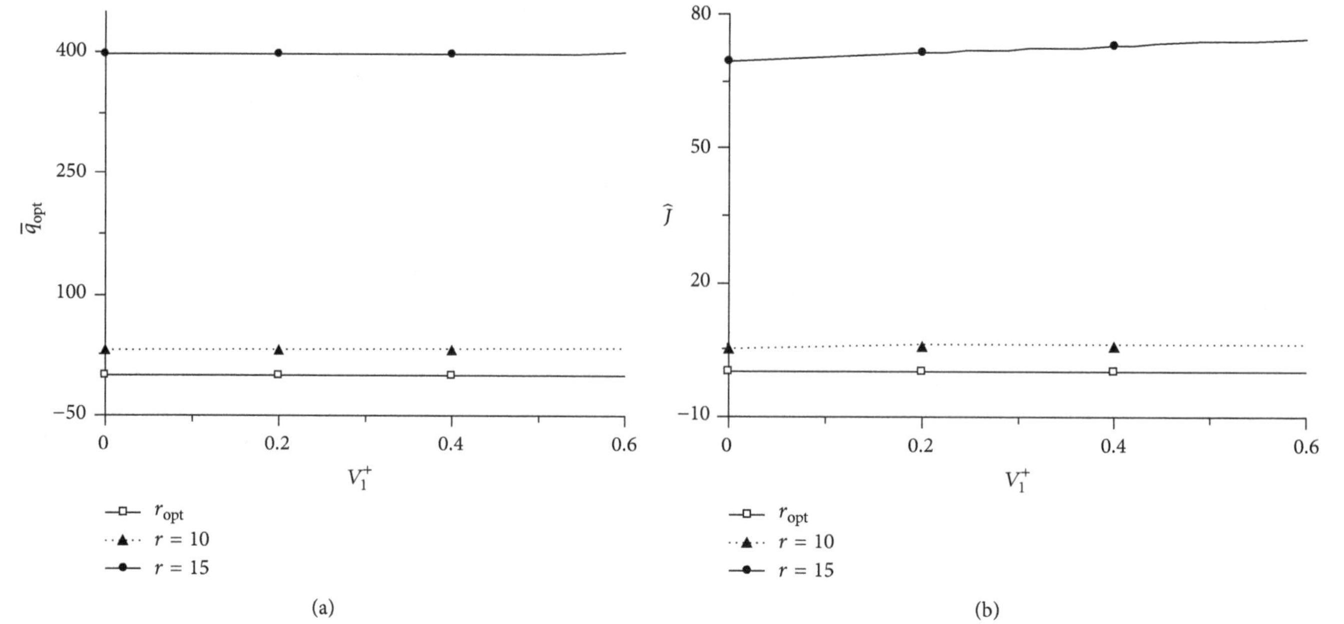

FIGURE 5: Effect of the thickness ratio r and volume fraction V_1^+ on curves of the control force $\overline{q}_{\text{opt}}$ and the energy $\widehat{J}$ for SSSS plates with $p = 4$ and $a/b = 1$.

7. Conclusion

A structural and control optimization procedure is presented to minimize the vibrational response of FG plates. A first-order plate theory including the normal strain and shear deformation effects is used. Optimal level of closed-loop control force, plate thickness, and the volume fractional index are determined. The influence of the normal strain effect on the accuracy of the obtained results is illustrated. It is found that the inclusion of the normal stain effect is required, so that neglecting the normal stain effect leads to errors reaching more than 22% for square plates with homogeneity index $p > 1$. The presented results show that the present control and design procedure has a high efficiency in improving the performance of the FG plates.

Appendices

A. The Definition of the Parameters in (6a) and (6b)

The stiffness coefficients A_{ij}, B_{ij}, and D_{ij} in (6a) and (6b) are

$$\left(A_{ij}, B_{ij}, D_{ij}\right) = \int_{-h/2}^{h/2} a_{ij} \left(\frac{1}{h^2}, \frac{12x_3}{h^4}, \frac{144x_3^2}{h^6}\right) dx_3 \qquad (i, j = 1, 2, 6),$$

$$\left(A_{ii}\right) = \int_{-h/2}^{h/2} \frac{9}{4h^2} a_{ii} \left(1 - \left(\frac{x_3}{h/2}\right)^2\right)^2 dx_3, \quad (i = 4, 5),$$

$$\left(A_{i3}, B_{i3}, \overline{B}_{i3}, D_{i3}\right) = \int_{-h/2}^{h/2} a_{i3} \left(\frac{3}{2h^2}, \frac{30x_3}{h^4}, \frac{18x_3}{h^4}, \frac{360x_3^2}{h^6}\right) \times \left(1 - \left(\frac{x_3}{h/2}\right)^2\right) dx_3, \quad (i, j = 1, 2),$$

$$\left(A_{33}, B_{33}, D_{33}\right) = \int_{-h/2}^{h/2} a_{33} \left(\frac{9}{4h^2}, \frac{45x_3}{h^4}, \frac{900x_3^2}{h^6}\right) \times \left(1 - \left(\frac{x_3}{h/2}\right)^2\right)^2 dx_3,$$

$$\begin{aligned}\left(F_{i3}, K_{i3}\right) = \int_{-h/2}^{h/2} a_{i3} \left(\frac{1}{h}, \frac{12x_3}{h^3}\right) &\times \left\{\frac{q_1}{4}\left[\left(\frac{x_3}{h/2}\right)^2 - \left(\frac{x_3}{h/2}\right) - 2\right] \times \left[1 + \left(\frac{x_3}{h/2}\right)\right]\right. \\ &\left. + \frac{q_2}{4}\left[\left(\frac{x_3}{h/2}\right)^2 + \left(\frac{x_3}{h/2}\right) - 2\right] \times \left[1 - \left(\frac{x_3}{h/2}\right)\right]\right\} dx_3, \quad (i = 1, 2),\end{aligned}$$

$$(F_{33}, K_{33}) = \int_{-h/2}^{h/2} a_{33}\left(\frac{3}{2h}, \frac{30x_3}{h^3}\right) \times \left(\frac{q_1}{4}\left[\left(\frac{x_3}{h/2}\right)^2 - \left(\frac{x_3}{h/2}\right) - 2\right]\left(1 + \frac{x_3}{h/2}\right) + \frac{q_2}{4}\left[\left(\frac{x_3}{h/2}\right)^2 + \left(\frac{x_3}{h/2}\right) - 2\right] \times \left(1 - \frac{x_3}{h/2}\right)\right) dx_3,$$

$$H_1 = \int_{-h/2}^{h/2} \frac{3a_{55}}{2h}\left[1 - \left(\frac{x_3}{h/2}\right)^2\right] \times \left\{\frac{P_1^+}{4}\left[1 - 3\left(\frac{x_3}{h/2}\right)\right]\left[1 + \left(\frac{x_3}{h/2}\right)\right] - \frac{P_1^-}{4}\left[1 + 3\left(\frac{x_3}{h/2}\right)\right]\left[1 - \left(\frac{x_3}{h/2}\right)\right]\right\} dx_3,$$

$$H_2 = \int_{-h/2}^{h/2} \frac{3a_{44}}{2h}\left[1 - \left(\frac{x_3}{h/2}\right)^2\right] \times \left\{\frac{P_2^+}{4}\left[1 - 3\left(\frac{x_3}{h/2}\right)\right]\left[1 + \left(\frac{x_3}{h/2}\right)\right] - \frac{P_2^-}{4}\left[1 + 3\left(\frac{x_3}{h/2}\right)\right]\left[1 - \left(\frac{x_3}{h/2}\right)\right]\right\} dx_3,$$

$$S_j = \frac{F_{j3}}{q(x_1, x_2, t)}, \quad S_{j+3} = \frac{K_{j3}}{q(x_1, x_2, t)}, \quad (j = 1, 2, 3). \tag{A.1}$$

B. The Definition of the Parameters in (12) and (14)

The coefficients $U_{1i}, U_{2i}, \Psi_{1i}, \Psi_{2i}, W_{0i}, W_{1i}, W_{2i}, Q_i$ in (12) are

$$U_{11} = A_{11}^* e_3 + A_{66}^* e_1,$$

$$U_{21} = (A_{12}^* + A_{66}^*) e_1,$$

$$\Psi_{11} = B_{11}^* e_3 + B_{66}^* e_1,$$

$$\Psi_{21} = (B_{12}^* + B_{66}^*) e_1,$$

$$U_{12} = (A_{12}^* + A_{66}^*) e_4,$$

$$U_{22} = A_{66}^* e_4 + A_{22}^* e_6,$$

$$\Psi_{12} = (B_{12}^* + B_{66}^*) e_4,$$

$$\Psi_{22} = B_{22}^* e_6 + B_{66}^* e_4,$$

$$U_{14} = B_{11}^* e_3 + B_{66}^* e_1,$$

$$U_{24} = (B_{12}^* + B_{66}^*) e_1,$$

$$\Psi_{14} = f_{11}^* e_3 + f_{66}^* e_1 - A_{55}^* e_2,$$

$$\Psi_{24} = (f_{12}^* + f_{66}^*) e_1,$$

$$U_{15} = (B_{12}^* + B_{66}^*) e_4,$$

$$U_{25} = B_{66}^* e_4 + B_{22}^* e_6,$$

$$\Psi_{15} = (f_{12}^* + f_{66}^*) e_4,$$

$$\Psi_{25} = f_{66}^* e_4 + f_{22}^* e_6 - A_{44}^* e_5,$$

$$U_{16} = -A_{13}^* e_7, \qquad U_{26} = -A_{23}^* e_8,$$

$$\Psi_{16} = -b_{13}^* e_7, \qquad \Psi_{26} = -b_{23}^* e_8,$$

$$U_{17} = -2B_{13}^* e_7, \qquad U_{27} = -2B_{23}^* e_8,$$

$$\Psi_{17} = \left(\frac{h^2 A_{55}^*}{20} - 2f_{13}^*\right) e_7,$$

$$\Psi_{27} = \left(\frac{h^2 A_{44}^*}{20} - 2f_{23}^*\right) e_8,$$

$$W_{04} = -A_{55}^* e_2, \qquad W_{11} = \alpha A_{13}^* e_2,$$

$$W_{21} = 2\beta B_{13}^* e_2, \qquad \Psi_{13} = A_{55}^* e_7,$$

$$\Psi_{23} = A_{44}^* e_8, \qquad W_{05} = -A_{44}^* e_5,$$

$$W_{12} = \alpha A_{23}^* e_5, \qquad W_{22} = 2\beta B_{23}^* e_5,$$

$$W_{23} = \frac{\beta h^2 (A_{55}^* e_7 + A_{44}^* e_8)}{20},$$

$$W_{07} = \frac{h^2 (A_{55}^* e_7 + A_{44}^* e_8)}{20},$$

$$W_{14} = \alpha b_{13}^* e_2,$$

$$W_{24} = \beta\left(2f_{13}^* - \frac{A_{55}^* h^2}{20}\right) e_2,$$

$$\overline{W}_3 = I_0 e_9, \qquad W_{15} = \alpha b_{23}^* e_5,$$

$$W_{25} = \beta\left(2f_{23}^* - \frac{A_{44}^* h^2}{20}\right)e_5,$$

$$\overline{W}_6 = I_1 e_9, \qquad W_{16} = -\alpha A_{33}^* e_9,$$

$$W_{26} = -2\beta B_{33}^* e_9, \qquad \overline{W}_7 = I_2 e_9,$$

$$W_{17} = -2\alpha B_{33}^* e_9,$$

$$W_{27} = \beta\left[\frac{h^4\left(A_{55}^* e_7 + A_{44}^* e_8\right)}{400} - 4f_{33}^* e_9\right],$$

$$U_{13} = U_{23} = W_{01} = W_{02} = W_{13} = W_{06} = 0,$$

$$Q_1 = \left(A_{11}^* S_1 + A_{12}^* S_2 + A_{13}^* S_3 + B_{11}^* S_4 + B_{12}^* S_5 + B_{13}^* S_6\right)e_2,$$

$$Q_2 = \left(A_{12}^* S_1 + A_{22}^* S_2 + A_{23}^* S_3 + B_{12}^* S_4 + B_{22}^* S_5 + B_{23}^* S_6\right)e_5,$$

$$Q_4 = \left(B_{11}^* S_1 + B_{12}^* S_2 + B_{13}^* S_3 + f_{11}^* S_4 + f_{12}^* S_5 + f_{13}^* S_6\right)e_2,$$

$$Q_5 = \left(B_{12}^* S_1 + B_{22}^* S_2 + B_{23}^* S_3 + f_{12}^* S_4 + f_{22}^* S_5 + f_{23}^* S_6\right)e_5,$$

$$Q_6 = -\left(A_{13}^* S_1 + A_{23}^* S_2 + A_{33}^* S_3 + b_{13}^* S_4 + b_{23}^* S_5 + B_{33}^* S_6 + h\right)e_9,$$

$$Q_7 = -2\left(B_{13}^* S_1 + B_{23}^* S_2 + B_{33}^* S_3 + f_{13}^* S_4 + f_{23}^* S_5 + f_{33}^* S_6 + \frac{9h^2}{40}\right)e_9,$$

$$Q_3 = -e_9,$$

$$\Delta_1 = \Delta_{41}\Psi_{13} + \Delta_{51}\Psi_{23} + \Delta_{71}W_{23} - \Delta_0 W_{03},$$

$$\Delta_2 = \Delta_{42}\Psi_{13} + \Delta_{52}\Psi_{23} + \Delta_{72}W_{23} + \Delta_0 e_9,$$

$$\Delta_3 = \Delta_0\overline{W}_3 - \Delta_{43}\Psi_3 - \Delta_{53}\Phi_3 - \Delta_{73}W_{23},$$

$$\Delta_0 = \left[P_{11}\ \ P_{22}\ \ P_{33}\ \ P_{44}\ \ P_{55}\ \ P_{66}\right],$$

$$\Delta_{11} = \left[\Gamma\ \ P_{22}\ \ P_{33}\ \ P_{44}\ \ P_{55}\ \ P_{66}\right],$$

$$\Delta_{12} = \left[\zeta\ \ P_{22}\ \ P_{33}\ \ P_{44}\ \ P_{55}\ \ P_{66}\right],$$

$$\Delta_{21} = \left[P_{11}\ \ \Gamma\ \ P_{33}\ \ P_{44}\ \ P_{55}\ \ P_{66}\right],$$

$$\Delta_{22} = \left[P_{11}\ \ \zeta\ \ P_{33}\ \ P_{44}\ \ P_{55}\ \ P_{66}\right],$$

$$\Delta_{41} = \left[P_{11}\ \ P_{22}\ \ \Gamma\ \ P_{44}\ \ P_{55}\ \ P_{66}\right],$$

$$\Delta_{42} = \left[P_{11}\ \ P_{22}\ \ \zeta\ \ P_{44}\ \ P_{55}\ \ P_{66}\right],$$

$$\Delta_{51} = \left[P_{11}\ \ P_{22}\ \ P_{33}\ \ \Gamma\ \ P_{55}\ \ P_{66}\right],$$

$$\Delta_{52} = \left[P_{11}\ \ P_{22}\ \ P_{33}\ \ \zeta\ \ P_{55}\ \ P_{66}\right],$$

$$\Delta_{61} = \left[P_{11}\ \ P_{22}\ \ P_{33}\ \ P_{44}\ \ \Gamma\ \ P_{66}\right],$$

$$\Delta_{62} = \left[P_{11}\ \ P_{22}\ \ P_{33}\ \ P_{44}\ \ \zeta\ \ P_{66}\right],$$

$$\Delta_{71} = \left[P_{11}\ \ P_{22}\ \ P_{33}\ \ P_{44}\ \ P_{55}\ \ \Gamma\right],$$

$$\Delta_{72} = \left[P_{11}\ \ P_{22}\ \ P_{33}\ \ P_{44}\ \ P_{55}\ \ \zeta\right],$$

$$\Delta_{13} = \left[\chi\ \ P_{22}\ \ P_{33}\ \ P_{44}\ \ P_{55}\ \ P_{66}\right],$$

$$\Delta_{23} = \left[P_{11}\ \ \chi\ \ P_{33}\ \ P_{44}\ \ P_{55}\ \ P_{66}\right],$$

$$\Delta_{43} = \left[P_{11}\ \ P_{22}\ \ \chi\ \ P_{44}\ \ P_{55}\ \ P_{66}\right],$$

$$\Delta_{63} = \left[P_{11}\ \ P_{22}\ \ P_{33}\ \ P_{44}\ \ \chi\ \ P_{66}\right],$$

$$\Delta_{53} = \left[P_{11}\ \ P_{22}\ \ P_{33}\ \ \chi\ \ P_{55}\ \ P_{66}\right],$$

$$\Delta_{73} = \left[P_{11}\ \ P_{22}\ \ P_{33}\ \ P_{44}\ \ P_{55}\ \ \chi\right],$$

$$P_{11} = \left[U_{11}\ \ U_{12}\ \ U_{14}\ \ U_{15}\ \ U_{16}\ \ U_{17}\right]^T,$$

$$P_{22} = \left[U_{21}\ \ U_{22}\ \ U_{24}\ \ U_{25}\ \ U_{26}\ \ U_{12}\right]^T,$$

$$P_{33} = \left[\Psi_{11}\ \ \Psi_{12}\ \ \Psi_{14}\ \ \Psi_{15}\ \ \Psi_{16}\ \ \Psi_{17}\right]^T,$$

$$P_{44} = \left[\Psi_{21}\ \ \Psi_{22}\ \ \Psi_{24}\ \ \Psi_{25}\ \ \Psi_{26}\ \ \Psi_{27}\right]^T,$$

$$P_{55} = \left[W_{11}\ \ W_{12}\ \ W_{14}\ \ W_{15}\ \ W_{16}\ \ W_{17}\right]^T,$$

$$P_{66} = \left[W_{21}\ \ W_{22}\ \ W_{24}\ \ W_{25}\ \ W_{26}\ \ W_{27}\right]^T,$$

$$\Gamma = \left[0\ \ 0\ \ W_{04}\ \ 0\ \ W_{06}\ \ W_{07}\right]^T,$$

$$\zeta = \left[Q_1\ \ Q_2\ \ Q_4\ \ Q_5\ \ Q_6\ \ Q_7\right]^T,$$

$$\chi = \left[0\ \ 0\ \ 0\ \ 0\ \ \overline{W}_6\ \ \overline{W}_7\right]^T. \tag{B.1}$$

The coefficients K_i $(i = 1, 2, \ldots, 6)$ in (14) are

$$K_1 = L_1 e_2 + L_2 e_5 + L_3 e_7 + L_4 e_8 + L_5 e_9 + L_6 e_{10} + L_7 e_{11} + L_8 e_{12} + L_9 e_{13},$$

$$K_2 = L_{10} e_2 + L_{11} e_5 + L_{12} e_7 + L_{13} e_8 + L_{14} e_9 + L_{15} e_{10} + L_{16} e_{11} + L_{17} e_{12} + L_{18} e_{13},$$

$$K_3 = L_{19}e_2 + L_{20}e_5 + L_{21}e_7 + L_{22}e_8 + L_{23}e_9 + L_{24}e_{10} + L_{25}e_{11} + L_{26}e_{12} + L_{27}e_{13},$$

$$K_4 = L_{28}e_2 + L_{29}e_5 + L_{30}e_9,$$

$$K_5 = L_{31}e_2 + L_{32}e_5 + L_{33}e_9,$$

$$K_6 = L_{34}e_2 + L_{35}e_5 + L_{36}e_9,$$

$$(e_1, e_2, e_3) = \int_0^b \int_0^a \left(X_{1,x_1} X_{2,x_2x_2}, X_{1,x_1} X_2, X_{1,x_1x_1x_1} X_2\right) \times X_{1,x_1} X_2 dx_1 dx_2,$$

$$(e_4, e_5, e_6) = \int_0^b \int_0^a \left(X_{1,x_1x_1} X_{2,x_2}, X_1 X_{2,x_2}, X_1 X_{2,x_2x_2x_2}\right) \times X_1 X_{2,x_2} dx_2 dx_2,$$

$$(e_7, e_8, e_9, e_{10}) = \int_0^b \int_0^a \left(X_{2,x_2x_2} X_2, X_1 X_{2,x_2x_2}, X_1 X_2, X_{1,x_1x_1} X_{2,x_2x_2}\right) \times X_1 X_2 dx_1 dx_2,$$

$$(e_{11}, e_{12}, e_{13}) = \int_0^b \int_0^a \left(X_{1,x_1}^2 X_{2,x_2}^2, X_1^2 X_{2,x_2x_2}^2, X_{1,x_1x_1}^2 X_2^2\right) dx_1 dx_2,$$

$$L_1 = \left(\left(h^4 \Delta_{71}^2 + 400\Delta_{41}^2 + 400\Delta_0^2 + 40\Delta_{41}^2 + 40h^2 \Delta_{41}\Delta_{71} - 40h^2 \Delta_0 \Delta_{71}\right) A_{55}\right) \left(800\Delta_0^2\right)^{-1},$$

$$L_2 = \left(\left(h^4 \Delta_{71}^2 + 400\Delta_{51}^2 + 400\Delta_0^2 + 40\Delta_{51}^2 + 40h^2 \Delta_{51}\Delta_{71} - 40h^2 \Delta_0 \Delta_{71} - 800h^2 \Delta_0 \Delta_{51}\right) A_{44}\right) \left(800\Delta_0^2\right)^{-1},$$

$$L_3 = \left(1600 f_{13} \Delta_{41}\Delta_{71} + 1600 B_{13} \Delta_{11}\Delta_{71} + 800 b_{13} \Delta_{41}\Delta_{61} + 800 A_{13} \Delta_{11}\Delta_{61}\right) \left(800\Delta_0^2\right)^{-1},$$

$$L_4 = \left(1600 f_{23} \Delta_{51}\Delta_{71} + 1600 B_{23} \Delta_{21}\Delta_{71} + 800 b_{23} \Delta_{51}\Delta_{61} + 800 A_{23} \Delta_{21}\Delta_{61}\right) \left(800\Delta_0^2\right)^{-1},$$

$$L_5 = \frac{\left(1600 f_{33} \Delta_{71}^2 + 1600 B_{33} \Delta_{61}\Delta_{71} + 400 A_{33} \Delta_{61}^2\right)}{800\Delta_0^2},$$

$$L_6 = \left(800 f_{12} \Delta_{41}\Delta_{51} + 800 A_{12} \Delta_{11}\Delta_{21} + 800 B_{12} \Delta_{11}\Delta_{51} + 800 B_{12} \Delta_{41}\Delta_{21}\right) \left(800\Delta_0^2\right)^{-1},$$

$$L_7 = \frac{m_1 A_{66} + m_2 B_{66} + m_3 f_{66}}{800\Delta_0^2},$$

$$L_8 = \frac{\left(400 f_{22} \Delta_{51}^2 + 800 B_{22} \Delta_{51}\Delta_{21} + 400 A_{22} \Delta_{21}^2\right)}{800\Delta_0^2},$$

$$L_9 = \left(40 A_{11} \Delta_{11}^2 + 800 B_{11} \Delta_{41}\Delta_{11} + 400 f_{11} \Delta_{41}^2 - 800 A_{55} \Delta_0 \Delta_{41}\right) \left(800\Delta_0^2\right)^{-1},$$

$$L_{10} = \left(\left(40h^2 \Delta_0 \Delta_{72} - 40h^2 \Delta_{41}\Delta_{72} + 800\Delta_0\Delta_{42} - 800\Delta_{41}\Delta_{42} - 40h^2 \Delta_{42}\Delta_{71} - 2h^4 \Delta_{71}\Delta_{72}\right) A_{55}\right) \left(800\Delta_0^2\right)^{-1},$$

$$L_{11} = \left(\left(40h^2 \Delta_0 \Delta_{72} - 40h^2 \Delta_{52}\Delta_{71} + 800\Delta_0\Delta_{52} - 800\Delta_{51}\Delta_{52} - 40h^2 \Delta_{51}\Delta_{72} - 2h^4 \Delta_{71}\Delta_{72}\right) A_{44}\right) \left(800\Delta_0^2\right)^{-1},$$

$$L_{12} = \frac{m_4 \Delta_0 \Delta_{11} + m_5 \Delta_0 \Delta_{41} + m_6}{800\Delta_0^2},$$

$$L_{13} = \frac{m_7 \Delta_0 \Delta_{21} + m_8 \Delta_0 \Delta_{51} + m_9}{800\Delta_0^2},$$

$$L_{14} = \frac{m_{10} \Delta_0 \Delta_{61} + m_{11} \Delta_0 \Delta_{71} + m_{12}}{800\Delta_0^2},$$

$$L_{15} = \frac{m_{13} A_{12} + m_{14} B_{12} + m_{15} f_{12}}{800\Delta_0^2},$$

$$L_{16} = \frac{m_{16} A_{66} + m_{17} B_{66} + m_{18} f_{66}}{800\Delta_0^2},$$

$$L_{17} = \left(-800\left(\Delta_{21}\Delta_{22} A_{22} + \left(\Delta_{51}\Delta_{22} + \Delta_{52}\Delta_{21}\right) B_{22} + \Delta_{51}\Delta_{52} f_{22}\right)\right) \left(800\Delta_0^2\right)^{-1},$$

$$L_{18} = \left(-800\left(\Delta_{11}\Delta_{12} A_{11} + \left(\Delta_{41}\Delta_{12} + \Delta_{42}\Delta_{11}\right) B_{11} + \Delta_{41}\Delta_{42} f_{11}\right)\right) \left(800\Delta_0^2\right)^{-1},$$

$$L_{19} = \frac{\left(400\Delta_{42}^2 + 40h^2 \Delta_{42}\Delta_{72} + h^4 \Delta_{72}^2\right) A_{55}}{800\Delta_0^2},$$

$$L_{20} = \frac{\left(400\Delta_{52}^2 + 40h^2\Delta_{52}\Delta_{72} + h^4\Delta_{72}^2\right) A_{44}}{800\Delta_0^2},$$

$$L_{21} = \frac{m_{19}\Delta_0\Delta_{12} + m_{20}\Delta_0\Delta_{42} + m_{21}}{800\Delta_0^2},$$

$$L_{22} = \frac{m_{22}\Delta_0\Delta_{22} + m_{23}\Delta_0\Delta_{52} + m_{24}}{800\Delta_0^2},$$

$$L_{23} = \frac{-\left(m_{10}\Delta_0\Delta_{62} + m_{11}\Delta_0\Delta_{72} + m_{25}\right)}{800\Delta_0^2},$$

$$L_{24} = \left(800\left(\Delta_{12}\Delta_{22}A_{12} + \left(\Delta_{12}\Delta_{22} + \Delta_{42}\Delta_{22}\right)B_{11} + \Delta_{42}\Delta_{52}f_{12}\right)\right)\left(800\Delta_0^2\right)^{-1},$$

$$L_{25} = \frac{m_{26}A_{66} + m_{27}B_{66} + m_{28}f_{66}}{800\Delta_0^2},$$

$$L_{26} = \frac{400\left(\Delta_{22}^2A_{22} + 2\Delta_{22}\Delta_{52}B_{22} + f_{22}\Delta_{52}^2\right)}{800\Delta_0^2},$$

$$L_{27} = \frac{400\left(\Delta_{12}^2A_{11} + 2\Delta_{12}\Delta_{42}B_{22} + f_{11}\Delta_{42}^2\right)}{800\Delta_0^2},$$

$$m_1 = 400\Delta_{11}^2 + 800\Delta_{11}\Delta_{21} + 400\Delta_{21}^2,$$

$$m_2 = 800\Delta_{51}\Delta_{21} + 800\Delta_{51}\Delta_{11} + 800\Delta_{41}\Delta_{21} + 800\Delta_{41}\Delta_{11},$$

$$m_3 = 400\Delta_{51}^2 + 400\Delta_{41}^2 + 800\Delta_{41}\Delta_{51},$$

$$m_4 = 400\left(A_{11}S1 + A_{12}S2 + A_{13}S3 + B_{11}S4 + B_{12}S5 + B_{13}S6\right),$$

$$m_5 = 400\left(B_{11}S1 + B_{12}S2 + b_{13}S3 + f_{11}S4 + f_{12}S5 + f_{13}S6\right),$$

$$m_6 = -800\left[\left(\Delta_{62}\Delta_{11} + \Delta_{61}\Delta_{12}\right)A_{13} + \left(\Delta_{62}\Delta_{41} + \Delta_{61}\Delta_{42}\right)b_{13} + 2\left(\Delta_{72}\Delta_{11} + \Delta_{71}\Delta_{12}\right)B_{13} + 2\left(\Delta_{42}\Delta_{71} + \Delta_{41}\Delta_{72}\right)f_{13}\right],$$

$$m_7 = 400\left(A_{12}S1 + A_{22}S2 + A_{23}S3 + B_{12}S4 + B_{22}S5 + B_{23}S6\right),$$

$$m_8 = 400\left(B_{12}S1 + B_{22}S2 + b_{23}S3 + f_{12}S4 + f_{22}S5 + f_{23}S6\right),$$

$$m_9 = -800\left[\left(\Delta_{62}\Delta_{21} + \Delta_{61}\Delta_{22}\right)A_{23} + \left(\Delta_{62}\Delta_{51} + \Delta_{61}\Delta_{52}\right)b_{23} + 2\left(\Delta_{72}\Delta_{21} + \Delta_{71}\Delta_{22}\right)B_{23} + 2\left(\Delta_{52}\Delta_{71} + \Delta_{51}\Delta_{72}\right)f_{23}\right],$$

$$m_{10} = 400\left(A_{13}S1 + A_{23}S2 + A_{33}S3 + b_{13}S4 + b_{23}S5 + B_{33}S6 + \frac{h}{2}\right),$$

$$m_{11} = 800\left(B_{13}S1 + B_{23}S2 + B_{33}S3 + f_{13}S4 + f_{23}S5 + f_{33}S6 + \frac{h^2}{10}\right),$$

$$m_{12} = -800\left[\Delta_{61}\Delta_{62}A_{33} + 2\left(\Delta_{61}\Delta_{72} + \Delta_{71}\Delta_{62}\right)B_{33} + 4\Delta_{71}\Delta_{72}f_{23}\right],$$

$$m_{13} = -800\left(\Delta_{11}\Delta_{22} + \Delta_{12}\Delta_{21}\right),$$

$$m_{14} = -800\left(\Delta_{41}\Delta_{22} + \Delta_{42}\Delta_{21} + \Delta_{52}\Delta_{11} + \Delta_{51}\Delta_{12}\right),$$

$$m_{15} = -800\left(\Delta_{41}\Delta_{52} + \Delta_{42}\Delta_{51}\right),$$

$$m_{16} = -800\left(\Delta_{11}\Delta_{12} + \Delta_{11}\Delta_{22} + \Delta_{21}\Delta_{12} + \Delta_{21}\Delta_{22}\right),$$

$$m_{17} = -800\left(\Delta_{41}\Delta_{12} + \Delta_{41}\Delta_{22} + \Delta_{42}\Delta_{11} + \Delta_{42}\Delta_{21} + \Delta_{51}\Delta_{12} + \Delta_{51}\Delta_{22} + \Delta_{52}\Delta_{11} + \Delta_{52}\Delta_{21}\right),$$

$$m_{18} = -800\left(\Delta_{41}\Delta_{42} + \Delta_{41}\Delta_{52} + \Delta_{42}\Delta_{51} + \Delta_{51}\Delta_{52}\right),$$

$$m_{19} = -400\left(A_{11}S1 + A_{12}S2 + A_{13}S3 + B_{11}S4 + B_{12}S5 + B_{13}S6\right),$$

$$m_{20} = -400\left(B_{11}S1 + B_{12}S2 + b_{13}S3 + f_{11}S4 + f_{12}S5 + f_{13}S6\right),$$

$$m_{21} = +800\left[\Delta_{12}\Delta_{62}A_{13} + \Delta_{62}\Delta_{42}b_{13} + 2\Delta_{12}\Delta_{72}B_{13} + 2\Delta_{42}\Delta_{72}f_{13}\right],$$

$$m_{22} = -400\left(A_{12}S1 + A_{22}S2 + A_{23}S3 + B_{12}S4 + B_{22}S5 + B_{23}S6\right),$$

$$m_{23} = -400\left(B_{12}S1 + B_{22}S2 + b_{23}S3 + f_{12}S4 + f_{22}S5 + f_{213}S6\right),$$

$$m_{24} = 800\left[\Delta_{22}\Delta_{62}A_{23} + \Delta_{62}\Delta_{52}b_{23} + 2\Delta_{22}\Delta_{72}B_{13} + 2\Delta_{52}\Delta_{72}f_{23}\right],$$

$$m_{25} = -800\left[\Delta_{62}^2 A_{33} + 2\Delta_{62}\Delta_{72}B_{33} + 4\Delta_{72}^2 f_{33} + \mu\Delta_0^2\right],$$

$$m_{26} = 400\Delta_{12}^2 + 800\Delta_{12}\Delta_{22} + 400\Delta_{22}^2,$$

$$m_{27} = 800\Delta_{52}\Delta_{12} + 800\Delta_{52}\Delta_{22} + 800\Delta_{42}\Delta_{12} + 800\Delta_{42}\Delta_{12},$$

$$m_{28} = 400\Delta_{42}^2 + 800\Delta_{42}\Delta_{52} + 400\Delta_{52}^2. \tag{B.2}$$

C. The Definition of the Parameters in (23)

The coefficients a_i $(i = 1, 2, \ldots, 18)$ of (23) are

$$a_1 = 4K_6 l^6,$$

$$a_2 = 4K_6 l^4\left(lK_2 + 2\omega^2 K_3\right),$$

$$a_3 = K_6 l^2\left(lK_2 + 2\omega^2 K_3\right)^2,$$

$$a_4 = -4K_3^3 l^2,$$

$$a_5 = -4K_3^3\left(lK_2 + 2\omega^2 K_3\right),$$

$$a_6 = K_3^3\left(4K_1K_3 - K_2^2\right),$$

$$a_7 = 4K_6 l^6, \qquad a_8 = -8K_3K_6 l^4,$$

$$a_9 = -K_3 l^2\left(K_3^2 + K_2K_6 l - K_3K_5 l\right),$$

$$a_{10} = 4K_3^2K_6 l^2,$$

$$a_{11} = 4K_3^2\left(2K_3^2 + K_2K_6 l - K_3K_5 l\right),$$

$$a_{12} = K_3^2\left(4K_4K_3^2 + K_2^2K_6 - 2K_2K_3K_5\right),$$

$$a_{13} = 4K_3^4,$$

$$a_{14} = 2K_6 l^4\left(lK_2 + 2\omega^2 K_3\right),$$

$$a_{15} = a_1, \qquad a_{16} = \frac{1}{2}a_8,$$

$$a_{17} = -2K_3 l^2\left(2K_3^2 + 2K_2K_6 l - K_3K_5 l + 2K_3K_6\omega^2\right),$$

$$a_{18} = -K_3\left(lK_2 + 2\omega^2 K_3\right)\left(2K_3^2 + K_2K_6 l - K_3K_5 l\right). \tag{C.1}$$

Conflict of Interests

The authors declare that there is no conflict of interests regarding the publication of this paper.

References

[1] M. Yamanouchi, M. Koizumi, T. Hirai, and I. Shiota, "Functionally gradient materials forum," in *Proceeding of the 1st International Symposium on Functionally Gradient Materials (FGM '90)*, pp. 197–202, Sendai, Japan, 1990.

[2] J. N. Yang and T. T. Soong, "Recent advances in active control of civil engineering structures," *Probabilistic Engineering Mechanics*, vol. 3, no. 4, pp. 179–188, 1988.

[3] R. K. Miller, S. F. Masri, T. J. Dehghanyar, and T. K. Caughey, "Active vibration control of large civil structures," *Journal of Engineering Mechanics*, vol. 114, no. 9, pp. 1542–1570, 1988.

[4] A. M. Reinhorn and G. D. Manolis, "Recent advances in structural control," *Shock & Vibration Digest*, vol. 21, no. 1, pp. 3–8, 1989.

[5] A. E. Bryson Jr., "Optimal control—1950 to 1985," *IEEE Control Systems*, vol. 16, no. 3, pp. 26–33, 1996.

[6] H. J. Sussmann and J. C. Willems, "300 years of optimal control: from the brachystochrone to the maximum principle," *IEEE Control Systems Magazine*, vol. 17, no. 3, pp. 32–44, 1997.

[7] A. Jajarmi, N. Pariz, A. Vahidian Kamyad, and S. Effati, "A highly computational efficient method to solve nonlinear optimal control problems," *Scientia Iranica*, vol. 19, no. 3, pp. 759–766, 2012.

[8] R. V. Grandhi, "Structural and control optimization of space structures," *Computers & Structures*, vol. 31, no. 2, pp. 139–150, 1989.

[9] J. M. Sloss, J. C. Bruch Jr., and I. S. Sadek, "A maximum principle for nonconservative self-adjoint systems," *IMA Journal of Mathematical Control and Information*, vol. 6, no. 2, pp. 199–216, 1989.

[10] S. Adali, J. C. Brunch Jr., I. S. Sadek, and J. M. Sloss, "Simultaneous design/control optimization of symmetric cross-ply laminates hybrid construction," *Engineering Optimization*, vol. 17, pp. 123–140, 1991.

[11] S. Adali, M. Walker, and V. E. Verijenko, "Multiobjective optimization of laminated plates for maximum prebuckling, buckling and postbuckling strength using continuous and discrete ply angles," *Composite Structures*, vol. 35, no. 1, pp. 117–130, 1996.

[12] J. M. Sloss, I. S. Sadek, J. C. Bruch Jr., and S. Adali, "Design/control optimization of cross-ply laminates under buckling and vibration," *Journal of Aerospace Engineering*, vol. 5, no. 1, pp. 127–137, 1992.

[13] I. S. Sadek, J. M. Sloss, S. Adali, and J. C. Bruch Jr., "Integrated design and control of laminated hybrid plates with dynamic response and buckling objectives," *Journal of Sound and Vibration*, vol. 163, no. 1, pp. 57–66, 1993.

[14] M. E. Fares, Y. G. Youssif, and A. E. Alamir, "Optimal design and control of composite laminated plates with various boundary conditions using various plate theories," *Composite Structures*, vol. 56, no. 1, pp. 1–12, 2002.

[15] M. E. Fares, Y. G. Youssif, and M. A. Hafiz, "Structural and control optimization for maximum thermal buckling and minimum dynamic response of composite laminated plates," *International Journal of Solids and Structures*, vol. 41, no. 3-4, pp. 1005–1019, 2004.

[16] M. E. Fares, Y. G. Youssif, and M. A. Hafiz, "Optimal control of the dynamic response of an anisotropic plate with various boundary conditions," *Mechanics Research Communications*, vol. 28, no. 5, pp. 525–534, 2001.

[17] M. E. Fares, Y. G. Youssif, and A. E. Elshoraky, "Non-linear design and control optimization of composite laminated plates with buckling and postbuckling objectives," *International Journal of Non-Linear Mechanics*, vol. 41, no. 6-7, pp. 807–824, 2006.

[18] P. Dash and B. N. Singh, "Buckling and post-buckling of laminated composite plates," *Mechanics Research Communications*, vol. 46, pp. 1–7, 2012.

[19] A. Lal, K. R. Jagtap, and B. N. Singh, "Post buckling response of functionally graded materials plate subjected to mechanical and thermal loadings with random material properties," *Applied Mathematical Modelling*, vol. 37, no. 5, pp. 2900–2920, 2013.

[20] M. E. Fares, M. K. Elmarghany, and D. Atta, "An efficient and simple refined theory for bending and vibration of functionally graded plates," *Composite Structures*, vol. 91, no. 3, pp. 296–305, 2009.

[21] R. Bellman, *Dynamic Programming*, Princeton University Press, 1957.

[22] T. Mori and K. Tanaka, "Average stress in matrix and average elastic energy of materials with misfitting inclusions," *Acta Metallurgica*, vol. 21, no. 5, pp. 571–574, 1973.

Permissions

All chapters in this book were first published in IJAE, by Hindawi Publishing Corporation; hereby published with permission under the Creative Commons Attribution License or equivalent. Every chapter published in this book has been scrutinized by our experts. Their significance has been extensively debated. The topics covered herein carry significant findings which will fuel the growth of the discipline. They may even be implemented as practical applications or may be referred to as a beginning point for another development.

The contributors of this book come from diverse backgrounds, making this book a truly international effort. This book will bring forth new frontiers with its revolutionizing research information and detailed analysis of the nascent developments around the world.

We would like to thank all the contributing authors for lending their expertise to make the book truly unique. They have played a crucial role in the development of this book. Without their invaluable contributions this book wouldn't have been possible. They have made vital efforts to compile up to date information on the varied aspects of this subject to make this book a valuable addition to the collection of many professionals and students.

This book was conceptualized with the vision of imparting up-to-date information and advanced data in this field. To ensure the same, a matchless editorial board was set up. Every individual on the board went through rigorous rounds of assessment to prove their worth. After which they invested a large part of their time researching and compiling the most relevant data for our readers.

The editorial board has been involved in producing this book since its inception. They have spent rigorous hours researching and exploring the diverse topics which have resulted in the successful publishing of this book. They have passed on their knowledge of decades through this book. To expedite this challenging task, the publisher supported the team at every step. A small team of assistant editors was also appointed to further simplify the editing procedure and attain best results for the readers.

Apart from the editorial board, the designing team has also invested a significant amount of their time in understanding the subject and creating the most relevant covers. They scrutinized every image to scout for the most suitable representation of the subject and create an appropriate cover for the book.

The publishing team has been an ardent support to the editorial, designing and production team. Their endless efforts to recruit the best for this project, has resulted in the accomplishment of this book. They are a veteran in the field of academics and their pool of knowledge is as vast as their experience in printing. Their expertise and guidance has proved useful at every step. Their uncompromising quality standards have made this book an exceptional effort. Their encouragement from time to time has been an inspiration for everyone.

The publisher and the editorial board hope that this book will prove to be a valuable piece of knowledge for researchers, students, practitioners and scholars across the globe.

List of Contributors

Xiangwei Bu, Yan Zhao and Guangjun He
Air and Missile Defense College, Air Force Engineering University, Xi'an 710051, China

Qiong Wang
Science College, Air Force Engineering University, Xi'an 710051, China

Minghui Yao, Li Ma and Wei Zhang
Beijing Key Laboratory of Nonlinear Vibrations and Strength of Mechanical Structures, College of Mechanical Engineering, Beijing University of Technology, Beijing 100124, China

Xueshu Liu
School of Automotive Engineering, Dalian University of Technology, Dalian 116024, China

Junhao Chang, Yuxing Yang, Yiqi Wang, Yongjie Bao and Hang Gao
2School of Mechanical Engineering, Dalian University of Technology, Dalian 116024, China

Sen Hu, Xianzhen Huang, Yimin Zhang and Chunmei Lv
School of Mechanical Engineering and Automation, Northeastern University, Shenyang 110819, China

C. C. Cui
State Key Laboratory of Complex Electromagnetic Environment Effects on Electronics and Information System, Luoyang, China
Department of Mechanics, Sun Yat-sen University, No. 135 Xingang Road West, Guangzhou 510275, China

S. X. Xie
Basic Education Department, Jiangxi Industry Polytechnic College, Qinshanhu Road 1988, Nanchang, China

X. C. Huang
Basic Education Department, Jiangxi Industry Polytechnic College, Qinshanhu Road 1988, Nanchang, China

J. K. Liu and Y. M. Chen
Department of Mechanics, Sun Yat-sen University, No. 135 Xingang Road West, Guangzhou 510275, China
Department of Mechanical Engineering, Ganzhou Institute of Technology, 88 Loutiling Rd, Ganzhou 341000, Jiangxi, China

Yulin Wang, Shengjing Tang and Jie Guo
School of Aerospace Engineering, Beijing Institute of Technology, Beijing 100081, China

Wei Shang
System Design Institute of Hubei Aerospace Technology Academy, Hubei, 430040, China

Sina Hamzehlouia and Kamran Behdinan
Department of Mechanical and Industrial Engineering, University of Toronto, Toronto, ON, Canada

Fei Liu and Yang Gao
Department of Geomatics Engineering, University of Calgary, Calgary, AB, Canada

Panfeng Huang, Lu Chen, Bin Zhang, Zhongjie Meng and Zhengxiong Liu
National Key Laboratory of Aerospace Flight Dynamics, Northwestern Polytechnical University, Xi'an, Shaanxi 710072, China
Research Center for Intelligent Robotics, School of Astronautics, Northwestern Polytechnical University, Xi'an, Shaanxi 710072, China

Tae-Yong Park and Hyun-Ung Oh
Space Technology Synthesis Laboratory, Department of Aerospace Engineering, Chosun University, 375 Seosuk-dong, Dong-gu, Gwangju 501-759, Republic of Korea

Joo-Yong Jung
Spaceborne SAR R&D Lab, LIG Nex1, Yongin, Republic of Korea

Rudolf Andoga, Ladislav Főző, Jozef Judičák, Róbert Bréda, Stanislav Szabo, Róbert Rozenberg and Milan Džunda
Faculty of Aeronautics, Technical University of Košice, Rampová 7, 041 21 Košice, Slovakia

Shengjiang Yang
Institute of Precision Guidance and Control, Northwestern Polytechnical University, Shaanxi 710072, China
Beijing Aerospace Technology Institute, Beijing 100074, China

Jianguo Guo and Jun Zhou
Institute of Precision Guidance and Control, Northwestern Polytechnical University, Shaanxi 710072, China

Jinshuo Hu and Hongbin Gu
College of Civil Aviation, Nanjing University of Aeronautics and Astronautics, Nanjing 211100, China

Zhichao Chen and Tao Zhang
Changchun Institute of Optics, Fine Mechanics and Physics, Chinese Academy of Sciences, Changchun 130033, China
University of Chinese Academy of Sciences, Beijing, China

Jin Zhou, Jianjiang Zeng, Jichang Chen and Mingbo Tong
College of Aerospace Engineering, Nanjing University of Aeronautics & Astronautics, Nanjing, China

M. E. Fares, M. Kh. Elmarghany and Doaa Atta
Department of Mathematics, Faculty of Science, Mansoura University, Mansoura 35516, Egypt

Index